Multidisciplinary know-how for smart-textiles developers

The Textile Institute and Woodhead Publishing

The Textile Institute is a unique organisation in textiles, clothing and footwear. Incorporated in England by a Royal Charter granted in 1925, the Institute has individual and corporate members in over 90 countries. The aim of the Institute is to facilitate learning, recognise achievement, reward excellence and disseminate information within the global textiles, clothing and footwear industries.

Historically, The Textile Institute has published books of interest to its members and the textile industry. To maintain this policy, the Institute has entered into partnership with Woodhead Publishing Limited to ensure that Institute members and the textile industry continue to have access to high calibre titles on textile science and technology.

Most Woodhead titles on textiles are now published in collaboration with The Textile Institute. Through this arrangement, the Institute provides an Editorial Board, which advises Woodhead on appropriate titles for future publication and suggests possible editors and authors for these books. Each book published under this arrangement carries the Institute's logo.

Woodhead books published in collaboration with The Textile Institute are offered to Textile Institute members at a substantial discount. These books, together with those published by The Textile Institute that are still in print, are offered on the Woodhead web site at: www.woodheadpublishing.com. Textile Institute books still in print are also available directly from the Institute's website at: www.textileinstitutebooks.com.

A list of Woodhead books on textile science and technology, most of which have been published in collaboration with The Textile Institute, can be found towards the end of the contents pages.

We are always happy to receive suggestions for new books from potential editors. To enquire about contributing to our Textiles series, please send your name, contact address and details of the topic/s you are interested in to sarah.lynch@woodheadpublishing.com. We look forward to hearing from you.

The Woodhead team responsible for publishing this book
Commissioning Editor: Kathryn Picking
Publications Coordinator: Adam Davies
Project Editor: Sarah Lynch
Editorial and Production Manager: Mary Campbell
Production Editor: Adam Hooper
Freelance Project Manager: Annette Wiseman
Copyeditor: Jo Egré
Proofreader: Clare Dobson
Cover Designer: Terry Callanan

Woodhead Publishing Series in Textiles: Number 139

Multidisciplinary know-how for smart-textiles developers

Edited by

Tünde Kirstein

The Textile Institute

WOODHEAD PUBLISHING

Oxford Cambridge Philadelphia New Delhi

Published by Woodhead Publishing Limited in association with The Textile Institute
Woodhead Publishing Limited,
80 High Street, Sawston, Cambridge CB22 3HJ, UK
www.woodheadpublishing.com
www.woodheadpublishingonline.com

Woodhead Publishing, 1518 Walnut Street, Suite 1100, Philadelphia,
PA 19102-3406, USA

Woodhead Publishing India Private Limited, G-2, Vardaan House,
7/28 Ansari Road, Daryaganj, New Delhi – 110002, India
www.woodheadpublishingindia.com

First published 2013, Woodhead Publishing Limited

British Library Cataloguing in Publication Data
A catalogue record for this book is available from the British Library.

Library of Congress Control Number: 2012954112

ISBN 978-0-85709-342-4 (print)
ISBN 978-0-85709-353-0 (online)
ISSN 2042-0803 Woodhead Publishing Series in Textiles (print)
ISSN 2042-0811 Woodhead Publishing Series in Textiles (online)

The publisher's policy is to use permanent paper from mills that operate a sustainable forestry policy, and which has been manufactured from pulp which is processed using acid-free and elemental chlorine-free practices. Furthermore, the publisher ensures that the text paper and cover board used have met acceptable environmental accreditation standards.

Typeset by RefineCatch Limited, Bungay, Suffolk
Printed and bound in the UK by the MPG Books Group

Contents

Contributor contact details

(* = main contact)

Editor and Chapter 1

Dr Tünde Kirstein
TechPublish
Römerfeldstrasse 1
8623 Wetzikon
Switzerland

E-mail: info@techpublish.ch

Professor Dr Lieva Van Langenhove
Department of Textiles
Ghent University
Technologiepark 907
9052 Zwijnaarde
Belgium

E-mail: Lieva.VanLangenhove@ugent.be

Chapter 2

Dr Anne Schwarz*
Institut für Textiltechnik
RWTH Aachen University
Otto-Blumenthal-Straße 1
52074 Aachen
Germany

E-mail: Anne.Schwarz@ita.rwth-aachen.de

Chapter 3

Professor Maksim Skorobogatiy
Department of Engineering Physics
Ecole Polytechnique de Montréal
C.P. 6079, succ. Centre-ville
Montréal
Québec H3C 3A7
Canada

E-mail: maksim.skorobogatiy@polymtl.ca

Chapter 4

Dr Songmin Shang* and Dr Wei Zeng
Institute of Textiles and Clothing
The Hong Kong Polytechnic University
Hung Hom
Kowloon
Hong Kong
China

E-mail: shang.songmin@polyu.edu.hk

Chapter 5

Dr Cédric Cochrane* and Dr Aurélie Cayla
ENSAIT / GEMTEX
2 Allée Louise et Victor Champier
F-59100 Roubaix
France

E-mail: cedric.cochrane@ensait.fr; aurelie.cayla@ensait.fr

Chapter 6

Professor Maksim Skorobogatiy
Department of Engineering Physics
Ecole Polytechnique de Montréal
C.P. 6079, succ. Centre-ville
Montréal
Québec H3C 3A7
Canada

E-mail: maksim.skorobogatiy@polymtl.ca

Chapter 7

Dr Julian Eichhoff and A. Hehl*
Institut für Textiltechnik of RWTH Aachen University
Otto-Blumenthal-Str. 1
52074 Aachen
Germany

E-mail: julian.eichhoff@ita.rwth-aachen.de; julian.eichhoff@gmx.de; achim.hehl@ita.rwth-aachen.de

Chapter 8

Christoph Zysset*, T. Kinkeldei, N. Münzenrieder and G. Tröster
ETH Zurich
Switzerland

E-mail: zysset@ife.ee.ethz.ch

K. Cherenack
Philips Research Eindhoven
The Netherlands

Chapter 9

Dr Corrado Carta* and Professor Frank Ellinger
Dresden University of Technology
Germany

E-mail: corrado.carta@tu-dresden.de; frank.ellinger@tu-dresden.de

Professor Arved Hübler and Dr Georg Schmidt
Chemnitz University of Technology
Germany

E-mail: arved.huebler@mb.tu-chemnitz.de; georg.schmidt@mb.tu-chemnitz.de

Dr Jörg Zapf
Siemens AG
Munich
Germany

E-mail: joerg.zapf@siemens.com

Professor Gerhard Tröster
ETH Zurich
Switzerland

E-mail: troester@ife.ee.ethz.ch

Anja Talo
Enfucell Oy
Finland

E-mail: anja.talo@enfucell.com

Dionyssis Kozakis
Data Control Systems Ltd
Athens
Greece

E-mail: dkozakis@dcs.com.gr

Dimitris Vassiliadis
Exoduss A.E.
Athens
Greece

E-mail: dvas@exodussa.com

Rita Paradiso
Smartex S.R.L.
Pisa
Italy

E-mail: rita@smartex.it

Martin Krebs
Varta Microbatteries GmbH
Germany

E-mail: martin.krebs@varta-microbattery.com

Markus Scharber
Konarka Technologies GmbH
Germany

E-mail: mscharber@konarka.com

Markus Tuomikoski
Teknologian Tutkimuskeskus VTT
Finland

E-mail: markus.tuomikoski@vtt.fi

Chapter 10

Dr Ivo Locher
SEFAR AG
Hinterbissaustrasse 12
CH-9410 Heiden
Switzerland

E-mail: Ivo.Locher@sefar.ch

Chapter 11

Professor Stephen Beeby*, Zhuo Cao and Ahmed Almusallam
Electronics and Computer Science
University of Southampton
SO17 1BJ
UK

E-mail: spb@ecs.soton.ac.uk

Chapter 12

Dr Daniel Roggen* and G. Tröster
ETH Zurich
Switzerland

E-mail: Daniel.roggen@ife.ee.ethz.ch

A. Bulling
University of Cambridge
UK

Chapter 13

Awa Garlinska and Andreas Röpert*
Interactive Wear AG
Starnberg
Germany

E-mail: awa.garlinska@interactive-wear.com; andreas.roepert@interactive-wear.com

Chapter 14

Stijn Ossevoort
Department of Design and Art
Lucern University of Applied Sciences and Arts
Sentimatt 1/Dammstrasse
6003 Lucerne
Switzerland

E-mail: stijn.ossevoort@hslu.ch

Chapter 15

Dr Shirley Coyle* and Professor Dermot Diamond
CLARITY: Centre for Sensor Web Technologies
National Centre for Sensor Research
Dublin City University
Glasnevin
Dublin 9
Ireland

E-mail: shirley.coyle@dcu.ie; dermot.diamond@dcu.ie

Chapter 16

Manfred Wagner, Dipl.-Phys
Daimler AG
Hanns-Klemm-Strassee 45
71034 Boeblingen
Germany

E-mail: Manfred.mw.wagner@daimler.com

Chapter 17

Dr Axel Ritter
ritter architekten
Telegrafenstrasse 11
53474 Bad Neuenahr-Ahrweiler
Germany

E-mail: info@ritter-architekten.com

Woodhead Publishing Series in Textiles

1 **Watson's textile design and colour Seventh edition**
Edited by Z. Grosicki
2 **Watson's advanced textile design**
Edited by Z. Grosicki
3 **Weaving Second edition**
P. R. Lord and M. H. Mohamed
4 **Handbook of textile fibres Volume 1: Natural fibres**
J. Gordon Cook
5 **Handbook of textile fibres Volume 2: Man-made fibres**
J. Gordon Cook
6 **Recycling textile and plastic waste**
Edited by A. R. Horrocks
7 **New fibers Second edition**
T. Hongu and G. O. Phillips
8 **Atlas of fibre fracture and damage to textiles Second edition**
J. W. S. Hearle, B. Lomas and W. D. Cooke
9 **Ecotextile '98**
Edited by A. R. Horrocks
10 **Physical testing of textiles**
B. P. Saville
11 **Geometric symmetry in patterns and tilings**
C. E. Horne
12 **Handbook of technical textiles**
Edited by A. R. Horrocks and S. C. Anand
13 **Textiles in automotive engineering**
W. Fung and J. M. Hardcastle
14 **Handbook of textile design**
J. Wilson
15 **High-performance fibres**
Edited by J. W. S. Hearle
16 **Knitting technology Third edition**
D. J. Spencer
17 **Medical textiles**
Edited by S. C. Anand
18 **Regenerated cellulose fibres**
Edited by C. Woodings
19 **Silk, mohair, cashmere and other luxury fibres**
Edited by R. R. Franck

20 **Smart fibres, fabrics and clothing**
Edited by X. M. Tao
21 **Yarn texturing technology**
J. W. S. Hearle, L. Hollick and D. K. Wilson
22 **Encyclopedia of textile finishing**
H-K. Rouette
23 **Coated and laminated textiles**
W. Fung
24 **Fancy yarns**
R. H. Gong and R. M. Wright
25 **Wool: Science and technology**
Edited by W. S. Simpson and G. Crawshaw
26 **Dictionary of textile finishing**
H-K. Rouette
27 **Environmental impact of textiles**
K. Slater
28 **Handbook of yarn production**
P. R. Lord
29 **Textile processing with enzymes**
Edited by A. Cavaco-Paulo and G. Gübitz
30 **The China and Hong Kong denim industry**
Y. Li, L. Yao and K. W. Yeung
31 **The World Trade Organization and international denim trading**
Y. Li, Y. Shen, L. Yao and E. Newton
32 **Chemical finishing of textiles**
W. D. Schindler and P. J. Hauser
33 **Clothing appearance and fit**
J. Fan, W. Yu and L. Hunter
34 **Handbook of fibre rope technology**
H. A. McKenna, J. W. S. Hearle and N. O'Hear
35 **Structure and mechanics of woven fabrics**
J. Hu
36 **Synthetic fibres: nylon, polyester, acrylic, polyolefin**
Edited by J. E. McIntyre
37 **Woollen and worsted woven fabric design**
E. G. Gilligan
38 **Analytical electrochemistry in textiles**
P. Westbroek, G. Priniotakis and P. Kiekens
39 **Bast and other plant fibres**
R. R. Franck
40 **Chemical testing of textiles**
Edited by Q. Fan
41 **Design and manufacture of textile composites**
Edited by A. C. Long
42 **Effect of mechanical and physical properties on fabric hand**
Edited by H. M. Behery
43 **New millennium fibers**
T. Hongu, M. Takigami and G. O. Phillips

44 **Textiles for protection**
Edited by R. A. Scott
45 **Textiles in sport**
Edited by R. Shishoo
46 **Wearable electronics and photonics**
Edited by X. M. Tao
47 **Biodegradable and sustainable fibres**
Edited by R. S. Blackburn
48 **Medical textiles and biomaterials for healthcare**
Edited by S. C. Anand, M. Miraftab, S. Rajendran and J. F. Kennedy
49 **Total colour management in textiles**
Edited by J. Xin
50 **Recycling in textiles**
Edited by Y. Wang
51 **Clothing biosensory engineering**
Y. Li and A. S. W. Wong
52 **Biomechanical engineering of textiles and clothing**
Edited by Y. Li and D. X-Q. Dai
53 **Digital printing of textiles**
Edited by H. Ujiie
54 **Intelligent textiles and clothing**
Edited by H. R. Mattila
55 **Innovation and technology of women's intimate apparel**
W. Yu, J. Fan, S. C. Harlock and S. P. Ng
56 **Thermal and moisture transport in fibrous materials**
Edited by N. Pan and P. Gibson
57 **Geosynthetics in civil engineering**
Edited by R. W. Sarsby
58 **Handbook of nonwovens**
Edited by S. Russell
59 **Cotton: Science and technology**
Edited by S. Gordon and Y-L. Hsieh
60 **Ecotextiles**
Edited by M. Miraftab and A. R. Horrocks
61 **Composite forming technologies**
Edited by A. C. Long
62 **Plasma technology for textiles**
Edited by R. Shishoo
63 **Smart textiles for medicine and healthcare**
Edited by L. Van Langenhove
64 **Sizing in clothing**
Edited by S. Ashdown
65 **Shape memory polymers and textiles**
J. Hu
66 **Environmental aspects of textile dyeing**
Edited by R. Christie
67 **Nanofibers and nanotechnology in textiles**
Edited by P. Brown and K. Stevens

68 **Physical properties of textile fibres Fourth edition**
W. E. Morton and J. W. S. Hearle
69 **Advances in apparel production**
Edited by C. Fairhurst
70 **Advances in fire retardant materials**
Edited by A. R. Horrocks and D. Price
71 **Polyesters and polyamides**
Edited by B. L. Deopura, R. Alagirusamy, M. Joshi and B. S. Gupta
72 **Advances in wool technology**
Edited by N. A. G. Johnson and I. Russell
73 **Military textiles**
Edited by E. Wilusz
74 **3D fibrous assemblies: Properties, applications and modelling of three-dimensional textile structures**
J. Hu
75 **Medical and healthcare textiles**
Edited by S. C. Anand, J. F. Kennedy, M. Miraftab and S. Rajendran
76 **Fabric testing**
Edited by J. Hu
77 **Biologically inspired textiles**
Edited by A. Abbott and M. Ellison
78 **Friction in textile materials**
Edited by B. S. Gupta
79 **Textile advances in the automotive industry**
Edited by R. Shishoo
80 **Structure and mechanics of textile fibre assemblies**
Edited by P. Schwartz
81 **Engineering textiles: Integrating the design and manufacture of textile products**
Edited by Y. E. El-Mogahzy
82 **Polyolefin fibres: Industrial and medical applications**
Edited by S. C. O. Ugbolue
83 **Smart clothes and wearable technology**
Edited by J. McCann and D. Bryson
84 **Identification of textile fibres**
Edited by M. Houck
85 **Advanced textiles for wound care**
Edited by S. Rajendran
86 **Fatigue failure of textile fibres**
Edited by M. Miraftab
87 **Advances in carpet technology**
Edited by K. Goswami
88 **Handbook of textile fibre structure Volume 1 and Volume 2**
Edited by S. J. Eichhorn, J. W. S. Hearle, M. Jaffe and T. Kikutani
89 **Advances in knitting technology**
Edited by K-F. Au
90 **Smart textile coatings and laminates**
Edited by W. C. Smith
91 **Handbook of tensile properties of textile and technical fibres**
Edited by A. R. Bunsell

92 **Interior textiles: Design and developments**
Edited by T. Rowe
93 **Textiles for cold weather apparel**
Edited by J. T. Williams
94 **Modelling and predicting textile behaviour**
Edited by X. Chen
95 **Textiles, polymers and composites for buildings**
Edited by G. Pohl
96 **Engineering apparel fabrics and garments**
J. Fan and L. Hunter
97 **Surface modification of textiles**
Edited by Q. Wei
98 **Sustainable textiles**
Edited by R. S. Blackburn
99 **Advances in yarn spinning technology**
Edited by C. A. Lawrence
100 **Handbook of medical textiles**
Edited by V. T. Bartels
101 **Technical textile yarns**
Edited by R. Alagirusamy and A. Das
102 **Applications of nonwovens in technical textiles**
Edited by R. A. Chapman
103 **Colour measurement: Principles, advances and industrial applications**
Edited by M. L. Gulrajani
104 **Fibrous and composite materials for civil engineering applications**
Edited by R. Fangueiro
105 **New product development in textiles: Innovation and production**
Edited by L.Horne
106 **Improving comfort in clothing**
Edited by G. Song
107 **Advances in textile biotechnology**
Edited by V. A. Nierstrasz and A. Cavaco-Paulo
108 **Textiles for hygiene and infection control**
Edited by B. McCarthy
109 **Nanofunctional textiles**
Edited by Y. Li
110 **Joining textiles: Principles and applications**
Edited by I. Jones and G. Stylios
111 **Soft computing in textile engineering**
Edited by A. Majumdar
112 **Textile design**
Edited by A. Briggs-Goode and K. Townsend
113 **Biotextiles as medical implants**
Edited by M. King and B. Gupta
114 **Textile thermal bioengineering**
Edited by Y. Li
115 **Woven textile structure**
B. K. Behera and P. K. Hari

116 **Handbook of textile and industrial dyeing. Volume 1: Principles, processes and types of dyes**
Edited by M. Clark
117 **Handbook of textile and industrial dyeing. Volume 2: Applications of dyes**
Edited by M. Clark
118 **Handbook of natural fibres. Volume 1: Types, properties and factors affecting breeding and cultivation**
Edited by R. Kozłowski
119 **Handbook of natural fibres. Volume 2: Processing and applications**
Edited by R. Kozłowski
120 **Functional textiles for improved performance, protection and health**
Edited by N. Pan and G. Sun
121 **Computer technology for textiles and apparel**
Edited by J. Hu
122 **Advances in military textiles and personal equipment**
Edited by E. Sparks
123 **Specialist yarn and fabric structures**
Edited by R. H. Gong
124 **Handbook of sustainable textile production**
M. I. Tobler-Rohr
125 **Woven textiles: Principles, developments and applications**
Edited by K. Gandhi
126 **Textiles and fashion: Materials design and technology**
Edited by R. Sinclair
127 **Industrial cutting of textile materials**
I. Vilumsone-Nemes
128 **Colour design: Theories and applications**
Edited by J. Best
129 **False twist textured yarns**
C. Atkinson
130 **Modelling, simulation and control of the dyeing process**
R. Shamey and X. Zhao
131 **Process control in textile manufacturing**
Edited by A. Majumdar, A. Das, R. Alagirusamy and V. K. Kothari
132 **Understanding and improving the durability of textiles**
Edited by P. A. Annis
133 **Smart textiles for protection**
Edited by R. A. Chapman
134 **Functional nanofibers and applications**
Edited by Q. Wei
135 **The global textile and clothing industry: Technological advances and future challenges**
Edited by R. Shishoo
136 **Simulation in textile technology: Theory and applications**
Edited by D. Veit
137 **Pattern cutting for clothing using CAD: How to use Lectra Modaris pattern cutting software**
M. Stott

138 **Advances in the dyeing and finishing of technical textiles**
M. L. Gulrajani

139 **Multidisciplinary know-how for smart-textiles developers**
Edited by T. Kirstein

140 **Handbook of fire resistant textiles**
Edited by F. Selcen Kilinc

141 **Handbook of footwear design and manufacture**
Edited by A. Luximon

142 **Textile-led design for the active ageing population**
Edited by J. McCann and D. Bryson

143 **Optimizing decision making in the apparel supply chain using artificial intelligence (AI): From production to retail**
W. K. Wong, Z. X. Guo and S. Y. S. Leung

144 **Mechanisms of flat weaving technology**
V. Choogin, P. Bandara and E. Chepelyuk

145 **Innovative jacquard textile design using digital technologies**
F. Ng and J. Zhou

146 **Advances in shape memory polymers**
J. Hu

147 **Clothing manufacture management: A systematic approach to planning, scheduling and control**
J. Gersak

148 **Anthropometry, apparel sizing and design**
D. Gupta and N. Zakaria

1

The future of smart-textiles development: new enabling technologies, commercialization and market trends

T. KIRSTEIN, TechPublish, Switzerland

DOI: 10.1533/9780857093530.1

Abstract: Smart-textiles development is entering a new era that is characterized by the convergence of different disciplines, such as electronics and polymer sciences. This chapter begins by reviewing the challenges in smart-textiles development and illustrating current trade-offs. Then, the chapter gives an overview of recent technological breakthroughs and shows how new enabling technologies from different disciplines help to overcome technological barriers. Finally, the chapter describes new approaches to the commercialization of smart textiles and reveals potential topics that will shape the future of smart textiles.

Key words: enabling technologies, trends in development, commercialization.

1.1 Introduction

A textile is a fascinating material. In various fabrication steps, millions of separate fibres are interlaced and modified and finally form a complicated structure with rather unpredictable properties. The resulting fibrous structures are highly inhomogeneous (electrical engineers might even say imprecise and chaotic) and therefore difficult to model. Knowing how to adjust textile process parameters in order to achieve a specific desired material behaviour is sometimes rather an art than science. However, the special structure of textiles is what makes them unique. No other material has this kind of drapability, lightness and porosity combined with robustness, strength and, last but not least, low manufacturing costs. A long time ago, textile producers discovered that textiles could provide more than just aesthetic and decorative features. The book *Extreme Textiles* shows amazing examples of recently developed high-performance textiles (McQuaid, 2005):

- textile containers that lift a weight of 12 tons;
- gloves that cannot be cut with knives or even razor wire;
- airbags that protect a space rover when landing on the surface of a foreign planet;
- racing cars (as well as boats, bicycles, etc.) that are made of carbon-fibre-reinforced polymers, that are lightweight, and can therefore achieve extremely high speeds.

If textiles can achieve such high-tech functionality, it is only logical to think about making them smart. The new research field of smart textiles arose about 15 years ago and initially focused on aerospace and military applications. After several years, researchers realized that smart functions in textiles also are attractive to civil and commercial products. However, the euphoria was dampened by a lack of success in the market. Now, smart-textiles research is finding new perspectives and directions. This chapter describes recent technological breakthroughs and analyses trends in smart-textiles development.

1.2 The technological trade-off between smartness and integration

1.2.1 Smartness of objects

The goal of every smart object is to make the lives of humans easier, safer and more comfortable. The smartness of an object refers to its ability to react to external stimuli. However, a clear definition is not easy as there are different levels of smartness. If an object property changes according to different environmental influences, this does not automatically make the object smart. After all, several mechanical, thermal, or electrical properties change depending, for example, on the surrounding climate. Smartness starts when the reaction of the object properties is not linear to the ambient changes and when this non-linearity brings benefits to the user. In the textile area, phase change materials are often called smart as they store heat in a hot environment and release heat when it gets cold. In contrast to 'normal' materials, smart materials change a specific property, like shape or aggregate state, rapidly and significantly. Smart materials are often called designed materials as their property changes can be tuned during fabrication. Common examples are:

- piezoelectric materials that can change shape when a voltage is applied;
- chromogenic materials that can change colour in response to electrical, optical or thermal stimuli;
- shape memory materials that can react with large deformations on temperature changes.

However, in such smart materials, the cause-and-effect relation is very simple: just one variable external stimulus is taken into account, and the reaction is always the same, regardless of any user needs.

A higher level of smartness can be achieved with electronic technology. In electronics, the non-linear behaviour of so-called 'active' electrical components like transistors is used to build programmable machines. These machines carry out logical operations and deliver a goal-oriented output, depending on multiple input parameters. Researchers have been improving the performance of computers and making them increasingly smarter. These advances have developed into the research field of artificial intelligence. In this area, smartness tries to imitate the human brain and involves planning, learning, reasoning and knowledge. This decision-making and problem-solving type of smartness can be called intelligence. 'An intelligent agent is a system that perceives its environment and takes actions that maximize its chances of success' (Russell and Norvig, 2003).

This enormous range of smartness often leads to confusion and controversy when the term 'smart textiles' is used. Certainly, for different applications, different levels of smartness are required. However, as a matter of principle, electronics is the only available technology that allows the generation of a goal-oriented output from multiple input parameters as well as user control. In addition, even the above-mentioned smart materials often do not function without electronics, as an electrical stimulus is required (e.g. to control piezoelectric actors). That means, no matter what level of smartness is envisioned, textiles have to learn from electronics. Textile technologies and materials sciences have to be combined with principles used in electrical engineering and measurement technology. For this reason, some experts prefer the term electronic textiles or e-textiles instead of smart textiles. However, this term is limiting as it excludes any kind of non-electronic smartness. In fact, the ultimate vision is smartness, and the use of electronics is just a means to an end.

1.2.2 Integration level

A combination of textile and smart functions in one product can be achieved on different integration levels. The first smart-textile products were inhomogeneous and consisted of textile and electronic parts that were rather attached to each other than really integrated. Over the years, smart-textiles developers have approached higher levels of integration. The shape and size of electronics were made more textile compatible so that they could be processed on textile machines and incorporated directly into fabrics. The integration of electronic functionality into yarns and fibres will be the next step. In this way, the electronics are not visible and do not disturb the desired textile look and feel. However, several problems need to be solved for this type of integration, and basic research still has a long way to go. The main challenges are the durability and reliability of the functions, as well as the scalability and industrialization of the fabrication processes.

1.2.3 Technological trade-off

So far, smart-textiles development has been characterized by a trade-off between smartness and level of integration. Most prototypes with smart functionality do not have a high level of integration, which means that several system parts – such as processing capabilities and energy supply – are placed in external devices. The few examples of 'monolithic' smart textiles only have a very limited functionality. This conflict can be illustrated with two examples of smart-window-shade concepts. Vili (2007) incorporated shape memory materials into textile window shades, in order to achieve an opening/closing of the woven structure depending on the surrounding temperature. This behaviour is sensitive only to the temperature, and it does not take into account other parameters such as the surrounding brightness, nor can it be influenced by the user. In contrast, the energy curtain (Fig. 1.1) is much more sophisticated and contains solar panels, batteries, LEDs and optical fibres. It can collect energy during the day and lights up the room during the night (Backlund, 2006). This system requires several electronic components and batteries but, in return, it offers a higher level of smartness.

Some researchers believe that textiles always will be just a part of a smart system where most of the processing power is outside the textiles. Others aim at pure textile systems where all the smartness is within the textile material. Their goal is to achieve a mergence of textile and electronic functionality within one

1.1 The energy curtain includes solar panels, batteries, LEDs and optical fibres. The user controls how much energy is collected during the day and how the room is lighted up during the night (Interactive Institute).

material. Either way, for both visions, experts from different disciplines have to join forces and push the boundaries. Textile engineers have to work together with material scientists, chemists, physicists, electrical engineers, designers and product developers from different branches. This kind of multidisciplinary cooperation is challenging because of the different ways of thinking, different languages and different approaches involved. However, in the last few years, tremendous changes have occurred because, in several disciplines, new enabling technologies have been established. Some of these new technologies have been known for a long time but can now be applied practically and combined for the first time. These major technological breakthroughs enable completely new approaches in smart-textiles development, as the next sections show.

1.3 New enabling technologies for smart textiles

In materials science, groundbreaking results have been achieved in the area of conductive polymers. Textile engineers have made substantial progress in the processing of substances and materials (e.g. nano-sized or non-polymer) that traditionally have not been used in textile technology. Especially relevant for smart textiles are new processes such as electrospinning, composite manufacturing and printing. In electrical engineering, the interest in flexible electronics and in organic electronics has drastically increased over the past few years. This trend perfectly matches the smart-textiles vision. Smart-textiles developers benefit from new electronic devices that are more and more textile compatible – which means flexible, organic, lightweight, low-cost and low power. Last but not least, advances in mobile electronics like the smart phone have a great influence on smart textiles. The following section provides an overview and examples of these recent technological achievements.

1.3.1 Textile technologies: towards nano and composite

Nanotechnology has been booming in many areas during the last few decades. The properties of nanoscale particles differ significantly from bulk material due to their high surface-area-to-volume ratio and their quantum mechanical effects. In the textile area, nanotechnologies are used to create fibres and fabrics with increased performance and added functionalities such as water-repellent, UV-protective, wrinkle-resistant, self-cleaning, anti-bacterial or anti-static abilities. There are already some nanotextiles on the market, mostly with dirt- or water-repellent properties. Nano-sized silver particles are used for their antimicrobial properties in a wide range of consumer products. Very promising nanomaterials for the future are based on carbon nanotubes, which possess extraordinary mechanical strength and high thermal and electrical conductivity.

The three major manufacturing technologies for nanotextiles are:

1. *Fabrication of nanofibres*: The best-established process for nanofibres fabrication is electrospinning. In this process, electrostatic forces are used to draw fluids into a fibrous form (Reneker and Fong, 2006).
2. *Surface modification of fibres and fabrics*: Technologies employed for applying nanocoatings include plasma treatment, atomic layer deposition (ALD), grafting of polymer chains, and self-assembly of monolayers, to name a few (Gorga, 2010).
3. *Filling of fibres with nanoparticles*: The embedding of nano-sized fillers, such as metal oxide or carbon into polymer fibres, leads to composite fibres with specific properties e.g. electrical conductivity.

The recently developed coaxial electrospinning technology enables the fabrication of composite nanofibres with a core-sheath structure or with embedded nanoparticles. Ko *et al.* (2008) demonstrate the possibilities of this so-called co-electrospinning process.

Bringing functional non-textile materials into fibrous forms is the great strength of nanotechnology. For this reason, nanotechnology plays an important role for smart textiles. First results show the potential of nanotechnology for creating sensor fibres as well as conducting and semiconducting fibres. Enz *et al.* (2010) produced temperature-sensitive fibres with electrospinning. Lima *et al.* (2011) presented an approach to spinning carbon nanotubes into yarns by scrolling carbon nanotube sheets. Sibinski *et al.* (2010) developed temperature-sensitive fibres with carbon nanotube coatings. Furthermore, ALD has been investigated as a method by which to produce conductive and semiconductive nanocoatings on fibres and fabrics (Jur, 2011).

1.3.2 Polymers: towards conducting and semiconducting

As far back as the 1970s, it was discovered that some polymers can be made electrically conductive. Until then, polymers were known for their electrically-insulating properties. In traditional polymer materials, valence electrons are bound in fixed and immobile sigma-bonds. It was thought that charge flow was only possible in materials like metal. However, some polymers (the conjugated polymers) possess a special molecular structure of alternating single and multiple bonds. In this structure, electrons can hop along the molecule chains if one or more electrons are removed or inserted. This process can occur in an oxidation or reduction process called 'doping.' With this invention, polymers can have electron mobility and imitate metal behaviour.

'Conducting polymers' have been synthesized before by mixing polymers with metal or graphite powders. However, in the case of this new class of materials, the conductivity was an intrinsic property. Therefore, the name intrinsically conducting polymers (ICPs) was recommended (Plieth, 2008). The main drawbacks of ICPs were their instability in air and their poor processability. It took several decades of basic research until the field of conductive polymers reached a transition to practical applications.

Now, this technology is also entering the textile world: Textile applications of conductive polymers range from anti-static fabrics or actuators for 'artificial muscles' (i.e. stimuli-responsive materials that react by modifying their shape or dimensions), wound dressing products with anti-bacterial properties or temperature, and optical- or strain-gauge sensors (Ferreras, 2011). Ferreras (2011) developed lightweight, low-cost strain sensors by coating fabrics with organic conductor films. Their fabrication method is compatible with printing technology.

Apart from the metal-like conductivity, the semiconductive properties of polymers have gained more and more importance. The variability of the conductivity, such as the ease with which the materials can be reversibly switched between their insulating and conducting forms, has enabled the development of organic field effect transistors (OFETs) (Inzelt, 2008; Plieth, 2008).

Electroluminescence in organic material has been recognized since the 1950s, but the drawback was the need to apply high voltages to create a light emission. The increased conductivity of modern conductive polymers has allowed them to generate higher amounts of light with lower voltages and has led to commercial applications in opto-electronic devices, such as organic light-emitting diodes (OLEDs) and photovoltaic cells (OPVs) (Kim J, 2011). Nowadays, we use OLED displays in everyday life in mobile devices. In the long term, the flexible nature of organic semiconductor materials will allow for the commercial development of opto-electronics into curved and flexible structures.

The tremendous impact of organic conductors on electronics is described in the next section. The technological progress in so-called plastic electronics is of great importance for smart textiles.

1.3.3 Electronics: towards large-area, flexible and printable

Miniaturization of electronics

The main driving force in the development of electronics always was the demand for miniaturization and performance increase. This goal was achieved by the integration of circuits onto semiconductor substrates. For these integrated circuits (ICs, microchips) single-crystal silicon is the semiconducting material most used (silicon wafer technology). The key transistor structures and the corresponding fabrication technology are called complementary metal-oxide semiconductors (CMOS). ICs have consistently migrated to smaller feature sizes over the years, allowing more circuitry to be packed on each chip. This high integration density not only enabled circuits with more functionality but also ones with higher computing power. In the last few decades, the integration has reached an amazing level. In the 1980s, the term 'very large-scale integration' (VLSI) was introduced and stood for thousands of transistors on one single chip. Today's ICs contain several billion transistors.

Large-area electronics

In the beginning of the 1990s, enormous interest for large-area electronics (also called macroelectronics) emerged. This trend was triggered by new requirements on displays. For several applications, a large size display is preferable, at best combined with a high resolution, a flat shape and a high image quality. Thus, a technology was needed for the active controlling of a high number of display pixels on a large area. Silicon wafer technology requires 'thick,' rigid and expensive silicon substrates and is therefore not suitable for large areas. Thin-film technology was the answer. In contrast to bulk silicon transistors (i.e. CMOS) thin-film transistors (TFTs) are made of amorphous silicon and can be deposited on non-semiconductor substrates (e.g. glass). The key benefit of thin-film technology is the ability to use low-cost, large-area and even transparent substrates. One further advantage is that amorphous silicon can be deposited at lower temperatures (below 500°C instead of ~1000°C). The result of this new technology was the expansion of flat-panel displays into numerous applications of our daily life, such as liquid crystal displays (LCDs) (Sun, 2011).

Organic electronics

The discovery of organic semiconductors opened up further possibilities for large-area electronics. The light-emitting properties of some organic semiconductors make them perfect for opto-electronic devices like displays. The development of OLEDs led to a new generation of large-area displays that consume less energy than LCDs and have lower manufacturing costs. The great advantage of OLEDs is that they emit light themselves and do not require a light-emitting backplane like LCDs. Today, OLEDs still are controlled with silicon-based TFTs. However, in the future, the switches could also be realized with transistors made of organic materials (e.g. organic field effect transistors, or OFETs). This would further reduce the thermal impact on the substrate as well as the manufacturing costs.

Flexible electronics

Parallel to the development of large-area electronics, another field of research has rapidly grown – namely flexible electronics. The main goals for the first flexible electronic devices were reductions in weight and size, and these were achieved by thinning of silicon-ICs (e.g. solar cells for satellites). Below a certain thickness, even bulk silicon chips become flexible. But, of course, this kind of bendability is not comparable to the drapability and elasticity typical of textiles. The next approach to making electronics flexible was to use thin-film technology on flexible substrates such as plastic or steel foils. This technology enabled, for example, a roll-to-roll fabrication of flexible thin-film solar cells

(Cheng and Wagner, 2009). The emergence of organic semiconductors had a major impact on flexible electronics, similar to the one it had on the aforementioned large-area electronics. The possibility to manufacture organic semiconductors at low temperatures is perfectly compatible with flexible substrates like plastic foils.

Printable electronics

A key manufacturing technology for organic electronics is printing. The most important benefits of printing are low-cost and large-scale fabrication. Several companies have developed printing technologies, for example, for OLEDs (DuPont, HP, Sony), for radio-frequency identification (RFID) and for solar cells. Jung *et al.* (2010) produced roll-to-roll printed RFID tags. Hübler *et al.* (2011) presented solar panels that are printed on paper. Enabled by nanotechnology, new printable inks have emerged that contain conductive particles. Das *et al.* (2011) evaluated novel printable materials. The roadmap of the Organic Electronics Association (Hecker, 2011) provides a good overview of technological developments in the fields of organic and printable electronics (Fig. 1.2). Two major challenges for the realization of organic devices are the limited lifetime of these devices and their sensitivity to atmospheric oxygen and humidity.

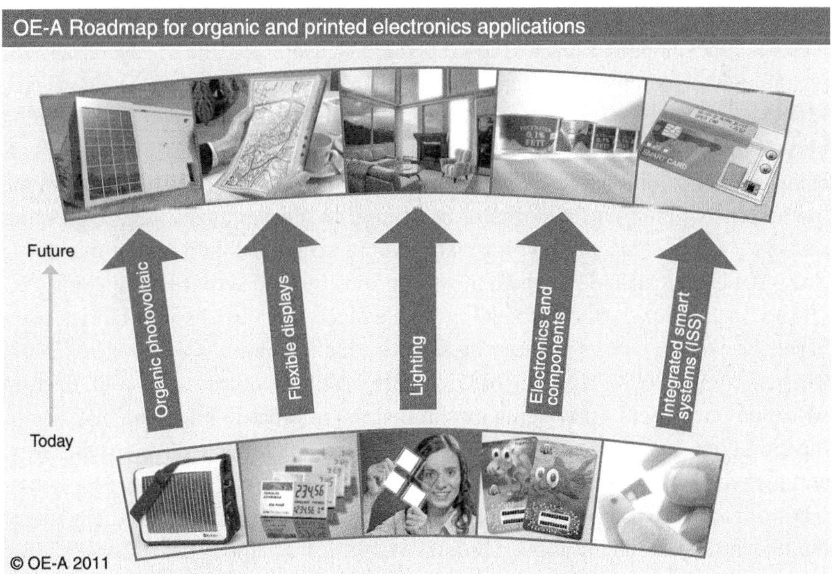

1.2 Roadmap of Organic Electronics Association (OE-A).

Stretchable electronics

From a textile point of view, one important question is how flexible electronics can become. Is it possible to make electronics not just bendable, but also stretchable and elastic? Some research groups already are working on different approaches to stretchable electronics. To date, the three main fabrication strategies for stretchable electronics are:

1. embedding ultra-thin silicon circuits in pre-stretched silicone rubber;
2. integrating thin-film circuits in an elastic, rubbery mesh;
3. distributing the electronic circuits on rigid platforms that are interconnected with stretchable metallization (Lacour, 2009).

Kim *et al.* (2011) recently reported the development of skin-like membranes that contain electronics. This kind of 'epidermal electronics,' as the researchers call it, can be attached to human skin like a tattoo.

Textile electronics

All the new electronic technologies that are moving towards large-area, flexible and printable have a huge potential to be adapted to textiles. In the following sections, the advances in implementing electronic technologies in textiles are shown.

Within the sixth framework programme of the European Union (2002–2006), several research projects in the area of smart textiles were carried out (cluster SFIT, which stands for smart fabrics, interactive textile). The first achievements were the development of textile-embedded sensors for the measurement of physiological parameters (e.g. within the projects WEALTHY, MyHeart, MERMOTH, OFSETH, BIOTEX, and PROETEX). Applications included personalized healthcare management (e.g. BIOTEX: Coyle, 2010) as well as smart garments for emergency operators (PROETEX: Curone, 2010). Most of the textile sensors monitor heart rate or breathing. In other projects, textile pressure sensors (Meyer *et al.*, 2006) and textile strain sensors (Mattmann *et al.*, 2008; Shyr, 2011) were also developed, mostly for posture and activity monitoring.

The first approaches to embedding whole electronic circuits into fabrics were based on woven fabrics with embedded conductive yarns (Cottet *et al.*, 2003; Locher *et al.*, 2004; Kirstein *et al.*, 2005). Silicon chips were mounted on so-called interposers (interfacing circuit boards) in order to adapt the distance of the chip connections to the larger distance of the conductive yarns. In this way, conductive yarns could be connected to chips and act as circuit paths.

Kim *et al.* (2010) applied printing technology to achieve more planar electronic circuits in fabrics. The European project MicroFlex aims to use standard printing techniques to realize micro-electromechanical systems (MEMS) and active films for sensing and actuating on fabrics.

The project STELLA demonstrated stretchable electronics for health monitoring devices and for car interior heating and lighting fabrics for the first time (Gonzalez, 2011). In 2010, two follow-up projects of STELLA started. The PLACE-it project aims to realize large-area and flexible opto-electronics systems, in particular light-emitting foils for interior lighting applications. The goal of the PASTA project is to create a conductive fibre with small silicon dyes that can be integrated into yarns. The envisioned applications range from monitoring applications for sportswear to interior textiles and industrial components. The project FLEXIBILITY aims to advance organic large-area electronics (OLAE). Multifunctional OLAE modules will be realized, such as printed, textile-integrated sound modules.

Several research groups are working towards the integration of electronics on the fibre and yarn levels. OFETs have been created on yarns and cylindrical shapes (Lee and Subramanian, 2005; Maccioni *et al.*, 2006; Hamedi, 2009). Mattana *et al.* (2011) developed transistors on cotton fibres. These examples show that electronic technologies are becoming more and more textile compatible in terms of shape, mechanical properties and fabrication processes.

1.3.4 Information technology: towards Internet and computing everywhere

The end of last century was characterized by the massive impact of information technology on our lives, on global society and on economics. The transition to an information society started when ICs revolutionized the world of electronics and led to the widespread use of personal computers. Technological advances made electronics smaller and, eventually, even mobile. In 1973, Motorola introduced the first mobile phone. Then, the Internet became a global network for everybody and enabled the easy and fast exchange of information. Today, the technological capacity to store, compute and share information is still increasing rapidly. In 2011, the projected amount of information created and replicated surpassed 1.8 zettabytes (1.8 trillion gigabytes) – and this figure will grow by an estimated factor of nine in just five years (Gantz, 2011). Hansmann *et al.* (2003) put it aptly when he wrote, 'Information is the new currency of the global economy.' The next step was the mergence of mobile telephony, personal computing and Internet services into multi-use handheld devices. The first mobile phones with wireless Internet connection were introduced in the late 1990s. Today, the so-called smart phone industry is booming. The smart phones run mobile operating systems and allow the installation of various software applications. According to estimates, by 2015, there will be at least 2 billion smart mobile devices in use globally (Kenney, 2011).

Researchers are now working on future mobile electronics that are wearable. Their goal is to create intelligent body-worn electronics that act as personal assistants. The research field of wearable computing belongs to the general vision of ambient intelligence. Experts believe that future electronics will surround us in a pervasive and ubiquitous manner:

> An ambient intelligence environment is capable of recognizing and responding to the presence of different individuals, working in a seamless, unobtrusive and often invisible way (Vasilakos, 2008).

New wireless technologies and distributed sensor networks enable this trend. Innovative sensor technologies allow the placement of sensors into everyday objects as well as close to the body. Barbaro (2010) illustrates how wearable technology has paved the way to a wide possibility of interesting biomedical applications, in particular in the field of individual physiological monitoring for diagnosis, therapy and rehabilitation. Even the on-body analysis of chemical substances, for example, in sweat is now possible.

Electronic devices are more and more distributed and the Internet is evolving from mobile towards embedded. Machine-to-machine (M2M) communication could be the next technological revolution after the mobile Internet. The vision behind M2M is the 'Internet of things,' that means embedding mobile Internet features into low-cost devices and connecting billions of everyday objects. 'Although this vision is not new, it is only now gathering momentum as ubiquitous connectivity is finally becoming a reality, and Moore's Law has driven device cost and size low enough to justify "smart devices everywhere" ' (Wu *et al.*, 2011). It can be expected that the massive interest in smart devices and ambient intelligence will push the development of smart textiles.

1.3.5 Consequences and impact on smart textiles

The previous sections revealed that the worlds of textiles and electronics are moving closer. On the one hand, electronics can increasingly have typical textile features such as flexibility, large-area and roll-to-roll processability. On the other hand textile technologies enable the processing of non-textile materials such as conductive particles in fibres or coatings. With these new enabling technologies, it is getting easier to combine the benefits of textile materials with electronic functionality. The realization of products that are smart and textile at the same time seems to be attainable. Some manufacturers even implemented automated assembly lines for the production of smart textiles. For example, the company TexTrace recently presented a manufacturing line for woven RFID labels. Nevertheless, it is still debatable when smart textiles will become widely commercialized.

Technological feasibility does not automatically guarantee commercial success. Yet several examples show that it sometimes takes a long time until the right application emerges and the breakthrough of a technology occurs. Several major technologies such as conductive polymers, electrospinning and thin-film electronics were discovered a long time ago and have only now broken through.

Electrospinning technology traces back into the last years of the 1800s but only reached a broad dissemination in the 1990s when the interest in nanotechnology increased (Reneker and Fong, 2006). Conductive polymers were introduced already in the 1970s with the discovery of polyacetylene and the process of oxidative doping (Plieth, 2008). However, only recently are conductive polymers rapidly gaining attraction in new applications such as organic LEDs and solar cells. Apparently, the emergence of new applications and the combination of methods from different disciplines can lead to an unexpected and rapid dissemination of a technology. The next section analyses new approaches and influencing factors in the commercialization of smart textiles.

1.4 New approaches in commercialization of smart textiles

The following aspects play a major role in commercializing smart textiles:

- What are the strengths of smart-textiles technologies?
- Are they predestined for specific application areas?
- Do smart textiles aim to replace existing products or satisfy new needs?

The following section evaluates strategies and chances in the smart-textiles business.

1.4.1 System-oriented development and focus on user needs

Healthcare has always been considered a key application for smart textiles. However, the first approaches of textile-based health monitoring failed mainly due to the difficulty of achieving certified medical reliability at reasonable costs. Now, the commercial focus is more on lifestyle monitoring. One major trend is sports monitoring. Partnerships are being formed between apparel companies and technology companies to develop and produce innovative ways of tracking, recording and sharing athletic performance. In 2010, Polar Electro teamed up with Nike to introduce the heart rate monitor Polar WearLink+. Recently, adidas gained commercial success with miCoach, a training tool including sensing shirts and football boots that measure performance data such as speed, number of sprints and distance. Nevertheless, the clothing and fashion market is rather difficult for smart textiles. For most fashion products, design and low cost are more important than sophisticated functionality. Even in the case of the athletic monitoring products, the profit comes largely from software selling rather than from the clothing itself.

Another popular application field for smart textiles is interior design. Philips is active in this area and offers luminous textile panels with integrated LEDs. Smart textiles provide a unique selling point for interior products because of their pleasant acoustic, haptic and optic properties in addition to their electronic functionality. Several textile manufacturers, designers and architects have developed ideas for smart furniture, interactive surfaces and car interiors. However, a piece of smart fabric is not marketable without a concept for the complete electronic system, including user interfaces, power supply, etc. Companies such as Philips can meet this need for system-oriented product development. Their luminous panels fulfil plug-and-play criteria and are delivered with standard connections for a power cable and an Ethernet cable. The user can install the system easily, upload and manage the dynamic content and even connect the system to building management systems. Similar to Philips, adidas and Nike also have access to their users and are experienced in developing end products. With this prerequisite, they manage to bring smart-textile systems to market. Yet for most of the textile manufacturers, system development is a critical point, as they normally supply just fibres and fabrics. For the breakthrough of smart-textile products, this barrier has to be overcome either by partnerships between component suppliers and end-product manufacturers or by the establishment of new 'system-building' companies.

Such new smart-system providers could come up with the required knowledge of the end users and their needs. The term 'system provider' also means that the focus is on finding solutions rather than on selling a specific type of product. The approach is not to ask 'What functionality can I put into a textile?' but rather 'What functionality is needed in a specific context, and which object could fulfil this need?' Depending on the task, several objects can contribute to smart services. In living spaces, vehicles or articles of daily use, some functions are best placed in textile surfaces, but others reasonably should be embedded in solid objects. The famous but unsuccessful smart jackets with embedded MP3 players demonstrate the failure of non-system-oriented product development. Well-engineered electronics like smart phones are hardly improved by textile gadgets like interfaces in sleeves or loud speakers in collars. This experience of unsuccessful smart garments also applies to technical textiles. Manufacturers of textiles for industrial applications and machines (i.e. belts, filters, etc.) are tempted to add value to the textile components by making them smart. However, a closer look at the whole system often reveals that sensors and actuators are much more easily embedded into non-textile parts of the system.

Focusing on the core competences of textiles helps to find suitable applications. Smart textiles are certainly predestined for large-area distributed sensing tasks. Apart from architecture, also branches such as civil engineering and agriculture deal with large surfaces. Within the European project POLYTECT, several smart geotextiles based on fibre-optic sensing systems have been developed, for example, for monitoring ground movements (Belli, 2009; Messervey, 2009). Could large-area textile sensor networks one day also monitor cultivated land, dykes or even glaciers? Some glaciers are actually covered with textiles in order to reduce melting (Fischer, 2011). The possibilities are endless, and focusing on system-oriented development might finally lead to the market success of smart textiles.

1.4.2 Do-it-yourself trend: chances for commercialization of smart textiles?

The do-it-yourself (DIY) movement started as a countermovement to the consumer culture many decades ago. The leading thought behind DIY was to move away from mass-produced goods towards unique handwork. There was a growing belief in the importance of the individual, in self-reliance and personal independence. The Internet has pushed this trend and has led to a boom in DIY culture. The web enables easy and global dissemination of DIY instructions. Online guides and YouTube tutorials are available for many topics such as home improvement, electronics, music production and fashion. On Internet platforms, such as Instructables.com, Techdiy.org or Makezine.com, the DIY community exchanges know-how and experience. In recent years, an explosive growth of online shops (e.g. Etsy.com) proved that DIY also has business potential. On these e-commerce pages, DIY products are sold from person to person.

DIY has already entered the smart textiles world as well. One major step has been the development of textile compatible electronic components that are accessible to a broad public and are easy to use. One example is the microcontroller board named LilyPad (Buechley *et al.*, 2008) by the High-Low Tech group of MIT. This circular-shaped flexible board was commercially introduced in 2007. It can be sewn to fabrics with conductive threads and connected to several sensors and actuators such as LEDs. It is based on the open-source computing platform Arduino and can be easily programmed. Following this development, new online shops, books, workshops and online databases emerged that are specialized on DIY smart textiles, for example:

- www.plugandwear.com (online shop, tutorials, workshops);
- www.kobakant.at/DIY (workshops, online database);
- DIY-books: *Fashion Geek* (Eng, 2009), *Fashioning Technology* (Pakhchyan, 2008) and *Switch Craft* (Lewis, 2008).

Figure 1.3 shows the result of a DIY workshop conducted by Kobakant: a T-shirt with a small solar-powered motor that moves an attached feather. Despite this popularity, some smart-textiles experts deride the DIY movement. From their point of view, handicraft and gadgets are not scientific, not marketable and economically irrelevant. Yet the smart-textiles area benefits from the DIY movement in many ways.

First, thanks to DIY, a broad public has gained access to smart-textiles technology and got to know its potential. This way, user acceptance of this new technology can be enhanced and concerns can be reduced. Furthermore, smart-textiles developers can learn from the DIY community about possible applications

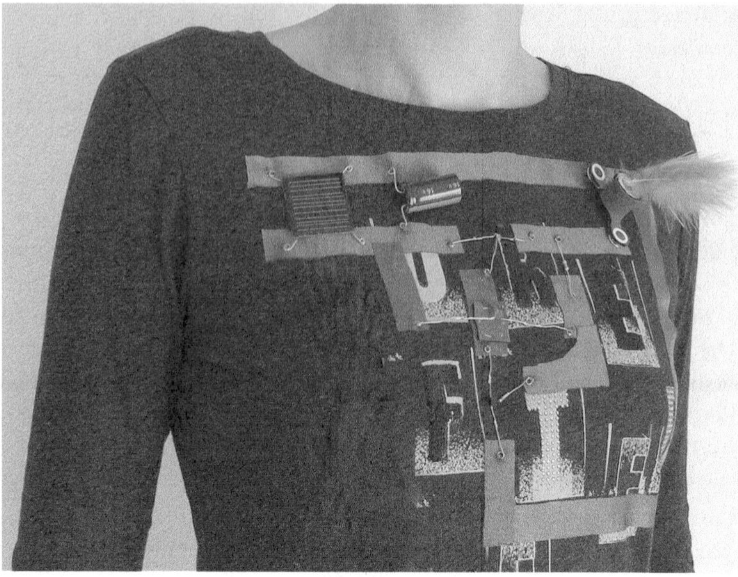

1.3 In this DIY solar shirt, a solar cell powered motor moves a feather (manufactured in a workshop of Mika Satomi, Kobokant).

and user needs. Researchers can use DIY products to perform real-world studies of the interactions between people and technology. Another positive effect of DIY toolkits is the spreading of smart-textiles technologies in the area of design, art and performance. It started with some individual fashion designers who experimented with smart textiles and expanded to well-known artists who have performed with interactive stage outfits. British musician Imogen Heap performed with musical gloves that enabled her to manipulate sounds and to compose on the fly with hand gestures. The light fashion clothes designer Moritz Waldemeyer developed illuminated video jackets for artists such as Take That and the Black Eyed Peas. Several exhibitions demonstrate the relevance of smart textiles to design (e.g. Prettysmarttextiles.com). Most smart textiles in the area of design and fashion are handmade masterpieces and are not comparable to marketable products. However, for the first time, they make publicly aware that smart textiles can be aesthetic and not just technical.

Finally, DIY has tremendous consequences for education. Now it is possible to teach smart-textiles topics not just in theory but also with hands-on experiments and workshops. In practical projects, students can experience the possibilities of this technology as well as the challenge of collaboration with other disciplines. Several schools and universities have carried out smart-textiles projects, where students with different backgrounds have worked together. One example is the e-motion project of www.design.udk-berlin.de/Modedesign/EMotion (the Berlin

University of the Arts (UdK), Germany). Another interdisciplinary student project is 'Ready-to-live,' which is a cooperation of the Wearable Computing Lab of the Swiss Federal Institute of Technology Zurich (ETH) and the Swiss Textile College (www.ready-to-live.ethz.ch/English). The participating students from the two disciplines of electrical engineering and fashion design seemed as different as day and night. However, they worked together with great enthusiasm and success. They appreciated the DIY approach and the possibility to get in touch with other ways of thinking. Figure 1.4 shows one of the interactive items of clothing that were presented at the final fashion show.

Smart-textiles projects are trend setting for future educational models – not just for university students but also for schoolchildren. Several research groups are exploring the potential of smart textiles in education. The EduWear project developed an educational low-cost construction kit for wearable interfaces (Katterfeldt *et al.*, 2009). The TeeBoard project resulted in an education-friendly construction platform for e-textiles (Ngai, 2010). These projects point out another important aspect of smart textiles in education: namely the potential to attract

1.4 In the fashion show 'Ready-to-live,' interactive clothes are presented. The clothes react to the movements of the wearer, e.g. with changing light effects. This student project was a cooperation of ETH Zurich and Swiss Textile College.

women to engineering and computer science. So far, these fields are mainly male-dominated. The integration of electronics and textiles create 'new spaces that historically have not been open to girls' (Kafai, 2010). Smart textiles provide the role of technology with a social context, which is an important component of attracting women to STEM (science, technology, engineering, and maths) (Lovell, 2010).

In conclusion, the DIY trend plays an essential role in the progress of smart textiles. The future of smart-textiles technology will be affected by end users who design and build their own products. Industrial companies can learn from the DIY community. Thus, DIY offers the opportunity to find new approaches to commercialization.

1.5 Future trends

The previous sections have shown the beginning of a new era of smart-textiles development: one that is characterized by novel enabling technologies and by the convergence of different disciplines such as electronics and polymer sciences. Yet in which direction will smart textiles develop? The next sections reveal potential topics that will shape the future of smart textiles. The possibilities of Web 2.0 and new applications with a focus on the global energy issue are interesting aspects of this future.

1.5.1 Web 2.0 and e-research

In recent years, the web has changed from being a publishing medium to a participative medium. New features of the so-called Social Web or Web 2.0 include tools and services that emphasize sharing of content among users and online collaboration (Murugesan, 2010). Some of the most widespread social software applications are:

- *Blogs*: The blog is a short term for web log and has its origin in online diaries. Nowadays a blog represents a kind of personal online journal that is frequently updated with comments, links, images, and other media pertaining to a given subject. The blog makes a statement and offers a space below for readers to comment and respond. Twitter is a microblogging service for posting short messages.
- *Wikis*: A wiki enables communities to write documents collaboratively. It is used as an online database for knowledge management, corporate intranets and group learning. One well-known example is the collaborative online encyclopaedia *Wikipedia*.
- *Social networks*: The idea behind platforms such as Facebook, Myspace, Google+ and LinkedIn is to share ideas, activities, events, and interests within individual social networks. According to Nielsen's *Social Media Report 2011*,

4 in 5 active Internet users visit social networks and blogs (Nielsen, 2011). Facebook is clearly the favoured social network in many countries.
- *Folksonomy* is also known as collaborative tagging. Tags act like keywords of a text and allow users to annotate and categorize content. Social websites use tagging to classify and find information.

Recently, the potential of Web 2.0 tools for business and marketing has been discovered by many companies. Several companies as well as non-profit organizations use social media for PR, branding and interaction with customers. For example, Nike has established a running community portal on Nikeplus.com, where users can monitor their sports performance. The web also has significant influence on research and technology development. Based on the progress of the web and of IT infrastructures a new generation of research has emerged, called e-research. E-research is characterized by the interaction between globally distributed teams of researchers. Virtual research communities share computer power and storage and collaborate across traditional boundaries of disciplines and locations (Wusteman, 2008). The social media support group interaction not only in research but also in industrial product development. Ahram *et al.* (2011) demonstrate the benefits of social networking as a collaborative tool for smarter product design. Social networking facilitates teamwork, collaborative online discussions, idea generation and peer review activities. The sharing of experiences enhances a team learning process.

Also in the smart-textiles world, researchers increasingly use Web 2.0 and e-research tools such as portals, blogs and online tutorials:

- SYSTEX is an EU-funded platform for Smart Textiles and Wearable Microsystems. Apart from workshops and conferences ('Smart Textiles Salon'), activities also include an online database (www.systex.org).
- Several universities and schools have established knowledge centres for smart textiles. One example is the Future Textiles initiative of the Danish design and management college TEKO (www.futuretextiles.dk). Seminars and competitions as well as exhibitions ('Pretty Smart Textiles') are organized.
- The Smart Textiles initiative of the Swedish School of Textiles of University of Borås aims at supporting textile companies and at promoting textile innovations (www.smarttextiles.se). Research is conducted in the Technology Lab and the Design Lab. A blog (www.stdl.se) provides information about projects and events such as the Ambience conference and summer camps.
- PhD students of the Textiles Futures Research Group, University of the Arts, London initiated a blog in order to engage in a dialogue about innovative textiles (textilefuturesphd.blogspot.com).
- Another interesting website is Fashioning Technology (www.fashioningtech.com). The author Syuzi Pakhchyan reports in her blog on the latest smart-textiles products and designs.

- +Plastic Electronics has a section about smart fabrics in their online and print magazine (www.plusplasticelectronics.com).
- Online shops and tutorials can be found at plugandwear (www.plugandwear.com) and SparkFun (www.sparkfun.com).
- Kobakant has an extensive online database of wearable technology (www.kobakant.at/DIY) and step-by-step instructions of DIY workshops.
- The blog of talk2myShirt (www.talk2myshirt.com) gives reviews and background information on textile-based wearable electronics and offers a weekly podcast (eTextile Lounge). Talk2myShirt is starting a new connection community platform that serves as an interaction place for experts of wearable electronics. It will provide an interactive whiteboard for the creation process and tools for the involvement of consumers.

These activities show that smart-textiles developers are moving towards online collaboration and interactive sharing of information. The Social Web helps to overcome the barriers between different disciplines and could lead to a revolutionary e-research approach in smart textiles.

1.5.2 Global energy issues

A sustainable energy supply is widely regarded as one of the most important intellectual and technological challenges of the twenty-first century. Climate change concerns and shortage of fossil fuel reserves are driving the development of alternative energy technologies. Energy from renewable resources such as sunlight, wind, rain, tides, and geothermal heat is becoming increasingly popular. An interesting aspect of renewable energy is the possibility of distributed electricity generation. In contrast to large centralized facilities for burning fossil fuel, renewable power sources can be small and decentralized. This allows electricity generation at the point of need, even in rural and remote regions. Small-scale wind generators and solar panels on the roofs of buildings are the first examples of this approach. The idea of autonomous electronic systems goes even one step further by powering devices directly with integrated miniaturized energy harvesters. Advances in MEMS technologies enable the embedding of energy harvesters in small wireless devices. Energy harvesting devices convert ambient energy such as vibrations, temperature gradients or solar power into electrical energy. The human body offers a variety of energy sources for wearable devices. Kinetic generators in wristwatches and shoes, for instance, capture energy from body movements.

This trend of distributed and embedded energy harvesting opens up tremendous possibilities for smart textiles. Textiles have the potential to act as energy suppliers, not just for electronic systems within the fabric but also for external devices:

- Several research groups are working on energy-harvesting microsystems. Researchers at the University of Southampton are building vibration-powered generators that could be integrated into clothing (Beeby, 2006).

- Konarka Technologies presented OPVs in a flexible wire format (Lee, 2009).
- Bedeloglu *et al.* (2010) examined the applicability of polymer-based solar cell materials onto thin polypropylene textile substrates and manufactured textile-compatible photovoltaic tapes.
- In the course of the EU-funded project DEPHOTEX, photovoltaic threads for automotive applications and clothes were developed (www.dephotex.com).
- Riaz (2011) and Wang (2008) have described concepts for piezoelectric nanogenerators.
- The architect Sheila Kennedy designs buildings with energy-harvesting textile surfaces (Soft House project, Fig. 1.5). In the non-profit initiative Portable Light Project, she provided photovoltaic textile kits for the developing world. Her goal was to enable local people to create and own energy-harvesting textile blankets, bags and clothing using available materials and traditional weaving and sewing techniques (Anonymous, 2008).
- Hu *et al.* (2010) recently demonstrated carbon nanotube coated textiles that can act as supercapacitor energy storage devices.
- Vatansever *et al.* (2011) used piezoelectric polymer fibres for energy scavenging from raindrops and wind.

1.5 The Soft House, designed by Kennedy & Violich Architecture, demonstrates how domestic infrastructure can become soft and flexible. The smart energy-harvesting textile cladding bends to respond to sun angle and twists to open views or create privacy.

These projects show the need for innovative energy-harvesting approaches. Smart textiles could play a pioneering role in meeting the energy demands of modern society. Their unique material properties make them perfect for new forms of drapable energy technology. With smart textiles, energy harvesting and storage could become soft and flexible. It could be distributed over large areas, as well as be embedded in fabrics surrounding the humans. Energy will be a key application area of future smart textiles and could possibly lead to a breakthrough of smart textiles.

1.6 Conclusion

This chapter has shown that technologies from different disciplines are moving towards similar goals: electronics are becoming soft and embedded, and textiles are becoming composite and functionalized from the macro- to the nano-level. These trends are opening up new perspectives in smart-textiles development. Some promising new approaches have been presented in this chapter, such as focusing on system-oriented product development, learning from the DIY movement, using Web 2.0 tools, as well as addressing global issues. These ideas could help and inspire researchers and companies to make the vision of smart textiles a reality.

1.7 References

Ahram T *et al.* (2011), 'Collaborative systems engineering and social networking approach to design and modelling of smarter products', *Behav Info Tech*, 30(1), 13–26.

Anonymous (2008), 'Kennedy's portable light wins tech award', *MIT TechTalk*, 53, 17 September, 7.

Backlund S (2006), 'STATIC! The aesthetics of energy in everyday things', *Proceedings of the Design Research Society International Conference*, Lisbon.

Barbaro M (2010), 'Active devices based on organic semiconductors for wearable applications', *IEEE Trans Info Tech Biomed*, 14(3), 758–766.

Bedeloglu A *et al.* (2010), 'Development of energy generating photovoltaic textile structures for smart applications', *Fibers Polym*, 11(3), 378–383.

Beeby S P (2006), 'Energy harvesting vibration sources for microsystems applications', *Meas Sci Technol*, 17, R175–R195.

Belli R (2009), 'Smart textiles for SHM of geostructures and buildings', *Proceedings of the SHMII-4*, Zurich, Paper No: 368.

Buechley L, Eisenberg M, Catchen J and Crockett A (2008), 'The LilyPad Arduino: using computational textiles to investigate engagement, aesthetics, and diversity in computer science education', in *Proceedings of the CHI 2008*, ACM Press, 423–432.

Cheng I and Wagner S (2009), 'Overview of flexible electronics technology', in Wong W S and Salleo A (eds.), *Flexible Electronics: Materials and Applications*, Series: Electronic Materials: Science and Technology, Springer, Berlin.

Cottet D, Grzyb J, Kirstein T and Tröster G (2003), 'Electrical characterization of textile transmission lines', *IEEE Trans Adv Pack*, 26(2), 182–190.

Coyle S (2010), 'BIOTEX: Biosensing textiles for personalised healthcare management', *IEEE Trans Info Tech Biomed*, 14(2), 364–370.

Curone D (2010), 'Smart garments for emergency operators: the ProeTEX project', *IEEE Trans Info Tech Biomed*, 14(3), 694–701.

Das R N (2011), 'Printable electronics: towards materials development and device fabrication', *Circuit World*, 37(1), 38–45.

Eng D (2009), *Fashion Geek*, North Light Books, F+W Media, Ohio.

Enz E *et al.* (2010), 'Electrospun microfibres with temperature sensitive iridescence from encapsulated cholesteric liquid crystal', *J Mater Chem*, 20, 6866–6872.

Ferreras L R (2011), 'Highly piezoresistive textiles based on a soft conducting charge transfer salt', *J Mater Chem*, 21, 637–640.

Fischer A (2011), 'Glaciers, snow and ski tourism in Austria's changing climate', *Annals Glaciol*, 52(58), 89–96.

Gantz J (2011), 'Extracting Value from Chaos', available from http://www.emc.com/leadership/programs/digital-universe.htm [Accessed 24 November 2011].

Gonzalez M (2011), 'Design and implementation of flexible and stretchable systems', *Microelect Reliab*, 51, 1069–1076.

Gorga R E (2010), 'Nanotechnology in textiles', *Textile World*, November–December.

Hamedi M (2009), 'Fiber-embedded electrolyte-gated field-effect transistors for e-textiles', *Adv Mater*, 21, 573–577.

Hansmann U *et al.* (2003), *Pervasive computing*, Springer, Berlin, p. 13.

Hecker K (2011), *Organic and Printed Electronics*, Fourth Edition. VDMA Verlag, Frankfurt.

Hu L *et al.* (2010), 'Stretchable, porous, and conductive energy textiles', *Nano Lett*, 10, 708–714.

Hübler A *et al.* (2011), 'Printed paper photovoltaic cells', *Adv Energy Mater*, 20, 1–5.

Inzelt G (2008), *Conducting Polymers: a new era in electrochemistry*, Springer, Berlin, pp. 1–6.

Jung M *et al.* (2010), 'All-printed and roll-to-roll-printable 13.56-MHz-operated 1-bit RF tag on plastic foils', *IEEE Trans Electron Dev*, 57 (3), 571–580.

Jur J S (2011), 'Atomic layer deposition of conductive coatings on cotton, paper, and synthetic fibers: conductivity analysis and functional chemical sensing using 'all-fiber' capacitors', *Adv Funct Mater*, 21, 1993–2002.

Kafai Y B (2010), 'Fröbel's forgotten gift: textile construction kits as pathways into play, design and computation', *Pro. IDC 2010*, 214–217.

Katterfeldt E *et al.* (2009), 'EduWear: smart textiles as ways of relating computing technology to everyday life', *Proc. ADC 2009*, 9–17.

Kenney M (2011), 'Structuring the smart phone industry: is the mobile internet OS platform the key?' *J Ind Compet Trade*, 11, 239–261.

Kim D *et al.* (2011), 'Epidermal electronics', *Science*, 333, 838–843.

Kim J (2011), 'Flexible OLEDs and organic electronics', *Semicond Sci Technol*, 26, 030301.

Kim Y (2010), 'Electrical characterization of screen-printed circuits on the fabric', *IEEE Trans Adv Pack*, 33 (1), 196–205.

Kirstein T, Cottet D and Grzyb J (2005), 'Wearable computing systems: electronic textiles', in Tao X, *Wearable Electronics and Photonics*, Woodhead Publishing Ltd, Cambridge.

Ko F K *et al.* (2008), 'Functional nanofibre: enabling material for the next generations smart textiles', *JFBI*, 1 (2), 81–92.

Lacour S (2009), 'Shapes of things to come', *C&I Magazine*, Society of Chemical Industry, No. 9.

Lee J B and Subramanian V (2005), 'Weave patterned organic transistors on fiber for E-textiles', *IEEE Trans Electron Dev*, 52 (2), 269–275.

Lee M R (2009), 'Solar power wires based on organic photovoltaic materials', *Science*, 324, 232–235.

Lewis A (2008), *Switch Craft: Battery-Powered Crafts to Make and Sew*, Potter Craft, The Crown Publishing Group, New York.

Lima M D *et al.* (2011), 'Biscrolling nanotube sheets and functional guests into yarns', *Science*, 331, 7 January, 51–55.

Locher I, Kirstein T and Tröster G (2004), 'Routing methods adapted to e-textiles', in *Proceedings of 37th Int Symposium on Microelectronics (IMAPS 2004)*, Long Beach CA, 14–18 November.

Lovell E (2010), 'An E-sewing tutorial for DIY learning', *Proc IDC 2010*, 230–233.

Maccioni M, Orgiu E, Cosseddu P, Locci S, and Bonfiglio A (2006), 'Towards the textile transistor: assembly and characterization of an organic field effect transistor with a cylindrical geometry', *Appl Phys Lett*, 89, 143 515–143 517.

Mattana G (2011), 'Organic electronics on natural cotton fibres', *Org Electron*, 12 (12), 2033–2039.

Mattmann C, Clemens F and Tröster G (2008), 'Sensor for measuring strain in textile', *Sensors (MDPI)*, 8(6), 3719–3732.

McQuaid M (2005), *Extreme Textiles*, Princeton Architectural Press, New York.

Messervey T B (2009), 'Industrial smart material applications: the Polytect Project and its expansion into the IMS Program', *Proc SHMII-4*, Zurich.

Meyer J, Lukowicz P and Tröster G (2006), 'Textile pressure sensor for muscle activity and motion detection', *Proc. ISWC 2006*, Montreux, Switzerland, 11–14 October.

Murugesan S (2010), 'Web X.0: a road map', in Murugesan S, *Handbook of Research on Web 2.0, 3.0, and X.0: Technologies, Business, and Social Applications*, Information Science Reference, 1–11.

Ngai G (2010), 'i*CATch: a scalable, plug-n-play wearable computing framework for novices and children', *Proc. CHI 2010*, 443–452.

Nielsen (2011), 'The Social Media Report 2011', Available from http://blog.nielsen.com/nielsenwire/social [Accessed 28 November 2011].

Pakhchyan S (2008), *Fashioning Technology: A DIY Intro to Smart Crafting*, O'Reilly Media, California.

Plieth W (2008), 'Intrinsically conducting polymers', in Plieth W, *Electrochemistry for Materials Science*, Elsevier: Amsterdam, pp. 323–363.

Reneker D H and Fong H (2006), 'Polymeric nanofibers: Introduction', in Reneker D H *et al.*, *Polymeric Nanofibers*, ACS Symposium Series, American Chemical Society: Washington, DC.

Riaz M (2011), 'Study of the piezoelectric power generation of ZnO nanowire arrays grown by different methods', *Adv Funct Mater*, 21, 628–633.

Russell S J and Norvig P (2003), *Artificial Intelligence: A Modern Approach*, Second Edition, Prentice Hall, Upper Saddle River, NJ, p. 34.

Shyr T (2011), 'The effect of tensile hysteresis and contact resistance on the performance of strain-resistant elastic-conductive webbing', *Sensors*, 11, 1693–1705.

Sibinski M *et al.* (2010), 'Flexible temperature sensors on fibers', *Sensors*, 10, 7934–7946.

Sun J (2011), 'Materials for printable, transparent, and low-voltage transistors', *Adv Funct Mater*, 21, 29–45.

Vasilakos V (2008), 'Special issue: ambient intelligence', *Information Sciences*, 178, 585–587.

Vatansever D *et al.* (2011), 'An investigation of energy harvesting from renewable sources with PVDF and PZT', *Smart Mater Struct*, 20, 055019.

Vili C Y F C (2007), 'Investigating smart textiles based on shape memory materials', *Textile Research Journal*, 77(5), 290–300.

Wang Z L (2008), 'Towards self-powered nanosystems: from nanogenerators to nanopiezotronics', *Adv Funct Mater*, 18, 3553–3567.

Wu G *et al.* (2011), 'M2M: from mobile to embedded internet', *IEEE Comm Mag*, April, 36–43.

Wusteman J (2008), 'Virtual research environments: what is the librarian's role?' *J Librarianship Info Sci*, 40 (2), 67–70.

Part I
Materials

2

Types and processing of electro-conductive and semiconducting materials for smart textiles

A. SCHWARZ, RWTH Aachen University, Germany
and L. VAN LANGENHOVE, Ghent University, Belgium

DOI: 10.1533/9780857093530.1.29

Abstract: The latest technological advances in new materials and devices enabled textile solutions to wearable systems. These solutions require electro-conductive textile materials as a basic component. This chapter first discusses key terms used in the context of electro-conductive textiles. It then addresses different conductive and semiconductive materials and their production methods that are applied in textiles.

Key words: electronic textiles, conductive and semiconductive yarns, production methods, coating technology.

2.1 Introduction

For a long time, the textile industry has used metallic yarn in weaving and knitting for aesthetic and decorative purposes. The first known conductive fabric was silk organza, which consisted of two types of yarns: a plain silk yarn running in the warp direction and, as the weft, a silk yarn wrapped in thin copper foil. Copper provided a shiny and reflective appearance; the silk core gave tensile strength. These yarns have been woven in India over the last century utilizing other metals such as silver and gold. In the 1920s and 1930s, metallic yarns were also used for ecclesiastical and courtly robes to give them the appearance of 'golden clothes'. In the 1950s, Lurex®, made from a thin strip of aluminium, entered the market. It could be added to a number of conventional fibres and yarns to create metallic fabrics, which were very popular at that time (Braddock *et al.*, 1999).

However, let us turn away from the aesthetic and decorative purposes of metallic yarns in clothing and take a look at the functional aspects that metallic filaments and yarns can fulfil. In more technically-oriented areas, conductive materials have been in use for anti-static and shielding purposes for the last 30 years. However, the conductivity needed for these purposes is rather low, which is explained in Section 2.2. With the increasing interest in smart textiles, the demand on highly conductive textile materials has also increased. They are used, for instance, in sensors, as transmission lines or as heating elements.

Besides using metals, such as copper, silver, stainless steel or aluminium, in fibre or filament form or incorporated into yarns, other types of conductive materials are now used, such as conductive polymers, conductive coatings and inks. In Sections 2.3 and 2.4, a range of conductive and semiconductive textile materials is discussed.

Conductivity can be introduced at different levels in a textile material:

- on a fibre, yarn or fabric level;
- during production; or
- applied as after-treatments.

The different possibilities are further discussed in Section 2.5.

2.2 Electro-conductive and semiconductive materials

2.2.1 What is electrical conductivity?

This section focuses on terms used to describe electrical properties of materials, such as electrical current, conductivity, resistance and resistivity, starting with the former term. Electrical current means charges in motion through an object. Depending on the matter, this charge is transported either by electrons (in a metal conductor), by ions (in an electrolyte) or by a combination of both, electrons and ions (in a plasma). Here, only electrons as charge carriers in solid conductors, such as metal filaments, are addressed, as they are of the most interest in the context of smart textiles.

To constitute an electrical current, there must be a flow of charge through an object along a potential difference that corresponds to an electrical voltage. Without a potential difference, no current will flow. This phenomenon can be depicted with the help of a copper wire (Fig. 2.1(a)). If we consider a closed-loop copper wire, free electrons will travel through it in all directions. Although electrons are available, no electrical force acts on them, as they are all at the same potential, and thus there is no current. If a battery is inserted into the loop (Fig. 2.1(b)), the conducting loop will no longer be at a single potential. Electrical fields act inside the copper wire, causing the electrons to move and thus establishing a current (Halliday *et al.*, 2001).

Conventional electrical current is defined by the movement of positive charge carriers in an electrical field (called drift movement). However, in metallic solids, such as copper wires, charge carriers are electrons, which are negative and therefore flow in the opposite direction. As illustrated in Fig. 2.1(b), the flow of current is indicated by an arrow associated with the symbol i to indicate the direction of the conventional current flow.

The electrical current that flows through an object is dependent on the material it is made of. For instance, if the same potential difference between the ends of copper wires or glass rods is applied, having the same geometry, different currents result. The object's resistance towards the flow of current through it plays a role.

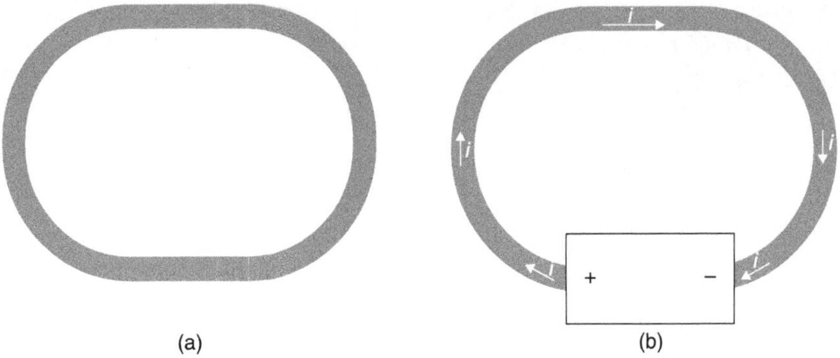

2.1 (a) A closed loop of copper wire; the wire is at one potential; (b) An electrical potential difference is caused between the ends of the copper wire due to a battery to which the ends are connected. The battery produces an electric field within the wire and causes charges (current *i*) to move around in one direction.

The resistance between any two points of a conductor is determined by applying a potential difference V between those points and measuring the current I that results. The resistance R is:

$$R = \frac{V}{I} \qquad [2.1]$$

The resistance is expressed in Ohm (Ω). It can be seen in the formula, that for a given potential difference, the greater the resistance is (to the current), the smaller the current. This formula is often referred to as Ohm's Law. However, it should be noted that resistance can be linear or non-linear. Only linear resistance obeys Ohm's Law, which states that the current is directly proportional to the potential difference between the two ends of the copper wire.

It is often necessary to take a general view and deal with materials instead of particular objects, because if the dimensions of the object change, the resistance will also change. However, a material's ability to oppose or conduct the flow of electrical current should be definable regardless of its dimension or shape. In this case, the potential difference V across a particular resistor should be regarded as an electrical field $\vec{E}$ at a point in a resistive material. It should not be focused on the current I through the resistor, but on current density $\vec{J}$ at the point in question. Instead of the resistance R of an object, it then needs to be referred to the resistivity ρ of the material:

$$\rho = \frac{E}{J} \qquad [2.2]$$

Combining the SI units for E and J, ohm-metre (Ωm) results as unit of ρ:

$$\frac{unit(E)}{unit(J)} = \frac{\frac{V}{m}}{\frac{I}{m^2}} = \frac{V}{I} m = \Omega\text{m} \qquad [2.3]$$

We often speak of the conductivity σ of a material. This is simply the reciprocal of its resistivity, resulting in:

$$\sigma = \frac{1}{\rho} \qquad [2.4]$$

If we know the resistivity of a material, such as copper, we can calculate the resistance of a length of copper wire. If the current density is uniform throughout the wire, the electrical field and the current density will be constant for all points within the wire and will have the values:

$$E = \frac{V}{L} \quad and \quad J = \frac{I}{A} \qquad [2.5]$$

where A is the cross-sectional area of the wire (in m^2) and L is its length (in m). The two formulas can be combined to give:

$$\rho = \frac{E}{J} = \frac{\frac{V}{L}}{\frac{I}{A}} \qquad [2.6]$$

However, V/I is the resistance R, which allows recasting the formula as:

$$R = \rho \frac{L}{A} \quad or \quad \rho = R \frac{A}{L} \qquad [2.7]$$

However, this formula, also known as the Pouillet formula, can only be applied to a homogenous isotropic material of uniform cross section. We can also give this formula a more 'textile character', as the cross-sectional area A of a fibre or yarn can be determined in different ways:

- It is possible to calculate the cross-sectional area applying a geometrical formula, when the dimension of the electro-conductive part in a fibre or yarn is known.
- The calculation can also be based on the linear density of the metallic fibre or yarn used and the electro-conductive material density of this fibre or yarn, respectively. By knowing the linear density of the fibre or yarn expressed in g/10 000 m (dtex), as well as its density ρ_D in g cm^{-3}, its diameter D can be calculated (Morton and Hearle, 2008) by:

$$D = \sqrt{\frac{400 * dtex}{\pi \rho_D}} \quad [2.8]$$

D is expressed in μm.

Applying the Pouillet formula, the resistivity can then be calculated by:

$$\rho = R \frac{\left[\left(\sqrt{\frac{400 * dtex}{\pi \rho_D}}\right) 10^{-a}\right]^2 \pi}{4} L^{-1} \quad [2.9]$$

expressed in $\Omega\ mm^2 m^{-1}$.

However, in the case of fibres and yarns, difficulties often appear when the cross-sectional area is included in the formula, as it assumes that the material is homogeneous. Hence, parameters, such as fineness, density and measured resistance per unit length, are considered to be constant along the yarn's axis. Since yarns are composed of several individual fibres or multiple filaments, they are far from homogeneous. This implies that the cross-sectional area A can vary along their length. Consequently, the variances may be too high to result in reliable results.

Against these considerations, the most reliable approach is also the most straightforward, that is to relate the electrical resistance R_y of a fibre or yarn to its length L:

$$R_y = \frac{R}{L} \quad [2.10]$$

where R_y is related to length L and is expressed in $\Omega\ cm^{-1}$.

The same difficulties appear when trying to express the resistivity of a conductive fabric, whether it be woven, knitted, braided, printed or a non-woven. Due to the textile's inhomogeneity and anisotropy, it is difficult to relate the resistance to the volume of the conductive area. Hence, it is often advantageous to express the surface resistivity of the fabric instead of its volume resistivity. The surface resistivity only takes into account the distance (D) between the two measuring electrodes and their contact length (L) with the sample, which are equal (Fig. 2.2). The physical dimensions, such as thickness and diameter of the conductive layer or yarn, are not considered.

The resistivity is therefore the quotient of the potential gradient, in V/m, and the current per unit of electrode length, A/m. Since the four ends of the measuring electrodes define a square, the lengths in the quotient cancel and surface resistivities are reported in ohms, although it is common to see the more descriptive unit of ohms per square (Ω/□) (IEC, ASTM).

Before turning to the differences between conductors, semiconductors and insulators, a tabular summary of the above discussed quantities, namely voltage, current, resistance, resistivity and conductivity, is presented in Table 2.1.

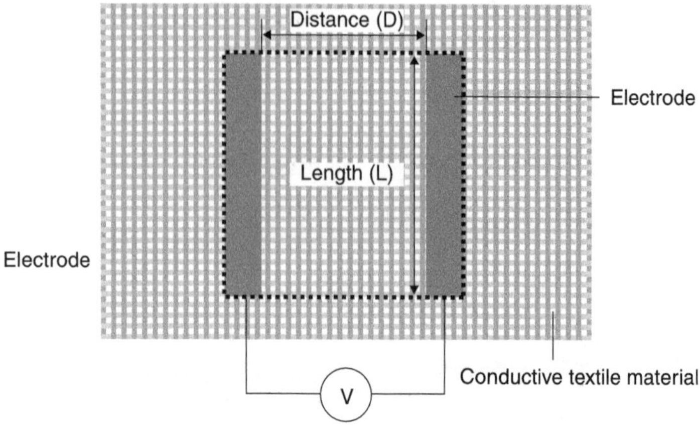

2.2 Set-up for surface resistivity measurements.

Table 2.1 Overview of quantities used in electronics around electric current

Quantity	Symbol	Unit of measure	Abbreviation
Voltage	V or E	Volt	V
Current	I	Ampère	A
Resistance	R	Ohm	Ω
Resistivity	ρ	Ohm metre	Ω m
Conductivity	σ	Siemens per metre	S m^{-1}

2.2.2 What is the difference between conductors, semiconductors and insulators?

The differences between conductors, semiconductors and insulators can be roughly explained when looking at their resistivities (Table 2.2). A conductor is a material that allows an electrical current to pass easily. Hence, it has a low value of resistivity. An insulator is just the opposite of a conductor. It opposes electrical current to flow and thus has a high resistivity. Semiconductors have electrical properties somewhere in the middle, between those of a conductor and an insulator.

The question arises as to why some materials allow the electrical current to flow easily and some others do not. The question can be answered when looking at the energies of the electrons that are present in a material. In a conductor, such as a copper wire, there are plenty of electrons that, roughly speaking, are only loosely held in place within the molecules and require only a little energy to become free. An electrical field applied across the conductor can supply that

Table 2.2 Different materials and their resistivity

Material	Resistivity ρ (Ω m)
Typical metals	
Silver	$1.62\ 10^{-8}$
Copper	$1.69\ 10^{-8}$
Gold	$2.44\ 10^{-8}$
Aluminium	$2.75\ 10^{-8}$
Nickel	$6.93\ 10^{-8}$
Iron	$9.68\ 10^{-8}$
Platinum	$10.6\ 10^{-8}$
Stainless steel	$72\ 10^{-8}$
Typical semiconductors	
Silicon, pure	$2.5\ 10^{3}$
Silicon, n-type[1]	$8.7\ 10^{-4}$
Silicon, p-type[2]	$2.8\ 10^{-3}$
Typical insulators	
Glass	$10^{10}–10^{14}$
Polyamide	10^{16}
Polyester	10^{17}

[1] Pure silicone doped with phosphorus impurities to charge carrier density of $10^{23}m^{-3}$

[2] Pure silicone doped with aluminium impurities to charge carrier density of $10^{23}m^{-3}$

energy. The field would not only free these loosely-held electrons but would also propel them along the wire. Thus, the electrons travel together through the material forming an electron drift. Generally speaking, most metals are good electrical conductors (Halliday *et al.*, 2001; Tutorials. 2011a,b).

In an insulator, all or most of the electrons are firmly locked into place within the molecules and significantly greater energy would be required to free them so they could move and participate in an electrical current. Any reasonable electrical field applied to the insulator cannot supply enough energy to move the electrons and hence, no current will flow giving these materials their isolating properties. Insulators play an important role within an electrical circuit as they prevent short circuits. Glass, epoxy glass resins and most polymers, also in textile form, are good insulators.

A semiconductor is like an insulator, except that the energy required to free some electrons is less. It has very few free electrons because its atoms are closely grouped together. However, by adding certain impurities or dopants to the semiconductive structure, charge carriers can be supplied that are very loosely held within the material and thus are easily movable. Charge carriers can be additional electrons or removed electrons, which act like a movable positive charge. Moreover, by controlling the doping of a semiconductor, the density of charge carriers that can participate in a current can be controlled, and thereby its electrical properties (Halliday *et al.*, 2001).

2.3 Electro-conductive materials and their properties

The variety of electro-conductive materials has significantly increased in recent years. The choice of which material to use depends to a large extent on its properties. Next to the material's electro-conductive properties, its chemical characteristics, such as reactivity to humidity, air or fluids, and its mechanical behaviour are relevant parameters during processing and for the performance of the final product.

The chemical characteristics of an electro-conductive material, especially in the case of metals, are best described when talking about corrosion. Corrosion means, in general, oxidation of metals due to an oxidant such as oxygen being present in the surrounding air. Hence, a metal oxide is formed on the surface of a metal. In most cases, the oxide of a metal has worse electro-conductive properties compared to the pure metal. Stainless steel, for instance, is covered with a chromium oxide layer after some time. This protective, non-conductive layer determines the high contact resistance of stainless steel surfaces. On a copper surface, to give another example, a layer of copper(I) and copper(II) oxide can be formed – surface corrosion processes that seriously affects its electro-conductive performance. Silver is an exceptional metal. Although an oxide film is built up over time on the silver's surface, the metal is not compromised in its electro-conductive behaviour, because silver in oxide form possesses the same electro-conductive properties as in its pure metal form. Noble metals, such as gold or palladium, possess a surface that is remarkably resistant to oxidation. Corrosion may also appear in polymers. However, in this context, the term degradation is most likely used.

As aforementioned, the mechanical properties of an electro-conductive material is another important aspect when choosing the right material. It is known that metal filaments compromise good electro-conductive properties with weak mechanical characteristics. They can hardly be stretched or stressed without breakage or at least a change of their electrical resistance. Oppositely, conventional polymeric yarns possess excellent elongation and recovery properties. Typical stress–extension curves are illustrated in Fig. 2.3.

When textiles are combined with electro-conductive materials, it is expected that the elastic behaviour of the conductive material is enhanced. This is important not only for comfort but also for processability. Electro-conductive yarns or metal filaments integrated into textiles need to be stressed during production and to be stretched during use. Thus their fragility will pose a major problem for their application in textiles, and their non-elasticity will decrease their service life. A good compromise can be achieved when combining metal filaments with elastic

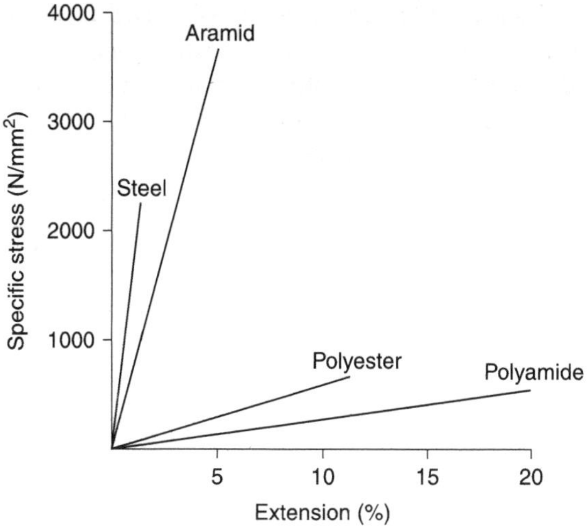

2.3 Typical stress–extension curves of steel, aramid, polyester and polyamide.

yarns, that is, a metal filament is wound around an elastic core yarn. While a metal filament is almost inextensible, an elastomeric yarn has a high extensibility. The combined yarn levels out the strength of the metal filament and the extension of the elastic core yarn (Schwarz *et al.*, 2011). The following discussion is subdivided by the origin of conductive materials, starting with classical metals in Section 2.4.1. Carbon and carbon-based materials are addressed in Section 2.4.2. Finally, in Section 2.4.3, we close with conductive polymers.

2.4 Metals

Metals are one of the most important industrial materials. They are usually classified into ferrous materials and non-ferrous metals. Both types are used in the textile world. However, their characteristics most likely define their application area in textiles. This section gives an overview of the most widely applied metals in combination with textiles and their properties.

2.4.1 Ferrous materials – stainless steel

In general, steel has a moderate electrical conductivity compared to other metals and is most often used for antistatic purposes in carpets and protective clothing. Stainless steels are high-alloy steels and have good corrosion resistance because they contain relatively large amounts of chromium (at least 10%) (Design, 2011). Nonetheless, as mentioned above, when exposed to air and humidity, stainless steels are covered with a thin layer of chromium oxide, which reduces their surface conductivity. Another issue to consider is that nickel is also present in stainless steel, causing a risk of allergenic reactions when applied on the human skin. Furthermore, the electrical contact resistance of stainless steel has to be critically assessed. When, for instance, being woven into a structure, interlacing stainless steel filaments exhibit a very high contact resistance, which does not allow the current to spread uniformly on the structure. Consequently, local hot spots and burns, when being applied on the skin, are very likely to appear (Banaszczyk *et al.*, 2009). Stainless steels are available as pure filaments, yarns and blended with polymeric fibres (Bekaert, 2011). As they are very heavy (generally 66% heavier than aluminium), stiff, and have a rough and abrasive surface handle, they may cause damage during processing, such as weaving and knitting. Examples of typical stainless steel yarns are depicted in Fig. 2.4.

2.4.2 Non-ferrous metals – copper, silver, gold, aluminum, nickel

Copper (Cu)

Copper possesses excellent electrical and thermal conductivity. It oxidizes in air, or more precisely with atmospheric oxygen, forming a copper oxide layer on its surface. This oxide layer prevents a further bulk corrosion and it is therefore often reported that copper has a good resistivity to corrosion (Association, 2011). However, its oxide layer is badly conducting, with the application of copper where surface conductivity is very important. On the contrary, in applications requiring materials with good electro-conductivity where inertness[1] is not so crucial, copper is a good choice. This includes textile transmission lines and textile heating elements. Furthermore, researchers have demonstrated the good antimicrobial properties of copper, which is a very important characteristic for touch surfaces such as textiles (Grzybowski and Trafny, 1999; Michels, 2006; Gould *et al.*, 2009). Disregarding its application as

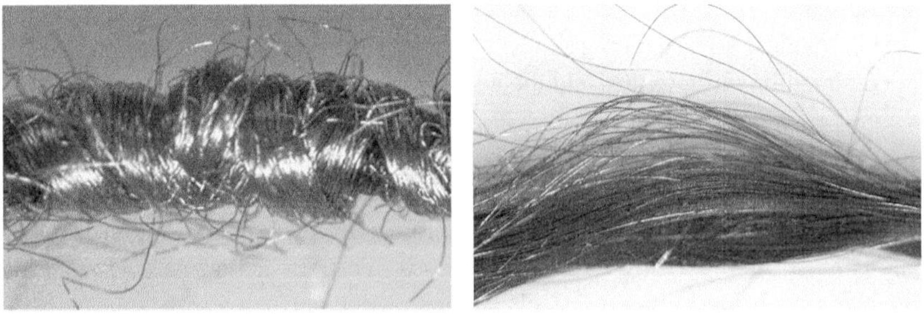

(a) (b)

2.4 Stainless steel yarn forms: (a) stainless steel staple yarn twisted around a core yarn; (b) stainless multifilament yarn.

electro-conductive components, it can also be used as an antimicrobial agent, for instance in healthcare environments.

Due to its malleability and formability, copper is extruded in wire forms having a diameter of a few micrometres, which is compatible with use in textiles (Fig. 2.5(a)). Copper is also applied as a thin layer onto textile surfaces (Gimpel *et al.*, 2003; Scheibner *et al.*, 2003; Westbroek *et al.*, 2006; Wei *et al.*, 2008a, Fibres, 2011; Schwarz *et al.*, 2012) (Fig. 2.5(b)).

Silver (Ag)

Silver can be regarded as a direct competitor to copper, as it also offers very high electrical and thermal conductivity among all metals (even higher than copper) and the lowest contact resistance, possesses antimicrobial properties and oxidizes when exposed to air. In contrast to copper, silver oxide is as conductive as pure silver and hence it is often applied in textiles to shape electrodes having direct contact to the human body and the environment (Cömert *et al.*, 2008). Due to its ability to protect against bacterial and fungal infection, its application in textiles has been extensively studied during the last 25 years (Tweden *et al.*, 1997; Pollini *et al.*, 2009; Landsdown, 2010, Filipowska *et al.*, 2011).

Silver is also very malleable and ductile and is therefore available in thin wire form that can be incorporated into textiles (Fig. 2.6(a)). Also, it is applied as coating (Fig. 2.6(b)) or printing onto textile structures (Jiang *et al.*, 2005, 2010; Scholz *et al.*, 2005; Statex – Ericson *et al.*, 2004; Swicofil 2011; Kazani *et al.*, 2012).

Gold (Au)

In history, gold has been predominately used to advertise affluence and wealth due to its rarity and permanence. Today, researchers and scientists take advantage

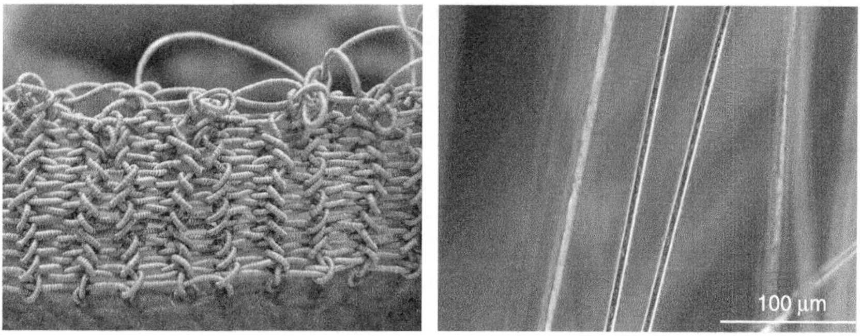

(a) (b)

2.5 Textile-compatible copper: (a) copper wires are twisted around elastic yarns and knitted; (b) copper-coated para-aramid yarn.

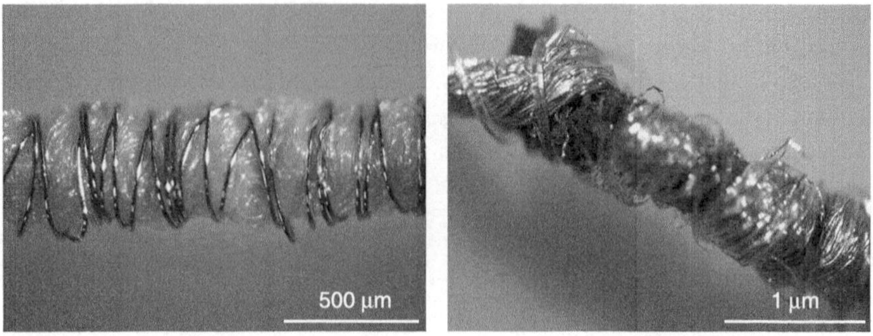

2.6 Textile compatible silver: (a) silver wires twisted around a rubber core yarn; (b) silver- coated Zylon® yarn.

of gold's outstanding chemical and physical properties, as it offers corrosion resistance, very good thermal and electrical conductivity (only silver and copper are better electrical conductors, but are not corrosion resistant). It has high ductility and excellent wear resistance (Corti and Holliday, 2004). Above that, it has a good bio-compatibility and therefore it is applied in versatile areas in medicine, as its surface cannot be changed via oxidation or other influences (Holliday, 2007). Furthermore, it does not cause allergenic reactions of the skin and the electrical properties do not change under the influence of body liquids such as sweat or urine (Council, 2011). Thus, gold is an ideal, but expensive, material to use as a sensing material measuring body parameters, such as heart rate or sweat rate.

In combination with textiles, gold foils, cut into thin strips, were used 90 years ago to decorate robes to give them a precious appearance. Up to today, it is still used as a colorant in textiles (Johnston and Kelly, 2011). Besides these decorative purposes, gold is also applied on yarns in the form of a thin film to produce highly electro-conductive yarns and fabrics (Fig. 2.7). Various methods have already been explored, such as electroless deposition (Pinto *et al.*, 2003; Matsuo *et al.*, 2007; Banaszczyk *et al.*, 2009; Schwarz *et al.*, 2010a), electroplating (Gimpel *et al.*, 2003; Cammarata *et al.*, 2011) and magnetron sputtering (Scholz *et al.*, 2005).

Aluminium (Al)

Aluminium is a very light, ductile and soft metal used particularly in applications where weight is a decisive factor, such as the automotive and airplane industry. Even applied as a very thin film, it is impervious to vapour and light, reflecting approximately 92% of visible light. It is therefore often used in the food and packaging industry. Beyond that, aluminium is also known for its good

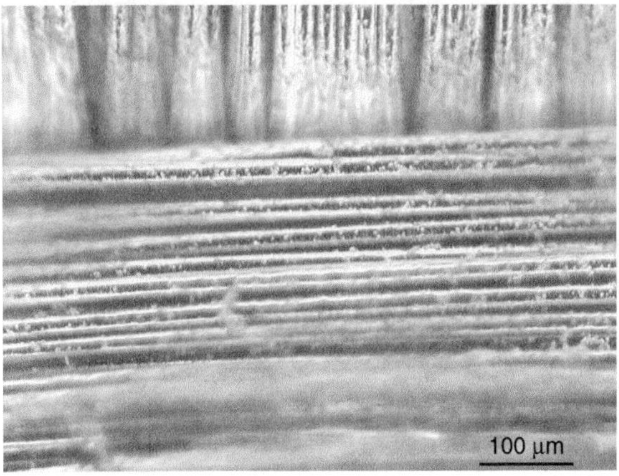

2.7 Gold-coated woven fabric through electroless deposition.

thermal- and electro-conductivity, having 60% the conductivity of copper (Ostermann, 2007). As with copper and silver, a thin layer of aluminium oxide forms on its surface when exposed to air. Forming a passive layer, the oxide protects aluminium from further corrosion. However, the oxide has very poor electro-conductivity, impairing the surface conductivity of the metal. When used close to the body, like many other textiles, it should be also taken into account that its corrosion resistance is often greatly reduced when many aqueous salts are present (e.g. sweat). Thin aluminium mono- and multi-filaments having diameters of at least 8 μm can be found on the market (Draht, 2011; Imattec 2011). Also, research has been conducted on printing with aluminium-based pastes on textiles (Yip *et al.*, 2009b) and coating textiles with thin layers of aluminium through flame spraying (Voyer *et al.*, 2008).

Nickel (Ni)

When people hear the word nickel, probably they first think about the allergenic reaction it can cause on the skin when used in jewellery. Hence, the amount of nickel that is allowed in products, which come into contact with human skin, is regulated by the European Union (Commission). Nickel belongs to the group of heavy metals and its chemical properties are similar to those of iron. At room temperature it is resistant to air, water and most chemicals. Like iron, it is ferromagnetic at room temperature. Due to its ductility it can be easily extruded in wire form. It is often applied as a coating to protect against corrosion and wear.

In combination with textiles, nickel is predominantly applied for electromagnetic shielding (Jiang and Guo, 2011, Perumalraj and Dasaradan, 2011). It is mostly applied as a thin layer on textile surface through electroless deposition (Pinto

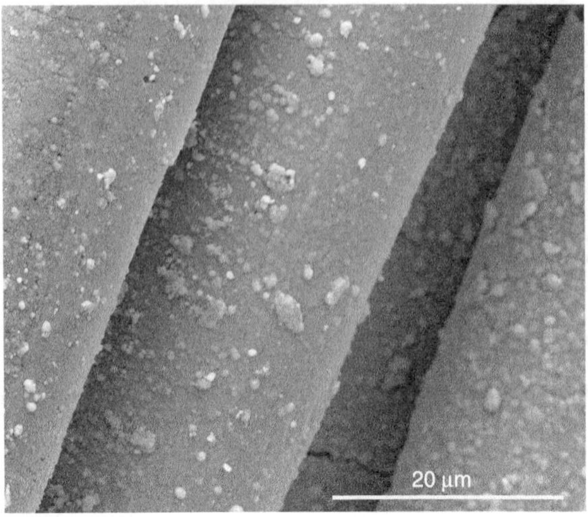

2.8 Nickel-coated polyester fibres through electroless deposition.

et al., 2003; Yuen *et al.*, 2007; Gan *et al.*, 2008; Jiang *et al.*, 2008; Association, 2011) and electroplating (Morin, 1982). An example of a nickel-coated yarn is illustrated in Fig. 2.8. Furthermore, the nickel titanium alloy, also known as *Nitinol*, has gained interest in the textile world due to its shape memory effect. In wire form, Nitinol is processed in textile stents (Riepe *et al.*, 2002; Berreklouw *et al.*, 2007) and in fashion items (Berzowska and Coelho, 2005; Vili, 2007).

2.5 Carbon: carbon black (CB), graphite and carbon nanotubes (CNT)

Carbon materials are well-known for their structural applications in lightweight composites in aerospace and automotive applications. Recently, they have been increasingly investigated for electrical applications, including electromagnetic reflection, heating and sensing. Carbon materials are typically built up in layers. As they only conduct in the in-plane direction, their electrical properties depend greatly on the degree of preferred orientation of the carbon layers. This degree can be influenced during production by applying heat treatments and aligning carbon units, such as carbon fibres in a composite structure. Hence, carbon can exhibit a range of electrical conductivity, which when compared to metals, is still rather low. Carbon is usually obtained by the pyrolysis of hydrocarbon gases and oils. Depending on their production parameters, different forms of carbon called allotropes exist. Next to amorphous carbon (carbon black), other allotropes of carbon are graphite and fullerene, such as carbon nanotubes.

2.5.1 Carbon black (CB)

In amorphous carbon, the arrangement of carbon atoms is non-crystalline and irregular. It is the main constituent of carbon black (CB), a typical filler material for polymers and polymeric fibres, as it provides electro-conductivity. There is a wide variety of CB products. Differences include particle size and particle size distribution, specific surface area, surface structure, amount of agglomeration, moisture content, and minor contaminants, such as metals, sulphur, oxygen and hydrogen (Peters, 1998). Compared to graphite and carbon nanotubes, CB is the least electro-conductive carbon material as it is amorphous.

In textiles, CB is either applied as filler material in polymers or as a layer. For the former, CB particles have been dispersed into different polymer solutions (Flandin *et al.*, 2001; Mallette *et al.*, 2001; Huang, 2002; Heiser *et al.*, 2004; Hwang *et al.*, 2007b). Subsequent extrusion (Hitchcock *et al.*, 1999; Jimenez *et al.*, 2009) or electrospinning (Kim and Yang, 2003; Pedicini and Farris, 2004; Hwang *et al.*, 2007b; Tiwari *et al.*, 2008; Schiffman *et al.*, 2011) produces fibres and fibre webs with electrical properties depending on loading of CB in the polymer solution and the percolation threshold[2], which is reported to be in the range of 2 to 12 vol.% (Hwang *et al.*, 2007b). Normally, the electrical resistance ranges between 10^5 and 10^7 Ohm/cm. The same effect can be achieved by applying CB coatings or films onto textile structures (Jin *et al.*, 2005; Koncar *et al.*, 2009; Xue *et al.*, 2011). In order to improve their electro-conductive properties, additional metallic particles or thin metallic layers are applied in or on CB containing fibrous and textile substrates (Chakravarthi *et al.*, 2011; Jalali *et al.*, 2011). Due to their high electrical resistance, CB-filled polymer composites are most often applied for antistatic, shielding and heating purposes (Koncar *et al.*, 2009). Their suitability as sensors and electrodes has also been explored (Chung, 2004; Xue *et al.*, 2011).

2.5.2 Graphite

Diamond,[3] graphite and fullerene are crystalline forms of carbon, in which each atom bonds with its neighbour forming a regular, repeating pattern. Graphite is the most common crystalline allotrope, with a layered, planar structure. In each layer, also referred to as graphene, each carbon atom forms three links with its neighbours, forming hexagonal rings. As carbon atoms have four electrons to share, each carbon atom can still bond to another carbon atom of another layer holding the material together (Sparrow, 1999).

The so-called flexible graphite is a textile-compatible form of graphite, as it is a flexible sheet made by compressing a collection of graphite flakes without a binder. Due to its microstructure involving graphite layers that are preferentially parallel to the surface of the sheet, flexible graphite has a good electrical and thermal conductivity in the plane of the sheet. It is particularly explored for shielding purposes and as a heating element (Chung, 2004).

2.5.3 Carbon nanotubes (CNT)

Another allotrope is fullerene, which has the form either of a hollow sphere, ellipsoid or cylinder. Fullerenes are similar in structure to graphite, but they may also contain pentagonal (or sometimes heptagonal) rings (Thostenson *et al.*, 2001). Typical fullerenes are carbon nanotubes, which are widely applied in composites. A carbon nanotube is produced by rolling a sheet at a specific angle. The combination of the rolling angle and radius specifies the nanotube properties, for example, whether the individual nanotube is a metal or semiconductor. Both, single-walled carbon nanotubes (SWCNT) and multi-walled carbon nanotubes (MWCNT) exist (Chou *et al.*, 2010).

Carbon nanotubes are applied in three different forms in textiles:

1. carbon nanotube fibres;
2. filler material in polymeric fibres and filaments;
3. surface layers on textile structures.

Carbon nanotube fibres

The development of continuous fibres based on carbon nanotubes has recently gained a lot of interest. The most applied technologies include the spinning of carbon nanotubes from solutions (Bhattacharyya *et al.*, 2003) and gels (Li *et al.*, 2004; Koziol *et al.*, 2007; Chae *et al.*, 2007, 2009), as well as the dry spinning from multi-walled nanotubes grown on a substrate (Jiang *et al.*, 2002; Zhang *et al.*, 2007). These technologies are explained in more detail in Section 2.5. Carbon nanotubes are known for their high axial strength and stiffness. Fibres with a greater strength than para-aramid fibres can be produced, depending on the production technologies and process parameters (Koziol *et al.*, 2007; Chou *et al.*, 2010).

Carbon nanotubes as filler material in polymeric fibres and filaments

Carbon nanotubes are used as fillers in polymeric solutions to extrude or spin fibres and filaments. Researchers have explored, for example, the melt spinning of carbon nanotube loaded polypropylene (PP) (Kearns and Shambaugh, 2002; Bhattacharyya *et al.*, 2003), poly(methyl methacrylate) (PMMA) (Haggenmueller *et al.*, 2000), polycarbonate (PC) (Sennett *et al.*, 2003; Potschke *et al.*, 2005; Thomas *et al.*, 2006), polyimide (PI) (Siochi *et al.*, 2004), polyamide (PA) (Sandler *et al.*, 2004), poly(ethylene terephthalate) (PET) (Luo *et al.*, 2006), polyurethane (Chen *et al.*, 2006), and shape memory polyurethane (Hu and Meng, 2008). Furthermore, some attempts were made to produce carbon nanotube loaded fibres through twin-screw extrusion with polyethylene and polyurethane (Chen *et al.*, 2006).

The electro-conductive properties of such fibres depend on a range of factors, including the purity of the nanotubes, the homogeneity of the dispersion in the polymeric host matrix, its loading and the aspect ratio between length and diameter for the connectivity between single nanotubes (Chung, 2004, Chou *et al.*, 2010). However, the achievable electro-conductivity of these yarns is limited when comparing them to carbon nanotube fibres.

Carbon nanotubes as surface layers on textile structures

Different technologies are employed to create carbon nanotube layers on textile materials. Chemical vapour deposition, for instance, was used to grow carbon nanotubes on woven fabrics (Baughman *et al.*, 2004; Veedu *et al.*, 2006; Zhang *et al.*, 2007). Another method that was applied in research is dye-printing of carbon nanotubes on polyester yarns (Zhang *et al.*, 2007; Fugetsu *et al.*, 2009). The achieved electrical resistance of the yarn by the latter method is $8\ 10^5\ \Omega$/cm.

2.6 Intrinsically conductive polymers (ICP)

Most polymers are insulating and so are not used in electrical, antistatic or electromagnetic shielding applications. However, in the second half of the last century a new group of polymers was discovered, known as intrinsically or inherently conducting polymers (ICP) or electroactive polymers (EAP), with the intrinsic ability to conduct electricity. A polymer can only conduct electrical charges if it has a conjugated structure, meaning that the chain of carbon–carbon bonds consists of alternating single (σ) and double (π) bonds. By doping, which is a process by which the polymer is either oxidized or reduced to create charge carriers, charge carriers can move along the carbon chain, exhibiting a conductive behaviour. The conductivity of such conjugated polymers is reported to be as high as 10^5 S/cm, which is close to that of copper with 10^6 S/cm (Ramakrishnan, 1997). These polymers include, among others, polyacetylene, polypyrrole (PPy), poly-3,4-ethylendioxythiophene (PEDOT) and polyaniline (PANi) (Fig. 2.9).

The electrical conductivity of the polymers can be varied by the amount of dopant that is added. Hence, the polymers can be tuned to have semiconductive properties as well as metallic conduction. Furthermore, they have the advantage, due to their nature, that they are lightweight and flexible. However, they are not resistant to high temperatures (e.g. 135°C maximum for PANi) and they become instable over time, because of a low resistance to humidity, oxygen and temperature.

In combination with textiles, PPy and PANi are the main polymers of choice and are usually applied as thin layers (Fig 2.10). Different textile materials have been used to produce, for instance, PPy-coated textiles obtained by chemical reaction, such as polyester (Kuhn *et al.*, 1995; Sparavigna *et al.*, 2008), polyamide (Wu *et al.*, 2005; Ferrero *et al.*, 2006), cotton (Bhat *et al.*, 2006) and wool (Kaynak

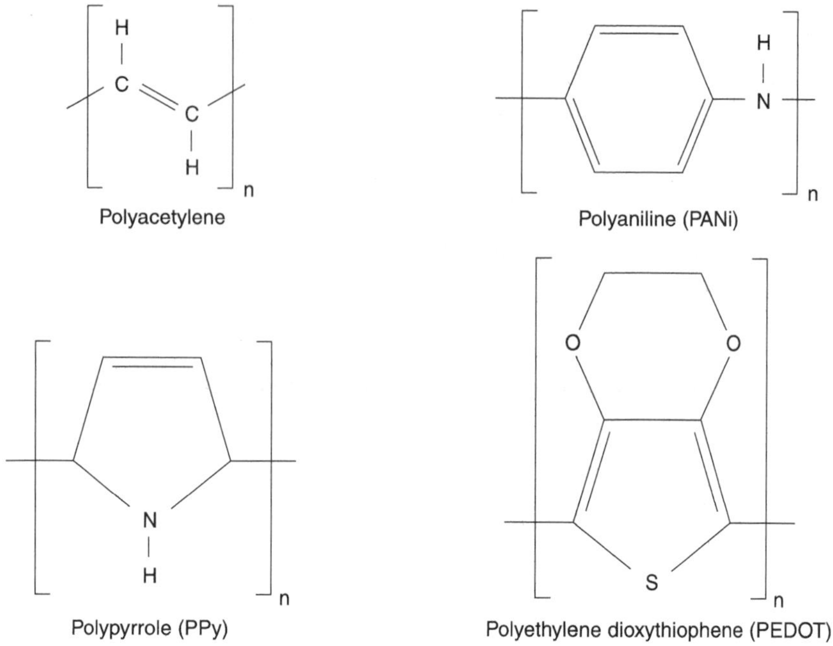

2.9 Structures of selected conductive polymers.

2.10 Polypyrrole-coated para-aramid fibres.

et al., 2002). Different methods are followed in chemical polymerization. While often a two-step polymerization is the method of choice (Lee and Hong, 2000; Wu *et al.*, 2005; Bhat *et al.*, 2006; Ferrero *et al.*, 2006), *in situ* polymerization (Reynolds *et al.*, 1998; Lee and Hong, 2000) and emulsion polymerization (Yanumet *et al.*, 2004) are also applied. PANi has been applied as a coating on polyester and polyamide fabrics via *in situ* polymerization (Oh *et al.*, 1999; Kim *et al.*, 2004). Moreover, two-step polymerization of aniline on fabrics is commonly used as well (Oh *et al.*, 2001).

Furthermore, conducting polymers are also inkjet- and screen-printed onto textiles (Calvert *et al.*, 2007, 2008, Hu *et al.*, 2011). Other possible techniques include the extrusion of conducting polymer filaments (Wang *et al.*, 2000) and the electrospinning of conducting polymer nanofibres (Henrich, 1996; Huang, 2002; Dai *et al.*, 2005; Pinto *et al.*, 2011). The application of conducting polymers in textiles ranges from sensors (Hwang *et al.*, 2007b) and dielectric layers in fibrous transistors (Krucinska *et al.*, 2011) to electromagnetic shielding (Kaynak *et al.*, 2006) and heating devices (Bhat *et al.*, 2006).

2.7 Semiconductive materials and their properties

Section 2.2 briefly outlined that a semiconductor has an intermediate electrical conductivity between that of a conductor and an insulator. The ability to conduct electrical conductivity can be selectively controlled by introducing impurities, called dopants, into the material. In general, a distinction is made between dopants that act as electron donors and electron acceptors. The former group of dopants contributes extra electrons and is therefore called *n*-type (*n* stands for negative). The latter dopant group withdraws electrons from the material, leaving a 'hole', and is referred to as *p*-type (*p* stands for positive). The addition of dopant is minuscule with a concentration of the order of 1 part in a 100 million (HyperPhysics, Hub, 2011).

In general, a distinction is made between inorganic and organic semiconductors. Inorganic ones typically comprise silicon, carbon, metal oxides and compound semiconductors (Rogers *et al.*, 2008). Their electrical performance is usually high and they have an excellent stability. However, their processing, for instance as thin films, requires temperatures that exceed the glass transition temperature and/or thermal decomposition temperatures of polymeric materials. Attempts are currently made to overcome the high processing temperatures of silicon and hence to tailor it for thin film applications. Organic semiconductors, on the contrary, can be processed by low temperature deposition or printing techniques. The research efforts to improve their electrical performance are promising but the results obtained are still lower than those achievable with inorganic semiconductors (Sun and Rogers, 2007). The following sections summarize the most commonly applicable semiconductors for textile structures.

2.7.1 Silicon

Common semiconducting materials are crystalline solids, such as silicon. In pure form it has such a high resistivity that it is an insulator. However, its resistivity can be greatly reduced by doping, which is achieved by adding impurities, such as phosphorous (to create an *n*-type semiconductor) or aluminium and boron (to create a *p*-type semiconductor), to a melt and subsequently allowing the melt to solidify into the crystal. Silicon exists with various levels of crystallinity, from amorphous (a-Si:H) to nanocrystalline (nc-Si), microcrystalline (μc-Si) and polycrystalline (pc-Si). Currently, low temperature processes, such as radiofrequency plasma-enhanced chemical vapour deposition (RF-PECVD) and hot-wire chemical vapour deposition (HW-CVD), are explored for all silicon types (Conde *et al.*, 2001; Wertheimer *et al.*, 2006; Sun and Rogers, 2007; Jeong *et al.*, 2010). In textiles, plasma-enhanced chemical vapour deposition has been applied to deposit amorphous silicon on Kapton fibres (Bonderover *et al.*, 2003).

2.7.2 Carbon

In general, carbon can be used as *n*-type and *p*-type semiconductor. A single graphite sheet, for instance, exhibits typically a semiconductive behaviour. Furthermore, as reported in the previous section, carbon nanotubes can be tuned to either possess conductive or semiconductive properties by imparting small differences in their diameter and chirality, such as the sheet direction in which the graphite sheet is rolled to form a nanotube cylinder (Baughman *et al.*, 2002).

The most common method to obtain semiconductive carbon-based textiles is to extrude composite fibres using carbon as filler material. Semiconductive properties can be achieved by the right amount of carbon filler content, the orientation of the carbon filler in the matrix and the contact resistance of neighbouring carbon components (Chou *et al.*, 2010). CB, to give one example, has been used as filler material in a thermoplastic resin to extrude or spin semiconductive fibres (Hwang *et al.*, 2007a, SmartFiber, Thostenson *et al.*, 2001). Above that, carbon nanotubes are applied as a coating on polyester fibres (Kuraray, 2011).

2.7.3 Metal oxides

Oxides of transition metals provide another class of inorganic semiconductors. These oxides, such as zinc oxide (ZnO), aluminium-doped zinc oxide (AZO) or tin-doped indium oxide (TIO), are ubiquitous in electronics and can be easily combined with textiles as they show in general a good compatibility with plastic substrates (Carcia and McLean). Especially zinc-oxide has drawn considerable attention as it offers a high chemical stability, electrical conductivity and optical transparency in the visible range (Wei *et al.*, 2009a). Usually dopants are used in a metal oxide to modify or improve its properties. Hence, AZO is an *n*-type semiconductor exhibiting a wide range of conductivities and conductivity changes under different environmental conditions allowing to sense gases (Suchea *et al.*, 2007). Like undoped zinc oxide, it is attractive due to low cost and non-toxicity (Wei *et al.*, 2009a). Indium doped metal oxides, on the contrary, are associated with high cost and limited availability. However, they are still one of the most widely used oxides due to their optical transparency, convenient processability and electrical properties (Henrich, 1996).

In general, metal oxides can be applied as thin films either by sputtering or solution processes onto textiles (Fig. 2.11). Magnetron sputtering of zinc, aluminium-doped zinc oxide and TIOs, for instance, is explored by different research groups on various textile substrates, such as polyester non-wovens (Wei *et al.*, 2007), PP non-wovens (Wei *et al.*, 2008c, 2009b) and polyamide nanofibres (Wei *et al.*, 2010). Solution processes include sol-gel coating of nanofibres with ZnO (Wei *et al.*, 2008b) and pulsed electrodeposition on polyamide filaments (Schlettwein *et al.*, 2009).

2.7.4 Polymeric and short organic semiconductors

Polymeric and short organic (oligomeric) semiconductors are an attractive alternative to inorganic semiconductors, especially when applied in textiles, due to their good mechanical flexibility, low-temperature processability and inherent compatibility with other polymers (Rogers *et al.*, 2008). In order to exhibit semiconductive properties, organic semiconductors require a highly conjugated π-system. Π-conjugated organic oligomers and polymers can function either as p-type or n-type semiconductor in which, respectively, the majority of carriers are holes or electrons. As a consequence, n-type materials are characterized by a high

2.11 ZnO-coated fibre surface through electroless deposition.

electron affinity and *p*-type by a low ionization potential. Most successful organic semiconductors are *p*-type, because of their stability in air and relative high mobility. *n*-type semiconductors have, in most cases, a tendency to react with air and moisture under operating conditions and have relatively low field-effect mobilities, making them less suitable for practical applications (Bao, 2004).

A distinction between organic semiconductors is typically made based on their size. Short organic semiconductors comprise polycyclic aromatic compounds, such as anthradithiophene (e.g. α-sexi-thiophene) and acenes (e.g. pentacene). Common polymeric semiconductors are polyacetylene, including PPy and PANi, poly(3-hexylthiophene) (P3HT) and polythienylenevinylene. Some examples are shown in Fig. 2.12.

With regard to textiles, the most explored oligomeric semiconductor is pentacene. Macconi *et al.* (2006), for instance, demonstrated the evaporation of pentacene on a cylindrical metal fibre. However, the main disadvantage is the poor solubility of pentacene molecules in organic solvents at room temperature (Afzali *et al.*, 2002). Despite its price, high-vacuum vapour deposition of pentacene is therefore still the most used technique for its deposition onto surfaces (Lee *et al.*, 2010). Another disadvantage of pentacene is that it is relatively unstable in oxidation conditions (Maliakal *et al.*, 2004). Hence, much work has been focused on using a soluble precursor allowing the fabrication of solution-cast unsubstituted pentacene (Wurthner *et al.*, 2006; Kwon *et al.*, 2008) or to permanently functionalize the structure with substituents. This allows the processing of pentacene with low temperature and cost technologies, such as drop casting, dip coating or inkjet printing. Hence, soluble TIPS-pentacene has been explored as a dip-coated layer on a PI-copper-coated polyester fibre (Genabet, 2010) (Fig. 2.13). In the area of polymeric semiconductors, PANi (Fryczkowski *et al.*, 2005), PPy (Corporation, 2011, Wallace *et al.*, 2008) and PEDOT (Skrifvars *et al.*, 2011) are mostly explored in textiles, as semiconductive properties can be achieved with the right amount and type of dopant.

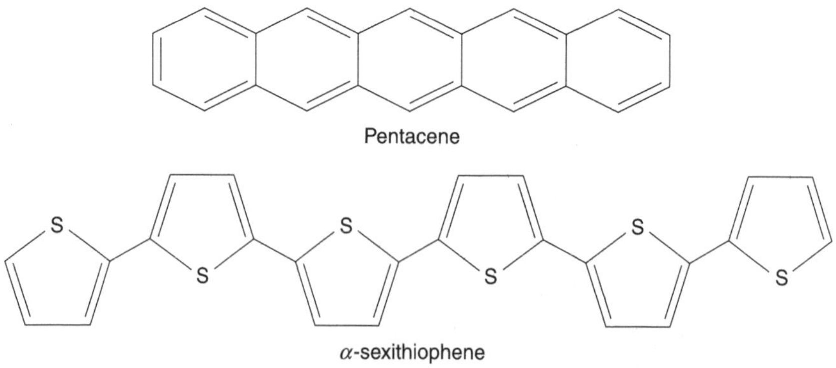

2.12 Structures of selected oligomeric semiconductors.

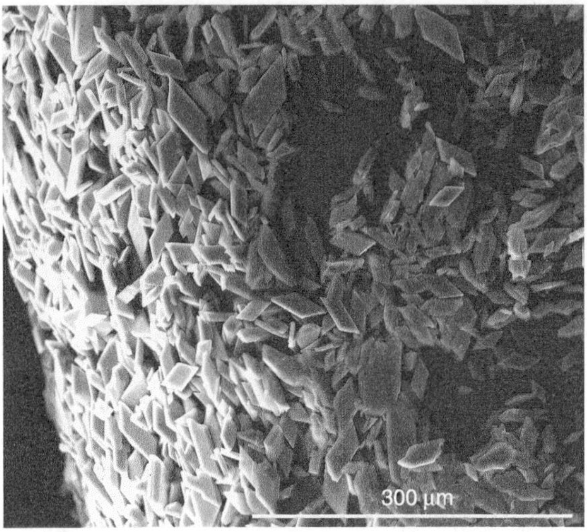

2.13 TIPS-pentacene dip coated on a polyester filament surface.

2.8 Processing electro-conductive and semiconductive materials into textile structures

As discussed in Section 2.3, conductive textile materials can be classified into two general categories, namely intrinsically conductive textile structures and textile structures that are specially treated to gain conductivity. Materials of these two groups can also be distinguished by their production technologies (Table 2.3). This section summarizes the most common production technologies applied to produce intrinsically conductive fibres, yarns and fabrics, as well as textile structures that used an after-treatment to obtain conductive properties.

Table 2.3 Overview of production technologies to obtain conductive textile structures

	Fibre level	Yarn level	Fabric level
During production	Spinning, wire drawing, bundled wire drawing, cutting production method, melt extraction, melt spinning, wet spinning	Spinning, twisting	Weaving, knitting braiding with conductive filaments
After-treatments	Coating, ink-jet printing	Coating	Coating, ink-jet printing, screen printing

2.8.1 Intrinsically conductive fibres and yarns

In general, intrinsically conductive fibres and yarns are conductors without the addition of a conductive substance. These fibres are pure metals, such as nickel, stainless steel, silver, aluminium and copper, a metal alloy or carbon. Recently, the development of intrinsically conductive polymer fibres has also been reported, described at the end of this section. Different production methods are used to produce intrinsically electro-conductive fibres, among them wire drawing, bundled wire drawing, cutting production method, melt spinning and wet spinning.

Wire drawing

The conventional process to produce metal fibres is wire drawing, a mechanical production process. This process is characterized by its various drawing steps, called coarse, medium, fine and carding trains. The drawing die, used to draw the fibre, consists of a steel mount with a core out of ceramics, carbide or diamond. The initial diameter of the metal wire varies, depending on the material. For copper, for example, it is usually 8 mm, while for iron it is 5 mm. After drawing, the wire is annealed in steel at temperatures ranging between 600 and 900°C. Subsequently, they are quenched. The fine metal wire is then wrapped onto a revolving wire drawing cylinder.

The wires produced in this way are used today in metal spun yarns, which are produced by wrapping the wire around a core yarn or are used to be processed further, for example by the bundling wire drawing process (Association, 2011).

Bundle wire drawing

Another mechanical metal fibre production process is the bundle wire drawing procedure. In this process, 1000 to 2000 conventional drawn wires are bundled into one strand and are wrapped by a thin metal cylinder.

The thin cylinder is used to enable further drawing of the whole bundle, because the drawing of a single wire would lead to wire breakage. As the cylinder material is chemically less stable than the wires, it can be removed by dissolving the whole system in an acidic bath. Finally, the process results in individual and parallel fibres of adjustable diameter between 4 to 25 μm. These can afterwards be spun to multifilament or broken into staple fibres with a length of 50 to 150 mm (Association, 2011).

The company Bekaert Fibre Technologies credits a US patent with the description of the bundle drawing production process of metal fibres. Their product range includes BEKINOX® VS, a 100% metal fibre in silver form or bulk, and BEKINOX® VN out of 100% stainless steel that can be applied in antistatic textiles, for signal and power transfer, cut resistant fabrics, heat resistant sewing and thermal conductivity. Therefore, the yarn is suitable for application in the sector of protective clothing, but also in fashion for the metal look (Bekaert, 2011).

Cutting production method

Besides using the bundle drawing procedure, Bekaert also produces metal fibres by shaving them endlessly off the edge of a rolled metal foil. With this process, it is possible to obtain several fibres per cutting section. Subsequently, the metal fibres are bundled into fibre skeins and wound onto a bobbin (Bekaert, 2011).

These fibres are branded under the name Bekinit®. In contrast to bundle-drawn fibres, they have a rectangular cross section. In addition, they are thicker, composed of more crimp and are less expensive due to a shorter production processing time. It has to be noted that the bundle-drawing process takes approximately six months.

Melt spinning

Metal fibres may also be produced by thermal methods based on processing the fibres through direct extrusion from the melt. The challenge of metal melt spinning compared to melt spinning of glass or polymers is that the melt viscosity of metals is approximately 100 times lower than that of glass and polymers. Thus, melt fractures can easily arise and instead of fibres, only droplets are formed. In the following, we refer to some methods for metal fibre production by using a coagulation bath.

In the Taylor process, a metal rod is utilized, which is placed into a glass tube. Both the material and the glass tube need to possess a comparable melting point to draw them at the same time. When heating the whole system, the metal rod melts and the glass tube softens. In this state, the glass tube can be drawn and forms a very thin hollow glass thread, with the metal as the filler material. The glass casing is afterwards removed from the metal, either by crushing with a roller or by ultrasound or chemical separation, and the result is a very fine metal fibre (diameter $<50\,\mu m$).

Continuous production of metallic fibres is achieved by melt spinning in a rotating fluid. During this process, metals or their alloys are molten and overheated in a gaseous atmosphere at high temperature and pressure. The liquid surface is charged with an inert gas excess pressure. Due to the high pressure, the melt can be pressed through a nozzle, after which the fibre is submerged as a free jet at a defined angle in a laminar flowing cooling agent. The solidifying fibre is accelerated through the rotation and through the density difference between the metal and the cooling agent at the inside of the drum, where it cools down to the temperature of the cooling agent. Extraction takes place either continuously or discontinuously, depending on the batch size.

A further way to produce melt spun metal fibres is the usage of the melt extraction process. In this procedure, a water-cooled extraction roller rotates in the melt batch. The roller usually comprises a multiple blade with V-shaped edges with which the mass is solidified and short fibres are pulled out. A magnetic field is created at the top of the melt batch to prevent waves occurring at the melt's surface. With this procedure, it is possible to process materials such as aluminium and tin fibres that cannot be produced by conventional methods due to their brittleness.

Wet spinning of carbon nanotubes from solutions

Researchers have explored the wet spinning process of lyotropic liquid crystalline suspension of SWCNTs. In this process, a dispersion of carbon nanotubes is dissolved in sulphuric acid. After being stored in a piston, the mixture is subsequently extruded through a spinneret into the air before entering a coagulation bath containing diethyl ether, aqueous sulphuric acid or water. The fibres are washed and dried before collecting them on a take-up bobbin (Ericson *et al.*, 2004).

2.8.2 Specially-treated conductive and semiconductive fibres, yarns and fabrics

Various coating techniques are currently in use to provide fibres, yarns and fabrics with conductive properties. This section gives an overview of the most applied techniques.

Electroless deposition

An electroless deposition is a process in which a metal is deposited onto a surface by an autocatalytic reduction. In the deposition solution, a metal salt (often in the form of a complex ion) and an appropriate reducing agent are dissolved.

To be of practical use, the reaction between dissolved metal species and the reducing agent must continuously occur only on the surface areas where metal deposition is desired and not in the bulk of the solution. Hence, electroless deposition is often referred to as autocatalytic deposition, because the deposited metal acts as catalytic surface for further deposition of the metal (Schwarz *et al.*, 2011).[4]

In the field of textiles, electroless deposition is mostly applied to polymeric fibres, yarns and fabrics. Various research groups deposited copper (Cu) onto fibres and fabrics (O'Connor and Capistran, 1986; Kim *et al.*, 2001; Oh *et al.*, 2004; Westbroek *et al.*, 2006; Gan *et al.*, 2008; Guo *et al.*, 2009; Jiang *et al.*, 2009; Schwarz *et al.*, 2009).

Other metals that can be applied in this way include nickel (Ni) (Pinto *et al.*, 2003; Yuen *et al.*, 2007) and silver (Ag) (Jiang *et al.*, 2005). Moreover, Cu-Ni alloys were deposited by electroless deposition on polyester fabrics (Gan *et al.*, 2008).

Silver-coated yarns have already been introduced to the market by different companies. For example, *X-Static*™ by Sauquoit Industries (2011) and *Shieldex*®

by Statex (2011) are both silver-coated polyamide yarns, either based on staple fibres or filaments.

Electroplating

Electroplating is a deposition process using electrical current to obtain a metallic layer on a sample's surface. The sample is immersed into an electrolyte solution and used as a cathode. The anode is made of the depositing material, which is dissolved in the depositing solution in the form of metal ions. These cations travel through the solution and are reduced to their metallic form on the cathode surface, resulting in a metallic layer. As the sample to be coated acts as cathode, it needs to be electro-conductive prior to the electroplating process.

In textiles, different conductive structures have been utilized to apply an additional metal or metal oxide layer to the surface. For instance, carbon fibres have been used to electroplate nickel and silver onto the surface (Morin, 1982). The additional metallic layer ensures a good bonding strength between the carbon fibre used for reinforcement and the matrix material in a composite structure.

Another approach was followed by the Textile Research Institute of Thuringia-Vogtland (TITV Greiz). Researchers used polyamide filaments that were coated with a thin silver layer as the base material, which were normally used for antistatic purposes and thus possessed a low conductivity. In a subsequent electrochemical treatment, the yarns could be modified with different metals, ranging from gold, to platinum, copper and silver (Gimpel *et al.*, 2003; Scheibner *et al.*, 2003). These yarns are registered under the name *ELITEX®* yarns.

The same base material was used to electroplate zinc oxide on the yarn's surface (Schlettwein *et al.*, 2009). The resulting semi-conductive yarn could then be used for the development of textile-based photovoltaics.

Physical vapour deposition (PVD)

In physical vapour deposition, a metal is evaporated in a vacuum. The vacuum allows metal vapour particles to travel directly to the target sample, where they condense back to a solid state. This condensation is achieved by cooling the sample surface, thus drawing energy from the metal particles as they arrive, allowing them to form a solid layer at the sample's surface.

This process, capable of uniformly depositing coatings on fibres, was implemented in various research labs to create metal-coated fibres by depositing metallic alloys (Subramanian *et al.*, 1998; Suzuki and Umehara, 1999) as well as metals, such as copper (Wei *et al.*, 2008a, Bula *et al.*, 2006) and silver (Scholz *et al.*, 2005, Hegemann *et al.*, 2009) on a fibre's surface. It is a technique that was also used to coat ceramic fibres with a metal layer, to ensure good shear strength in ceramic matrix composites (Miller *et al.*, 2001).

Chemical vapour deposition (CVD)

Chemical vapour deposition of films and coatings involve the chemical reactions of gaseous reactants on or near the vicinity of a heated substrate surface. This atomistic deposition method can provide pure materials with structural control at atomic scales (Choy, 2003).

CVD is a commonly-used method to coat textile structures with layers of conductive polymers (Malinauskas, 2001). Fibrous materials, prepared from chemically-modified polyacrylonitrile (PAN), which possesses ion-exchange properties, were charged by sorption with iron chloride ($FeCl_3$), provided from an aqueous or etheric solution. Then they were treated with pyrrole vapour in vacuum yielding PPy-coated composite material (Cvetkovska *et al.*, 1996).

Cotton yarns were coated with $FeCl_3$ and exposed to pyrrole vapour, resulting in a coating with a layer of PPy (Tan and Ge, 1996). Similarly, cotton yarns were saturated with ammonium persulphate, and then exposed to aniline vapour, resulting in polymerization on the cotton surface (Tan and Ge, 1998).

Conductive aramid and PPy composite fibres (Cho and Jung, 1997), as well as wool and PPy composite yarns (Kaynak *et al.*, 2007), were obtained by vapour-phase polymerization of pyrrole using $FeCl_3$ as an oxidant. Furthermore, PPy-poly(p-phenylene terephthalamide) composite fibres have been prepared by continuous vapour phase polymerization using $FeCl_3$ as an oxidant (Xu *et al.*, 1995). The technique was also applied to deposit thin silicon films onto PI substrates, which were subsequently integrated into a textile to shape a piezo-resistive sensor array (Alpuim *et al.*, 2009).

Lately, researchers have reported the idea of a PPy coating technique for fabrics using a screen-printing method, followed by vapour phase polymerization (Tsang *et al.*, 2007). An emulsion containing $FeCl_3$ was screen-printed onto the fabric, which was then transferred to a desiccator containing pyrrole for the deposition. After the deposition process, a layer of PPy had formed on the fabric surface.

Solution polymerization of conductive polymers

Conductive polymers, such as PPy, PANi or polythiophene, can also be applied onto textile surfaces through a solution polymerization. Here, their monomers are present in a polymerization solution. The solution also contains an oxidizing agent to start the polymerization of the monomers (Malinauskas, 2001).

In combination with textiles, mostly PPy and PANi are the polymers of choice. Different textile materials have been used to produce PPy-coated textiles obtained by chemical reaction, such as polyester (Kuhn *et al.*, 1995, Sparavigna *et al.*, 2008), polyamide (Wu *et al.*, 2005; Ferrero *et al.*, 2006), cotton (Bhat *et al.*, 2006) and wool (Kaynak *et al.*, 2002). Different methods are followed in chemical polymerization. While often a two-step polymerization is the method of choice (Lee and Hong, 2000), *in situ* polymerization (Reynolds *et al.*, 1998; Lee and Hong, 2000) and emulsion polymerization (Yanumet *et al.*, 2004) are also applied.

PANi has been applied as a coating on polyester and polyamide fabrics via *in situ* polymerization (Oh *et al.*, 1999; Kim *et al.*, 2004). Moreover, two-step polymerization of aniline on fabrics is also commonly used (Oh *et al.*, 2001). The chemical coating with conducting polymers (Scholz *et al.*, 2005; Shahidi *et al.*, 2007), or the reinforcement of polymeric materials with conductive compounds via spinning (Kumar *et al.*, 2002), are possible techniques.

Spin coating, drop casting and dipcoating

Spin coating (Park *et al.*, 2007; Kim Y H, 2007; Kim *et al.*, 2009), solution-drop casting (Reese, 2004, Park *et al.*, 2007, 2009) and dip coating (Park *et al.*, 2007) are popular ways of direct solution deposition. The basic working principle for all three is the same. By dropping or casting solutions to a certain rate, thin films can be obtained on a substrate when the solvent evaporates. In spin coating, a solution is deposited on a flat substrate and spun at a determined rate, usually in the order of magnitude of 10^3 rpm (Park *et al.*, 2007). For drop casting, the solution is deposited as a droplet on the substrate and the solvent slowly evaporates. Dip coating is a process in which the substrate is immersed in a liquid and lifted out of the solution at a preset speed controlled by a continuous motor. A part of the solution will remain on the substrate, which will form the coated film with a certain thickness after the solvent has evaporated.

These three techniques are frequently used to apply semiconductive layers onto different substrates, among them textiles. Dip coating, for instance, was applied to deposit TIPS-pentacene onto polyamide filaments (Van Genabet *et al.*, 2011).

Screen printing

Screen printing is a traditional way of printing that is applied in both the textile as well as electronic industry, as it is a low-cost and simple process. In this printing technique, the image to be printed is photographically transferred to a very fine fabric screen; the non-printing areas are coated with an impermeable substance and the fabric serves as a stencil. Ink is forced through the unblocked mesh by moving a squeegee across it. Hence, the ink is transferred to the underlying substrate.

Different research groups applied screen printing to cover a fabric with a conductive layer. Usually the applied inks contained silver particles and aluminium powder (Yip *et al.*, 2009a; Kazani *et al.*, 2012).

Ink-jet printing

Being developed in the early 1950s, ink-jet printing is a young technology compared to screen printing. In general, we can distinguish between two different ink-jet printing principles: continuous and drop-on-demand. In continuous ink-jet printing, the charged drops are allowed to fly directly onto the media, while the uncharged drops are deflected into a gutter for recirculation. Depending on the mechanism used in the drop formation process, the drop-on-demand technology can be categorized into four major methods: thermal, piezoelectric, electrostatic and acoustic ink-jet.

Two distinct ways are followed to print conductive patterns onto textile substrates. Either conductive ink-jet inks contain a dispersion of metallic nanoparticles in an aqueous or organic vehicle. A major challenge is to formulate suitable inks. The inks must contain the appropriate precursors and a carrier vehicle. In addition, they may contain various binders, dispersants and adhesion promoters, depending on the nature of the precursor and the type of fabric.

Another approach is the two-step printing of a water-soluble metal salt and a reducing agent on the textile forming the metal *in situ* (Li *et al.*, 2009).

2.9 Future trends

From a textile point of view, the overall aim for smart textiles is to convert all related components, such as sensors, actuators, transmission lines, etc., into 100% textile materials. To achieve this aim, we have to face a large challenge from a technological point of view, as concepts, materials and treatments must be made appropriate for use in, on or as textile materials.

Basic components that will be needed for all smart-textile components are electro-conductive yarns and materials. To be fully applicable, research will have to concentrate on the properties' improvement of these materials. Hence, focus will be directed to increase the washability and textile character, as well as conductivity of yarns.

Yet there is another trend visible in that electro-conductive yarns still exhibit rather passive electronic functions and there is a great need to move one step ahead, such as by developing yarns that carry out an electronic function themselves. One possible function may be switching or amplifying, which can be realized by a transistor (Maccioni *et al.*, 2006; Schwarz *et al.*, 2010b; Krucinska *et al.*, 2011). Another possible approach may be to develop photovoltaic fibres, yarns and textiles to collect and store energy from sunlight (Ghosh *et al.*, 2006; Schlettwein *et al.*, 2009, Bedeloglu *et al.*, 2010).

2.10 Sources of further information and advice

Table 2.4 gives an overview of producers that currently offer electro-conductive filaments and yarns on the market.

An interesting link may be the wiki information database on flexible, organic and large area electronics (Association, 2011), which gives an overview on materials and processes that are currently applied in the world of flexible electronics, into which we may cluster electro-conductive yarns.

Table 2.4 Overview of producers of conductive yarns and textiles

Material	Companies Wires/filaments	Companies Yarns
Copper	Elektrisola Feindraht Drahtwerk Waidhaus	Syscom Advanced Materials R.Stat
Silver	Elektrisola Feindraht	Statex Sauquoit Industries TITV Greiz (Elitex) Syscom Advanced Materials Textronics Swiss Shield
Gold		Microcroax Syscom Advanced Materials
Aluminium	Imattec Gutmann	
Nickel	Western Metal Materials	Electro Fiber Technologies Syscom Advanced Materials
Stainless steel	Leoni Nippon Seisen	Bekaert Textiles Schoeller Epitropic Fibres R.Stat Créafibers IMATTEC International TibTech Innovation
Carbon black	SmartFiber	Kuraray Teijin Monofilament Ascend Performance Materials Shakespeare Conductive Fibers Unitika Fibers Barnet
Carbon nanotubes		Kuraray and Mitsui
Polypyrrole		Sterling Fibers Econyx Corporation

2.11 Notes

1 Inertness means not chemically reactive with other elements and hence forming few or no chemical compounds.

2 The percolation threshold is the minimal amount of filler that needs to be added to create a conductive path in between the fillers.

3 Diamond is not addressed in this chapter as it is an isolator.

4 An electroless deposition process is truly autocatalytic when the rate of deposition is more or less independent of thickness. That is given when thickness of the deposited metal is linear with respect to deposition time at a given temperature.

2.12 References

Afzali A, Dimitrakopoulos C D and Breen T L (2002), 'High-performance, solution-processed organic thin film transistors from a novel pentacene precursor', *J Amer Chem Soc*, 124, 8812–8813.

Alpuim P, Marins E S, Rocha P F, Trindade I G, Carvalho M A and Lanceros-Mendez S (2009), 'Ultra-sensitive shape sensor test structures based on piezoresistive doped nanocrystalline silicon', *Vacuum*, 83, 1279–1282.

Association, C D (2011), 'The Copper Advantage: A Guide to Working With Copper and Copper Alloys', available from http://www.copper.org/ [accessed 20 September 2011].

Association, O E (2011) *oe-a-wiki*, available from http://www.oe-a-wiki.org/ [accessed 31 October 2011].

ASTM. ASTM D257.

Banaszczyk J, De Mey G, Anca A, Schwarz A and Van Langenhove L. (2009), '*Contact resistance investigation between stainless steel electroconductive yarns*', *Proceedings* of the *2009 MIXDES-16th International Conference Mixed Design of Integrated Circuits and Systems (MIXDES)* Łódź, Poland.

Bao Z N (2004), 'Designing organic materials for thin film transistors', *Abs Papers Amer Chem Soc*, 228, U215–U216.

Baughman R H, Zakhidov A A and De Heer W A (2002), 'Carbon nanotubes: the route toward applications', *Science*, 297, 787–792.

Baughman R H, Zhang M and Atkinson K R (2004), 'Multifunctional carbon nanotube yarns by downsizing an ancient technology', *Science*, 306, 1358–1361.

Bedeloglu A, Demir A, Bozkurt Y and Sariciftci N S (2010), 'A photovoltaic fiber design for smart textiles', *Textile Res J*, 80, 1065–1074.

Bekaert (2011), 'Stainless Steel Fibers', available from http://www.bekaert.com/ [accessed 20 September 2011].

Berreklouw E, Vogel B, Fischer H, Weinberg F, Van Rijk-Zwikker G L, *et al.* (2007), 'Fast sutureless implantation of mechanical aortic valve prostheses using Nitinol attachment rings: Feasibility in acute pig experiments', *J Thor Cardiovas Surg*, 134, 1508–1512.

Berzowska J and Coelho M (2005), 'Kukkia and Vilkas: kinetic electronic garments', *Proceedings of the Ninth IEEE International Symposium on Wearable Computers*, Osaka, Japan.

Bhat N V, Seshadri D T, Nate M M and Gore A V (2006), 'Development of conductive cotton fabrics for heating devices', *J Appl Polym Sci*, 102, 4690–4695.

Bhattacharyya A R, Sreekumar T V, Liu T, Kumar S, Ericson L M, *et al.* (2003), 'Crystallization and orientation studies in polypropylene/single wall carbon nanotube composite', *Polymer*, 44, 2373–2377.

Bonderover E, Wagner S and Suo Z G (2003), 'Amorphous silicon thin film transistors on Kapton fibers', *Elect Unconv Subs-Electrotext Giant-Area Flex Circ*, 736, 109–114.

Braddock S E and O'Mahony M (1999), *Techno Textiles: Revolutionary Fabrics for Fashion and Design*. London, Thames and Hudson.

Bula K, Koprowska J and Janukiewicz J (2006), 'Application of cathode sputtering for obtaining ultra-thin metallic coatings on textile products', *FibText E Eur*, 14, 75–79.

Calvert P, Sawhney A, Agrawal A, Lo T C, Patra P K and Chen C H (2007), 'Soft-structured sensors and connectors by inkjet printing', *Aatcc Rev*, 7, 42–46.

Calvert P, Duggal D, Patra P, Agrawal A and Sawhney, A. (2008), 'Conducting polymer and conducting composite strain sensors on textiles', *Mol Cryst Liq Crys*, 484, 657–668.

Cammarata V, Little B K, Li Y F, Broughton R and Mills G. (2011), 'Metallization of Kevlar fibers with gold', *Acs Appl Mat Interf*, 3, 1965–1973.

Carcia P F and Mclean R S 'Oxide semiconductors for flexible electronics', *DuPont Research and Development*, available from http://www.electrochem.org/dl/ma/206/pdfs/0977.pdf [accessed 8 January 2013]

Chae H G, Minus M L, Rasheed A and Kumar S (2007), 'Stabilization and carbonization of gel spun polyacrylonitrile/single wall carbon nanotube composite fibers', *Polymer*, 48, 3781–3789.

Chae H G, Choi Y H, Minus M L and Kumar S (2009), 'Carbon nanotube reinforced small diameter polyacrylonitrile based carbon fiber', *Comp Sci Tech*, 69, 406–413.

Chakravarthi D K, Khabashesku V N, Vaidyanathan R, Blaine J, Yarlagadda S, *et al.* (2011), 'Carbon fiber-bismaleimide composites filled with nickel-coated single-walled carbon nanotubes for lightning-strike protection', *Adv Funct Mat*, 21, 2527–2533.

Chen W, Tao X M and Liu Y Y (2006), 'Carbon nanotube-reinforced polyurethane composite fibers', *Comp Sci Tech*, 66, 3029–3034.

Cho J W and Jung H (1997), 'Electrically conducting high-strength aramid composite fibres prepared by vapour-phase polymerization of pyrrole', *J Mat Sci*, 32, 5371–5376.

Chou T W, Gao L, Thostenson E T, Zhang Z and Byun J H (2010), 'An assessment of the science and technology of carbon nanotube-based fibers and composites', *Comp Sci Tech*, 70, 1–19.

Choy K L (2003), 'Chemical vapour deposition of coatings', *Prog Mat Sci*, 48, 57–170.

Chung D D L (2004), 'Electrical applications of carbon materials', *J Mat Sci*, 39, 2645–2661.

Cömert A, Honkala M, Aydogan B, Vehkaoja A, Verho J and Hyttinen J (2008), 'Comparison of different structures of silver yarn electrodes for mobile monitoring', *IFMBE Proc*, 22, 1204–1207.

Commission E (2004), Commission Directive 2004/96/EC, in: Commission E (Ed.).

Conde J P, Alpuim P, Boucinha M, Gaspar J and Chu V (2001), 'Amorphous and microcrystalline silicon deposited by hot-wire chemical vapor deposition at low substrate temperatures: application to devices and thin-film microelectromechanical systems', *Thin Solid Films*, 395, 105–111.

Corti C W and Holliday R J (2004), 'Commercial aspects of gold applications: from materials science to chemical science', *Gold Bull*, 37, 20–26.

Council W G (2011), available from http://www.gold.org/technology/why_gold/#conductivity [accessed 20 September 2011].

Cvetkovska M, Grchev T and Obradovic T (1996), 'Electroconductive polymer composite: Fibrous carrier-polypyrrole', *J Appl Polym Sci*, 60, 2049–2058.

Dai L, Aussawasathien D and Dong J H (2005), 'Electrospun polymer nanofiber sensors', *Synth Met*, 154, 37–40.

Design M (2011), available from http://machinedesign.com/article/comparing-stainless-steel-and-other-metals-1209 [accessed 20 September 2011].

Draht G A (2011), available from http://www.gutmann-wire.com/newsite/de/index.html [accessed 20 September 2011].

EEonyx Corporation. *Eeonyarn*™ (2011), available from http://www.eeonyx.com/eeonyarn.php [accessed 20 September 2011].

Elektrisola Feindraht (2011), available from http://www.textile-wire.ch/ [accessed 20 September 2011].

Ericson L M, Fan H, Peng H Q, Davis V A, Zhou W, *et al.* (2004), 'Macroscopic, neat, single-walled carbon nanotube fibers', *Science*, 305, 1447–1450.

Ferrero F, Napoli L, Tonin C and Varesano A (2006), 'Pyrrole chemical polymerization on textiles: Kinetics and operating conditions', *J Appl Poly Sci*, 102, 4121–4126.

Fibers M C Amberstrand (2011), Syscom Advanced Materials, Amberstrand available from http://www.metalcladfibers.com/amberstrand-fiber/ [accessed 21 September 2011].

Filipowska B, Rybicki E, Walawska A and Matyjas-Zgondek E (2011), 'A new method for the antibacterial and antifungal modification of silver finished textiles', *Fib Text E Eur*, 19, 124–128.

Flandin L, Hiltner A and Baer E (2001), 'Interrelationships between electrical and mechanical properties of a carbon black-filled ethylene-octene elastomer', *Polyme*r, 42, 827–838.

Fryczkowski R, Rom M and Fryczkowski B (2005), 'Polyester fibres finished with polyaniline', *Fib Text E Eur*, 13, 141–143.

Fugetsu B, Akiba E, Hachiya M and Endo M (2009), 'The production of soft, durable, and electrically conductive polyester multifilament yarns by dye-printing them with carbon nanotubes', *Carbon*, 47, 527–530.

Gan X, Wu Y, Liu L, Shen B and Hu W (2008), 'Electroless plating of Cu–Ni–P alloy on PET fabrics and effect of plating parameters on the properties of conductive fabrics', *J Alloys Comp*, 455, 308–313.

Genabet B V (2010), 'Synthesis and characterisation of copper, polyimide and TIPS-pentacene layers in the development of a solution processed fibrous transistor', MSc thesis, Ghent University.

Ghosh T, Dhawan A and Muth J F (2006), 'Formation of electrical circuits in textile structures', in: Mattila H R (Ed.), *Intelligent Textiles and Clothing*. Abington, Woodhead Publishing in Textiles.

Gimpel S, Moehring U, Mueller H, Neudeck A and Scheibnr, W (2003), 'The galvanic and electrochemical modification of textiles', *Band – und Flechtindustrie*, 40, 115–121.

Gould S W J, Fielder M D, Kelly A F, Morgan M, Kenny J and Naughton D P (2009), 'The antimicrobial properties of copper surfaces against a range of important nosocomial pathogens', *Ann Microbiol*, 59, 151–156.

Grzybowski J and Trafny E A (1999), 'Antimicrobial properties of copper-coated electroconductive polyester fibers', *Polimery w medycynie*, 29, 27–33.

Guo R H, Jiang S Q, Yuen C W M and Ng M C F (2009), 'Surface characterization of electroless copper plated polyester fibres', *Surf Eng*, 25, 101–105.

Haggenmueller R, Gommans H H, Rinzler A G, Fischer J E and Winey K I (2000), 'Aligned single-wall carbon nanotubes in composites by melt processing methods', *Chem Phys Lett*, 330, 219–225.

Halliday D, Resnick R and Walker J (2001), *Fundamentals of Physics*. Hoboken, John Wiley and Sons.

Hegemann D, Amberg M, Ritter A and Heuberger M (2009), 'Recent developments in Ag metallised textiles using plasma sputtering', *Mat Tech*, 24, 41–45.

Heiser J A, King J A, Konell J P and Sutter L L (2004), 'Electrical conductivity of carbon filled nylon 6,6', *Adv Polym Tech*, 23, 135–146.

Henrich V E (1996), *The Surface Science of Metal Oxides*. Cambridge, Cambridge University Press.

Hitchcock G R, Akisanya A R and Thompson D S (1999), 'The mechanical response and anti-extrusion characteristics of fibre-filled elastomers', *Proc Inst Mech Eng Part L: J Mats – Design App*, 213, 37–46.

Holliday R (2007), 'Going for gold', *Mat World*, 15, 39–40.

Hu B, Li D P, Ala O, Manandhar P, Fan Q G, *et al.* (2011), 'Textile-based flexible electroluminescent devices', *Adv Funct Mats*, 21, 305–311.

Hu J L and Meng Q H (2008), 'Self-organizing alignment of carbon nanotube in shape memory segmented fiber prepared by in situ polymerization and melt spinning', *Comp Pt A: Appl Sci Manu*, 39, 314–321.

Huang J C (2002), 'Carbon black filled conducting polymers and polymer blends', *Adv Polym Tech*, 21, 299–313.

Hub B (2011), available from http://www.brighthub.com/engineering/clectrical/articles/41694.aspx [accessed 20 September 2011].

Hwang J, Muth J and Ghosh T (2007a), 'Electrical and mechanical properties of carbon-black-filled, electrospun nanocomposite fiber webs', *J Appl Polym Sci*, 104, 2410–2417.

Hwang J, Muth J and Ghosh T (2007b), 'Electrical and mechanical properties of carbon-black-filled, electrospun nanocomposite fiber webs', *J Appl Polym Sci*, 104, 2410–2417.

Hyperphysics (2011), available from http://hyperphysics.phy-astr.gsu.edu/hbase/solids/dope.html#c3 [accessed 20 September 2011].

IEC, IEC 60093

IMATTEC (2011), available from http://www.imattec.com/imattec_012.htm [accessed 20 September 2011].

Industries S (2011), available from http://www.sauquoit.com [accessed 20 September 2011].

Jalali M, Dauterstedt S, Michaud A and Wuthrich R (2011), 'Electromagnetic shielding of polymer-matrix composites with metallic nanoparticles', *Comp Part B: Eng*, 42, 1420–1426.

Jeong G H, Lee B J and Yu H Y (2010), 'Controlled synthesis of monolayer graphene toward transparent flexible conductive film application', *Nanos Res Lett*, 5, 1768–1773.

Jiang K L, Li Q Q and Fan S S (2002), 'Nanotechnology: spinning continuous carbon nanotube yarns: carbon nanotubes weave their way into a range of imaginative macroscopic applications', *Nature*, 419, 801–801.

Jiang S, Guo R, Yuen C and Ng M (2009), 'An alternative process for electroless copper plating on polyester fabric', *J Mat Sci: Mat Elect*, 20, 33–38.

Jiang S Q, Newton E, Yuen C W M and Kan C W (2005), 'Chemical silver plating and its application to textile fabric design', *J Appl Polym Sci*, 96, 919–926.

Jiang S Q, Kan C W, Yuen C W M and Wong W K (2008), 'Electroless nickel plating of polyester fiber', *J Appl Polym Sci*, 108, 2630–2637.

Jiang S X, Qin W F, Guo R H and Zhang L (2010), 'Surface functionalization of nanostructured silver-coated polyester fabric by magnetron sputtering', *Surf Coat Tech*, 204, 3662–3667.

Jiang S X and Guo R H (2011), 'Electromagnetic shielding and corrosion resistance of electroless Ni-P/Cu-Ni multilayer plated polyester fabric', *Surf Coat Tech*, 205, 4274–4279.

Jimenez L, Rocha A M, Aranberri I, Covas J and Catarino A P (2009), 'Electrically conductive mono filaments for smart textiles', in: Vincenzini P P R (Ed.), *Smart Textiles*. Firanza, Italy, Trans Tech Publications Inc.

Jin X, Xiao C F, An S L, Wang Y Y, Jia G X, *et al.* (2005), 'Studies on electrically conductive textile fiber made by a new coating method', *Fibres Polym Mat*, 1 and 2, 193–196.

Johnston J H and Kelly F M (2011), 'Colored and functional silver nanoparticle-wool fiber composites', *Acs Appl Mat Interf*, 3, 1083–1092.

Kaynak A, Wang L J, Hurren C and Wang X G (2002), 'Characterization of conductive polypyrrole coated wool yarn', *Fib Polym*, 3, 24–30.

Kaynak A, Hakansson E and Amiet A (2006), 'Electromagnetic shielding properties of polypyrrole/polyester composites in the 1–18 GHz frequency range', *Synth Met*, 156, 917–925.

Kaynak A, Najara S S and Foitzik R C (2007), 'Conductive wool yarns by continuous vapour phase polymerization of pyrrole', *Synth Met*, 157, 1–4.

Kazani I, Hertleer C, De Mey G, Schwarz A, Guxho G and Van Langenhove L (2012), 'Electrical conductive textiles through screen printing', *Fib Text E Eur*, 1, 57–63.

Kearns J C and Shambaugh R L (2002), 'Polypropylene fibers reinforced with carbon nanotubes', *J Appl Polym Sci*, 86, 2079–2084.

Kim B, Koncar V, Devaux E, Dufour C and Viallier P (2004), 'Electrical and morphological properties of PP and PET conductive polymer fibers', *Synth Met*, 146, 167–174.

Kim C and Yang K S (2003), 'Electrochemical properties of carbon nanofiber web as an electrode for supercapacitor prepared by electrospinning', *Appl Phys Lett*, 83, 1216–1218.

Kim E A, Han E G and Oh K W (2001), 'Electromagnetic interference shielding effectiveness of electroless Cu-plated PET fabrics', *Synth Met*, 123, 469–476.

Kim J, Jeong J, Cho H D, Lee C, Kim S O, *et al.* (2009), 'All-solution-processed bottom-gate organic thin-film transistor with improved subthreshold behaviour using functionalized pentacene active layer', *J Phys D: Appl Phys*, 42, 115107. 6 pp.

Kim Y H, Han S M and Han M K (2007), 'Influence of solvent on the film morphology, crystallinity and electrical characteristics of triisopropylsilyl pentacene OTFTs', *J Elect Soc*, 154, H995–H998.

Koncar V, Cochrane C, Lewandowski M, Boussu F and Dufour C (2009), 'Electro-conductive sensors and heating elements based on conductive polymer composites', *Intern J Cloth Sci Tech*, 21, 82–92.

Koziol K, Vilatela J, Moisala A, Motta M, Cunniff P, *et al.* (2007), 'High-performance carbon nanotube fiber', *Science*, 318, 1892–1895.

Krucinska I, Urbaniak-Domagala W, Skoneczna M, Cosseddu P and Bonfiglio A (2011), 'Possibility of the application of low temperature plasma for the deposition of a polypyrrole insulating layer to construct a textile-based organic field effect transistor,' *Fib Text E Eur*, 19, 78–83.

Kuhn H H, Child A D and Kimbrell W C (1995), 'Toward real applications of conductive polymers', *Synth Met*, 71, 2139–2142.

Kumar S, Doshi H, Srinivasarao M, Park J O and Schiraldi D A (2002), 'Fibers from polypropylene/nano carbon fiber composites', *Polymer*, 43, 1701–1703.

Kuraray (2011), available from http://www.kuraray.co.jp/en/release/2007/pdf/071009_02.pdf [accessed 20 September 2011].

Kwon J H, Seo J H, Shin S I, Kim K H, Choi D H, *et al.* (2008), 'A 6,13-bis(Triisopropyls ilylethynyl) pentacene thin-film transistor using a spun-on inorganic gate-dielectric', *IEEE Trans Electr Dev*, 55, 500–505.

Landsdown A (2010), *Silver in Healthcare: Its Antimicrobial Efficacy and Safety in Use*. Cambridge, RSC Publishing.

Lee H S and Hong J (2000), 'Chemical synthesis and characterization of pyrrole coated on porous membranes and its electrochemical stability', *Synth Met*, 113, 115–119.

Lee S D, Bae J H, Park J, Keum C M, Kim W H, *et al.* (2010), 'Thermal annealing effect on the crack development and the stability of 6,13-bis(triisopropylsilylethynyl)-

pentacene field-effect transistors with a solution-processed polymer insulator', *Org Electr*, 11, 784–788.

Li D P, Sutton D, Burgess A, Graham D and Calvert P (2009), 'Conductive copper and nickel lines via reactive inkjet printing', *J Mat Chem*, 19, 3719–3724.

Li Y L, Kinloch I A and Windle A H (2004), 'Direct spinning of carbon nanotube fibers from chemical vapor deposition synthesis', *Science*, 304, 276–278.

Luo G H, Li Z F, Wei F and Huang Y (2006), 'Microstructure of carbon nanotubes/PET conductive composites fibers and their properties', *Comp Sci Tech*, 66, 1022 1029.

Maccioni M, Orgiu E, Cosseddu P, Locci S and Bonfiglio A (2006), 'Towards the textile transistor: Assembly and characterization of an organic field effect transistor with a cylindrical geometry', *Appl Phys Lett*, 89, 143 515–143 517.

Maliakal A, Raghavachari K, Katz H, Chandross E and Siegrist T (2004), 'Photochemical stability of pentacene and a substituted pentacene in solution and in thin films', *Chem Mats*, 16, 4980–4986.

Malinauskas A (2001), 'Chemical deposition of conducting polymers', *Polymer*, 42, 3957–3972.

Mallette J G, Quej L M, Marquez A and Manero O (2001), 'Carbon black-filled PET/HDPE blends: effect of the CB structure on rheological and electric properties', *J Appl Polym Sci*, 81, 562–569.

Matsuo M, Ishikawa H, Chen Q Y, Bin Y Z and Komatsu K (2007), 'High electrical conductivity and high corrosion resistance fibers with high modulus and high strength prepared by electroless plating of gold on the surface of poly (p-phenylene benzobisoxazole) (PBO)', *J Mat Sci*, 42, 7772–7779.

Michels H T (2006), 'Anti-microbial characteristics of copper', *ASTM Standardization News*, October.

Miller J H, Liaw P K and Landes J D (2001), 'Influence of fiber coating thickness on fracture behaviour of continous woven Nicalon fabric-reinforced silicon-carbide matrix ceramic composites', *Mat Sci Eng A: Struct Mat Prop Microstr Proc, A*, 317, 49–58.

Morin L G (1982), 'Yarns and tows comprising high strength metal coated fibers, process for their production, and articles made therefrom', *US patent 4661403*.

Morton W E and Hearle J W S (2008), *Physical Properties of Textile Fibres*. Cambridge, Woodhead Publishing Ltd.

O'Connor M and Capistran J D (1986), *'Oxidation-resistant metal coatings', US patent application 4900618, 1990*.

Oh K W, Hong K H and Kim S H (1999), 'Electrically conductive textiles by in situ polymerization of aniline', *J Appl Polym Sci*, 74, 2094–2101.

Oh K W, Kim S H and Kim E A (2001), 'Improved surface characteristics and the conductivity of polyaniline-nylon 6 fabrics by plasma treatment', *J Appl Polym Sci*, 81, 684–694.

Oh K W, Kim S H and Bahk J H (2004), 'Electrochemically synthesized polypyrrole and Cu-plated nylon/spandex for electrotherapeutic pad electrode', *J Appl Polym Sci*, 91, 4064–4071.

Ostermann F (2007), *Anwendungstechnologie Aluminium*. Berlin, Springer.

Park S K, Jackson T N, Anthony J E and Mourey D A (2007), 'High mobility solution processed 6,13-bis(triisopropyl-silylethynyl) pentacene organic thin film transistors', *Appl Phys Lett*, 91, 063514, 3 pp.

Park S K, Mourey D A, Han J I, Anthony J E and Jackson T N (2009), 'Environmental and operational stability of solution-processed 6,13-bis(triisopropyl-silylethynyl) pentacene thin film transistors', *Org Electr*, 10, 486–490.

Pedicini A and Farris R J (2004), 'Thermally induced color change in electrospun fiber mats', *J Polym Sci Part B: Polym Phys*, 42, 752–757.

Perumalraj R and Dasaradan B S (2011), 'Electroless nickel plated composite textile materials for electromagnet compatibility', *Indian J Fib Text Res*, 36, 35–41.

Peters S (1998), *Handbook of Composites*. London, Chapman and Hall.

Pinto N J, Da Silva A N R, Fachini E, Carrion P, Furlan R and Ramos I (2003), 'Electroless deposition of thin metallic films on polymer fibers prepared via electrospinning', *Abs Papers Amer Chem Soc*, 226, U402–U402.

Pinto N J, Rivera D, Melendez A, Ramos I, Lim J H and Johnson A T C (2011), 'Electrical response of electrospun PEDOT-PSSA nanofibers to organic and inorganic gases', *Sens Actuat B: Chem*, 156, 849–853.

Pollini M, Russo M, Licciulli A, Sannino A and Maffezzoli A (2009), 'Characterization of antibacterial silver coated yarns', *J Mat Sci: Mat Med*, 20, 2361–2366.

Potschke P, Brunig H, Janke A, Fischer D and Jehnichen D (2005), 'Orientation of multiwalled carbon nanotubes in composites with polycarbonate by melt spinning', *Polymer*, 46, 10 355–10 363.

Ramakrishnan S (1997), 'Conducting polymers: from a laboratory curiosity to the market place', *Resonance*, November.

Reese C (2004), 'Organic thin film transistors', *Mat Today*, 7(9), 20–27.

Reynolds J R, Kincal D, Kumar A and Child A D (1998), 'Conductivity switching in polypyrrole-coated textile fabrics as gas sensor', *Synth Met*, 92, 53–56.

Riepe G, Heintz C, Kaiser E, Chakfe N, Morlock M, *et al.* (2002), 'What can we learn from explanted endovascular devices?' *Eur J Vas Endovas Surg*, 24, 117–122.

Rogers J A, Baca A J, Ahn J H, Sun Y G, Meitl M A, *et al.* (2008), 'Semiconductor wires and ribbons for high-performance flexible electronics', *Angewandte Chemie: Intern Edn*, 47, 5524–5542.

Sandler J K W, Pegel S, Cadek M, Gojny F, Van Es M, *et al.* (2004), 'A comparative study of melt spun polyamide-12 fibres reinforced with carbon nanotubes and nanofibres', *Polymer*, 45, 2001–2015.

Scheibner W, Feustel M, Moehring U, Hofmann J and Linz T (2003), 'Textile electrical connecting cables', *Narrow Fab Braid Ind*, 40, 76–78.

Schiffman J D, Blackford A C, Wegst U G K and Schauer C L (2011), 'Carbon black immobilized in electrospun chitosan membranes', *Carb Polym*, 84, 1252–1257.

Schlettwein D, Rudolph M, Loewenstein T, Arndt E, Zimmermann Y and Neudeck A (2009), 'Pulsed electrodeposition of porous ZnO on Ag-coated polyamide filaments', *Phys Chemistry Chemical Phys*, 11, 3313–3319.

Scholz J, Nocke G, Hollstein F and Weissbach A (2005), 'Investigations on fabrics coated with precious metals using the magnetron sputter technique with regard to their anti-microbial properties', *Surf Coat Tech*, 192, 252–256.

Schwarz A, Hakuzimana J, Westbroek P and Van Langenhove L (2009), 'How to equip para-aramid yarns with electro-conductive properties', in: *Proceedings of the Sixth International workshop on wearable and implantable body sensor networks*, San Francisco, IEEE Computer Society, pp. 278–281.

Schwarz A, Hakuzimana J, Kaczynska A, Banaszcyck J, Westbroek P, *et al.* (2010a), 'Gold coated yarns through electroless deposition', *Surf Coat Tech*, 204, 1412–1418.

Schwarz A, Cardoen J, Bruneel E, Hakuzimana J, Westbroek P. *et al.* (2010b), 'Steps towards a textile based transistor: development of the gate and insulating layer', *Text Res J*, 80, 1738–1746.

Schwarz A, Kazani I, Cuny L, Hertleer C, Ghekiere F, *et al.* (2011), 'Comparative study on the mechanical properties of elastic, electro-conductive hybrid yarns and their input materials', *Text Res J*, 81(16), 1713–1723.

Schwarz A, Hakuzimana J, Westbroek P, Mey G, Priniotakis G, *et al.* (2012), 'A study on the morphology of thin copper films on para-aramid yarns and their influence on the yarn's electro-conductive and mechanical properties', *Text Res J*, 82(15), 1587–1596.

Sennett M, Welsh E, Wright J B, Li W Z, Wen J G and Ren Z F (2003), 'Dispersion and alignment of carbon nanotubes in polycarbonate', *Appl Phys A: Mat Sci Proc*, 76, 111–113.

Shahidi S, Ghoranneviss M, Moazzenchi B, Anvari A and Rashidi A (2007), 'Aluminum coatings on cotton fabrics with low temperature plasma of argon and oxygen', *Surf Coat Tech*, 201, 5646–5650.

Siochi E J, Working D C, Park C, Lillehei P T, Rouse J H, *et al.* (2004), 'Melt processing of SWCNT-polyimide nanocomposite fibers', *Comp Part B: Eng*, 35, 439–446.

Skrifvars M, Bashir T and Persson N K (2011), 'Surface modification of conductive PEDOT coated textile yarns with silicone resin', *Mat Tech*, 26, 135–139.

Smartfiber (2011), available from http://www.smartfiber.de/index.php?option=com_contentandview=articleandid=16andItemid=97 [accessed 20 September 2011].

Sparavigna A, Avloni J, Lau R, Ouyang M, Florio L and Henn A R (2008), 'Polypyrrole-coated nonwovens for electromagnetic shielding', *J Ind Text*, 38, 55–68.

SPARROW, G. 1999. *The Elements: Carbon*. Corbis, Benchmark Books.

Statex (2011), available from http://www.statex.biz [accessed 20 September 2011].

Subramanian P R, Krishnamurthy S, Keller S T and Mendiratta M G (1998), 'Processing of continuously reinforced Ti-alloy metal matrix composities (MMc) by magnetron sputtering', *Mat Sci Eng: A*, 244, 1 10.

Suchea M, Christoulakis S, Katsarakis N, Kitsopoulos T and Kiriakidis G (2007), 'Comparative study of zinc oxide and aluminum doped zinc oxide transparent thin films grown by direct current magnetron sputtering', *Thin Solid Films*, 515, 6562–6566.

Sun Y G and Rogers J A (2007), 'Inorganic semiconductors for flexible electronics', *Adv Mat*, 19, 1897–1916.

Suzuki T and Umehara H (1999), 'Pitch-based carbon fiber microstructure and texture and compatibility with aluminum coated using chemical vapor deposition', *Carbon*, 37, 47–59

Swicofil (2011), available from http://www.swicofil.ch/ [accessed 20 September 2011].

Tan S N and Ge H L (1996), 'Investigation into vapour-phase formation of polypyrrole', *Polymer*, 37, 965–968.

Tan S N and Ge H L (1998), 'Investigation into vapour formation of polyaniline', *New Polym Mat*, 5, 169–176.

Thomas E L, Fornes T D, Baur J W and Sabba Y (2006), 'Morphology and properties of melt-spun polycarbonate fibers containing single- and multi-wall carbon nanotubes', *Polymer*, 47(5), 1704–1714.

Thostenson E T, Ren Z and Chou T W (2001), 'Advances in science and technology of carbon nanotubes and their composite: a review', *Comp Sci Tech*, 61, 1899–1912.

Tiwari M K, Yarin A L and Megaridis C M (2008), 'Electrospun fibrous nanocomposites as permeable, flexible strain sensors', *J Appl Phys*, 103(4), 044305.

Tsang J, Leung S, Tao X M, Yuen M and Xue P (2007), 'Effect of fabrication temperature on strain-sensing capacity of polypyrrole-coated conductive fabrics', *Polym Intern*, 56, 827–833.

Tutorials E (2011a), *Semiconductor basics*, available from http://www.electronics-tutorials.ws/ [accessed 20 September 2011].

Tutorials E (2011b), *Voltage, Current and Resistance*, available from http://www.electronics-tutorials.ws/ [accessed 20 September 2011].

Tweden K S, Cameron J D, Razzouk A J, Bianco R W, Holmberg W R, *et al.* (1997), 'Silver modification of polyethylene terephthalate textiles for antimicrobial protection', *Asaio J*, 43, M475–M481.

Van Genabet B, Schwarz A, Bruneel E, Rambausek L, Van Driessche I and Van Langenhove, L (2011), 'Synthesis and characterization of copper, polyimide and TIPS-pentacene layers for the development of a solution processed fibrous transistor', *AIP Adv*, 1(4), 042119.

Veedu V P, Cao A Y, Li X S, Ma K G, Soldano C, *et al.* (2006), 'Multifunctional composites using reinforced laminae with carbon-nanotube forests', *Nat Mat*, 5, 457–462.

Vili Y Y F C (2007), 'Investigating smart textiles based on shape memory materials', *Text Res J*, 77, 290–300.

Voyer J, Schulz P and Schreiber M (2008), 'Electrically conductive flame sprayed aluminum coatings on textile substrates', *J Therm Spray Tech*, 17, 818–823.

Wallace G G, Foroughi J, Spinks G M and Whitten P G (2008), 'Production of polypyrrole fibres by wet spinning', *Synthc Met*, 158, 104–107.

Wang H L, Romero R J, Mattes B R, Zhu Y T and Winokur M J (2000), 'Effect of processing conditions on the properties of high molecular weight conductive polyaniline fiber', *J Polym Sci Pt B: Polym Phys*, 38, 194–204.

Wei Q F, Deng B Y, Yan X and Gao W D (2007), 'AFM characterization of nonwoven material functionalized by ZnO sputter coating', *Mat Char*, 58, 854–858.

Wei Q F, Yu L, Wu N and Hong S (2008a), 'Preparation and characterization of copper nanocomposite textiles', *J Ind Text*, 37, 275–283.

Wei Q F, Shao D F, Zhang L W, Cai Y B and Jiang S D (2008b), 'Surface functionalization of carbon nanofibers by sol-gel coating of zinc oxide', *Appl Surf Sci*, 254, 6543–6546.

Wei Q F, Yu L Y, Hou D Y and Huang F (2008c), 'Surface characterization and properties of functionalized nonwoven', *J Appl Polym Sci*, 107, 132–137.

Wei Q F, Xu Y and Wang Y (2009a), *Textile Surface Functionalization by Physical Vapor Deposition (PVD)*. Abington, Woodhead Publishing in Textiles.

Wei Q F, Shao D F, Deng B Y and Xu Y (2009b), 'Comparative studies of polypropylene nonwoven sputtered with ITO and AZO', *J Appl Polym Sci*, 114, 1813–1819.

Wei Q F, Wang H F, Xu Y and Deng B Y (2010), 'Surface functionalization of polymer nanofibers by ITO sputter coating', *J Coat Tech Res*, 7, 511–514.

Wertheimer M R, Macqueen L A, Zikovsky J, Dennler G, Latreche M and Czeremuszkin G (2006), 'PECVD of nanocrystalline Si layers on high-Tg polymer substrates', *Plasma Proc Polym*, 3, 58–65.

Westbroek P, Gasana E, Hakuzimana J, De Clerck K, Priniotakis G, *et al.* (2006), 'Electroconductive textile structures through electroless deposition of polypyrrole and copper at polyaramide surfaces', *Surf Coat Tech*, 201, 3547–3551.

Wu J, Zhou D, Too C O and Wallace G G (2005), 'Conducting polymer with Lycra', *Synthc Met*, 155, 698–701.

Wurthner F, Schmidt R, Gottling S, Leusser D, Stalke D and Krause A M (2006), 'Highly soluble acenes as semiconductors for thin film transistors', *J Mat Chem*, 16, 3708–3714.

Xu C C, Wang P and Bi X T (1995), 'Continuous vapor-phase polymerization of pyrrole: 1. Electrically conductive composite fiber of polypyrrole with poly(p-phenelyne terephtalamide)', *J Appl Polym Sci*, 58, 2155–2159.

Xue P, Tao X M and Park K H (2011), 'Electrically conductive fibers/yarns with sensing behavior from PVA and carbon black', in: Key Engineering Materials (KEM) (Ed.) *Fracture and Strength of Solids Vii, Pts 1 and 2*. Trans Tech Publications, Switzerland.

Yanumet N, Lekpittaya P, Grady B P and O'rear E A (2004), 'Resistivity of conductive polymer-coated fabric', *J Appl Polym Sci*, 92, 2629–2636.

Yip J, Jiang S and Wong C (2009a), 'Characterization of metallic textiles deposited by magnetron sputtering and traditional metallic treatment', *Surf Coat Tech*, 204, 380–385.

Yuen C W M, Jiang S Q, Kan C W and Tung W S (2007), 'Influence of surface treatment on the electroless nickel plating of textile fabric', *Appl Surf Sci*, 253, 5250–5257.

Zhang X, Li Q, Tu Y, Li Y, Coulter J Y, *et al.* (2007), 'Strong carbon-nanotube fibers spun from long carbon-nanotube arrays', *Small*, 3, 244–248.

3

Optical fibers for smart photonic textiles

S. GORGUTSA, J. BERZOWKSA and
M. SKOROBOGATIY, Ecole Polytechnique
de Montréal, Canada

DOI: 10.1533/9780857093530.1.70

Abstract: This chapter starts with an overview of conventional photonic textiles based on total internal reflection (TIR) fibers and briefly discusses light extraction from such fibers. It then concentrates on the novel type of photonic textiles fabricated using plastic photonic band gap (PBG) Bragg fibers. Next, the reflective properties of PBG fibers are studied, which enable active control of fiber color by balancing the emitted and reflected light. Because PBG plastic fibers are thin and highly flexible, they can be woven using a computerized Jacquard loom. Finally, examples of various technologies are demonstrated.

Key words: photonic displays, electronic textiles, photonic bandgap fibers, Jacquard loom, fiber design and fabrication, Bragg reflectors.

3.1 Introduction to photonic textiles

Driven by consumer demand for a unique appearance, increased performance and multi-functionality of woven textiles, smart textiles have become an active area of current research. Applications of smart textiles include:

- interactive clothing for sports, hazardous occupations and military personnel;
- industrial textiles with integrated sensors for monitoring or signage to display information;
- fashion accessories and apparel with a unique and variable appearance.

Major advances in textile functionality can only be achieved through further development of its fundamental element – a fiber. In this chapter we discuss the potential of photonic band gap (PBG) Bragg fibers in photonic textiles. Among newly discovered functionalities we highlight the real-time color-changing capability of PBG fiber-based textiles, with potential applications in dynamic signage and environmentally-adaptive coloration.

As suggested by their name, photonic textiles integrate light emitting or light processing elements into a mechanically flexible matrix of a woven material, so that appearance or other properties of such textiles can be controlled or investigated. Practical implementation of photonic textiles is through integration of specialty optical fibers during the weaving process of textile manufacturing. This approach is natural as optical fibers, being long threads of sub-millimeter diameter, are geometrically and mechanically similar to regular textile fibers and, therefore, are suitable for processing in the same way. Various applications of photonic textiles have been researched, including large-area structural health monitoring and wearable sensing, large-area illumination and clothes with unique esthetic appearance, and flexible and wearable displays.

Optical fibers embedded into woven composites have been used for in-service structural health monitoring and stress–strain monitoring of industrial textiles and composites (Uskokovic, 1999; Uskokovic *et al.*, 2001; D'Amato, 2002; Kuang and Cantwell, 2003; Kojovic *et al.*, 2005; Zivkovic *et al.*, 2005). Integration of optical fiber-based sensor elements into wearable clothes allows real-time monitoring of bodily and environmental conditions, which is of importance to various hazardous civil occupations and the military. Examples of such sensor elements can be optical fibers with chemically or biologically activated claddings for biochemical detection (El-Sherif *et al.*, 2000), Bragg gratings and long-period gratings (Ghosh *et al.*, 2005) for temperature and strain measurements, as well as microbending-based sensing elements for pressure detection (Zheng *et al.*, 2006). Advantages of optical fiber sensors over other sensor types include resistance to corrosion and fatigue, a flexible and lightweight nature, immunity to electricity and magnetism (E&M) interference, and ease of integration into textiles.

Flexible displays based on emissive fiber textiles have recently received considerable attention due to their potential applications in wearable advertisement and dynamic signage (Hatcher, 2002; Koncar, 2005). However, it has been noted that such displays are natural 'attention grabbers' and might not be suitable for applications that do not require constant user awareness (Wakita and Shibutani, 2006). An alternative to such displays are the so-called ambient displays, based on non-emissive or, possibly, weakly emissive elements. In such displays, color change is typically achieved in the light reflection mode via variable spectral absorption of chromatic inks. Color or transparency changes in such inks can be thermally (Berzowska, 2004; Wakita and Shibutani, 2006) or electrically activated (Hardake and Gregory, 2003). An ambient display normally blends in with the environment, while the display presence is recognized only when the user becomes aware of it. It is argued that in such ambient displays, comfort, esthetics and information streaming are the easiest to combine.

Apart from photonic textiles, a vast body of research has been conducted to understand and design the light scattering properties of synthetic non-optical fibers. Prediction of the color of an individual fiber based on the fiber absorption and reflection properties was discussed by Sokkar *et al.* (1992). Prediction of textile appearance due to multi-fiber redirection of light was addressed by Rubin *et al.* (1997), Grasso *et al.* (1997) and Schuster *et al.* (2003). It was also established that the shape of the individual fibers comprising a yarn bundle has a major effect on the appearance of the resultant textile (Maekawa *et al.*, 1984; Rubin *et al.*, 1994; Sirikasemlert and Tao, 1999; Yamaguchi and Takanabe, 2001; Zhang *et al.*, 2003), including textile brightness, glitter, color, etc. The use of synthetic fibers with non-circular cross-sections, or microstructured fibers containing air voids

(Rubin, 1999) running along their length, became one of the major product differentiators in the yarn manufacturing industry.

Recently, a novel type of optical fibers, called photonic crystal fibers (PCFs), were introduced. In cross section, such fibers contain either periodically arranged micron-sized air voids (Knight *et al.*, 1998; Morikawa *et al.*, 2006) or a periodic sequence of micron-sized layers of different materials (Hart *et al.*, 2002; Benoit *et al.*, 2003; Dupuis *et al.*, 2007). Unsurprisingly, when such fibers are illuminated transversally, spatial and spectral distribution of scattered light is complex. The fibers appear colored due to optical interference effects in the microstructured region of each fiber. By varying the size and position of the fiber structural elements, we can, in principle, design fibers of unlimited unique appearances. Thus, starting with transparent colorless materials and by choosing transverse fiber geometry correctly, we can design the fiber color, translucence and iridescence. This has several manufacturing advantages, namely, color agents are no longer necessary for the fabrication of colored fibers, and the same material combination can be used for the fabrication of fibers with very different designable appearances.

Moreover, fiber appearance is very stable over time, as it is defined by the fiber geometry rather than by chemical additives such as dyes, which are prone to fading. In addition, some PCFs guide light using the photonic bandgap effect rather than total internal reflection (TIR). Intensity of side-emitted light can be controlled by choosing the number of layers in the microstructured region surrounding the optical fiber core. Such fibers always emit a certain color sideways without the need of surface corrugation or microbending, thus promising considerably better fiber mechanical properties compared to TIR fibers adapted for illumination applications. In addition, by introducing into the fiber microstructure materials whose refractive index could be changed through external stimuli (e.g. liquid crystals at a variable temperature), spectral position of the fiber bandgap (color of the emitted light) can be varied at will (Busch and John, 1999; Fudouzi and Xia, 2003; Larsen *et al.*, 2003). Finally, as we demonstrate in this chapter, PCFs can be designed that reflect one color when side illuminated, while emitting another color while transmitting the light. By mixing the two colors, we can either tune the color of an individual fiber, or change it dynamically by controlling the intensity of the launched light. This provides new opportunities for the development of photonic textiles with adaptive coloration, as well as wearable fiber-based color displays.

So far, application of PCFs in textiles was only demonstrated in the context of distributed detection and emission of mid-infrared radiation (wavelengths in the 3–12 mm range) for security applications (Hart *et al.*, 2002; Benoit *et al.*, 2003); there the authors used photonic crystal Bragg fibers made of chalcogenide glasses, which are transparent in the mid-IR range. However, proposed fibers were of limited use for textiles operating in the visible (wavelengths in the 0.38–0.75 mm range) due to high absorption of chalcogenide glasses, and a dominant orange-metallic color of the chalcogenide glass. In the visible spectral range, in principle, both silica (Vienne *et al.*, 2004; Russell, 2006) and polymer-based PBG fibers (Argyros *et al.*, 2001) are now available and can be used for textile applications. However, at this point, the cost of textiles based on such fibers is prohibitively high, as the price of such fibers ranges in hundreds of dollars per meter due to the complexity of their fabrication. We believe that acceptance of PCFs by the textile industry can only become possible if much cheaper fiber fabrication techniques are used. Such techniques can be either extrusion-based (Mignanelli *et al.*, 2007), or should involve only simple processing steps requiring limited process control. To this end, our group has developed all-polymer PBG Bragg fibers (Gao *et al.*, 2006; Dupuis *et al.*, 2007) using layer-by-layer polymer deposition, as well as polymer film co-rolling techniques, which are economical and suitable for industrial scale-up.

In Section 3.2, we briefly discuss light extraction from TIR fibers and then concentrate on the novel type of photonic textiles fabricated in our group using plastic photonic band gap (PBG) Bragg fibers (Section 3.3); their manufacturing (Section 3.4), reflective properties (Section 3.5) and behavior under ambient light (Section 3.6). Finally, we discuss possible applications of the photonics textiles, such as the design of architectural animated wall panels, interactive garment, interactive paper surfaces, along with manufacturing of a civil worker safety garment operating at low light (night) conditions and interactive and sensing textiles responsive to mechanical influences (Section 3.7).

3.2 Total internal reflection (TIR) fiber-based photonic textiles

Traditional fiber optic textiles use commercial TIR optical fibers developed for the telecommunication industry. Such fibers are made of transparent materials, such as glass or polymers, and optimized for low loss information transmission. TIR optical fibers embedded into woven composites have been applied for in-service structural health monitoring and stress–strain monitoring of industrial textiles and composites (Uskokovic *et al.*, 1999; D'Amato, 2002; Kuang and Cantwell, 2003). Integration of optical fiber-based sensor elements into wearable apparel allows real-time monitoring of bodily and environmental conditions, which is important to various hazardous civil occupations, including the security and military sectors. Examples of such sensor elements can be optical fibers with biologically- or chemically-activated claddings for biochemical detection (El-Sherif *et al.*, 2000), Bragg gratings and long-period gratings (Ghosh *et al.*, 2005) for temperature and strain measurements, as well as microbending-based sensing elements for pressure detection (Zheng *et al.*, 2006). Advantages of the optical fiber sensors over other sensor types include resistance to corrosion and fatigue, flexible and lightweight nature, immunity to E&M interference, and ease of integration into textiles.

TIR optical fibers modified to emit light sideways (Spigulis *et al.*, 1997) have been used to produce emissive fashion items (Lumigram), as well as backlighting panels for medical and industrial applications (Selem *et al.*, 2007; Lumitex). To implement such emissive textiles, common silica (Spigulis *et al.*, 1997) or plastic (Harlin *et al.*, 2003) optical fibers are used, in which light extraction is achieved through corrugation of the fiber surface or fiber microbending. Moreover, specialty fibers capable of transverse lasing (Balachandran *et al.*, 1996; Shapira *et al.*, 2006) have been demonstrated, with additional applications in security and target identification.

3.2.1 Total internal reflection fibers

The key functionality of a standard optical fiber is efficient guiding of light from an optical source to a detector. To date, all photonic textiles are made using TIR optical fiber which, by its nature, confines light very efficiently in the fiber core. Due to considerations of commercial availability and cost, silica glass-based telecommunication grade fibers are often used, however they are even less suitable for photonic textiles, as such fibers are designed for ultra-low loss transmission with virtually undetectable side leakage. Thus the main problem for the photonic textile manufacturers becomes the extraction of light from the optical fibers. In TIR fibers, light is guided via consecutive reflections of light at the fiber/air cladding interface. Only the rays within the cone defined by the fiber numerical aperture are guided along the fiber, while the rays with steeper angles of propagation leak out within small (typically sub-cm) propagation distances from the source.

3.2.2 Extraction of light from total internal reflection fibers

Light extraction from the core of a TIR fiber is typically accomplished by introducing perturbations at the fiber core/cladding interface. The two most frequently used methods to realize such perturbations are macro-bending of optical fibers by the threads of a supporting fabric (Fig. 3.1(a)), or scratching of the fiber surface to create light scattering defects (Fig. 3.1b).

The principal disadvantage of the macro-bending approach is high sensitivity of scattered light intensity on the value of a bend radius. In particular, insuring that the fiber is sufficiently bent with constant bending radii throughout the whole textile is challenging. If uniformity of the fiber bending radii is not assured, then only a part of a textile featuring tightly bent fibers will be lit up. This technical problem becomes especially acute in the case of wearable photonic textiles in which the local textile structure is prone to changes due to variable force loads during wear, resulting in 'patchy' looking non-uniformly luminescing fabrics. Moreover, optical and mechanical properties of the commercial silica fibers degrade irreversibly when the fibers are bent into tight bends (bending radii of

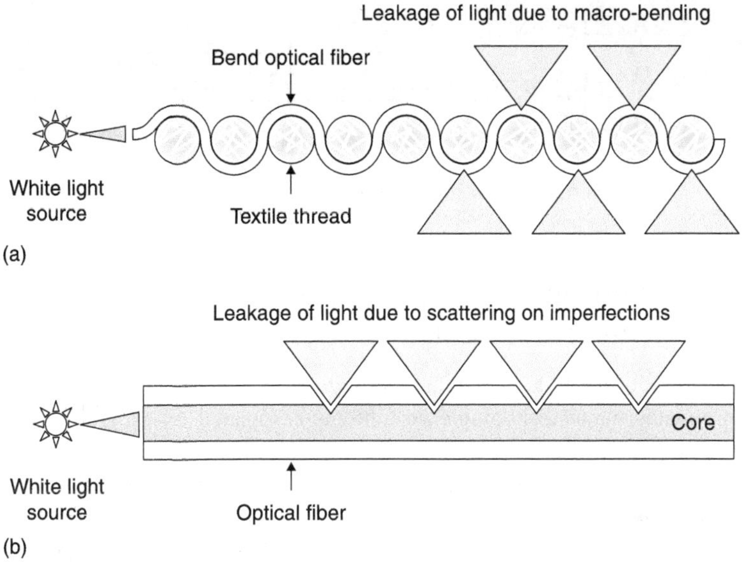

3.1 Light extraction from TIR optical fibers: (a) microbending in TIR fibers; (b) surface corrugation in TIR fibers.

several mm), which are necessary for efficient light extraction, thus resulting in somewhat fragile textiles. The main disadvantage of the scratching approach is that mechanical or chemical methods used to roughen the fiber surface tend to introduce mechanical defects into the fiber structure, thus resulting in weaker fibers prone to breakage. Moreover, due to the random nature of mechanical scratching or chemical etching, such post-processing techniques tend to introduce a number of randomly located very strong optical defects, which result in almost complete leakage of light at a few singular points, making the photonic textile appearance unappealing.

3.2.3 Dynamic color manipulation using total internal reflection fibers

To adjust the overall color of the traditional TIR fibers, we have to individually control the intensities of light launched into the fiber. Usually three different light sources of red, green and blue (the so-called RGB) are used (Fig. 3.2). The main disadvantage of this technique is that relative intensities of the three light sources have to be monitored and maintained constantly over time to avoid color drift. In the event of failure of one of the light sources, it has to be replaced and the fiber color has to be recalibrated by adjusting the relative intensities of all three sources.

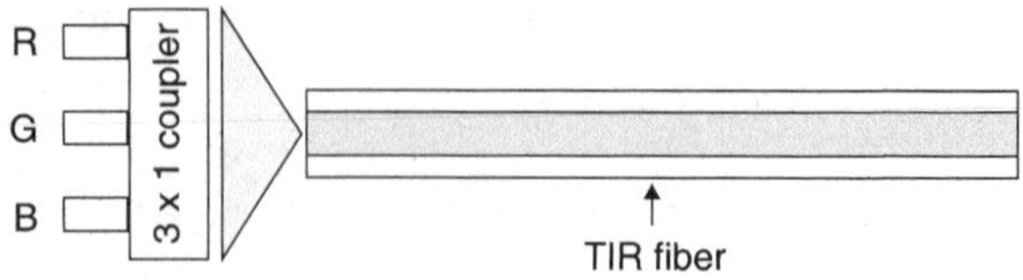

3.2 Schematic of the color manipulation in TIR fibers.

3.3 Photonic bandgap (PBG) fiber-based photonic textiles

In this section, we demonstrate that usage of the PBG fibers in place of TIR fibers eliminates many technological problems associated with light extraction from optical fibers, while also allowing additional functionalities. Among others, PBG fibers enable static multi-color flexible displays requiring only white light sources, semi-transparent flexible displays, as well as passive optical textiles that change their color as a function of the intensity of ambient light. Although the following discussion holds for any PBG fiber, we structure our presentation around hollow and solid core PBG Bragg fibers that we fabricate in our laboratory (Fig. 3.3(a) and (b)).

3.3.1 Photonic bandgap fibers

Hollow core Bragg fiber (Fig. 3.3(a)) consists of an air-filled core surrounded by a periodic sequence of high and low refractive index layers forming a so-called Bragg reflector (Johnson *et al.*, 2001). The distinguishing feature of such a reflector is the presence of bandgaps – spectral regions of high reflector efficiency caused by the interference effects inside a periodic multilayer. In a Bragg fiber, light with frequency inside a reflector bandgap can be effectively confined in the fiber hollow core through reflections from a surrounding periodic reflector. The effective refractive index of a core guided mode is typically somewhat smaller than that of the air filling the core. In practice, due to the finite number of layers in the multilayer, there is always tunneling and leakage of light sideways across the reflector. By changing the number of reflector layers, we can control the leakage rate. Thus, for energy transmission applications, we need to increase the number of reflector layers to suppress radiation loss, while for illumination applications, we need to choose a relatively small number of reflector layers to allow sizable sideways irradiation. Another characteristic feature of bandgap guidance is wavelength filtering.

Especially when launching white light into a PBG fiber, only a particular color defined by the bandgap will be guided, while all the other colors will be irradiated out of the waveguide after a few cm of propagation. The key advantage of the hollow core PBG fiber technology for photonic textiles is that such fibers can emit

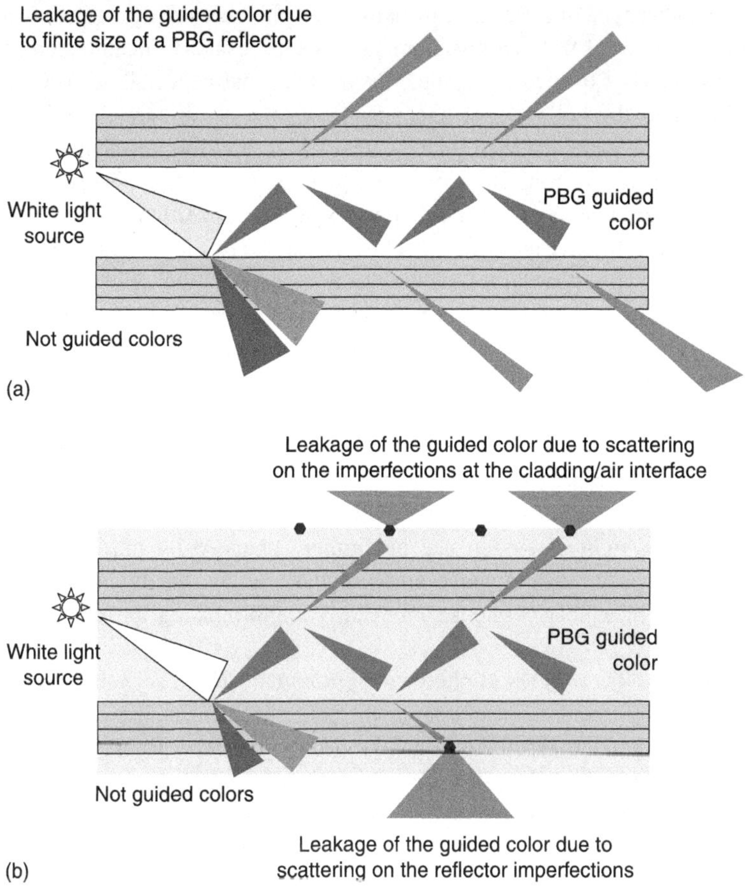

3.3 Light extraction from PBG optical fibers: (a) leaky modes in straight hollow core PBG fibers; (b) leaky modes in straight low refractive index solid core PBG fibers with scatterers at the fiber/air interface. Also the core of the fiber has the same refractive index as the low refractive index layer.

guided radiation sideways without the need of any mechanical deformations. Moreover, the emission rate and the color of irradiated light can be controlled by varying the number of layers in the reflector and the reflector layer thicknesses, respectively. Currently, commercial hollow core fibers are available both for the visible-near IR (Russell, 2006) and mid-IR ranges (Johnson *et al.*, 2001). Due to challenges in fabrication of hollow core fibers, their cost ranges in hundreds of dollars per meter. As hundreds of meters of fiber are needed for even a small sample, this makes photonic textile applications on the basis of such fibers economically impractical.

Another type of fiber that can be used for photonic textiles is a solid core PBG fiber (Fig. 3.3(b)). Such fibers are significantly easier to manufacture, therefore their cost is expected to be much lower than that of a hollow core PBG fiber. Similar to the hollow core fiber, geometry of a solid core PBG fiber features periodic reflector made of a sequence of high and low refractive index layers. The reflector surrounds a low refractive index core, typically made of the same material as the low refractive index layers of a periodic reflector. Guidance in the low refractive index core of a fiber is enabled by the bandgap of a surrounding reflector, with non-guided colors leaking through the reflector and into the fiber cladding. The effective refractive index of a core guided mode in a solid core PBG fiber is somewhat smaller than that of a core refractive index, while larger than the refractive index of air. Therefore, light that leaks from the fiber core into the cladding will be contained in the fiber cladding material. In this respect, the fiber exhibits an overall TIR guidance and no sideways radiation of light is expected. In practice, due to scattering on the imperfections in the geometry and materials of a fiber, light is always partially irradiated outside the fiber. Moreover, light extraction can be further enhanced by introducing scattering centers in the form of nano-powders that can be easily integrated into the structure of a fiber, without significantly compromising the fiber's mechanical properties.

In Plate I(a) we present photos of approximate 20 cm sections of various solid core Bragg fibers fabricated from the same preform, while drawn to different outside diameters (see Plate I in color section between pages 90 and 91). Under injection of white light, the fibers are glowing uniformly along their lengths, which reach distinct colors defined by the bandgaps of their corresponding Bragg reflectors. Moreover, even in the absence of guided light, the fibers appear colored when externally illuminated (Plate I(b)), while remaining semi-transparent due to transparency of plastics used in the fiber fabrication.

3.3.2 Understanding colors of photonic band-gap fibers

In our laboratory, we fabricate solid and hollow core PBG Bragg fibers using layer-by-layer deposition of polymer films, as well as co-rolling of commercial and custom-made polymer films around the core mandrel (Gao *et al.*, 2006; Dupuis *et al.*, 2007). Photos of a typical solid core fiber preform and a resultant fiber are presented in Fig. 3.4(a). For fabrication of Bragg fibers, we mainly use two material combinations (the high – n_h and low – n_l refractive index layers), which are polystyrene (PS)/poly(methyl methylacrylate) (PMMA) and polycarbonate (PC)/poly(vinylene difloride) (PVDF), featuring the refractive index contrasts of $n_h/n_l = 1.6/1.48$ and $n_h/n_l = 1.58/1.4$, respectively. To describe guided states in such fibers, we typically start with finding the bandgaps of a Bragg reflector. In Fig. 3.4(b), we present a typical band diagram (frequency versus the propagation constant, which is defined by the ratio of the amplitude at

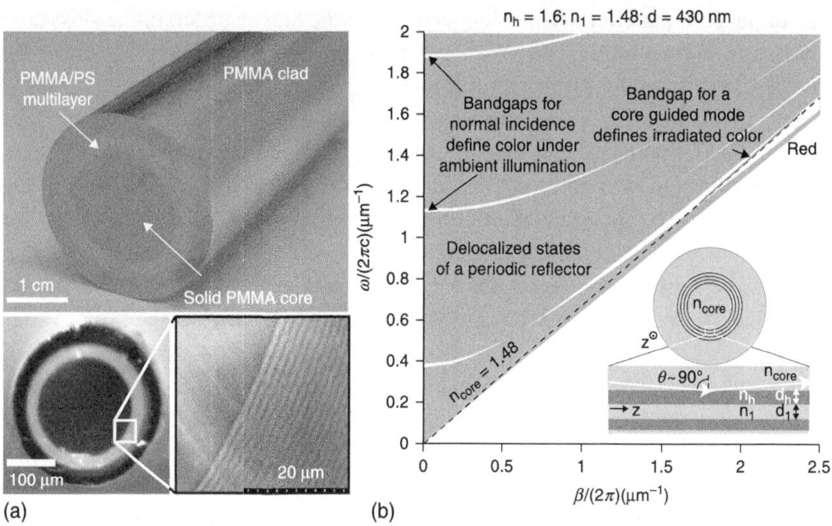

3.4 (a) Solid core plastic Bragg fiber preform and a resultant fiber. (b) Band diagram of the modes of a solid PMMA core Bragg fiber with a PMMA/PS reflector. Colors of the emitted and reflected light from the Bragg fibers are determined by the positions of the fiber reflector band gaps. Direction Z corresponds to the central axis of the fiber and η is the incident angle on the core-cladding interface.

the source of the wave to the amplitude at a certain distance) of the guided modes of an infinite planar periodic reflector fabricated from PMMA/PS and having layers of equal thicknesses $d_h = d_l = d = 430\,\text{nm}$.

Gray regions in the band diagram describe states delocalized over the whole periodic reflector. Such states are efficiently irradiated out of the fiber on the imperfections at the fiber/air interface. Thus, when launching white light into the fiber (Plate I(a)), states delocalized over the whole fiber cross section are typically irradiated after the first few cm of propagation. Clear regions in Fig. 3.4(b) define regions of phase space where no delocalized states exist inside the periodic reflector; these are the reflector bandgaps. Bragg reflector can therefore, confine light in the fiber core if the frequency and angle of incidence (propagation constant) of guided light falls into the reflector bandgap. As the core size of a Bragg fiber is very large (compared to the wavelength of operation), light propagation inside the fiber core can be envisioned as a sequence of consecutive bounces of rays traveling at shallow angles with respect to the core/reflector interface.

The effective refractive index of such rays is close, while somewhat smaller than that of a core material. Dispersion relation of the Gaussian-like fundamental core guided mode (shown in Fig. 3.4(b) as a solid red curve), therefore, appears inside the Bragg reflector bandgap, and is positioned somewhat above the light line of the core material. Dispersion relations of the higher-order, higher-loss core modes (not shown in Fig. 3.4(b)) are positioned further above the light line of the core material, while propagation of such modes within the fiber core is characterized by steeper incidence angles onto the core/reflector interface. The color guided by the fiber core is, therefore, defined by the spectral region corresponding to the intersection of the core material light line with the reflector bandgap. Spectral position of a reflector bandgap (guided color) can be varied at will by changing the thicknesses of the reflector layers, with thicker layers resulting in bandgaps positioned at longer wavelengths. Practically, layer thicknesses are varied by drawing the same preform to fibers of various diameters.

It is important to note that, although the bandgap position is determined solely by the geometry of a reflector, the color of guided light is rather determined by the intersection of the light line of a fiber core material with the reflector bandgap (solid line in Fig. 3.4(b)). From basic theory of the low refractive index-contract fibers (Skorobogatiy, 2005), it follows that the center wavelength λ_c of the reflector bandgap is given by:

$$\frac{\lambda_c}{2} = d_h\sqrt{n_h^2 - n_c^2} + d_l\sqrt{n_l^2 - n_c^2} \qquad [3.1]$$

where n_c is the core refractive index, $n_{h,l}$ are the refractive indices of the high and low refractive index layers in the fiber Bragg reflector, while $d_{h,l}$ are the corresponding layer thicknesses. As follows from Fig. 3.4(b) and Eq. 3.1, we can actively change the color of the guided light by either changing the thickness of the reflector layers, or by changing the value of the core refractive index. The former can be implemented by stretching the fiber. The latter can be implemented by filling the hollow core of a PBG Bragg fiber with a material whose refractive index can be changed by varying certain environmental parameters such as temperature or electric field. One class of such materials is liquid crystals that have already been successfully applied to tune bandgap positions in various PBG systems (Busch and John, 1999; Fudouzi and Xia, 2003; Larsen *et al.*, 2003).

In addition, even with no light traveling inside a fiber, while only under the ambient (external) illumination, the PBG Bragg fibers still appear colored (Plate I(b)). Typically, the fiber color in the far field is determined by the reflection properties of the fiber Bragg reflector under the normal light incidence ($\beta = 0$). From Fig. 3.4(b), it is clear that for low refractive index-contrast all-polymer Bragg fibers, the bandgap of a reflector at normal incidence is, generally, located at a different spectral position than the reflector bandgap that supports the core guided mode. Therefore, the fiber color under ambient illumination is different from the fiber color due to irradiation of the core guided light. This provides the interesting opportunity to adjust the overall fiber color by controlling the relative intensities of the ambient and propagating light. Finally, we note that when operating within higher-order bandgaps of the fiber Bragg reflector, the color of guided light can be both of higher or lower frequency than the color of the fiber under ambient illumination.

3.3.3 Color-changing effects under variable ambient illumination

One of the unique properties of the PBG fiber-based textiles is their ability to change their overall color by mixing the color of reflected ambient light with the color of the irradiated guided light. Consider the schematic of the hollow core PBG fiber shown in Plate II(a). As was detailed in the previous section, when launching white light into such a fiber, only the color guided by the reflector bandgap will propagate along the fiber, while all the other colors will be lost to radiation in the first few cm of propagation. Due to the finite number of layers in the reflector, guided color will slowly leak out of the fiber core and scatter onto the imperfections at the fiber/air interface, thus resulting in fiber coloration. However, under ambient illumination, and in the absence of guided light, such fibers will again be colored (Plate II(b)). However, this is due to the reflection of a particular color by the reflector bandgap at close to normal angles of incidence of the ambient light (Plate II(a)). As was demonstrated in the previous section, color of the reflected ambient light is, generally, different from the color of the irradiated core-guided light. Therefore, when both the ambient illumination and guided light are present, the overall color of the fiber will be determined by mixing of the two colors in the radiation far field. This opens up the interesting possibility of actively controlling the fiber color by controlling the relative intensities of the ambient and guided light.

In Plate II(b), we present a practical demonstration of a color-mixing concept. In this experiment, four fibers were suspended in air parallel to each other. The picture at the bottom left of Plate II(b) was taken in the dark with light going through the fiber, showing red coloration of the fiber. The picture at the top left of Plate II(b) was taken under laboratory illumination with no light going through the fiber, showing green coloration of the fiber. When allowing both ambient illumination and guided light (middle picture in Plate II(b)), the resulting appearance of the fibers in the near field is a collection of green and red stripes. When defocusing the microscope from the fiber surface (right picture in Plate II(b)) we obtain, effectively, an image of the fibers in the far field, which reveals a resultant yellow color of the fiber bundle. Note that the backgrounds always appear black, even in the photos captured under the laboratory illumination. This is because in taking the images of fibers we have used a 5× microscope. As the fibers were suspended in air, there was no reflective background in the field of view of the microscope, thus resulting in the black background. Finally, in Plate II(c), we show a collection of several randomly chosen lit fibers under ambient illumination, with both the emitted and reflected colors clearly visible.

3.3.4 Red green blue yarns for dynamic color manipulation

To implement a PBG fiber-based textile capable of changing its emissive color, we can employ RGB yarns. In particular, consider a braid made of three Bragg fibers having red (R), green (G) and blue (B) emissive colors (Plate III(a)). The resultant color emitted by a RGB yarn can be adjusted by controlling the intensities of white light launched into the individual R, G or B fibers. In principle, we can use a single white light source with an unbalanced 1×3 splitter, as shown in Plate III(b). The advantage of this approach is that the color emitted by such a textile will be very stable over time and largely independent of fluctuations in the intensity of a light source. Moreover, in contrast with TIR fibers, in case of failure of a white light source, we have to simply substitute it with a source of comparable emission spectrum with no re-calibration of the fiber color being required. Similarly, in the reflection mode, textile color can be designed by mixing in the proper proportion of fibers of different reflected colors. Moreover, fibers having the same reflected color (e.g. pink in Plate II(c)) can have different emissive colors (e.g. cyan, blue and pink in Plate II(c)). This provides the opportunity of designing a monochromatically colored textile under ambient illumination which is, nevertheless, capable of having any resultant color through emission of R, G and B colors. Such color-changing textiles could find their use in uniforms, signage and machine vision. As no chemical colorants are used in the fabrication of the PBG fiber-based textiles, and as only white light sources are necessary to illuminate them, such textiles can prove to be more stable over time and easier to maintain than their traditional counterparts.

3.4 Photonic textile manufacturing

Photonic textiles can be fabricated using conventional weaving techniques, but some important aspects arise due to the mechanical properties of the optical fibers. To clarify this question, we would like to compare two textile prototypes fabricated in our lab.

3.4.1 Hand-woven prototypes

Photos of the first 15 cm × 15 cm textile prototype, which was created by weaving the Bragg fibers into the matrix of white silk, are presented in Plate IV. About 200 m of continuous colored Bragg fiber was created in a single draw. Fiber preform was fabricated by co-rolling of the two commercial 50 μm thick PMMA and PS films around 1 in diameter PMMA rod. The fiber diameter was varied in the range 100 mm to 500 μm during the production run, to obtain the fibers of different colors. Fiber propagation loss for the fabricated batch was in the 10 to 20 dB/m range (higher loss for smaller diameter fibers) in the visible part of the spectrum, while the fiber numerical aperture (NA) in the visible part of the spectrum varied between 0.15 and 0.38, with higher NA for the larger diameter fibers.

The PBG textile prototype was handwoven by K. Schicker at the University of the Arts, London, using our fibers and a Dobby loom. Specific weave structures were used to place as much of the optical fiber as possible on the white silk matrix, or ground cloth, so that the fibers could emit and reflect the maximum amount of light. As it is standard in textile manufacturing, several rows of tightly packed threads at the textile sides were used to hold the textile and fibers firmly together. Fibers on one end of a textile were extended outside a sample for further connection and light launching. Thus, the fabricated photonic textile appeared colored and reflective when externally illuminated. When examined closer, the textile showed colored stripes made of fibers with similar diameters and coloration. Finally, when examined even closer, we distinguished a white silk matrix visible through the semi-transparent colored fibers.

When launching white light into a textile sample (Plate V), we note that light propagates easily through the coupling section containing mostly straight fibers, resulting in a collection of brightly lit fibers of every color. At the textile edge, the fibers are held by rows of tightly-packed silk threads causing thinner fibers to experience strong macro-bending, and resulting in extensive light loss. In contrast, thicker fibers, which are less prone to bending, persist inside a textile and light it up through emission of guided color.

In Plate VI, we present photos of a second 20 cm × 20 cm textile prototype, which was created by weaving the PBG Bragg fibers into the matrix of black silk. Before weaving, individual fibers were roughly sorted by their two colors in reflection and emission modes. Therefore, both unlit and lit textiles appear to be made of wide bands of distinct colors. Fibers for the second prototype were made to have larger diameters to avoid the problem of excessive light loss due to textile induced fiber bending. In particular, fiber preform was fabricated by co-rolling around a PMMA rod of the home-extruded PMMA and PS films, each of less than 20 μm thickness. As we started with thinner reflector layers in the preform, we needed smaller drawdown ratios (compared to that of the first prototype) to result in the same size of the reflector layers, as in the case of the first prototype. As a consequence, fibers in the second prototype were significantly larger and more mechanically robust than the fibers in the first prototype, while having similar colors both in reflection and emission modes. Overall, about 200 m of continuous fiber was produced. Fiber diameter was varied in the range 300 to 600 μm during the production run to obtain fibers of different colors. Fiber propagation loss for the fabricated batch was in the range of 5 to 10 dB/m (higher loss for smaller diameter fibers) in the visible part of the spectrum.

The second sample was also handwoven on a Dobby loom; however, unlike the first prototype, care was taken not to create tightly packed threads at the textile sides in order to avoid an excess bending induced loss at the textile input and output edges. Fibers on one end of a textile were extended outside of a sample for further connection and light launching. Thus, fabricated photonic textile appears colored when externally illuminated (Plate VI(a)). When examined closer, the textile exhibited colored bands made of fibers with similar diameters and coloration. When launching white light into a textile sample (Plate VI(b)), light flows easily over the textile input edge, exhibiting a number of brightly lit wide bands of distinct colors.

In Plate VI(c), we compare the appearances of various textile patches under ambient illumination, with or without the light going through the fibers. Note that the same patch looks different, depending on whether the textile is lit or unlit. This opens up the possibility of controlling the resultant textile color by balancing the intensities of the guided and ambient light.

To couple light into the textile fibers they are batched into two groups, each one butt-coupled to one of the two fiber bundles coming out of the 100 W Edmund Optics halogen lamp source. The picture is taken under the normal ambient illumination in the laboratory. The textile is lit, and the white light source is powerful enough so that the colors emitted by the textile are dominating the textile colors due to ambient illumination, even under standard laboratory illumination.

Finally, we note that as power leaks out of the fiber, the optical power carried along the fiber decreases exponentially with traveled distance, thus resulting in non-uniform brightness of a photonic textile. This problem is common to all fiber-based textiles. The simplest solution is to ensure that textile dimension is much smaller than the characteristic decay length of light in the fiber. In principal, we can also resort to fabrication of non-uniform fibers to compensate for the decay in the transmitted light intensity by forcing more light to escape closer to the fiber end.

3.4.2 Prototypes fabricated with a Jacquard loom

The appearance of photonic textile can be dramatically varied by using different weave patterns, such as the ones enabled by a computerized Jacquard loom. A Jacquard loom allows the weaver to individually address each warp thread so as to create complex weave structures, including double weaves. Double weave is a type of woven cloth in which two warps and two sets of weft yarns are interwoven to form a two-layered cloth. We used a white cotton warp. We also used a white cotton weft and a PBG fiber weft to create individual illuminated designs in the textile display. All the textile samples and prototypes that follow were woven by J Berzowska at Concordia University, Montréal, Canada.

To systematically study the emission and reflection from PBG fiber textiles, we prepared a textile sample with a three-leaf pattern, each using a variety of weft face double weave structures. We wove the pattern twice, using two different diameters of PBG fibers, 500 and 300 microns. The first sample (Plate VII), was woven with thicker PBG fibers, which used a double weave where the warp yarns were separated into two layers, so as to bring the PBG fibers to the surface in the leaf pattern and weave them in the back layer in the area around the leaf. In the leaf featured in Plate VII, the PBG fiber is woven in a 1/7 twill structure in the front layer, while the cotton yarn is woven in a 2/2 pocket weave structure in the back layer. The two layers are connected by reversing the positions of these two layers outside the leaf pattern, but are not connected inside the leaf. This is also known as pocket cloth or pocket weave, since it creates pockets between the two layers. The leaf structure features relatively long sections of straight fiber (1–2 cm) placed on the surface of white cotton. When reflecting daylight, textile sections that have Bragg fibers show a warm red color and have a 'sparkly' appearance (Plate VII).

3.5 Reflective properties of photonic bandgap textiles under ambient illumination

PBG fibers are highly reflective and have the appearance of colored metallic mirrors. A straight PBG fiber exposed to ambient light can 'sparkle' when the incidence angle and the angle of observation are properly chosen (Plate VII). The sparkling effect can be enhanced by introducing bends into the fiber.

To study textile reflection properties under controlled conditions, we used artificial illumination provided by a Luxeon cool-white light-emitting diode (LED) source (LXK2-PW14-T00 model) placed 60 cm from the textile surface at a 45-degree inclination (Plate VIII(a)). We positioned the photo camera 60 cm from the center of the sample and took images at 15, 30, 45, 60, 75, and 90-degree inclinations with respect to the textile surface. From the photos, we see that the reflection of ambient light from this textile sample is very directional, and that the observed reflection is the strongest when the angle of observation matches the angle of incidence. Note from Plate VIII(a) that the color of the reflected LED light appears white, in strong contrast to the color of the reflected daylight (Plate VIII(b)), which appears red. Moreover, when inspecting the same textile sample under magnification in the LED light (see the top part in Plate VIII) and in daylight (see the top part in Plate VII), we clearly see that the individual strands are colored in both cases, while the overall hues are different. The main reason for the difference in perceived appearance of the same textile under different illumination conditions is caused by the spectral differences of the two illumination sources. Thus, a cool-white spectrum of an LED source features the main maximum in the blue region and a secondary much smaller maximum in the green-red region, while the spectrum of daylight has a much stronger red component.

3.6 Animated photonic bandgap textiles using mixing of ambient and emitted light

To explore variations in textile appearance produced by changes in the relative intensities of the ambient illumination and the emitted radiation, we performed the following experiment. The sample under study is the textile panel shown in Plate VIII(a). It was illuminated with a cool-white LED at a 45-degree angle with intensity I_a, and at the same time light was launched into the textile with intensity I_e. In Plate IX(a), we show a sequence of photos that were taken at a fixed intensity of ambient light $I_a = const$ and a variable intensity of light launched into the textile $I_e = 0 \rightarrow \max$. We observed that when the intensity of the emitted light is small, the overall appearance of a textile panel is determined by the reflected light. When intensities of the emitted and reflected lights are comparable, the textile appearance becomes strongly affected (compare the top and bottom pictures in Plate IX(a)). Another experiment reflected in a series of photos in Plate IX(b) was performed by keeping the light launched into the textile at the maximal intensity, while gradually reducing the intensity of the ambient light. Note the remarkable dynamics both in color, warmth, iridescence, etc. of the textile provoked by changes in the illumination conditions.

3.7 Potential applications of photonic bandgap textiles

We have shown that PBG fibers can be designed to reflect one color when side illuminated with ambient light and emit a different color when transmitting light. Moreover, we can dynamically change the color of an individual fiber by controlling the relative intensity of guided and ambient light. This allows for creative opportunities in the application of photonic textiles under changing ambient illumination conditions. In what follows, we present several photonic textiles prototypes designed for applications in animated architectural wall panels, interactive garments, interactive paper surfaces, worker ware operating in low visibility conditions, as well as in textiles that require distributed sensing of the mechanical influence.

3.7.1 Accents and novel lighting solutions

Appearance of the reflective textile can be varied dramatically by using different weave patterns to create individual illuminated designs in the textile display (Plate X). Such decorative elements can be then integrated into a garment or textile-based panels.

A woven modular architectural animated lighting wall panel, with photonic bandgap fibers, is shown in Plate XI. All the control electronics is integrated into the frame. The panel dynamically changes its appearance, depending on external light conditions.

The 'sparkle' panels presented in Plate XII change in response to the angle and intensity of light, both transmitted and reflected in their design, creating illusions of depth. Different layers of imagery and color are revealed. The tangled imagery contrasts the orthogonal weave structures and suggests a seaside landscape where the reflected and transmitted light collapses sea and grass, creating dynamic ripples.

In Plate XIII an interactive garment with integrated silk and photonic bandgap fiber woven strips were concealed within pleats is demonstrated. Custom control electronics, light sources and batteries were integrated into the textile (Plate XIII). The woven photonic elements create different patterns and textures when illuminated; light is reflected, transformed and animated within the pleats. This illumination is not a binary operation; it fades slowly and looks different from every angle. This kinetic light brings a performative aspect to the garment, which is quiet and subdued. The wearer can interact with this animation, concealing the luminous structure or revealing the luminosity by unfolding the pleats, creating an illusion of intimacy through artificial light.

As a first step in the development of large-scale interactive paper surfaces intended for architectural space, we have created paper samples integrating PBG fibers (Plate XIV). In order to research different design possibilities, the samples are created from a variety of paper pulps, of varying thickness, and contain fibers arranged in various ways. Another possible application of the PBG fibers incorporated into the paper lies in the field of secure documents.

3.7.2 Worker ware for low visibility conditions

Another scenario of using PBG textiles relates to worker ware operating in low visibility conditions (Plate XV). In this scenario, the worker is operating at night, and he/she is wearing an emissive textile panel (Plate XV). When no external illumination source is present, we detect a colorful emission from the garment. When a car (with headlights turned on) approaches from a distance, the textile appearance starts changing. The closer the car, the stronger the reflected light from the garment. As the overall appearance of a textile is strongly sensitive to the balance between intensities of emitted and reflected light (see previous section), we expect strong changes in the textile appearance under this scenario. In Plate XV, we demonstrate the results of an experiment in which a camera and a lit camera-mounted LED were placed at various distances from a lit photonic textile panel. All the pictures were taken with the same shutter speed and aperture.

When an LED and a camera are placed far from a textile panel ($L > 4.0$ m), the camera can only register light emitted from the photonic textile. When moving closer towards a textile panel, reflection of the LED light becomes more pronounced. At a distance $L \sim 1.5$ m, intensities of the reflected and emitted light become comparable, while at smaller distances the textile appearance is dominated by reflections. As the human eye is most sensitive to detection of changes in the appearance of objects, a PBG fiber-based photonic textile panel will be highly noticeable in the night time environment.

3.7.3 Strain sensing fabrics

Finally, we would like to discuss another interesting feature of plastic PBG fibers, which is the change in their optical properties under mechanical influence. We have previously mentioned that scaling of Bragg fiber geometry leads to the shift in the reflector bandgap position, and as a consequence, to the change in the fiber transmitted and reflected colors. We would expect that under mechanical strain, fiber dimensions would vary, thus having an impact on both the fiber appearance and transmission spectrum.

In order to verify this experimentally, a piece of Bragg fiber was attached to two rigid blocks, with one of the blocks (at the fiber output) being translated along the fiber axis using a linear stage, thus applying a measurable strain on a fiber sample (see inset in Plate XVI(a)). The linear stage at the output end also hosted a lens assembly for collimating the transmitted light (fiber output end placed at the focal plane of the lens). Optical transmission spectra were then acquired with the aid of a monochromator for various fiber elongations reaching up to 2:2 cm with an initial distance between the fiber clamps of 43 cm. During the experiments we made sure that changes in the recorded spectra were not due to translation of the set-up optics.

As fiber relative elongation progressively increases from 0 to 5%, the fiber transmission peak shifts towards shorter wavelengths, while attenuation increases. We presume that shift of the fiber transmission peak is due to spectral shift of the Bragg fiber bandgap. In particular, we expect that due to fiber strain the fiber transverse cross-section contracts proportionally, thus resulting in a blue shift of the Bragg fiber bandgap. Decrease in the transmitted power can be explained in two ways. First, the two rigid clamps used to secure the fiber in place prevent about 2 cm of fiber on each end to expand compared to the rest of the fiber. When strain is applied, the clamped fiber will have a bandgap positioned somewhat differently from the one of a suspended fiber. This will amount to increased loss as the transmission peak for a suspended fiber will be located closer to the bandgap edge of a clamped fiber. The net effect will be that of a transmission through a series of three somewhat mismatched narrow band filters. Another possible reason for the increase in fiber attenuation is the increase in the amount of the fiber structural defects due to applied strain.

Moreover, from the transmission data of Plate VXI(a), we can also deduce the dependence of the transmission peak wavelength with respect to the relative elongation of the Bragg fiber. Plate XVI(b) shows that transmission peak shift is linear, with the fiber elongation having a slope of 1:73 nm%. Thus, for the maximal fiber elongation of 5%, we observe 8:5 nm shift in the transmission peak position. Sensitivity of the fiber transmission and reflection spectra to the mechanical strain can lead to applications of such fibers as distributed sensors or indicators. Moreover, when fiber is left under strain for an extended time period, such a strain can become permanent in polymer fibers, thus allowing for tuning of the fiber color and transmission spectrum.

3.8 Conclusion

We have shown that under ambient illumination Bragg fibers appear colored due to optical interference in their microstructure. As Bragg fibers guide light in the low refractive index core by the photonic bandgap effect, they naturally emit sideways a portion of guided color without the need of mechanical perturbations, which is their key advantage over traditional TIR fibers. Moreover, we have demonstrated Bragg fibers that reflect one color when side illuminated, and emit another color while transmitting the light. We then showed that by controlling the relative intensities of the ambient and guided light, the overall fiber color can be varied. General implementation of the color-on-demand textiles using RGB yarns in the form of tri-fiber braids was discussed. It was established that another key advantage offered by PBG fibers in application to photonic textiles is stability of the emitted color over time. We then showed that emitted color of the PBG fiber can be tuned by applying strain. Besides, it was demonstrated that by controlling the relative intensities of the ambient and guided light the overall textile appearance (color, warmth, iridescence, etc.) can be strongly varied.

Finally, we have explored the possibility of using PBG fiber-based textiles in the design of architectural animated wall panels, interactive garments (from dresses to a worker safety garment), interactive paper surfaces and sensing textiles responsive to mechanical influence.

3.9 Acknowledgments

We would like to thank K Schicker for being instrumental in creating the first prototypes, B Gauvreau for his major contribution in the early research on PBG textiles, the Hexagram Research Center for giving us access to the Jacquard loom, Ronald Borshan for helping photograph the textiles, and Damir Cheremisov and Helene Gingras for weaving assistance.

3.10 References

Argyros A, Bassett I, Eijkelenborg M, Large M, Zagari J, *et al.* (2001), 'Ring structures in microstructured polymer optical fibres', *Opt Exp*, 9, 813–820.

Balachandran R M, Pacheco D P and Lawandy N M (1996), 'Photonic textile fibers', *Appl Optics*, 35, 91–1994.

Benoit G, Hart S D, Temelkuran B, Joannopoulos J D and Fink Y (2003), 'Static and dynamic properties of optical micro-cavities in photonic bandgap yarns', *Adv Mater*, 15, 2053–2056.

Berzowska J and Banasik A (2004), 'Very slowly animating textiles: shimmering flower', *Proceedings of the SIGGRAPH '04: AGM SIGGRAPH, Sketches*, New York, p. 34.

Busch K and John S (1999), 'Liquid-crystal photonic-band-gap materials: the tunable electromagnetic vacuum', *Phys Rev Lett*, 83, 967–970.

D'Amato E (2002), 'Stress-strain monitoring in textile composites by means of optical fibers', *Exp Tech Des Comp Mat*, 221, 245–253.

Dupuis A, Guo N, Gauvreau B, Hassani A, Pone E, *et al.* (2007), 'Guiding in the visible with 'colorful' solid-core Bragg fibers', *Opt Lett*, 32, 2882–2884.

El-Sherif M A, Yuan J M, MacDiarmid A (2000), 'Fiber optic sensors and smart fabrics', *J Intell Mat Syst Struct*, 11, 407–414.

Fudouzi H, Xia Y N (2003), 'Colloidal crystals with tunable colors and their use as photonic papers', *Langmuir*, 19, 9653–9660.

Gao Y, Guo N, Gauvreau B, Rajabian M, Skorobogata O, *et al.* (2006), 'Consecutive solvent evaporation and co-rolling techniques for polymer multilayer hollow fiber preform fabrication', *J Mater Res*, 21, 2246–2254.

Ghosh S, Amidei C and Furrow K (2005), 'Development of a sensor-embedded flexible textile structure for apparel or large area applications', *Indian J Fibre Text Res*, 30, 42–48.

Grasso M M, Hunn B D and Rewerts A M (1997), 'Effect of textile properties in evaluating a directional shading fabric', *Text Res J*, 67, 233–247.

Hardaker S S and Gregory R V (2003), 'Progress toward dynamic color-responsive "chameleon" fiber systems', *MRS Bull*, 28, 564–567.

Harlin A, Mäkinen M and Vuorivirta A (2003), 'Development of polymeric optical fibre fabrics as illumination elements and textile displays', *Autex Res J*, 3(10), 1–8.

Hart S D, Maskaly G R, Temelkuran B, Prideaux P H, Joannopoulos J D and Fink Y (2002), 'External reflection from omnidirectional dielectric mirror fibers,' *Science*, 296, 510–513.

Hatcher M (2002), 'France telecom debuts fiber screen', *Optics.org News*, 2 July.

Johnson S G, .Ibanescu M, Skorobogatiy M, Weiseberg O, Engeness T D, *et al.* (2001), 'Low-loss asymptotically single-mode propagation in large core OmniGuide fibers', *Opt Exp*, 9, 748.

Kojovic A, Zivkovic I, Brajovic L, Mitrakovic D and Aleksic R (2005), 'Low energy impact damage detection in laminar thermoplastic composite materials by means of embedded optical fibers', *Curr Res Adv Mat Proc*, 494, 481–486.

Koncar V (2005), 'Optical fiber fabric displays', *Opt Photon News*, 16, 40–44.

Knight J C, Birks T A, Russell R S J and Rarity J G (1998), 'Bragg scattering from an obliquely illuminated photonic crystal fiber', *Appl Opt*, 37, 449–452.

Kuang K S C and Cantwell W J (2003), 'Detection of impact damage in thermoplastic-based glass fiber composites using embedded optical fiber sensors', *J Thermoplas Comp* Mat, 16, 213–229.

Larsen T T, Bjarklev A, Hermann D S and Broeng J (2003), 'Optical devices based on liquid crystal photonic bandgap fibres', *Opt Exp*, 11, 2589–2596.

Maekawa I, Gunji T and Tsuboi T (1984), 'Study on optical properties of silk-like fabrics', *J Text Mach Soc Japan*, 30, 18–27.

Mignanelli M, Wani K, Ballato J, Foulger S and Brown P (2007), 'Polymer microstructured fibers by one-step extrusion', *Opt Exp*, 15, 6183–6189.

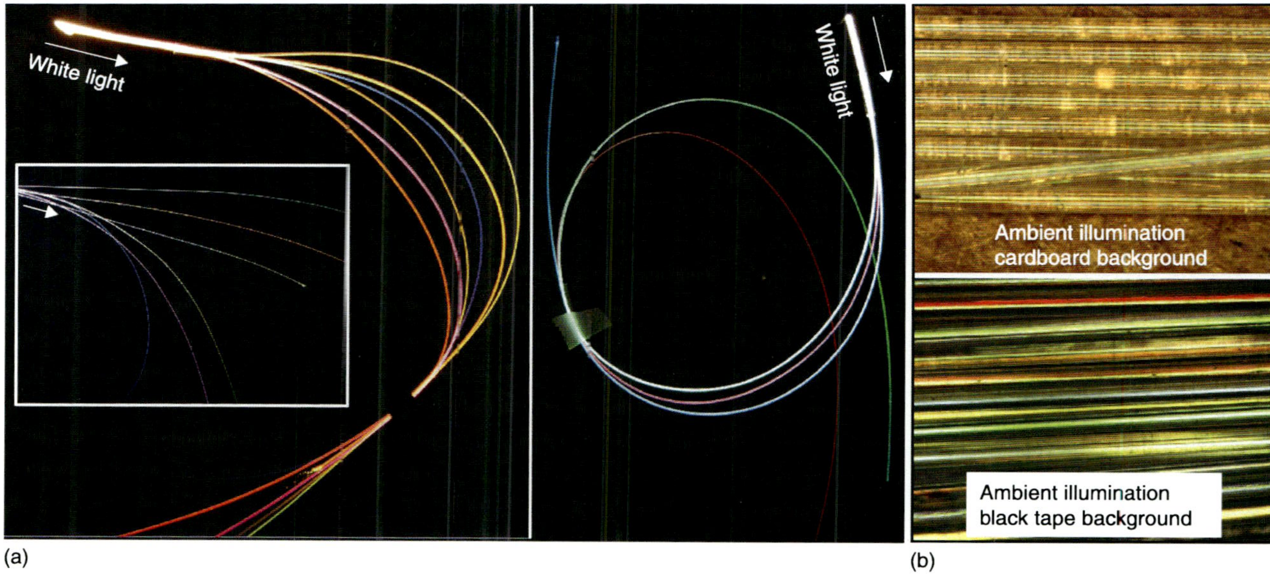

Plate I Colorful PBG Bragg fibers: (a) when launching white light into the Bragg fibers, after a few cm from the coupling end the fibers appear intensely colored. Color of an individual fiber is defined by the spectral position of the fiber reflector bandgap; (b) under ambient illumination, semi-transparent Bragg fibers appear colored. Fiber color in reflection of the ambient light can be different from the color, due to emission of guided light.

Plate X Swatches for wearable advertisement. Semi-automatic Jacquard loom was used to research into the most efficient ways of photonic textile fabrication.

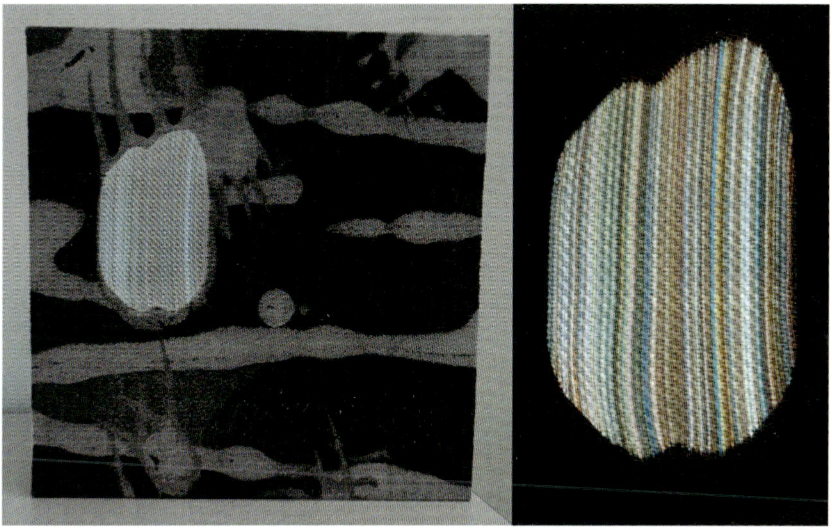

Plate XI Architectural lighting panel with dynamic appearance.

Plate XII Reconfigurable architectural lighting panel. The 'sparkle' panels change in response to the angle
and intensity of light, both transmitted and reflected in their design, creating illusions of depth.

Plate XIII An interactive garment with integrated silk and photonic band-gap fiber woven strips concealed within pleats. All the control electronics, light sources and batteries are integrated into textile.

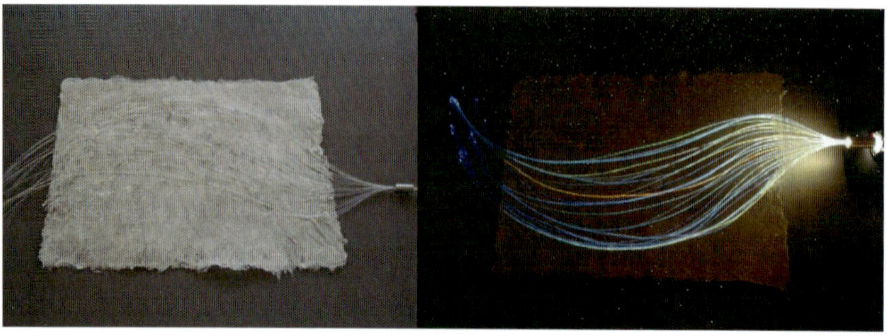

Plate XIV Integration of the PBG fibers into papers for illumination applications and for applications in secure documents.

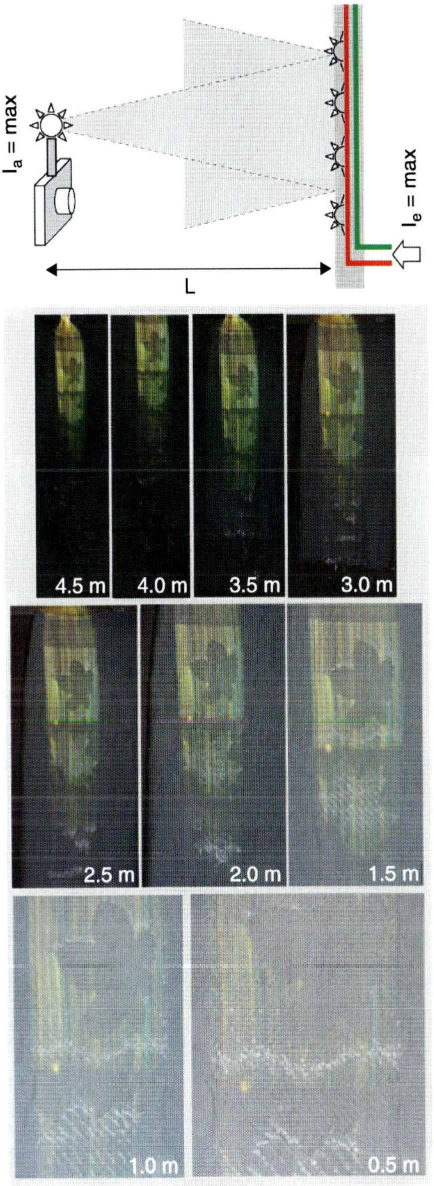

Plate XV Changes in the textile appearance under the scenario of night-time operation of an emissive textile being illuminated with the headlights of an approaching car. The emissive textile panel represents a worker operating at night. The worker's appearance changes considerably when a bright illumination source (car headlights) approaches.

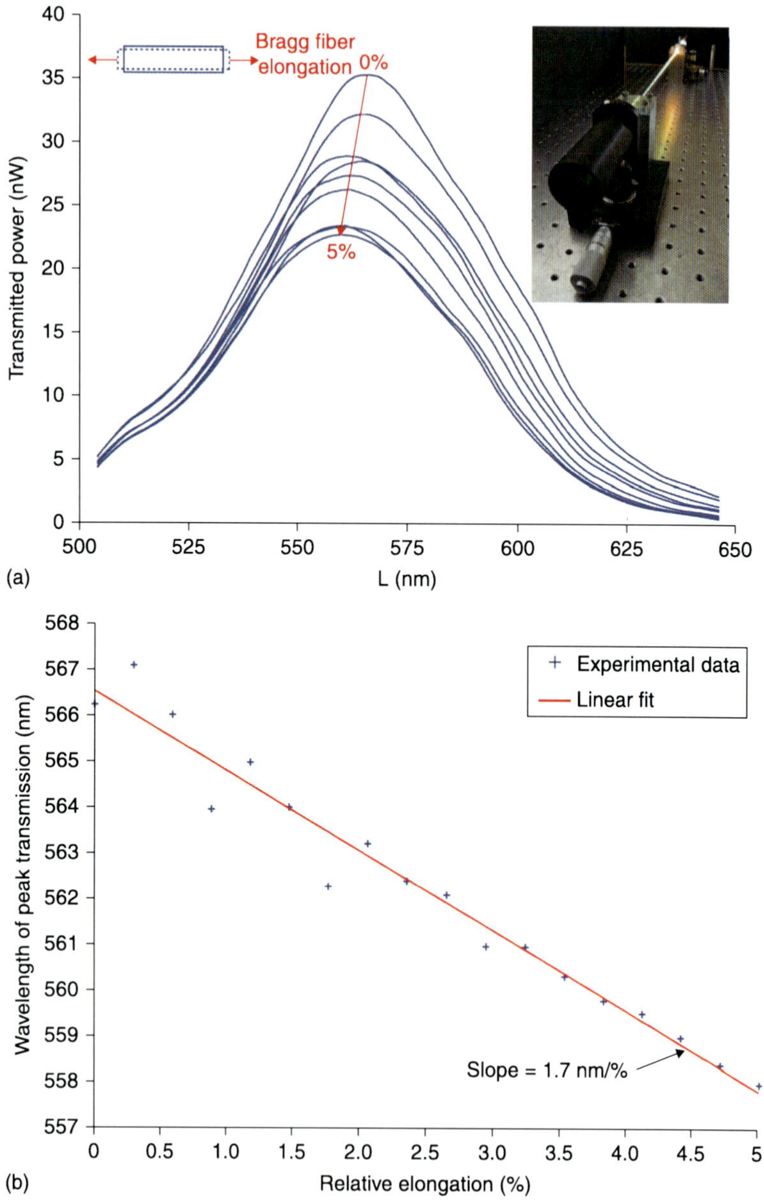

Plate XVI Stretching of plastic Bragg fibers: (a) transmission spectra of a Bragg fiber sample subjected to longitudinal stretching. Fiber elongation results in the shift of the fiber transmission peak to shorter wavelengths due to Bragg reflector thinning. Inset: photo of an experimental set-up. Collimated beam of a super-continuum white light source is injected into a Bragg fiber sample, which is clamped at both ends to the rigid support blocks; (b) linear shift of the transmission peak wavelength with respect to the Bragg fiber elongation with a slope of 1:73 nm%.

Morikawa K, Fujisawa T, Saitoh K and Koshiba M (2006), 'Transmission characteristics of laterally illuminated photonic crystal fibers', *IEICE Electr Exp*, 3, 70–73.

Rubin B (1999), 'Tailored fiber cross-sections', *Adv Mat*, 10, 1225–1227.

Rubin B, Kobsa H and Shearer S M (1994), 'Modeling the dependence of fabric reflectance on denier per filament', *Text Res J*, 64, 685–689.

Rubin B, Kobsa H and Shearer S M (1997), 'Prediction and verification of an iridescent synthetic fiber', *Appl Opt*, 36, 6388–6392.

Russell P St J (2006), 'Photonic crystal fibers', *J Lightware Tech*, 24, 4729–4749.

Schuster J, Trahan M, Heider D and Li W (2003), 'Influence of fabric ties on the performance of woven-in optical fibres', *Comp Pt A: Appl Sci Manu*, 34, 855–861.

Selem B, Rothmaier M, Camenzind M, Khan T and Walt H (2007), 'Novel flexible light diffuser and irradiation properties for photodynamic therapy', *J Biomed Opt*, 12, 034024.

Shapira O, Kuriki K, Orf N D, Abouraddy A F, Benoit G, *et al.* (2006), 'Surface-emitting fiber lasers', *Opt Exp*, 14, 3929–3935.

Sirikasemlert A and Tao X (1999), 'Effects of fabric parameters on specular reflection of single-jersey knitted fabrics', *Text Res J*, 69, 663–675.

Skorobogatiy M (2005), 'Efficient anti-guiding of TE and TM polarizations in low index core waveguides without the need of omnidirectional reflector', *Opt Lett*, 30, 2991.

Sokkar T Z N, Kabeel M A, Ramadan W A and Hamza A A (1992), 'A contribution to the study of color of fabrics', *Color Res Appl*, 17, 219–224.

Spigulis J, Pfafrods D, Stafekis M and Jelinska-Platace W (1997), 'The "glowing" optical fibre designs and parameters', *Proc SPIE*, 2967, 231–236.

Uskokovic P S, Miljkovic M, Krivokuca M, Putic S S and Aleksici R (1999), 'An intensity based optical fibre sensor for flexural damage detection in woven composites', *Adv Comp Lett*, 8, 55–58.

Uskokovic P S, Balac I, Brajovic L, Simic M, Putic S and Aleksic R (2001), 'Delamination detection in woven composite laminates with embedded optical fibers', *Adv Eng Mat*, 3, 492–496.

Vienne G, Xu Y, Jakobsen C, Deyerl H-J, Jensen J, Sorensen T, *et al.* (2004), 'Ultra-large bandwidth hollow-core guiding in all-silica Bragg fibers with nano-supports', *Opt Express*, 12, 3500–3508.

Wakita A and Shibutani M (2006), 'Mosaic textile: wearable ambient display with non-emissive color-changing modules', *Proceedings of the ACE '06, Advances in Computer Entertainment Technology*, p. 48.

Yamaguchi S and Takanabe H (2001), 'Fibers having fine concave and convex surface from silica hybrid polyester', *Sen-Iakkaishi*, 57, 111–119

Zhang H Q, Gao W D and Qiu H (2003), 'Retro-reflection of round fibers', *Text Res J*, 73, 965–970.

Zheng Y H, Pitsianis N P and Brady D J (2006), 'Non-adaptive group testing based fiber sensor deployment for multi-person tracking', *IEEE Sen J Color Res Appl*, 6, 490–494.

Zivkovic I, Brajovic L, Uskokovic P and Aleksic R (2005), 'Indentation damage detection in thermoplastic composite laminates by using embedded optical fibers', *J Adv Mat*, 37, 33–37.

4

Conductive nanofibres and nanocoatings for smart textiles

S. M. SHANG and W. ZENG,
The Hong Kong Polytechnic University, Hong Kong

DOI: 10.1533/9780857093530.1.92

Abstract: This chapter summarizes the recent application of nanotechnology in smart textiles. Two major technical applications, using nanosize fibre/fillers and employing specific nanocoating techniques to create smart textile, are reviewed. The existing functionality of textiles can be improved using nanotechnology, and it is also possible to manufacture smart textiles with entirely new properties or a combination of different functions. Nanotechnology can be used in textiles with a variety of methods. At the end of the chapter, the future development of nanotechnology for smart textiles is addressed.

Key words: nanotechnology; conductive nanofibre; conducting polymer; nanocoating technology; smart textiles.

4.1 Introduction

Nanotechnology is an emerging interdisciplinary technology and the role of nanotechnology in the textile industry has accelerated significantly during recent years. Its profound impact has been considered as a second industrial revolution. Fuelled by the rapid development of nanotechnologies, and driven by the need to increase the value of conventional textile products, intelligent or smart-textile applications have flourished. The application of nanotechnology in textiles aims to improve a material's functionality and give it new characteristics. Nanotechnology has versatile applications in the textile industry in manufacturing garments with wrinkle resistant finishes, flame retardant finishes, stain resistance, antimicrobial qualities, moisture management, ultraviolet (UV) protection, self-cleaning, water- and stain repellent and anti-odour properties, etc. A whole variety of novel smart textiles is already on the market. Examples of smart textiles include sportswear, healthcare and space technology applications.

4.2 Conductive nanofibres

4.2.1 Background

In recent decades, the application of conductive nanofibres in the creation of smart textiles has drawn much attention. Conductive nanofibres are viewed as key to the realization of molecular-scale electronics. The development and properties of nanofibres fabricated from conductive polymer or composites, carbon nanotubes (CNTs), carbon nanofibres and other conductive nanoparticles, have been the focus of much investigation.[1–3] The preparation of organic and/or inorganic materials from conductive nanofibres, which are functionalized via a nanoscale structuring process, is currently an area of intense international activity, both in fundamental and applied science.

Advantages of conductive nanofibres include the ability to be stretched, folded and deformed into complex curvilinear shapes, enabling many new applications that could not be achieved using conventional rigid conductors. Such flexibility means conductive nanofibres hold great potential and offer exciting opportunities, particularly in application to the arbitrary curved surfaces and movable parts of large-area electronics. Depending on the application, opportunities exist for the use of three-dimensional (3-D) systems (photonic bandgap materials), two-dimensional (2-D) systems (quantum well structures) or one-dimensional (1-D) systems (quantum wires, nanocables). Semi-ordered or disordered (non-woven) conductive nanofibres are of interest for such applications as filter media,[4] nanofibre-reinforced conductive composites,[5] solar and light sails and mirrors in space,[6] along with application in electronic textile systems (operating panels in textiles such as soft keyboards, displays, etc.), biomedical applications (scaffolds, tissue engineering bandages, drug release systems and sensors), and protective clothing aimed at biological and chemical protection.[7]

Furthermore, they offer the potential for application in smart and technical textiles, which are more efficient, lower in weight, and have more functions than conventional textiles. Their use in medical textiles, for example, can help improve the quality of life for elderly people, whilst the use of high-efficiency filters can improve environmental conditions in varied situations.

Textile materials that exhibit conductivity or serve an electronic or computational function are called electro-textiles. They can have a variety of functions, including electromagnetic shielding, antistatic, electronic applications, infrared absorption or protective clothing in explosive areas.[8–10] Tactile properties such as stretch, recovery, drape, shear and handling are particularly important for clothing, meaning that the conductive fibres used should be fine and fabrics should have a low weight.

These demands are inconsistent with the materials and geometries that are needed for electrical conductivity, as the incorporation of elements such as metal wires within textiles increases stiffness and reduces elasticity. Whilst this may be acceptable in outer garments like jackets, it would reduce the comfort of garments worn close to the body, such as underwear or shirts. In the longer term, an integration of electrical devices with the textile substrate is desired. Connections between components and interfaces are likely to be based either on textile transmitters utilizing modified textiles, such as fibres and yarns, or on the application of coatings and prints.

4.2.2 Carbon nanotubes

Carbon nanotubes (CNTs) are allotropes of carbon with a cylindrical nanostructure. CNTs, discovered in 1991 by Iijima, are ideal building blocks for nanotechnology because they have many unique properties including chemical inertness, extremely large surface area formation, high aspect ratio structure, super strength, high melting temperature in vacuum, super high thermal and electrical conductivity, and biocompatibility.[11] CNTs have been constructed with the length-to-diameter ratio of up to 132 000 000:1, which is significantly larger than any other material. On the basis of their high conductivity, transparency and mechanical stability, there is a rapid increase in commercial demand, with the novel properties of these cylindrical carbon molecules making them potentially useful for many applications in conductive and high-strength composite fibres, energy storage and energy conversion devices, sensors, and field emission displays.[12–16]

There are two main types of CNTs that can achieve high structural perfection: single-walled nanotubes (SWNTs) and multi-walled nanotubes (MWNTs). SWNTs consist of a cylindrical single sheet with a diameter between 1 and 3 nm and a length of several micrometers. They possess a cylindrical nanostructure formed by rolling up a single graphite sheet into a tube. MWNTs consist of a coaxial arrangement of concentric single nanotubes, like rings of a tree trunk, separated from one another by 0.34 nm. They usually have a diameter of about 2 to 20 nm. The production of SWNTs or MWNTs is highly dependent on the synthesis process and conditions.[17]

The production of MWNTs, which have made a promising appearance on the marketplace in recent times, dates back to early reports from Endo in the late 1970s. The MWNTs with multilayer graphitic tubular structures were catalytically grown by chemical vapour deposition (CVD), and could be further annealed to decrease their defect density. The dimensions of these were typically 10 to 100 nm in diameter, but occasionally smaller structures, including single-layer tubular structures, were also observed. In the early 1990s, Iijima *et al.* and others found that the high-temperature arc-discharge process could produce smaller, highly graphitized MWNTs. Then in 1993 came the discovery of SWNTs by the NEC and IBM groups. This discovery suggested that catalytic vapour deposition could be controlled to make single-layer tubular structures of graphene with diameters as small as 1 nm.

Since the discovery of fullerenes and CNTs, synthesis of carbon nanostructures has attracted a great deal of attention, both for basic scientific research and for industrial applications. Although rapid development has been made in CNT science, the diversification of CNT fabrication methods is still of great interest. The synthesis of well-defined CNTs with diameters of 50 to 100 nm has been a long-standing goal of materials chemistry. Recently, we successfully synthesized the CNTs via pyrolysis of the PPy nanotubes at 900 °C in a nitrogen atmosphere. This synthesis method of the as-prepared CNTs is a simple, easy way to produce nanotubes on a large scale. The SEM and TEM images of the as-prepared CNTs are shown in

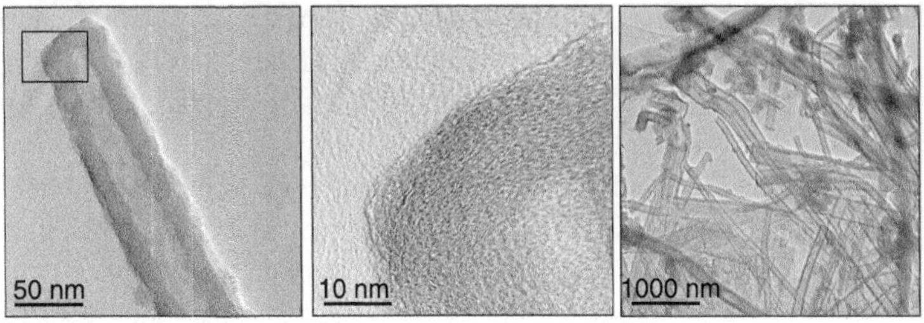

4.1 The SEM and TEM images of the prepared CNTs.[18] High-resolution transmission electron microscopy: (a) The TEM image of one carbon nanotube. (b) The HRTEM image of circled part of (a). (c) The TEM images of the carbon nanotubes. (Reprinted with permission from Elsevier.)

Fig. 4.1. The wall thickness of the CNT is about 15 nm, which is thinner than that of the original PPy nanotubes (20 nm). This size reduction originates from the weight loss and formation of a more compact structure in the carbonization process.[18]

Today, various nanotubes with a precisely controlled number of layers, such as SWNTs and double-wall nanotubes (DWNTs), can be prepared and a lot of progress has occurred in the controllable synthesis of this material. It is now possible to synthesize CNTs in large quantities that are uniformly doped with elements such as nitrogen, holding great promise for specific applications. The applications of nanotubes can be divided into the traditional (where nanotubes provide value addition compared to carbon fibres and other carbons) and non-traditional uses (where nanotube applications provide unique applications), often delineating the timeline expected for applications.

4.2.3 Carbon nanofibres

Carbon nanofibres belong to the family of covalent carbon nanomaterials and have similar conductivity and stability to CNTs. The main characteristic distinguishing carbon nanofibres from CNTs is the stacking of graphene sheets of varying shapes, producing more edge sites on the outer wall of carbon nanofibres than CNTs. This can facilitate the electron transfer of electroactive analytes. Their unique chemical and physical properties make carbon nanofibres exceptional candidates for next-generation on-chip interconnect materials, and promising as immobilization substrates.

Moreover, carbon nanofibres possess lower manufacturing costs in relation to CNTs, while maintaining large aspect ratios and high mechanical and electrical properties, making them particularly appealing in the design of the flexible circuits, electrodes, sensors and actuators used in large quantities in smart textiles. Recently, for example, carbon nanofibres have entered the textile industry for use in increasing the thermal comfort of garments. In such extreme applications of textiles as diving suits, ski wear and active sportswear, carbon nanofibre can impart outstanding properties to garments. Its high tensile strength, thermal insulating properties and low weight make it a popular choice for making sportswear that is both comfortable and protective.

There are several methods for the preparation of carbon nanofibres. Catalytic CVD, often including variants such as thermal and plasma-assistants, is the dominant technique for the fabrication of carbon nanofibres.[19–22] Gas-phase molecules are decomposed at high temperatures and carbon is deposited in the presence of a transition metal catalyst on a substrate. Here, subsequent growth of the fibre around the catalyst particles occurs.

More recently, several research efforts have been made to develop carbon nanofibres from electrospun precursors and to explore their potential applications.[23–25] Carbon fibres can be derived from the precursor material through thermochemical procedures. As a first step, the precursor should be oxidized, which means a thermal treatment at approximately 250 to 350°C. The base material undergoes cyclization, which is an exothermic process. The heat energy released can degrade or even burn fibres, making optimization of this step crucially important. In the next step, the material is heated in an inert atmosphere up to 800 to 1200°C.[26]

Carbon fibres can be heat treated at even higher temperatures, in a process called graphitization. Figure 4.2 shows the prepared carbon fibre with an average diameter of 220 nm, more than 95% of which is made of carbon atoms. The newly developed carbon nanofibres can be manufactured using current facilities, helping to minimize manufacturing costs. In addition, it is possible to manufacture high-purity carbon nanofibre, as no catalyst is used during the production process.

4.2.4 Nanotechnology in manufacturing conductive polymer or composite films

The development of conductive polymer or composite fibres for smart textiles has received great attention because of the unique optical, electronic, chemical and mechanical properties offered. Materials such as metal nanoparticle/nanowire (gold, silver and metal oxides), carbon black, CNTs and conductive polymers have been used to prepare conductive nanofibres. These materials are promising for a variety of applications, including flexible optical and electronic devices, and chemical and biological sensors,[27] leading to the publication of hundreds of research articles on the use of conductive nanofibre for smart textiles.

Conductive nanofibres can be obtained via a number of methods. Indium phosphide nanowires, for example, have been prepared using laser assisted catalytic growth,[28] with metallic-nanowire electrodeposited.[29] A one-step route for the fabrication of highly porous polyaniline nanofibre and DNA-templated assembly has also been proposed as a possible method of nanofibre and nanowire production.[30–31] All methods listed above allow for good process control. However,

4.2 SEM images showing the representative morphologies of (a) as-electrospun PAN nanofibre bundle, (b) stabilized PAN nanofibre bundle, (c) low temperature (1000°C) carbonized PAN nanofibre bundle, and (d) high temperature (2200°C) carbonized PAN nanofibre bundle.[26] (Reprinted with permission from Elsevier.)

all of them yield significantly short nanofibres and nanowires, with lengths of the order of several microns.

Since the discovery that conjugated conductive polymers can be made to conduct electricity through doping,[32] a tremendous amount of research work has been carried out in the field. Nanoscale π-conjugated organic molecules and polymers can be used for sensors, actuators, transistors, flexible electronic devices and field emission display in the textiles systems.[33–36] Different routes are used to prepare nanofibres of various conducting polymers. Polyaniline nanofibres were prepared via chemical polymerization of aniline.[37] Similarly, polypyrrole nanofibres were synthesized (60–100 nm in diameter) in the presence of p-hydroxy-azobenzene sulphonic acid as a functional dopant.[38] The nanofibres have high conductivity (120–130 S/cm) at room temperature and a photo-isomerization function that results from proton doping and isomerization of the azobenzene moiety.

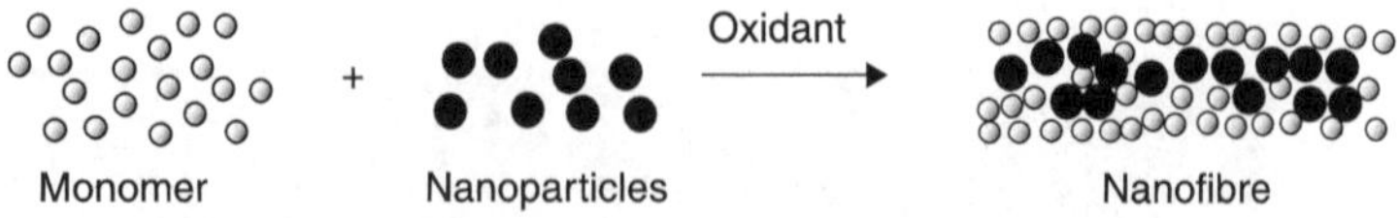

4.3 Formation of nanofibre.

Polymer/nanoparticle composite conductive nanofibres have also attracted the interest of a number of researchers, due to their synergistic and hybrid properties, derived from several components.[39–44] A simple representation of the formation of nanofibre is given in Fig. 4.3. The ease of processing an organic polymer, combined with the improved mechanical and electrical properties of nanoparticles, has led to the fabrication of many electro-devices. Figure 4.4 shows

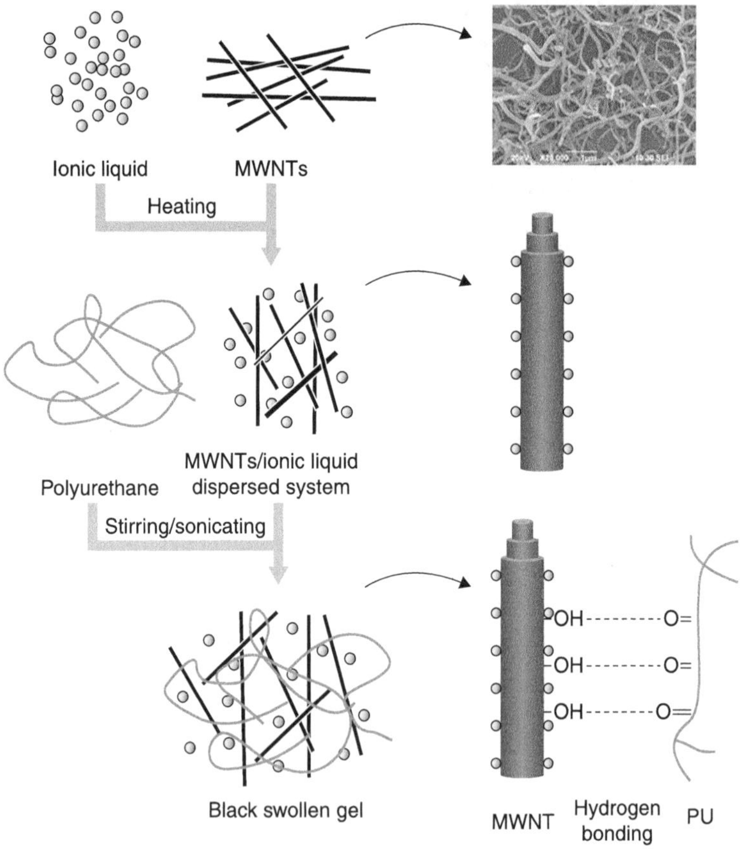

4.4 Schematic representation of the process to fabricate the MWNTs/PU composite.[45] (Reprinted with permission The Royal Society of Chemistry.)

the preparation of conductive composites from CNTs, which can be stably dispersed in a polyurethane matrix, and thus reveals potential applications in electronic devices.[45]

Another method used to prepare conductive nanofibres is electrospinning. Moreover, the electrospinning technology is currently the only method that allows the fabrication of continuous fibres with diameters from several micrometers to a few nanometers.[46–47] This method can be applied to both polymers and polymer/nanoparticle composites. Electrospun nanofibres with complex architectures, such as core–shell structure nanofibres or hollow nanofibres, can be produced using coaxial electrospinning methods. It is also possible to produce structures ranging from single nanofibres to highly ordered nanofibre arrangements.

Electrospinning is not only employed in university laboratories, but is also increasingly being applied in industry. The scope of applications is very broad, with potential in areas as diverse as filtration, opto-electronics, catalysis, medicine, and sensor and actuator technology. A typical electrospinning set-up (Fig. 4.5) consists of three basic components: a syringe with a metal spinning nozzle, a high-voltage supplier and a collector. During the electrospinning process, a high voltage in the range of 10 to 100 kV is used to create an electrically-charged jet of electrospun solution. The droplet of the polymer spun solution on the spinning nozzle is slowly stretched under the high voltage and becomes narrower. Through evaporation of solvents, the nanofibre will be formed on the collector.[48]

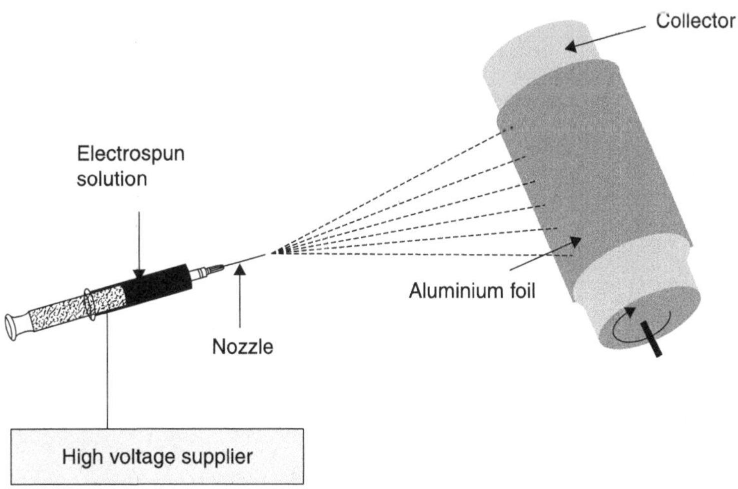

4.5 A schematic diagram showing the electrospinning apparatus used to prepare the nanofibres.

Our research group also successfully fabricated conductive MWNTs/polyurethane nanofibres using this process. Due to the electrospinning-induced alignment of CNTs in the polymer matrix, it provides an effective way to explore high conductivity, flexibility and reliability of electrodes. We employed the electrospinning method to prepare MWNTs/PU nanofibres with differing MWNT content (ranging from 0.1 wt.% to 10 wt.%). For improving dispersion of MWNTs in PU solution, MWNTs were first modified using ionic liquid, with the modified MWNTs expected to form a homogeneous dispersion in the PU solution because of their compatibility.

The SEM and TEM images of conductive nanofibres containing 10 wt.% MWNTs are presented in Fig. 4.6. As can be seen in the figure, uniform, highly smooth nanofibres were formed without the occurrence of bead defects for all experimental materials. More importantly, in previous reports many beads (>1 μm) appeared along the MWNTs/PU fibres as the amount of MWNTs increased, meaning that the MWNTs/PU composites containing more than 2 wt.% of MWNTs could not be electrospun into fibres.[49–50] Yet in this study, there was no formation of beads, and the solution with 10.0 wt.% MWNTs was successfully electrospun into nanofibres. The reason for this success may be that the MWNTs/PU electrospun solution possessed excellent conductivity and elasticity.[51]

When the MWNTs content reaches 10%, the conductivity is as high as 1.8 S/cm, indicating that excellent conductivity is obtained for the MWNTs/PU nanofibres (Fig. 4.7). Therefore, the inclusion of MWNTs led to a further increase in the conductivity of the MWNTs/PU nanofibres. The extraordinary conductivity of the MWNTs/PU nanofibres can also be applied to high-performance mechanical actuation systems, such as in flexible electrical artificial skin and integrated actuators.

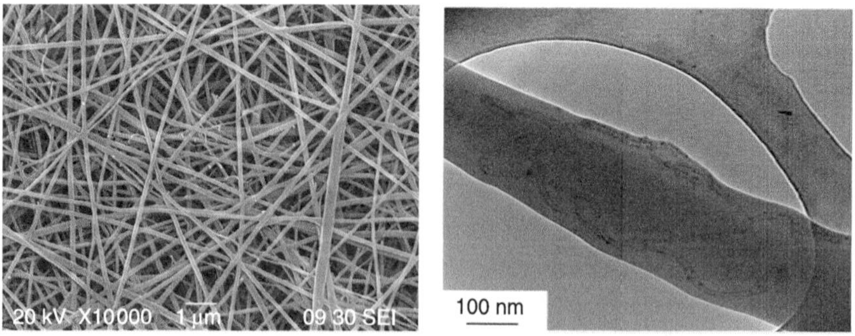

(a) (b)

4.6 (a) SEM image of conductive nanofibres with 10.0 wt.% MWNTs. (b) TEM micrograph of nanofibres with 10.0 wt.% MWNTs.

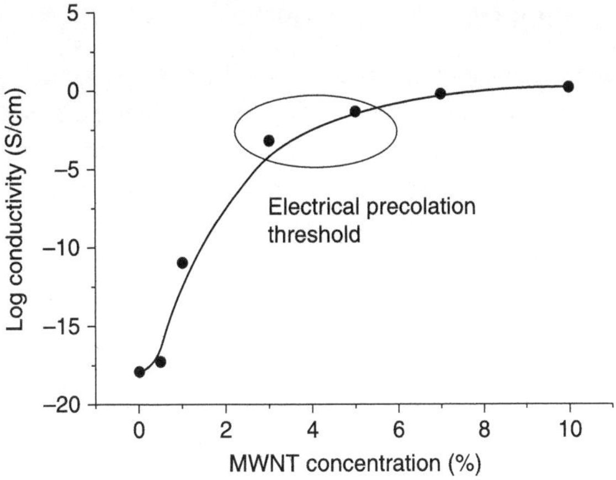

4.7 Conductivity of nanofibre containing different MWNTs ranging from 0 to 10 wt.%.

4.3 Conductive nanocoating

4.3.1 Background

Conductive nanocoating technologies are an alternative approach for improving the performance and functionality of textiles.[57] Conventional methods of adding various properties to textiles degrade through washing and wearing. However, a successful nanocoating technology can impart the desired functionalities and provide high durability fabrics. This is due to the large surface area-to-volume ratio and high surface energy of nanomaterials.

Conductive nanocoating with discrete molecules or conductive nanomaterials can be added with a specific orientation and trajectory to designated textile material sites, through thermodynamic, electrostatic or other technical approaches. This can produce enhanced or specifically targeted properties, due to the combined properties and/or structuring effects of the nanomaterials used. With advances in nanomaterials and processes, conductive nanocoating technology plays a crucial role in realizing new engineering breakthroughs and performance enhancement of smart textiles.[9]

A number of different approaches have been used to fabricate nanocoatings, including vapour deposition,[53] chemical reduction,[54] electrochemical deposition,[55] pulsed laser deposition[56] and molecular self-assembly.[57] Highly conformal nanoscale coatings on complex fibrous systems could enable advances in fibre-based batteries and capacitors, as well as photoactive electrochemical and sensing devices. If we can find a way to apply them to textiles, it would represent a cost-effective approach and framework for improving current and future types of electronic devices.

4.3.2 Vapour deposition technology

Development in vapour deposition techniques over the last two decades has led to the introduction of many advanced nanocoating technologies for smart textiles. Generally, vapour deposition refers to any process in which materials in a vapour state are condensed through condensation, chemical reaction or conversion to form a solid material. There are two categories of vapour deposition processes: physical vapour deposition (PVD) and CVD. Examples include single-layered and multilayered conductive nanocoatings, which feature very good electrical, mechanical and chemical properties.

Plasma- and ion-based vacuum coating techniques have been at the forefront of these new developments. They allow the coating/textile system to be designed in such a way that the combinations can perform in an optimal manner. Considerable progress has been made in understanding the theoretical aspects of mass transfer process, conductivity and mechanical mechanisms.

A more practical approach is to consider the requirements of a surface for different applications, that is, at the interface of the substrate, within the coating itself, and at its surface (Fig. 4.8). Multilayer coatings offer the possibility to design desired functionalities according to the application requirements. A functional nanocoating designed to use specific layers in achieving distinct specific properties is presented in Fig. 4.9. Such a concept can be achieved with just two layers; for example, an electroless nickel coating can be used as a load supporting layer for a hard PVD TiN or CrN coating, in order to improve abrasive wear performance and enhance corrosion resistance.[58] It is even possible to pretreat a substrate surface for the production of a hardened outer layer, which can then support a coating more effectively. Such approaches have been called hybrid or duplex coatings.[59]

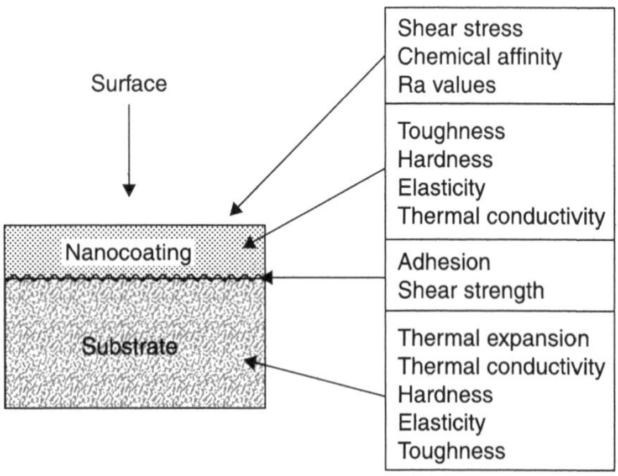

4.8 Property requirements in different zones of a coated surface.

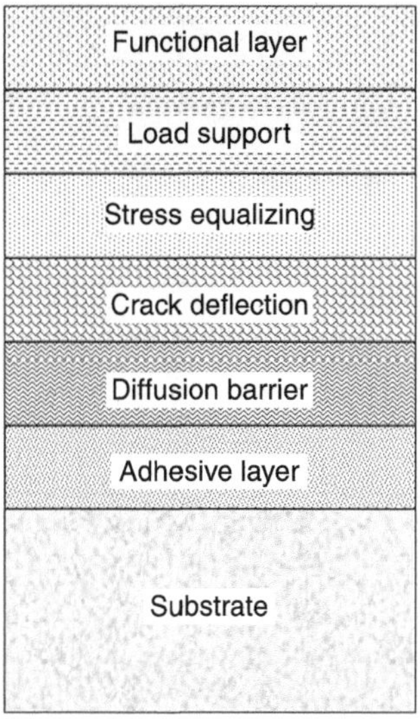

4.9 A typical functionally graded multilayer coating design to utilize specific layers for distinct properties.

Physical vapour deposition

Physical vapour deposition (PVD) is a process used to produce a metal vapour that can be deposited on electrically conductive materials as a thin, highly adhered pure metal or alloy coating. The process is carried out in a vacuum chamber at high vacuum (10–6 torr) using a cathodic arc source.[60] PVD methods use clean and dry vacuum deposition, in which the coating is deposited over the entire object simultaneously, rather than in localized areas. All reactive PVD hard coating processes combine:

- a method for depositing the metal;
- combination with an active gas, such as nitrogen, oxygen, or methane;
- plasma bombardment of the substrate to ensure a dense, hard coating.

PVD methods differ in the production of metal vapour and the details of plasma creation. The primary PVD methods are ion plating, ion implantation, sputtering and laser surface alloying.

Chemical vapour deposition

Chemical vapour deposition (CVD) is a technique whereby gaseous reactants can be deposited onto a substrate to form a stable solid film.[61] The sequence of events during a CVD reaction is shown in Fig. 4.10. The film morphology is mostly dependent on the surface morphology of the substrate, which is usually used as a template. However, through selection of the gaseous reactants and

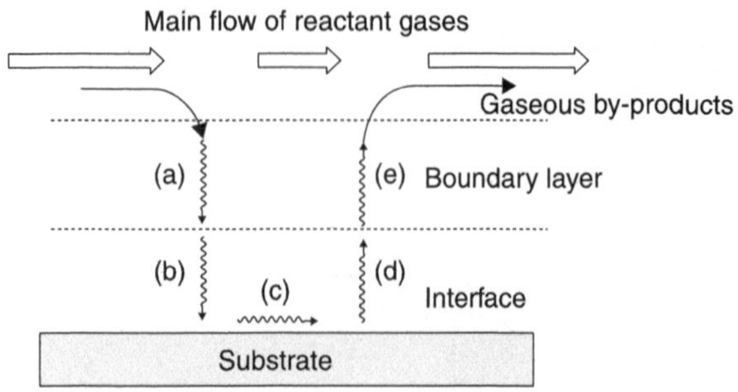

4.10 Sequence of events during CVD: (a) diffusion of reactants through boundary layer; (b) adsorption of reactants on substrate; (c) chemical reaction takes place; (d) desorption of adsorbed species; and (e) diffusion out of by-products through boundary layer.

adjustment of the reaction conditions, different morphologies can be realized on the substrate. There are great varieties of CVD processes, as shown in Table 4.1. Applications of CVD range from the fabrication of microelectronic devices to the deposition of protective coatings.[62–64] CVD generally has faster growth rates and better uniformity than the other deposition processes; it is more suitable for large-scale commercial applications. In a typical CVD process, reactant gases enter the reaction chamber at room temperature, and the gas mixture is then either heated as it approaches the deposition surface, heated radiatively, or placed upon a heated substrate. Depending on the process and operating conditions, the reactant gases may undergo homogeneous chemical reactions in the vapour phase before striking the surface.

4.3.3 Chemical reduction

Amongst the various solution-phase chemistry routes developed for the preparation of metal nanoparticles on the surface of the fibrous system, the reduction of metal salts is probably the most popular due to its simplicity, low-cost and ability to produce large quantities. To prepare the stable conductive metal nanoparticles via the chemical reduction method, it is important to choose appropriate stabilizer and reducing agents. In a typical process, we can choose the reducing agent and corresponding protective agent to obtain a uniform dispersion of products. The nanoparticle size and aggregation state of the conductive nanocoating are affected by various parameters, such as initial metallic ions and stabilizer concentrations, the molar ratios of reducing agent/metallic ions, and pH value. In an aqueous reaction medium, strong reducing agents such as hydrazine (N_2H_4) and sodium borohydride ($NaBH_4$), will directly affect the reaction rate and reduce very small nanoparticles. Furthermore, the high precursor concentration also influences the size distribution of the final product, because of the diffusion limit of protective agent onto the colloidal surface.[54, 74–75]

Table 4.1 Varieties of chemical vapour deposition processes

Atmospheric pressure chemical vapour deposition	CVD processes at atmospheric pressure.[65]
Plasma assisted chemical vapour deposition	CVD processes that utilize plasma to enhance chemical reaction rates of the precursors. Plasma assisted chemical vapour deposition processing allows deposition at lower temperatures, which is often critical in the manufacture of semiconductors.[66]
Combustion chemical vapour deposition	Combustion chemical vapour deposition process is an open-atmosphere, flame-based technique for depositing high-quality thin films and nanomaterials.[67]
Photochemical vapour deposition	Photochemical CVD uses the energy of photons to initiate the chemical reactions. Photo-dissociation of the chemical precursor involves the absorption of one or more photons resulting in the breaking of a chemical bond. The most common precursors for photo-assisted deposition are the hydrides, carbonyls, and the alkyls.[68]
Laser chemical vapour deposition	Laser chemical vapour depositions a derivative of CVD whereby the global heat source for the furnace is replaced with a localized spot heated by a laser. There are two types of laser CVD, pyrolytic and photolytic.[69]
Rapid thermal chemical vapour deposition	CVD processes that use heating lamps or other methods to rapidly heat the wafer substrate. Heating only the substrate rather than the gas or chamber walls helps reduce unwanted gas phase reactions that can lead to particle formation.[70]
Metalorganic chemical vapour deposition	CVD method of epitaxial growth of materials, especially compound semiconductors from the surface reaction of organic compounds or metal-organics and metal hydrides containing the required chemical elements.[71]
Chemical vapour infiltration	A chemical vapour deposition type process used for the preparation of ceramic matrix composites such as alumina-alumina, in which a chemical vapour consisting of $AlCl_3$-H_2-CO_2 is deposited onto porous alumina fibres or preforms. Can be used as a substitute for CVD.[72]
Chemical beam epitaxy	Chemical beam epitaxy forms an important class of deposition techniques for semiconductor layer systems, especially III-V semiconductor systems. This form of epitaxial growth is performed in an ultrahigh vacuum system. The reactants are in the form of molecular beams of reactive gases, typically as the hydride or a metalorganic.[73]

The reduction reaction of metallic ions is also sensitive to the solution's pH. Moreover, it may affect the product's morphology via the formation of certain species. More recently, Xue *et al.* (2008) prepared conductive textiles with silver nanoparticles via the chemical reduction method. The nanoparticle roughening effect of the fibre surface favours the construction of super-hydrophobic surfaces, while the compact coating of silver imparts not only the metallic feature to the fibres, rendering the textiles conductive, but also imbues antibacterial properties to the textiles. This method to multi-functionalize conventional textiles with one material is useful in the textile industry, and this strategy is expected to become a powerful platform for the fabrication of multi-functional materials.[76]

4.3.4 Electrochemical deposition

Electrochemical deposition is a unique technique in which a variety of materials can be processed, including polymers, metals and even ceramics. Electrochemical strategies offer important advantages and unique possibilities in the development of nanomaterials and nanostructures. A conductive nanocoating can be produced to any thickness from about 10 nm to more than 100 μm.[77–79] The deposition process can be most effectively controlled when used with a source of external electrical potential. However, the deposition process requires electrical contact to the substrate when immersed in the liquid bath. In any electrochemical deposition process, the surface of the substrate must have an electrically conducting coating before the deposition can be done. A schematic diagram of a typical set-up for electroplating is shown in Fig. 4.11.

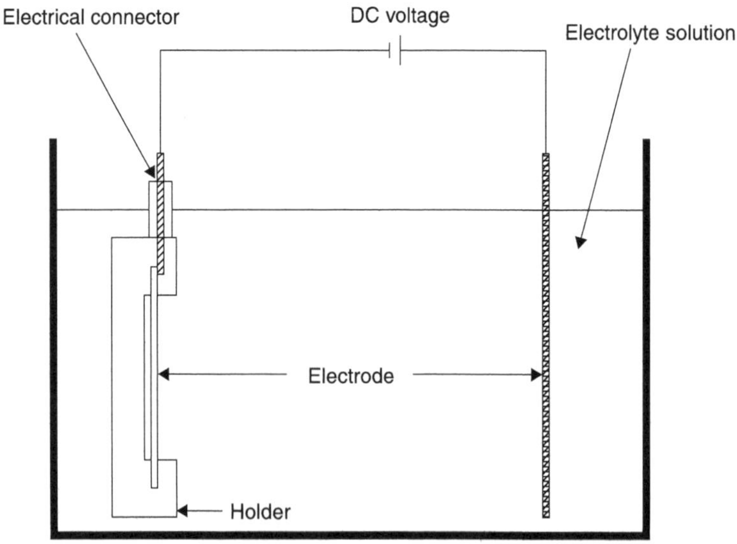

4.11 Typical set-up for electro-deposition.

Using this technique, it is possible to obtain a metal nanocoating in a single step process.[80] To obtain enhanced characteristics of any nanocoating, it is necessary to optimize the bath (electrolyte) composition, as well as the deposition parameters, and anode and cathode materials.

Similarly, optimized parameters and bath composition are paramount in the electrochemical deposition of any new coating or material. Studies have been carried out using electrochemical methods for the preparation of insulating fibres coated with conductive polymers, which could also be employed for electrodes.[81] Bhadani *et al.* (1997) prepared conductive fibres with cylindrical platinum electrodes.[82] In this study, insulating fibres were wound around the anode, and the polymerization was carried out in an electrolytic cell charged with a solution of aniline in aqueous HBF_4. The electrochemical deposition is an indirect process, since the substrate is not a conducting material, and polyaniline was deposited on poly(ethylene terephthalate) fabrics through an indirect electrochemical deposition. Polyaniline was formed on the anode and was adsorbed onto the fabric.[83]

In further studies, the conductive polymer of polypyrrole has been synthesized directly on conducting textiles of PES-Ppy/AQSA using electrochemical deposition, whilst polyester has been used as a base material due to the extensive literature on its behaviour.[84] It must be mentioned that only the polarity of the surface may have an effect on the adhesion of the conducting polymer. Moreover, the electrochemical synthesis of the conductive polymer takes place in the acidic medium, so the selected substrate should be resistant to damage by acids. For this purpose, the employment of natural fibres like cotton should be avoided, because acids damage such fibres.

As a result of the unique properties of electrodeposited nanomaterials, a number of industrial applications have emerged. The combination of increased hardness/wear resistance and reduced localized corrosion results, for example, in improved protective coating performance. Protective materials can be used as a hard surface on softer, less wear-resistant substrates. The magnetic and electrical properties also make electrodeposited nanomaterials attractive as soft magnets for high-efficiency transformers, power supplies and healthcare applications. Therefore next-generation smart textiles may also benefit from these excellent magnetic, electrical and wear properties.

4.3.5 Pulsed laser deposition technology

Pulsed laser deposition is a thin film deposition (specifically a PVD) technique, where a high power pulsed laser beam is focused inside a vacuum chamber to strike a target of the material that is to be deposited.[85] Conceptually and experimentally, pulsed laser ablation is an extremely simple technique; probably the simplest among all thin-film growth techniques. Figure 4.12 is a schematic diagram of a pulsed laser deposition system. It consists of a target holder and a substrate holder housed in a vacuum chamber. A high-power laser is used as an

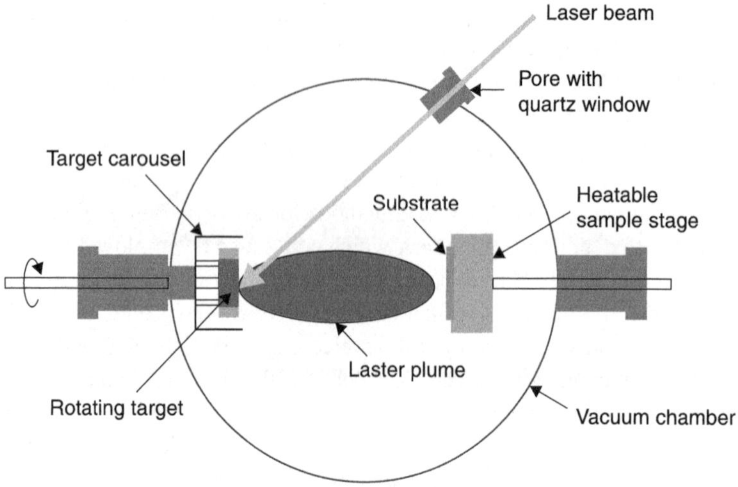

4.12 Schematic view of a PLD System.

external energy source to vapourize materials and to deposit thin films. A set of optical components is used to focus and raster the laser beam over the target surface. The decoupling of the vacuum hardware and the evaporation power source makes this technique extremely flexible. It is therefore easily adapted to different operational modes without the constraints imposed by using internally powered evaporation sources. The main advantages of pulsed laser deposition are:

- *Conceptually simple*: a laser beam vapourizes a target surface, producing a film with the same composition as the target.
- *Versatile*: many materials can be deposited in a wide variety of gases over a broad range of gas pressures.
- *Cost-effective*: one laser can serve many vacuum systems.
- *Fast*: high-quality samples can be grown reliably in 10 or 15 minutes.
- *Scalable*: as complex oxides move toward volume production.

Popescu *et al.* (2011) report on ZnO thin film or nanoparticle deposition on textile materials by pulsed laser deposition, which ensures the stoichiometric transfer of oxides from target to collector and a high adherence of the deposited structures to substrate.[86] A major advantage of pulsed laser deposition is related to its high versatility. By control of the deposition parameters, we can obtain nanocoating films with completely different morphology, structure and/or functionality.[87] Etchegoin *et al.* (2004) showed that the structures deposited by pulsed laser deposition can be controlled in terms of wetting behaviour.[88] Thus, ZnO nanocoating film deposited by pulsed laser deposition on Si became hydrophilic after exposure to UV. The structures became hydrophobic after storage in complete darkness for 10 days or after thermal treatment. This study might offer guidance for one step in the laser manufacturing of stable super-hydrophobic textile surfaces, for use in both everyday garments and medical clothing.

4.3.6 Self-assembly nanocoating

Self-assembly nanocoating is a versatile way to construct thin-film nanocomposites and to modify surfaces. Popular due to its simplicity and adaptability, this technology is most commonly used to fabricate interfaces involving multilayer ionic assemblies of polyelectrolytes.[89] In self-assembled nanocoating, target chemical molecules form a layer of nanosize thickness on the surface of textile materials. Additional layers can be added on the top of the existing ones, creating a nanolayered structure. Different self-assembled nanocoating approaches are being explored to confer special functions to textile materials.[90]

One technical approach is to use electrostatic attraction to self-assemble nanocoatings on textile materials, producing protective, self-cleaning and self-repairing functions.[91–92] This technique is most commonly used to fabricate an interface involving a multilayer ionic assembly of polyelectrolytes, because the thickness, homogeneity and sequence of these nanolayers can be precisely controlled by regulation of the molecular architecture, self-assembly and electrostatic interactions during the self-assembly process.[93–95] However, some ionic self-assemblies are of limited applicability, because of the weak interaction between the substrate and nanolayers.

Another kind of self-assembly nanocoating technology involves layer-by-layer assembly using covalent bonds to form multilayer grafts on the substrate. These approaches are much more durable, stable and practical compared with the ionic self-assemblies. As shown in Fig. 4.13, the design of chemical reactions between each layer is very important, as this affects the stability of the formed coating, as well as the feasibility of the process. Functionalized nanoparticles usually facilitate the production of rough surfaces on a substrate simply by controlling the nanoparticle layers. To enhance the interaction between the first layer and the substrate, the substrate should be activated using the surface modification or fictionalization technology. After layer-by-layer assembly, the outer surface contains residual functional groups, marked by the letters A and C shown in Fig. 4.13, which can be utilized to react with low-surface-energy agents on the surface of textiles.[96]

Xue *et al.* (2008) prepared super-hydrophobic surfaces on cotton textiles. They used layer-by-layer assembly nanocoating of amino- and epoxy-functionalized silica nanoparticles on epoxy-functional cotton textiles to generate a dual-size surface roughness and super-hydrophobic property. Current layer-by-layer assemblies offer many advantages, including easy availability of raw materials, simple fabrication, and large-scale area production of super-hydrophobic surfaces. These advantages make the present strategy suitable for real applications.[97]

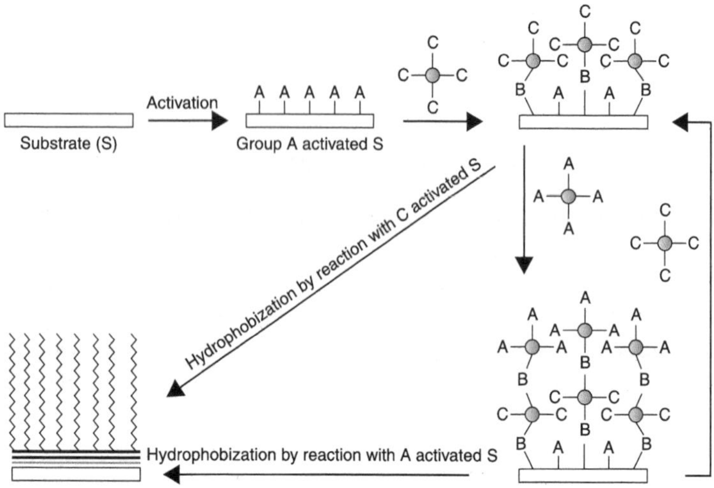

4.13 Steps used in the covalent layer-by-layer assembly of functional particles, with C groups and A groups on a substrate activated with A groups, for super-hydrophobic surfaces. B groups are formed by the reactions of A groups with C groups.[96] (Reprinted with permission from IOP Publishing Ltd.)

4.4 Application of nanotechnology in smart textiles

Initial nanotechnology influenced improvements to textiles will be relatively simple changes to existing technologies. However, the future holds the possibility of extremely complex solutions to the production, not only of smart textiles, but ones that are highly intelligent. This new generation of smart textile may incorporate nanomachines, nanosensors and even nanocomputers into their structure. This will enable them to respond directly to their environment rather than make simple changes caused by the environment. The development of nanotechnologies in textiles will have a two-fold focus:

1. upgrading existing functions and performances of textile materials; and
2. developing new smart textiles with unprecedented functions.

4.4.1 Upgrading existing functions and performances of textile materials

Nanotechnology in textile finishing

The impact of nanotechnology in the textile finishing area has led to increased innovation and research into new applications. Particular attention has been paid to refining and enhancing the controllability of chemical finishes. Ideally, discrete molecules or nanoparticles of a finishing agent can be brought individually to designated sites on textile materials, and set in a specific orientation and trajectory through thermodynamic, electrostatic or other technical approaches.

The main advantages of nanotechnology applications in textile finishing include:

- creating a carefree quality to wrinkle resistant fabric that minimizes stains;
- offering superior liquid repellence and providing additional wrinkle resistance;
- designing fabrics that breathe, yet remaining liquid and stain-repellent;
- developing creative modifications of fabric surface properties, such as permanent anti-static, antibacterial and UV protection.

One of the trends in the synthesis process is to pursue a nanoscale emulsification, through which finishing agent can be applied to textile materials in a more thorough, even and precise manner. The finishing agent can be emulsified into nano-micelles or wrapped in nanocapsules that can adhere to textile substrates more evenly. These advanced finishing agents set up an unprecedented level of textile performance, with greatly enhanced wrinkle resistant, moisture content, stain-resistant, anti-static and shrink-proof abilities.

For finishing purposes, another leading nanotechnology is chemical oxidative deposition, which involves the deposition of conducting polymers onto different kinds of fibres and textiles, resulting in composite materials with high-tensile strength and good thermal stability and conductivity. Furthermore, surface polymerization of conducting polymer fibres has the potential to increase conductivity almost 10-fold by decreasing the electrical resistivity.[98] Such coated polymeric composite materials can be used in electromagnetic interference shielding, microwave attenuation and dissipation of static electric charge. Hence, they can be particularly useful for military applications, for example in camouflage and stealth technology.

Nanoparticles, such as metal oxides, CNTs and ceramics, are also used in textile finishing to alter surface properties and impart textile functions. Nanosize particles have a large surface area, and by combining the nanoparticles with the organic and inorganic compounds, the surfaces of the fabrics treated with abrasion resistant, water repellent, UV, electromagnetic and infrared protection finishes can be appreciably modified. Fabric treated with nanoparticles TiO_2 and MgO, for example, replaces the active carbon-treated fabrics previously used as chemical and biological protective materials.[99–100]

The photo-catalytic activity of TiO_2 and MgO nanoparticles can break harmful and toxic chemicals and biological agents. During the finishing process, these nanoparticles can be pre-engineered to adhere to textile substrates using spray coating or electrostatic methods. Furthermore, finishing with conductive nanoparticles can convert fabrics into sensor-based materials. If conductive nanoparticles are incorporated into fabrics, the finished fabric can convert exerted mechanical forces into electrical signals, enabling the monitoring of bodily functions such as heart rhythm and pulse if worn next to skin.

4.4.2 Developing smart and intelligent textiles with unprecedented functions

The development of smart and intelligent textiles with unprecedented functions is particularly urgent from the standpoint of homeland security and advancement of technology. Key new textile functions to be developed include:

1. wearable solar cell and energy storage system;
2. sensors and information acquisition system;
3. healthcare and wound healing functions; and
4. self-cleaning, repairing and heating functions.

Wearable solar cell and energy storage

Nanotechnology has shown the possibility of fulfilling everyone's dream of producing cheap and clean energy through its strategic application. Its intersection with energy is going to change the way energy was hitherto generated, stored, transmitted, distributed and managed. In particular, nanotechnology is going to revolutionize the solar energy and energy storage sectors. Integration of flexible solar cells into clothing can provide power for portable electronic devices, offering a range of useful applications in a variety of consumer application sectors, including home and sports, leisure clothing, architecture textiles and the automotive industry. Applications envisioned include solar tents and parasols with the capability of energy generation.

Since 2000, design studies on the integration of solar cells into clothing have regularly been presented at fairs and exhibitions on smart textiles.[101] Although both consumers and the clothing industry seem to be very interested in the wearable solar cell, the advent of real products on the market has been hindered and delayed by the limited availability and performance of flexible solar cells. Amongst various technologies for flexible solar modules, photovoltaics is the most advanced way of providing electricity far from any mains supply.

However, this system suffers from the limits of ambient light intensity. Silicon thin films currently offer the most promising combination of flexibility and performance, especially under typical low-light conditions.[102] Ultra-thin pc-Si cells are ready for direct lamination with textiles, while high-efficiency c-Si transfer cells are suitable for add-on accessories or pocket integration. Low-loss charge controllers, textile wiring and specialized know-how on integration with garment production will pave the road to the first integrated photovoltaic products.

Wearable energy storage systems, such as the textile based micro-supercapacitor, have also been the subject of much research. Figure 4.14 shows a prototype of a high-efficiency fibre-based electrochemical micro-supercapacitor using ZnO nanowires as electrodes.[103] These fibre supercapacitors, which have great potential for scale-up, comprise of two electrodes that employ a flexible plastic wire and a Kevlar fibre as a substrate. Both wire and fibre are covered with arrays of high

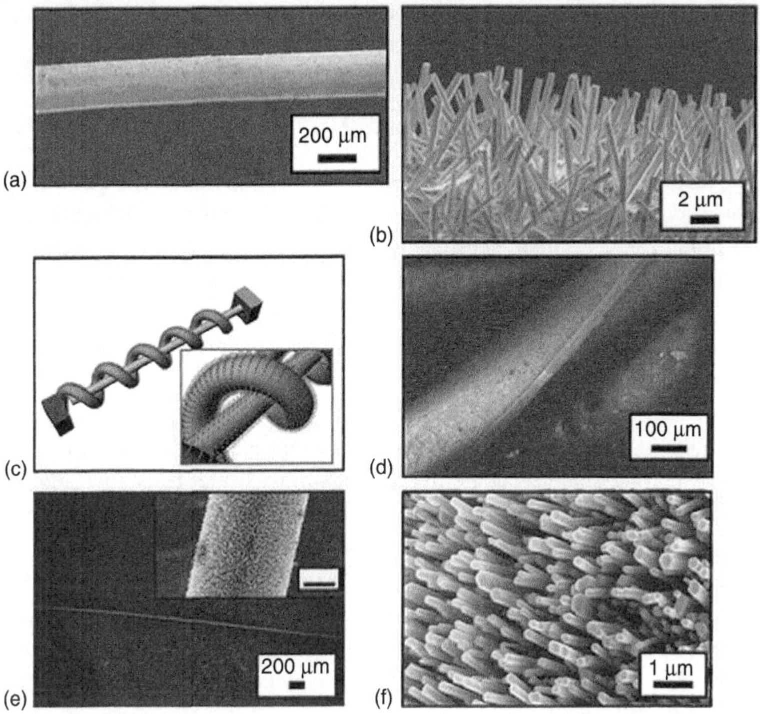

4.14 (a) Low-resolution SEM image of a gold-coated plastic wire covered with ZnO nanowires arrays. (b) Higher-magnification SEM image of the plastic wire, showing arrays of nanowires. (c) A fibre-based super-capacitor. (d) Low-resolution SEM image of fibre super-capacitors consisting of entangled Kevlar fibre and plastic wire. (e) Low-resolution SEM image of a Kevlar fibre covered with ZnO nanowires arrays. Inset: close-up; scale bar: 10 mm. (f) Higher-magnification SEM image of the fibre, showing ZnO nanowires arrays grown along the radial direction[103] (Reprinted with permission from Wiley-VCH Verlag GmbH & Co. KGaA, Weinheim.)

quality ZnO nanowires, grown by the hydrothermal method. ZnO nanowires on a Kevlar fibre were coated with a thin gold film to improve the charge–collection capacity.

Furthermore, employment of ZnO nanowires could provide exciting solutions to the future development of wearable energy devices. The fibre-based micro-supercapacitor would be large enough for use in self-powering nanosystems, such as a power shirt using piezoelectric ZnO nanowires grown radially around textile fibres. With lightweight, cheap cotton or fabrics and the existing textile fabrication infrastructure, large-scale energy storage with low material and processing costs is also feasible. When combined with inks of other functional materials, alternative types of electronics could be created, such as wearable solar cells and batteries.

These porous conductors with excellent properties will be found in a wide range of applications.

Sensors and information acquisition systems

The incorporation of sensors into new materials has emerged as a new response to needs in medical monitoring of physiological signals, including heart and respiration rates. Such sensory materials can be used for guided training, rehabilitation of athletes and assistance for emergency first-responders. As for real devices, ultimately most signals are being transformed into electric devices.[104–105] Electroconductive materials are consequently of utmost importance with respect to smart textiles.

As well as technical considerations, concepts, materials, structures and treatments, development must focus on appropriateness for use in or as a textile material. This includes such criteria as flexibility, water resistance and durability against deformation and radiation. Recent progress in flexible electronics has debunked many traditional barriers, including the lack of practical ways to incorporate electronics into textiles.[106–108] Several kinds of fabric or textile sensors have been investigated. As one example, a single-layer fabric pressure sensor array containing two series of orthogonally placed conductive yarns was fabricated, in which the conductive multifilament yarns serve as both the sensing and conducting elements.[109]

Smart electronic yarns and wearable fabrics for human bio-monitoring have also been investigated by Bong Sup Shim *et al.* (2008). In this report, cotton yarn has been coated with CNTs and polyelectrolytes.[110] When anti-albumin was incorporated into the CNT-cotton yarn, it became an e-textile biosensor that quantitatively and selectively detected albumin, the essential protein in blood. This method provides a fast, simple, robust, low-cost, and readily scalable process for making e-textiles.

Alternatively, a flexible and partially extensible textile based capacitive pressure sensor array has been reported by Sergio *et al.* (2003). That measures the capacitance variations between rows and columns of conductive fibres patterned on the two opposite sides of an elastic fabric when pressure is exerted.[111–112] A capacitance variation as low as 100 fF induced by a light pressure exerted over a 1 m^2 fabric surface can be detected. In a very recent study, a capacitive textile pressure sensor was used for classification of sitting postures, and the measurement error was reduced from 24 to 5% on average, by modelling the hysteresis provoked by compression of the spacer with the Preisach model.[113]

Compared with those in the reported literature, the fabric strain sensors have outstanding linearity with little strain rate effect, a working range up to 70%, excellent fatigue resistance (>100 000 cycles) and very small hysteresis.[114–115] The environmental stability of the fabric sensors is very good, and they are ideal candidates for measuring pressure on 3-D surfaces. Figure 4.15 presents a new

4.15 Fabric sensors for 3-D surface pressure mapping.

family of fabric based sensors for 3-D pressure mapping, which has been developed for man–machine interfaces and other engineering applications under this project. The technology has been licensed to AdvanPro Limited to produce such sensors commercially.

Healthcare and wound healing functions

Healthcare smart textiles play a key role in the development of bio-communicative clothes for ambulatory measurement and monitoring of vital physiological, kinematic and behavioural human parameters.[116] Integration of sensors, actuators and communication systems into woven or knitted textiles is now feasible, providing light and wearable user-friendly electronic systems capable of exchanging information with other health-related information systems.

Examples of wearable prototypes are the Georgia Tech Wearable Motherboard,[117] LifeShirt[118] and WEALTHY system.[119] The Georgia Tech Wearable Motherboard is currently being manufactured for commercial use under the name Smart Shirt. LifeShirt is commercially distributed by Vivometrics, Inc. and WEALTHY systems are being developed in research projects financed by the European Commission.

Foreseen applications in healthcare include medical monitoring in pharmaceutical trials, rehabilitation, obstetrics, geriatric care, drug delivery, mental health, and detection of sudden infant death syndrome. The next step could be the development and integration of new nanotechnology-based smart textiles with novel mechanical, electrical and/or optical properties.

Another important and growing area of the textile industry is focused on smart textiles for wound care. The extent of this growth is due to constant improvements in both textile technology and medical procedures. Figure 4.16 shows functions connected with the development of new applications of modern wound dressings.

There are a number of materials such as metals, metal oxides at nanoscale, biological materials such as enzymes, and drugs that can add functionality to a fibrous system. Such value-added fibrous systems can be used effectively in applications, including tissue engineering and biomaterials, drug delivery and protective clothing. As a degradable implant, achieving the optimum levels of mechanical and biological stability of nanofibre scaffolds, for instance, facilitates very important applications.[120]

Furthermore, wound dressings composed of an electrospun polyurethane nanofibrous membrane and silk fibroin nanofibres have been developed.[121–122] These electrospun materials are characterized by a wide range of pore size distribution, high porosity, and high surface area-to-volume ratio, which are favourable parameters for cell attachment, growth and proliferation. The porous structure is particularly important for fluid exudation from the wound, avoiding wound desiccation, and impairing exogenous microorganism infection.

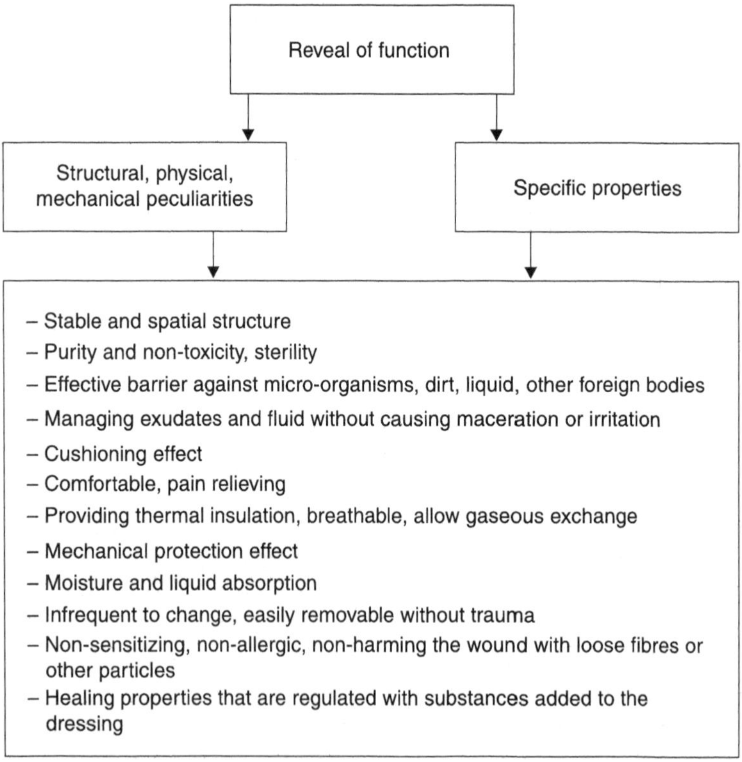

4.16 Functions of modern wound dressings.

Recent advancements in electrospinning enable the production of nanoscale and biodegradable polymer fibres, which have potential applications in wound care.[123] Alltracel Pharmaceuticals plc is developing such first-aid wound care products.[124] Micro-dispersed oxidized cellulose-based nanofibre wound care prototypes have been developed and are now at an advanced stage of technical evaluation. Here, micro-dispersed oxidized cellulose, capable of stopping bleeding, is currently being marketed as the leading ingredient brand within recognized wound care, oral care and relevant first aid brands worldwide.

Self-cleaning, repairing and heating functions

Self-cleaning surfaces based on photo-catalysis are an extremely promising nano-technological field, and are the focus of extensive research and development. The self-cleaning effect of a textile material can be obtained by a photo-catalytically active coating containing a photo-catalytically active oxide of a transition metal, such as titanium dioxide (TiO_2) and zinc oxide (ZnO).[125–126]

Self-cleaning coatings react with and decompose organic compounds or pollutants deposited on the textiles from the environment, under the effects of exposure to sunlight, utilizing UV radiation in particular. The organic pollutants are decomposed to simple inorganic compounds, such as CO_2 and H_2O, and are removed under the effects of heat, wind and/or rain. The efficiency of the self-cleaning coating is dependent upon the photo-catalytic activity of the metal oxide catalyst, which is directly proportional to the total surface area of the metal oxide particles.[127–128]

The Institute of Textiles and Clothing at The Hong Kong Polytechnic University has worked wonders with nano-technology, using it to develop a special fabric that can be made into self-cleaning clothes. The researchers found the use of a titanium dioxide coating to be an efficient nano-treatment for the textile industry. The titanium coating on fabrics acts as an anti-bacterial photo-catalyst that helps to break down carbon-based molecules. Once triggered by sunlight, clothing made of such fabrics will be able to remove dirt, pollutants and microorganisms by itself.

More recently, Wang *et al.* (2010) reported that a highly active $Au/TiO_2/SiO_2$ photo-catalyst can be obtained by embedding Au and Si species using sol-gel nanotechnology.[129] The as-prepared composites are porous nanoparticles with a size of less than 30 nm. A hydrophilic surface was achieved. The treated fabrics did not change color in comparison to the control, even after several months, and the durable UV blocking properties were still observed after 30 wash fastness tests. The outstanding self-cleaning durability was evidenced by the wash-fastness effect on decoloration of red wine and coffee stained cotton samples. Figure 4.17 presents the decoloration results of a red wine stain and a concentrated coffee stain, respectively, on control cellulosic fabrics, TiO_2-treated cellulosic fabrics

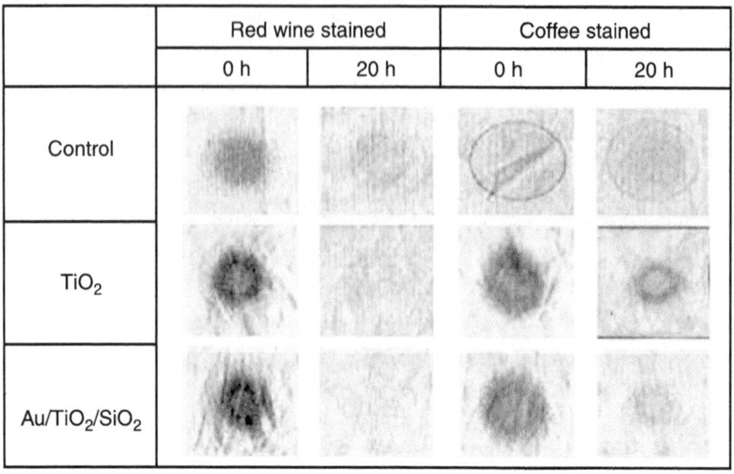

4.17 Degradation results of a red wine stain and a concentrated coffee stain.[129] (Reprinted with permission The American Chemical Society.)

and Au/TiO_2/SiO_2 coated white cellulosic fabrics under visible-light irradiation. The as-prepared structural material demonstrated a remarkable step forward in visible light self-cleaning treatment for flexible substrates, and has a high potential for commercialization.

Self-repairing smart textiles are used as a crack management strategy in building reinforced concrete. These polymer composites offer high potential for long-lived structural materials.[130] In the near future, self-repairing nanocoating for textiles will become a hot issue in the textile industry. Self-repairing nanocoating is introduced to protect the underlying textile substrate. However, the challenge is to find the appropriate microcapsules or nanocapsules for the desired specific applications.[131]

Self-repairing fibres have received a significant level of interest, primarily for inclusion in reinforced concrete as crack management technology. These self-repairing fibres have been added as discrete elements within the concrete matrix. The self-repairing fibres are primarily fluid-filled hollow capillaries that contain a bonding agent that, when released, slow or prevent the spread of a crack through the concrete matrix.[132–133] Self-repairing fibres have also been proposed as a strategy for addressing debonding events between the concrete matrixes and reinforcing bars.

The management of heat transfer through an exterior wall is an important aspect of thermal performance; one that is substantially compromised by air infiltration and exfiltration. Standard building practice, especially in residential construction, usually requires that a membrane be applied to the building volume to reduce the movement of air between the interior and exterior. Any discontinuities in this membrane may allow for the passage of air to and from the exterior. Self-repairing fibres, as an inclusion within the weave of an air/vapour barrier textile, will be studied as a strategy for passively sealing the miscellaneous discontinuities that arise during the application and lifetime of the membrane.

Self-heating smart textiles would offer wide-ranging applications in areas such as wearable equipment for military, medical care and outdoor purposes. Fabrics coated with or containing conjugated polymers have received considerable attention due to their possible applications in the areas of heat generation. It is known that a conjugated polymer can be synthesized by either an oxidative chemical or an electrochemical polymerization of its monomer in the liquid phase. Textile substrates can be easily covered with a conjugated polymer layer by immersing the fabric in a polymerization solution containing the conjugated monomers, an oxidant and a doping agent.

In our recent research, a shiny, smooth and adherent conducting PPy film coated PET fabric was successfully fabricated by a simple method of vapour-phase polymerization.[134] Results show that impregnated oxidant concentration, vapour polymerization temperature and time had significant influences on the final surface resistance of the coated fabric. Figure 4.18 demonstrates that the PPy-coated PET fabric generated heat when a DC voltage was applied. The effectiveness can be increased by reducing the surface resistance and improving the ability of the film to remain conductive.

Furthermore, we successfully fabricated another conductive fabric using the vapour-phase polymerization method. The flexible PET fabric was coated with a uniform poly(3,4-ethylenedioxythiophene) (PEDOT) film, producing a conductive fabric with a lowest resistance of only about $52\ \Omega\cdot\square^{-1}$. The PEDOT-coated fabric sample, with the surface resistance of $121\ \Omega\cdot\square^{-1}$, was heated to about 40 °C when subjected to 15 DC voltages.[135]

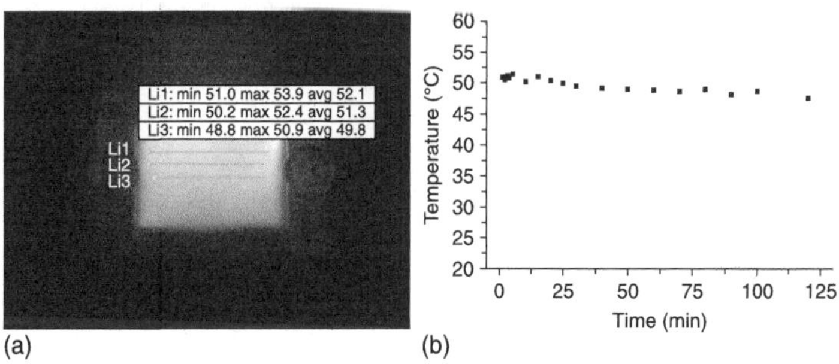

4.18 (a) Thermal images and temperature profiles of the selected three lines across the self-heating fabric. (b) Temperature changes of the self-heating fabric with time (applied DC voltage: 30 V).[134] (Reprinted with permission from Wiley-VCH Verlag GmbH & Co. KGaA, Weinheim.)

4.5 Future trends

The advent of nanotechnology has opened up a whole new vista of value-addition possibilities in the textile sector. The future success of nanotechnology in textile applications lies in areas where new functionalities are combined into durable, multifunctional textile systems without compromising the inherent favourable textile properties, including processability, washability, flexibility and softness. In the next few years, nanotechnology will doubtlessly penetrate every area of the textile industry. The exciting developments between nanotechnology, polymer science and the textile industry will certainly be a lasting and growing relationship, leading to exciting innovations.

Next generation nanomaterials will be designed, engineered and produced more efficiently thanks to advances in nanotechnology, bioinformatics, spectroscopy combinatorial chemistry, robotics and high-throughput screening. We will continuously seek to express a desired property in a nanomaterial, whatever the composition and structure, and will continue to develop strategies aimed toward building precise molecules with desired properties and functions. This will finally result in smart-textile materials with attributes of sensitivity, selectivity, shape ability, self-repair, self-diagnosis, self-recovery, self-tuning and switchability.

4.6 Sources of further information and advice

The following books are recommended for further reading:

- Nierstrasz V A and Cavaco-Paulo A (2010), *Advances in Textile Biotechnology*, Philadelphia, PA, Woodhead Publishing.
- Pan N and Sun G (2011), *Functional Textiles for Improved Performance, Protection and Health*, Philadelphia, PA, Woodhead Publishing.
- Tao X M (2001), *Smart Fibres, Fabrics and Clothing: Fundamentals and Applications'*, Cambridge, UK, Woodhead Publishing.
- Tao X M (2005), *Wearable Electronics and Photonics*, Cambridge, UK, Woodhead Publishing.

4.7 References

1. Al-Saleh M H and Sundararaj U (2009), 'A review of vapor grown carbon nanofiber/polymer conductive composites', *Carbon*, 47, 2–22.
2. Yao Y, Huo K F, Hu L B, Liu N A, Ha J J, *et al.* (2011), 'Highly conductive, mechanically robust, and electrochemically inactive TiC/C nanofiber scaffold for high-performance silicon anode batteries', *Acs Nano*, 5, 8346–8351.
3. Zhou J, Gao Q, Fukawa T, Shirai H and Kimura M (2011), 'Macroporous conductive polymer films fabricated by electrospun nanofiber templates and their electromechanical properties', *Nanotech*, 22, 275501, 8 pp.
4. Chen P, Liang H W, Lv X H, Zhu, H Z, Yao H B and Yu S H (2011), 'Carbonaceous nanofiber membrane functionalized by beta-cyclodextrins for molecular filtration', *Acs Nano*, 5, 5928–5935.

5. Mei Q L, Wang J H, Wang F L, Huang Z X, Yang X L and Wei T (2008), 'Conductive behaviors of carbon nanofibers reinforced epoxy composites', *J Wuhan Univ Technol*, 23, 139–142.
6. Yam P (2003), 'Light sails to orbit: NASA watches from the sidelines as cosmos 1, the first solar sail, goes up', *Sci Amer*, 289, 23–24.
7. Tao, X M (2001), *Textile Institute (Manchester England), Smart Fibres, Fabrics and Clothing*, Cambridge, UK, Woodhead Publishing.
8. Skrifvars M and Soroudi A (2008), 'Strategies for the preparation of conductive textile fibres for smart and functional textiles', *86th Textile Institute World Conference*, vol 1, Conference Proceedings.
9. Devaux E, Koncar M, Kim B, Campagne C, Roux, C, *et al.* (2007), 'Processing and characterization of conductive yarns by coating or bulk treatment for smart textile applications', *T I Meas Control*, 29, 355–376.
10. Bowman D and Mattes B R (2005), 'Conductive fibre prepared from ultra-high molecular weight polyaniline for smart fabric and interactive textile applications', *Synthetic Met*, 154, 29–32.
11. Ajayan P M and Iijima S (1992), 'Smallest carbon nanotube', *Nature*, 358, 23–23.
12. Kim J H, Lee K H, Overzet L J and Lee G S (2011). 'Synthesis and electrochemical properties of spin-capable carbon nanotube sheet/MnO(x) composites for high-performance energy storage devices', *Nano Lett*, 11, 2611–2617.
13. Liu K, Sun Y H, Lin X Y, Zhou R F, Wang J P *et al.* (2010), 'Scratch-resistant, highly conductive, and high-strength carbon nanotube-based composite yarns', *Acs Nano*, 4, 5827–5834.
14. Lee S W, Kim B S, Kim J, Hammond P T and Shao-Horn Y (2009), 'COLL 77-Layer-by-layer assembled multiwall carbon nanotube electrodes for electrochemical energy storage and conversion devices', *Abstr Pap Am Chem Soc*, 238.
15. Hwang J, Jang J, Hong K, Kim K N, Han J H, *et al.* (2011), 'Poly(3-hexylthiophene) wrapped carbon nanotube/poly(dimethylsiloxane) composites for use in finger-sensing piezoresistive pressure sensors', *Carbon*, 49, 106–110.
16. Kuznetzov A A, Lee S B, Zhang M, Baughman R H and Zakhidov A A (2010), 'Electron field emission from transparent multiwalled carbon nanotube sheets for inverted field emission displays', *Carbon*, 48, 41–46.
17. Wang J and Musameh M (2003), 'Carbon nanotube/teflon composite electrochemical sensors and biosensors', *Anal Chem*, 75, 2075–2079.
18. Shang S M, Yang X M and Tao M M (2009), 'Easy synthesis of carbon nanotubes with polypyrrole nanotubes as the carbon precursor', *Polymer*, 50, 2815–2818.
19. Thakur D B, Tiggelaar R M, Gardeniers J G E, Lefferts L and Seshan K (2009), 'Growth of carbon nanofiber coatings on nickel thin films on fused silica by catalytic thermal chemical vapor deposition: on the use of titanium, titanium-tungsten and tantalum as adhesion layers', *Surf Coat Tech*, 203, 3435–3441.
20. Longtin R, Fauteux C, Goduguchinta R and Pegna J (2007), 'Synthesis of carbon nanofiber films and nanofiber composite coatings by laser-assisted catalytic chemical vapor deposition', *Thin Solid Films*, 515, 2958–2964.
21. Denysenko I, Ostrikov K, Cvelbar U, Mozetic M and Azarenkov N A (2008), 'Carbon nanofiber growth in plasma-enhanced chemical vapor deposition', *J Appl Phys*, 104, 073301, 9 pp.
22. Shimoi N and Tanaka S (2008), 'Carbon nanofiber bundles grown by plasma enhanced chemical vapor deposition', *J Ceram Process Res*, 9, 437–439.

23. Seo D K, Jeun J P, Bin Kim H and Kang P H (2011), 'Preparation and characterization of the carbon nanofiber mat produced from electrospun pan/lignin precursors by electron beam irradiation', *Rev Adv Mater Sci*, 28, 31–34.
24. Kim C, Jeong Y I, Ngoc B Y N, Yang K S, Kojima M, *et al.* (2007), 'Synthesis and characterization of porous carbon nanofibers with hollow cores through the thermal treatment of electrospun copolymeric nanofiber webs', *Small*, 3, 91–95.
25. Ji L W, Yao Y F, Toprakci O, Lin Z, Liang Y Z, *et al.* (2010), 'Fabrication of carbon nanofiber-driven electrodes from electrospun polyacrylonitrile/polypyrrole bicomponents for high-performance rechargeable lithium-ion batteries', *J Power Sources*, 195, 2050–2056.
26. Zhou Z P, Lai C L, Zhang L F, Qian Y, Hou H Q, *et al.* (2009), 'Development of carbon nanofibers from aligned electrospun polyacrylonitrile nanofiber bundles and characterization of their microstructural, electrical, and mechanical properties', *Polymer*, 50, 2999–3006.
27. Ahuja R T and Kumar D (2009), 'Recent progress in the development of nano-structured conducting polymers/nanocomposites for sensor applications', *Sensor Actuat B: Chem*, 136, 275–286.
28. Wu S J, Venugopal R and Chen Y T (2005), 'Laser assisted catalytic growth of ZnS/CdSe core-shell and wire-coil nanowire heterostructures', *J Chin Chem Soc-Taip*, 52, 725–732.
29. Zhang B, Weng Y Y, Huang X P, Wang M, Peng R W, *et al.* (2009), 'Creating in-plane metallic-nanowire arrays by corner-mediated electrodeposition', *Adv Mater*, 21, 3576.
30. Li X, Tian S J, Ping Y, Kim D H and Knoll W (2005), 'One-step route to the fabrication of highly porous polyaniline nanofiber films by using PS-b-PVP diblock copolymers as templates', *Langmuir*, 21, 9393–9397.
31. Iwaura R, Ohnishi-Kameyama M and Shimizu T (2008), 'Nanofiber formation from sequence-selective DNA-templated self-assembly of a thymidylic acid-appended bolaamphiphile', *Chem Commun*, 5770–5772.
32. Shirakawa H, Louis E J, Macdiarmid A G, Chiang C K and Heeger A J (1977), 'Synthesis of electrically conducting organic polymers: halogen derivatives of polyacetylene', *J Chem Soc Chem Comm*, Ch. X, 578–580.
33. Jang S Y, Seshadri V, Khil M S, Kumar A, Marquez M, *et al.* (2005), 'Welded electrochromic conductive polymer nanofibers by electrostatic spinning', *Adv Mater*, 17, 2177–2180.
34. Kim B H, Kim M S, Park K T, Lee J K, Park D H, *et al.* (2003), 'Characteristics and field emission of conducting poly (3,4-ethylenedioxythiophene) nanowires', *Appl Phys Lett*, 83, 539–541.
35. Malinauskas A, Malinauskiene J and Ramanavicius A (2005), 'Conducting polymer-based nanostructurized materials: electrochemical aspects', *Nanotech*, 16, R51–R62.
36. Hatchett D W and Josowicz M (2008), 'Composites of intrinsically conducting polymers as sensing nanomaterials', *Chem Rev*, 108, 746–769.
37. Huang J X and Kaner R B (2004), 'Nanofiber formation in the chemical polymerization of aniline: a mechanistic study', *Angew Chem Int Edit*, 43, 5817–5821.
38. Huang K, Wan M X, Long Y Z, Chen Z J and Wei Y (2005), 'Multi-functional polypyrrole nanofibers via a functional dopant-introduced process', *Synth Met*, 155, 495–500.
39. Lin B C, Cheng S, Qiu L H, Yan F, Shang S M and Lu J M (2010), 'Protic ionic liquid-based hybrid proton-conducting membranes for anhydrous proton exchange membrane application', *Chem Mater*, 22, 1807–1813.

40. Yang X M, Li L, Shang S M and Tao X M (2010), 'Synthesis and characterization of layer-aligned poly(vinyl alcohol)/graphene nanocomposites', *Polymer*, 51, 3431–3435.
41. Yang X M, Tu Y F, Li L, Shang S M and Tao X M (2010), 'Well-dispersed chitosan/graphene oxide nanocomposites', *Acs Appl Mater Inter*, 2, 1707–1713.
42. Diao H B, Yan F, Qiu L H, Lu J M, Lu X H, *et al.* (2010), 'High performance cross-linked poly(2-acrylamido-2-methylpropanesulfonic acid)-based proton exchange membranes for fuel cells', *Macromol*, 43, 6398–6405.
43. Yang X M, Shang S M and Li L (2011), 'Layer-structured poly(vinyl alcohol)/graphene oxide nanocomposites with improved thermal and mechanical properties', *J Appl Polym Sci*, 120, 1355–1360.
44. Yang J H, Qiu L H, Liu B Q, Peng Y J, Yan F and Shang S M (2011), 'Synthesis of polymeric ionic liquid microsphere/pt nanoparticle hybrids for electrocatalytic oxidation of methanol and catalytic oxidation of benzyl alcohol', *J Polym Sci Pol Chem*, 49, 4531–4538.
45. Shang S M, Zeng W and Tao X M (2011), 'High stretchable MWNTs/polyurethane conductive nanocomposites', *J Mat Chem*, 21, 7274–7280.
46. Chronakis I S, Grapenson S and Jakob A (2006), 'Conductive polypyrrole nanofibers via electrospinning: electrical and morphological properties', *Polymer*, 47, 1597–1603.
47. Laforgue A and Robitaille L (2010), 'Production of conductive PEDOT nanofibers by the combination of electrospinning and vapor-phase polymerization', *Macromol*, 43, 4194–4200.
48. Li M, Li Z Q and Li Z (2008), 'Mechanism and electric field properties of the dual electrode mode electrospining', *Chem J Chinese U*, 29, 1041–1045.
49. Park S Y, Saeed K, Lee H J, Baek J B and Huh W S (2006), 'Preparation of electrospun nanofibers of carbon nanotube/polycaprolactone nanocomposite', *Polymer*, 47, 8019–8025.
50. Dubois C, Mazinani S and Ajji A (2010), 'Fundamental study of crystallization, orientation, and electrical conductivity of electrospun PET/carbon nanotube nanofibers', *J Polym Sci Pol Phys*, 48, 2052–2064.
51. Fukushima T and Aida T (2007), 'Ionic liquids for soft functional materials with carbon nanotubes', *Chem-Eur J*, 13, 5048–5058.
52. Li Y C, Mannen S, Morgan A B, Chang S C, Yang Y H, *et al.* (2011), 'Intumescent all-polymer multilayer nanocoating capable of extinguishing flame on fabric', *Adv Mater*, 23, 3926.
53. Kyo Y, Yadav A P, Nishikata A and Tsuru T (2011), 'Hydrogen entry behaviour of newly developed Al-Mg-Si coating produced by physical vapour deposition', *Corros Sci*, 53, 3043–3047.
54. Roy A, Srinivas V, Ram S and Rao T V C (2007), 'The effect of silver coating on magnetic properties of oxygen-stabilized tetragonal Ni nanoparticles prepared by chemical reduction', *J Phys-Condens Mat*, 19, 346220, 16 pp.
55. Hu R, Lin C J, Shi H Y and Wang Y H (2009), 'Electrochemical deposition mechanism of calcium phosphate coating in dilute Ca-P electrolyte system', *Mater Chem Phys*, 115, 718–723.
56. Lozovan A A, Frangulov S V and Chulkov D V (2010), 'Structure and composition analysis of multilayer nanocoating deposited onto inner surfaces of tubes by pulsed laser deposition', *J Surf Invest-X-Ray*, 4, 530–533.

57. de Villiers M M, Otto D P, Strydom S J and Lvov Y M (2011), 'Introduction to nanocoatings produced by layer-by-layer (LbL) self-assembly', *Adv Drug Del Rev*, 63, 701–715.
58. Yeldose B C and Ramamoorthy B (2008), 'Characterization of DC magnetron sputtered diamond-like carbon (DLC) nano coating', *Int J Adv Manu Tech*, 38, 705–717.
59. Springer R W and Catlett D S (1978), 'Structure and mechanical-properties of al-alxoy vacuum-deposited laminates', *Thin Solid Films*, 54, 197–205.
60. Zhang Y J, Zhang Q, Li Y B, Wang N L and Zhu J (2000), 'Coating of carbon nanotubes with tungsten by physical vapor deposition', *Solid State Commun*, 115, 51–55.
61. Goto T and Kimura T (2006), 'High-speed oxide coating by laser chemical vapor deposition and their nano-structure', *Thin Solid Films*, 515, 46–52.
62. Daves W, Krauss A, Behnel N, Haublein V, Bauer A and Frey L (2011), 'Amorphous silicon carbide thin films (a-SiC:H) deposited by plasma-enhanced chemical vapor deposition as protective coatings for harsh environment applications', *Thin Solid Films*, 519, 5892–5898.
63. Herasymchuk A I, Mazurenko E A, Vrochyns'kyi S L, Medvedev O M and Molyar O H (2006), 'Formation of technological protective coatings by chemical vapor deposition', *Mater Sci*, 42, 627–633.
64. Jiang S X, Qin W F, Tao X M, Zhang Z M, Yuen C W M, *et al.* (2011), 'Surface characterization of sputter silver-coated polyester fiber', *Fiber Polym*, 12, 616–619.
65. Vincent B, Gencarelli F, Bender H, Merckling C, Douhard B, *et al.* (2011), 'Undoped and in-situ B doped GeSn epitaxial growth on Ge by atmospheric pressure-chemical vapor deposition', *Appl Phys Lett*, 99, 152103, 3 pp.
66. Selvakumar J, Sathiyamoorthy D and Nagaraja K S (2011), 'Tetrakis(trimethylsilyl) silane: Temperature dependence of vapor pressure, kinetics, and silicon carbide thin films by plasma-assisted liquid injection chemical vapor deposition process', *Mater Chem Phys*, 129, 62–67.
67. Struppert T, Jakob A, Heft A, Grunler B and Lang H (2010), 'The use of silver(I)-2-[2-(2-methoxyethoxy)ethoxy]acetate as precursor in the deposition of thin silver layers on float glass by the atmospheric pressure combustion chemical vapor deposition process', *Thin Solid Films*, 518, 5741–5744.
68. Obata K, Sakamoto K and Kira M (2001), 'Novel polysilane synthesis using photochemical vapor deposition', *Macromole*, 34, 2739–2741.
69. Goto T, Kimura T and Tu R (2010), 'High-speed engineering ceramic coating by laser chemical vapor deposition', *Ceram Mat Comp Ener Envir Appl*, 210, 363–369.
70. Reynolds S, Vook D W and Gibbons J F (1986), 'Limited reaction processing: growth of iii-v epitaxial layers by rapid thermal metalorganic chemical vapor-deposition', *Appl Phys Lett*, 49, 1720–1722.
71. Kwon S Y, Sun Q, Kwak J, Seo H C and Han J (2011), 'Growth of cubic InN on GaP(100) with GaN buffer by metalorganic chemical vapour deposition', *J Phys D Appl Phys*, 44, 285403, 6 pp.
72. Benzinger W and Huttinger K J (1996), 'Chemical vapour infiltration of pyrocarbon: 1. Some kinetic considerations', *Carbon*, 34, 1465–1471.
73. Ratcliff C, Grassman T J, Carlin J A and Ringel S A (2011), 'High temperature step-flow growth of gallium phosphide by molecular beam epitaxy and metalorganic chemical vapor deposition', *Appl Phys Lett*, 99, 141905, 3 pp.

74. Zhang D M, Mrozek M F, Xie Y and Ben-Amotz D (2004), 'Chemical segregation and reduction of Raman background interference using drop coating deposition', *Appl Spectro*, 58, 929–933.
75. Caley W F, Kanary L E, Clarke D B and Didyk S H (1987), 'Reduction of chemical corrosion in transfer ladles by application of an oxide ceramic coating,' *Cim Bull*, 80, 96–96.
76. Xue C H, Chen J, Yin W, Jia S T and Ma J Z (2012), 'Superhydrophobic conductive textiles with antibacterial property by coating fibers with silver nanoparticles', *Appl Surf Sci*, 258, 2468–2472.
77. Wang F, Arai S, Morimoto S and Endo M (2004), 'Ni-fluorinated vapor growth carbon fiber (VGCF) composite films prepared by an electrochemical deposition process', *Electrochem Commun*, 6, 242–244.
78. Wang R X, Tao X M, Wang Y, Wang G F and Shang S M (2010), 'Microstructures and electrical conductance of silver nanocrystalline thin films on flexible polymer substrates', *Surf Coat Tech*, 204, 1206–1210.
79. Liu F J, Shang S M, Duan Y J and Li L (2011), 'Electrical and optical properties of polymer-Au nanocomposite films synthesized by magnetron cosputtering', *J Appl Polym Sci*, 123, 2800–2804.
80. Hirase R, Hasegawa M and Shirai M (2003), 'Conductive fibers based on poly (ethylene terephthalate)-polyaniline composites manufactured by electrochemical polymerization', *J Appl Polym Sci*, 87, 1073–1078.
81. Bhadani S N, SenGupta S K, Sahu G C and Kumari M (1996), 'Electrochemical formation of some conducting fibers', *J Appl Polym Sci*, 61, 207–212.
82. Bhadani S N, Kumari M, SenGupta S K and Sahu G C (1997), 'Preparation of conducting fibers via the electrochemical polymerization of pyrrole', *J Appl Polym Sci*, 64, 1073–1077.
83. Hirase R, Hasegawa M and Shirai M (2003), 'Conductive fibers based on poly (ethylene terephthalate)-polyaniline composites manufactured by electrochemical polymerization', *J Appl Polym Sci*, 87, 1073–1078.
84. Molina J, del Rio A L, Bonastre J and Cases F (2009), 'Electrochemical polymerisation of aniline on conducting textiles of polyester covered with polypyrrole/AQSA', *Eur Polym J*, 45, 1302–1315.
85. Zhang D (2006), *Pulsed Laser Deposition Dynamics and Thin Film Deposited onto Glass*, First Edition, Wuhan Shi.
86. Popescu A C, Duta L, Dorcioman G, Mihailescu I N, Stan G E, *et al.* (2011) 'Radical modification of the wetting behavior of textiles coated with ZnO thin films and nanoparticles when changing the ambient pressure in the pulsed laser deposition process', *J Appl Phys*, 110, 064321, 8 pp.
87. Socol G, Axente E, Ristoscu C, Sima F, Popescu A, *et al.* (2007), 'Enhanced gas sensing of Au nanocluster-doped or -coated zinc oxide thin films', *J Appl Phys*, 102, 083103, 6 pp.
88. Etchegoin P, Cohen L F, Hartigan H, Brown R J C, Milton M J T and Gallop J C (2004), 'Localized plasmon resonances in inhomogeneous metallic nanoclusters', *Chem Phys Lett*, 383, 577–583.
89. Yin M J, Gu B B, Zhao Q A, Qian J W, Zhang A, *et al.* (2011), 'Highly sensitive and fast responsive fiber-optic modal interferometric pH sensor based on polyelectrolyte complex and polyelectrolyte self-assembled nanocoating', *Anal Bioanal Chem*, 399, 3623–3631.

90. Ali S W, Joshi M and Rajendran S (2011), 'Novel, self-assembled antimicrobial textile coating containing chitosan nanoparticles', *Aatcc Rev*, 11, 49–55.
91. Jennings G K, Munro J C and. Laibinis P E (1999), 'Effects of film crystallinity on the protective properties of self-assembled monolayers of alkanethiols on copper', *Adv Mater*, 11, 1000–1003.
92. Skrzeszewska P J, Sprakel J, de Wolf F A, Fokkink R, Stuart M A C and van der Gucht J (2010), 'Fracture and self-healing in a well-defined self-assembled polymer network', *Macromole*, 43, 3542–3548.
93. Neves B R A, Salmon M E, Troughton E B and Russell P E (2001), 'Self-healing on OPA self-assembled monolayers', *Nanotech*, 12, 285–289.
94. Jee H G, Hwang H N, Han J H, Lim J, Shin H J, *et al.* (2007) 'Patterning of self-assembled pentacene nanolayers by extreme ultraviolet-induced three-dimensional polymerization', *Acs Nano*, 4, 4997–5002.
95. Yokota S, Matsuyama S K, Kitaoka T and Wariishi H (2007), 'Thermally responsive wettability of self-assembled methylcellulose nanolayers', *Appl Surf Sci*, 253, 5149–5154.
96. Xue C H, Jia S T, Zhang J and Ma J Z (2010), 'Large-area fabrication of superhydrophobic surfaces for practical applications: an overview', *Sci Technol Adv Mat*, 11.
97. Xue C H, Jia S T, Zhang J, Tian L Q, Chen H Z and Wang M (2008), 'Preparation of superhydrophobic surfaces on cotton textiles', *Sci Technol Adv Mat*, 9, 035008, 7 pp.
98. Li Z F and Ruckenstein E (2002), 'Strong adhesion and smooth conductive surface via graft polymerization of aniline on a modified glass fiber surface', *J Colloid Interf Sci*, 251, 343–349.
99. Pekakis P A, Xekoukoulotakis N P and Mantzavinos D (2006), 'Treatment of textile dyehouse wastewater by TiO_2 photocatalysis', *Water Res*, 40, 1276–1286.
100. Bandara J and Ranasinghe R A S S (2007), 'The effect of MgO coating on photocatalytic activity of SnO2 for the degradation of chlorophenol and textile colorants; the correlation between the photocatalytic activity and the negative shift of flatband potential of SnO_2', *Appl Catal A: Gen*, 319, 58–63.
101. Knittel D, Kontges M, Heinemeyer F and Schollmeyer E (2010), 'Coatings on textiles for Cu(In,Ga)Se(2) photovoltaic cell formation on textile carriers: preparation of Cu(In,Ga)Se(2) solar cells on glass-fiber textiles', *J Appl Polym Sci*, 115, 2763–2766.
102. Schubert M B and Werner J H (2006), 'Flexible solar cells for clothing', *Mater Today*, 9, 42–50.
103. Bae J, Song M K, Park Y J, Kim J M, Liu M L and Wang Z L (2011), 'Fiber supercapacitors made of nanowire-fiber hybrid structures for wearable/flexible energy storage', *Angew Chem Int Edit*, 50, 1683–1687.
104. Cochrane C, Koncar V, Lewandowski M and Dufour C (2007), 'Design and development of a flexible strain sensor for textile structures based on a conductive polymer composite', *Sensors-Basel*, 7, 473–492.
105. Shang S M, Zeng W, and Tao X M (2012), 'Investigation on the electrical response behaviors of multiwalled carbon nanotube/polyurethane composite in organic solvent vapors', *Sens Actuat: B Chem*, 166–167, 330–337.
106. Bahadir S K, Kalaoglu F, Thomassey S, Cristian I and Koncar V (2011), 'A study on the beam pattern of ultrasonic sensor integrated to textile structure', *Int J Cloth Sci Tech*, 23, 232–241.

107. Rai P, Kumar P S, Oh S, Kwon H, Mathur G N and Varadan V K (2011), 'Printable low-cost sensor systems for healthcare smart textile', *Nanosense Biosen Info-Tech Sens Syst*, 7980, 79800E.
108. Wei Y Y, Gao D Z, Li L and Shang S M (2011), 'Memory effect in polymer brushes containing pendant carbazole groups', *Polymer*, 52, 1385–1390.
109. Grant J (2005), 'Distributed sensing of composite over-wrapped pressure vessel using fiber Bragg gratings at ambient and cryogenic temperatures', *Smart Structures and Materials 2005: Smart Sensor Technology and Measurement Systems*, 5758, 201–208.
110. Shim B S, Chen W, Doty C, Xu C L and Kotov N A (2008), 'Smart electronic yarns and wearable fabrics for human biomonitoring made by carbon nanotube coating with polyelectrolytes', *Nano Lett*, 8, 4151–4157.
111. Sergio M, Manaresi N, Campi F, Canegallo R, Tartagni M and Guerrieri R (2003), 'A dynamically reconfigurable monolithic CMOS pressure sensor for smart fabric', *IEEE J Solid-St Circ*, 38, 966–975.
112. Sergio M, Manaresi N, Nicolini M, Gennaretti D, Tartagni M and Guerrieri R (2004), 'A textile-based capacitive pressure sensor', *Sensor Lett*, 2, 153–160.
113. Meyer J, Arnrich B, Schumm J and Troster G (2010), 'Design and modeling of a textile pressure sensor for sitting posture classification', *IEEE Sens J*, 10, 1391–1398.
114. Wang Y Y, Hua T, Zhu B, Li Q, Yi W J and Tao X M (2011), 'Novel fabric pressure sensors: design, fabrication, and characterization', *Smart Mater Struct*, 2.
115. Li Y, Cheng X Y, Leung M Y, Tsang J, Tao X M and Yuen C W M (2005), 'A flexible strain sensor from polypyrrole-coated fabrics', *Synthetic Met*, 155, 89–94.
116. G. Reimers G (2002), 'Textile hygiene in healthcare facilities', *Gesundheitswesen*, 64, 234–234.
117. http://www.gtwm.gatech.edu/
118. http://www.teeshirts.com/
119. http://www.wealthy-ist.com/
120. Zdebiak P and El Fray M (2008), 'Hybrid polymer/ceramic nanofibre scaffolds for soft tissue engineering', *Tissue Eng Pt A*, 14, 860–860.
121. Khil M S, Cha D I, Kim H Y, Kim I S and N. Bhattarai N (2003), 'Electrospun nanofibrous polyurethane membrane as wound dressing', *J Biomed Mater Res B*, 67B, 675–679.
122. Min B M, Lee G, Kim S H, Nam Y S, Lee T S and Park W H (2004), 'Electrospinning of silk fibroin nanofibers and its effect on the adhesion and spreading of normal human keratinocytes and fibroblasts *in vitro*', *Biomat*, 25, 1289–1297.
123. Petrulyte S (2008), 'Advanced textile materials and biopolymers in wound management', *Dan Med Bull*, 55, 72–77.
124. www.alltracel.com
125. Yuranova T, Mosteo R, Bandara J, Laub D and Kiwi J (2006), 'Self-cleaning cotton textiles surfaces modified by photoactive SiO_2/TiO_2 coating', *J Mol Catal A: Chem*, 244, 160–167.
126. Guo Z, Chen X, Li J, Liu J H and Huang X J (2011), 'ZnO/CuO hetero-hierarchical nanotrees array: hydrothermal preparation and self-cleaning properties', *Langmuir*, 27, 6193–6200.
127. Montazer M and Seifollahzadeh S (2011), 'Enhanced self-cleaning, antibacterial and uv protection properties of nano $TiO(_2)$ treated textile through enzymatic pretreatment', *Photochem Photobiol*, 87, 877–883.

128. Bedford N M and Steckl A J (2010), 'Photocatalytic self cleaning textile fibers by coaxial electrospinning', *Acs Appl Mater Inter*, 2, 2448–2455.
129. Wang R H, Wang X W and Xin J H (2010), 'Advanced visible-light-driven self-cleaning cotton by Au/TiO($_2$)/SiO($_2$) photocatalysts', *Acs Appl Mater Inter*, 2, 82–85.
130. Dry C M (1992), 'Smart materials which sense, activate and repair damage – hollow porous fibers in composites release chemicals from fibers for self-healing, damage prevention, and or dynamic control', *First European Conference on Smart Structures and Materials*, 1777, 367–371.
131. Li T T, Wang R and Liu X (2011), 'Synthesis of self-healing microcapsules via different heating profiles', *Manu Proc Sys*, 1–2, 1486–1490.
132. Woldesenbet E (2010), 'Self-healing in single and multiple fiber(s) reinforced polymer composites', *Epj Web Conf*, 6.
133. Bond I P, Trask R S and Williams H R (2008), 'Self-healing fiber-reinforced polymer composites', *MRS Bull*, 33, 770–774.
134. Shang S M, Yang X M, Tao X M and Lam S S (2010), 'Vapor-phase polymerization of pyrrole on flexible substrate at low temperature and its application in heat generation', *Polym Int*, 59, 204–211.
135. Yang X M, Shang S M, Li L, Tao X M and Yan F (2011), 'Vapor phase polymerization of 3,4-ethylenedioxythiophene on flexible substrate and its application on heat generation', *Polym Advan Technol*, 22, 1049–1055.

5

Polymer-based resistive sensors for smart textiles

C. COCHRANE and A. CAYLA,
University Lille Nord de France, ENSAIT / GEMTEX, France

DOI: 10.1533/9780857093530.1.129

Abstract: This chapter covers resistive sensors (electrical passive response), which are classified according to the external stimulus to which they respond: mechanical, chemical or temperature. Currently in the textile field, hard, heavy and rigid sensors are prohibited in favor of light and flexible ones, as most polymers are not electrically conductive. However, two families of compounds are suitable for use in sensor applications (i.e. having the mechanical behavior and electrical conductivity close to that of polymers): conductive polymer composites (CPC) and intrinsically conductive polymers (ICP). In this chapter, textile based resistive mechanical (elongation and compression), chemical (solvent and humidity) and temperature sensors are described.

Key words: resistive sensors, mechanical sensors, chemical sensors, temperature sensors, CPC, ICP.

5.1 Introduction

It is generally accepted that a sensor is a device that reversibly produces a measurable electrical response to a physical, non-electrical, stimuli. By analogy with electronic components, if A sensor consumes energy (but does not produce it) it will be named a 'passive sensor', otherwise it will be named an 'active sensor'. Active sensors produce an electrical charge or electrical potential, such as piezoelectric sensors, photocells or thermocouples. The field of passive sensors is wide and so split into specific families according to their electrical passive response, such as resistance, impedance, capacity, etc. (Fraden, 1996).

This chapter deals with resistive sensors only, those whose electrical resistance changes in response to external stimuli. Resistive sensors are classified according to the external stimulus to which they respond, mechanical, chemical and temperature being the most common stimuli. Apart from in the textile field, these sensors are not necessarily resistive: most temperature sensors are thermocouples (based on the Seebeck effect: Besançon, 1985) and some chemical/solvent sensors are based on an optical measurement (absorption spectrum of chemical species: Fraden, 1996). As far as possible in the textile field, hard, heavy and rigid materials are avoided in favor of light and flexible ones. Metals, ceramics and electronic components are therefore avoided in favor of polymer-based materials. Obviously, these stiff conductive materials (typically metal) can be formed into filament, or thread, and used as sensor structures. The main problem is that in most cases polymers are not electrically conductive. However, two families of compounds have the mechanical behavior close to that of polymers, while also being electrically conductive: conductive polymer composites (CPC) and intrinsically conductive polymers (ICP).

5.1.1 Conductive polymer composites (CPC) sensors materials

CPCs are based on a mix of polymer matrix and electrical conductive filler. The matrix can be a thermoplastic (polyester, polyamide, etc.) or thermoset (epoxy resin, polyurethane, etc.) polymer. The fillers mainly used are divided into three groups:

1. *carbon filler*: carbon fiber, graphite, carbon black (CB), carbon nanotube (CNT), fullerene, etc.;
2. *organic conductive filler*: powder of ICP, insulator particles coated with ICP, etc.;
3. *metal-based filler*: silver particles, glass particles coated by silver, copper sulphide, indium tin oxide (ITO), etc.

The borders of the final group are unclear, because it also contains electro-conductive ceramic compounds. However, the most commonly used fillers are carbon fillers, since they are stable (no oxidation under normal conditions), easy to use, lightweight and inexpensive (except CNT). The choice of the filler is often linked to the end use. The size and amount of filler required to achieve the desired properties (mechanical, rheological, conductivity) are often the limiting factors. For example, a filler content above 4% is rarely appropriate for melt spinning of nanocomposites. Indeed, the addition of a filler induces a dramatic increase in the viscosity, which hinders spinning to multifilaments (Devaux *et al.*, 2011).

The main parameter that influences the electrical properties of the CPC is the filler concentration. The transition from the insulating (polymer alone and polymer with low filler content) to the conductive state is obtained at a critical filler concentration (ϕ_C). This value, known as the percolation threshold, can be determined from the conductivity evolution as a function of the conductive filler (Fig. 5.1) and corresponds to the inflexion point (Kirkpatrick, 1973). This change in the electrical conductivity is explained by the creation of continuous conductive paths through the material (State III in Fig. 5.1, i.e. filler content above ϕ_C).

It is well-established that at the percolation concentration threshold the sensitivity of the sensor is optimal (Feller *et al.*, 2002(a,b), 2003). Indeed, external stimuli (e.g. mechanical, thermal or chemical) act mainly on the dimensions, therefore on the volume of the CPC sensor. Thus, by exposing a CPC sensor to a stimulus, it is the equivalent of increasing or decreasing its volume and thus diluting or concentrating the conductive fillers. When the volume of the fraction of filler artificially varies, the electrical conductivity changes following the percolation curve of the system being considered. CPC sensors based on this

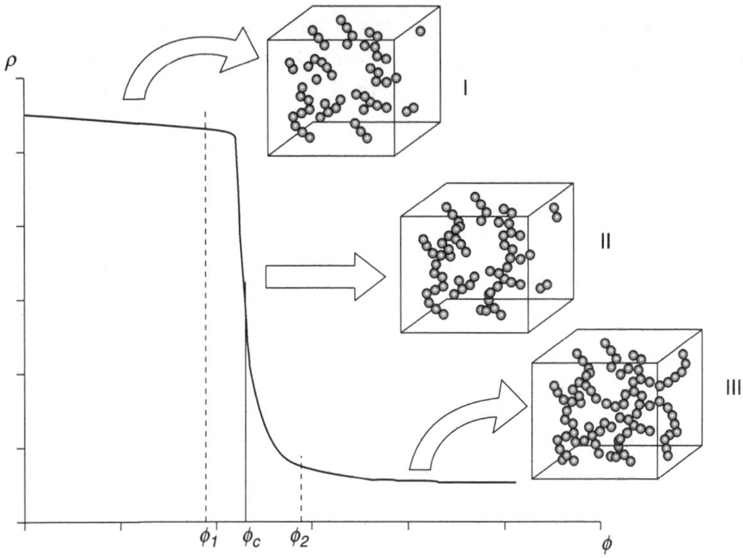

5.1 Evolution of conductivity (ρ) as a function of conductive filler content (ϕ). I-Insulator composite, II-composite in percolation range, III-conductive composite (Cochrane, 2011).

behavior are known as electrical percolation type *X* sensor (EPT*X*S), with X corresponding to the external stimulus (mechanical, temperature, chemical, solvent, moisture, etc.).

Often the material resistance in the percolation range (State II in Fig. 5.1) is too high to be measured by conventional electrical apparatus. Thus it is necessary to find a good compromise between the sensitivity (dependence of conductivity on filler content and maximum near electrical percolation concentration) and the global conductivity leading to a sufficiently low resistance for it to be easily measurable.

5.1.2 Intrinsically conductive polymers (ICP) sensors materials

Intrinsically conducting polymers (ICP), also called 'synthetic metals', are inherently conducting or semi-conducting in nature, due to the presence of a conjugated π electron in their molecular structure. A high level of conductivity can be achieved though oxidation–reduction reactions or acid-base treatments with a suitable dopant (doping). Polypyrrole (PPy), polythiophene (PT) and polyaniline (PANI) propose some of the best compromises between electrical conductivity, stability and processability. ICPs can be classified as (Stenger-Smith, 1998):

- *Oxidized* (*conducting*): the polymer form, which has electrons removed from the backbone, resulting in (radical) cations.

- *Neutral* (*insulating polymer*): the polymer form in its uncharged state. This can be referred to as a 'reduced' form of the oxidized polymer.
- *Reduced* (*conducting*): the polymer form, which has electrons added to the backbone, resulting in (radical) anions.

Conducting polymers are either made directly by electro- or oxidative-polymerization or polymerized and then oxidized chemically or electrochemically. Conductive polymers are provided both as solid compounds, liquid dispersions or solutions. The liquid versions can be easily applied onto a textile substrate by coating methods (Skrifvars, 2008).

5.1.3 Resistive sensors: general sensing principle

It is well-established that ordinary resistive sensors are governed by the law:

$$R = \rho \times \frac{L}{S} \tag{5.1}$$

where R is the electrical resistance of sample (Ω); ρ is the electrical resistivity of material (Ω.m); L is the distance between measurement electrode (m); and S is the section surface area of sample (m^2).

From Eq. 5.1, to obtain variation on the electrical resistance, it is necessary for the external stimulus to act either on the geometry of the sample (L and/or S) or/and on the electrical resistivity (ρ) of the material. Sensor geometry can be affected by mechanical stress or thermal expansion of the material. For most materials, electrical resistivity (ρ) is an intrinsic variable and depends only on temperature (Fraden, 1996), but for CPC and ICP, it is otherwise. For CPC materials, the matrix (polymer) plays a key role. The electrical resistivity of filler remains constant (temperature dependence only), while global electrical resistivity of CPC can change according to several external parameters such as stress, moisture, temperature, etc. External stimuli can act on the polymer matrix (swelling, crystallinity degree, etc.), on the filler (aspect ratio, electrical conductivity, etc.), on the interface matrix/filler or filler/filler (decohesion, contact resistance, etc.) and on the arrangement between matrix and filler (dispersion, agglomeration, segregation, etc.). For ICP materials, the oxidation state of polymer, the doping rate, and the quality of lateral contact between polymer chains strongly influence the global electrical conductivity.

5.2 Mechanical resistive sensors

Mechanical sensors are based on resistive materials or structures whose electrical resistance changes reversibly according to an applied stress (elongation or pressure). The term 'piezoresistive sensor' is commonly used. For mechanical resistive sensors, the sensing mechanism is based on at least one of the following three:

1. change in electrical resistance of textile structure (e.g. stretching or deformation of textile structure);
2. change in electrical resistance of material not affected by fillers disconnection;
3. change in electrical resistance of material affected by fillers disconnection.

In real-case examples, the three changes often occur simultaneously and so it can be difficult to identify the dominant mechanism.

Mechanical sensors are characterized by gauge factor (or sensitivity) (k, Eq. 5.2). k is the ratio of relative change in electrical resistance $\left(\frac{\Delta R}{R_0}\right)$ to the mechanical strain $\left(\varepsilon = \frac{\Delta L}{L_0}\right)$:

$$k = \frac{\Delta R}{R_0} \times \left(\frac{\Delta L}{L_0}\right)^{-1} \qquad [5.2]$$

where k is the gauge factor; ΔR is the change in strain gauge resistance (Ω); R_0 is the unstrained resistance of strain gauge (Ω); ΔL is the change in length (m); and L_0 is the length of unstrained strain gauge (m).

5.2.1 Elongation resistive sensors

Conventional strain gauges are resistive sensors, but they are used only for low elastic materials such as concrete or steel. However, one study shows that by a clever assembly in a 'curved beam', these gauges can be used to evaluate the rigid extension of plain weave (Heinrich and Saari, 1975). This solution is bulky and induces a very important local stiffening of textile. The need to develop resistive elongation sensors specific to textile applications (lightweight, flexible, elastic and washable, if possible) is important.

Sensors based on change in electrical resistance of textile structures

The first family of elongation resistive sensors consists of 'potentiometer per contact', where neither the material resistivity nor its geometry varies. There are changes in the overall macroscopic textile structure that influence electrical resistance. This kind of sensor is similar to conventional potentiometer sensors. One of the best examples to illustrate this family of elongation measurement systems is made of a stainless steel (non-elastic materials) yarn crocheted chain on elastic fabric. This type of sensor can be obtained by embroidering or sewing a conductive thread onto an elastic textile support (Fig. 5.2(a), Strazdiene *et al.*, 2007). When the stitches are stretched, the efficiency of the electrical contact between the conductive thread increases and then the electrical resistance of the device decreases (Fig. 5.2(b), Strazdiene *et al.*, 2007). Based on this system,

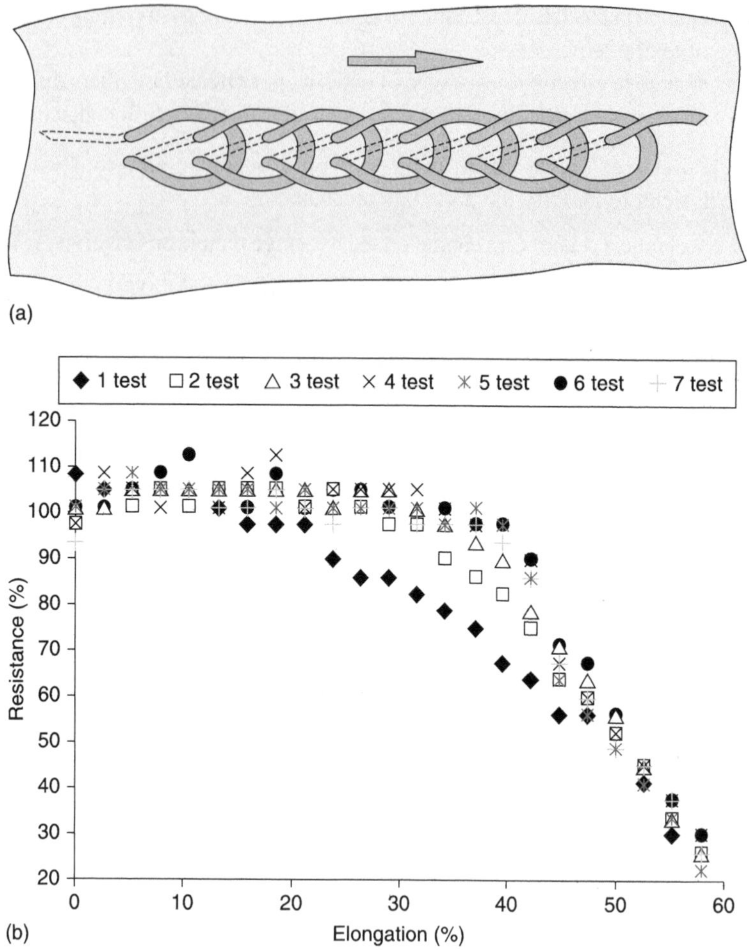

5.2 (a) Embroidering stretch sensor and (b) example of measurement (Strazdiene *et al.*, 2007).

dynamic anthropometric data can be measured in the range of 4 to 16% of elongation (Strazdiene *et al.*, 2007).

It is also possible to build these sensors out of knitted structures. Highly resistive yarns (or thread) can be knitted and placed into flat textiles (woven or another knit) to construct stretch sensors (Yoshikai *et al.*, 2009; Glazzard and Kettley, 2010). The main problem of these sensor types is their lifetime. After each elongation, the stitch sensor does not return to its initial configuration (electrical resistance). The repeatability is poor. The system has a hysteresis and becomes less sensitive with the number of uses.

Elastic narrow fabric, composed of Lycra fibres and polyamide coated carbon particles, can be used as elongation sensors. The materials and structures used are able to reduce hysteresis (Shyr *et al.*, 2011). For this kind of sensor, the conductive fibers are elastic but the detection of elongation is mainly a function of geometrical structure, not the elongation of the conductive fibre. Similarly, if a very thin-coated or printed conductive layer (CPC or ICP) only coats some fibers of the textile structure, global electromechanical behavior appears as a 'potentiometer per contact' rather than a sensor where electrical resistance varies with dimension (see below: 'Sensors based on change in electrical resistance of material non-affected by fillers disconnection'). These supported ICP sensors can be obtained by coating or inkjet printing (Wu *et al.*, 2005; Zhang *et al.*, 2006; Calvert *et al.*, 2008).

In all these cases, the number and efficiency of electrical contacts between the conductive yarn (pure conductive material fibre, fibre coated by conductive material or textile coated by conductive material) dominate the performance of the sensor. This results in a decrease of electrical resistance when efficiency of contact, therefore sensor length, increases. These sensors exhibit a negative gauge factor (k, Eq. 5.2), even if the bulk materials have classical piezoresistive behavior.

Sensors based on change in electrical resistance of materials non-affected by fillers disconnection

This second family of elongation resistive sensors is similar to conventional elongation sensors governed by Eq. 5.1, where only gauge dimension affects the electrical resistance (Fraden, 1996). Elongation induces almost entirely the geometrical change of the gauge. Elongation sensors generally have a high aspect ratio (>25), low thickness (e) and width (l), and high length (L) (i.e. rectangular cross section). Electrical resistance modification, resulting from elongation in the length direction, is mainly caused by length and not section variation (S = e × l). For this kind of gauge, the gauge factor (k) does not exceed 6 and is directly linked to the Poisson ratio of conductive material (Window, 1992).

This family includes the textile (or at least flexible) elongation sensors made of ICP, where macromolecular chains are randomly oriented (isotropic ICP), and CPC sensors that are filled with a large amount of conductive filler (i.e. filler content much higher than percolation concentration, ϕ_C). The electromechanical behavior of CPCs containing a large content of filler is governed mainly by Eq. 5.1: ρ is quasi non-dependent on elongation (during stretching, the ratio between disconnected conductive paths and connected conductive paths is low) and only dimensions of the sensor (L and S) varies.

ICP is already used in standard strain gauges when large deformation is required. ICP-based conductive fibers can be easily obtained by coating. In the case of PANI-coated PET (Kim *et al.*, 2004), the process produces electroconductive fibers with good mechanical properties (i.e. mechanical properties of core fiber) and flexibility, allowing knitting or weaving (Neelakandan and Madhusoothanan,

2010). Studies of the electromechanical properties of these products show that coated fibers can be used as a fibrous sensor (Kim *et al.*, 2006).

In general, the coating process (and sometimes inkjet printing) on textiles can easily produce ICP-based sensors, which will act as standard piezoresistive gauges. However, if the coating thickness is low, the structure (textile + coating) will tend to behave as a 'potentiometer per contact' and not as a classical gauge. Also thin coating of ICP produces anisotropic materials, whose piezoresistivity will depend on macromolecular organization. Similarly, behavior of pure ICP fibers (i.e. PANI) is due to the small diameter and stretching applied to the fiber during its production (Mattes, 2004; Bowman and Mattes, 2005). Due to this anisotropy, the resistance of pure PANI fiber decreases with applied load (negative gauge factor, k) (Fig. 5.3, Mattes, 2004). This can be explained by macromolecular organization of the PANI backbone. Stress increases the drawing of the backbone and quality of lateral contact between chains. The major problem of ICP mechanical sensors are their great sensibility compared to other environmental parameters (temperature, solvent, water, etc.) and their poor stability due to oxidation in comparison to CPC materials (i.e. stainless steel fiber or silver platted fiber).

Sensors based on change in electrical resistance of materials affected by fillers disconnection

This third family consists of Electrical Percolation Type Mechanical Sensors (EPTMS) and contains only CPC sensors. Elongation induces a geometrical change due to the aspect ratio (variation of L and/or S) and also the modification of electrical resistivity (ρ) of the materials. Indeed, if the conductive filler content is near, but above, the percolation concentration (ϕ_C), the electrical conductivity is very sensitive to volume change of the material due to stretching. A small volume change breaks a lot of conductive networks (disconnections), causing a large variation in the overall resistivity of the CPC. EPTMS exhibit a gauge factor (k, Eq. 5.2) far above that of conventional piezoresistive sensors (k below 6: Fraden, 1996) and sometimes comparable to semiconductor gauges (k between 40 and 200: Fraden, 1996). In addition, with CPC materials, conductive particles come close and conductive paths disconnect leading to conduction arising from hopping and tunnelling (Donnet *et al.*, 1993). Indeed, if the conductive particles are near, but are not in contact with each other, activation energy is required to produce a pair of charges (negative and positive) in which their migration is from site to site, resulting in conductivity.

The term 'hopping' refers to the sudden displacement of a charge carrier from one site to another neighbor. This hop usually includes the passage of a potential barrier and tunneling conduction (Dyre and Schroder, 2002). The main difference between a metallic conduction and hopping conduction (or tunneling) corresponds to the decrease of the metallic electrical conductivity with temperature, while the

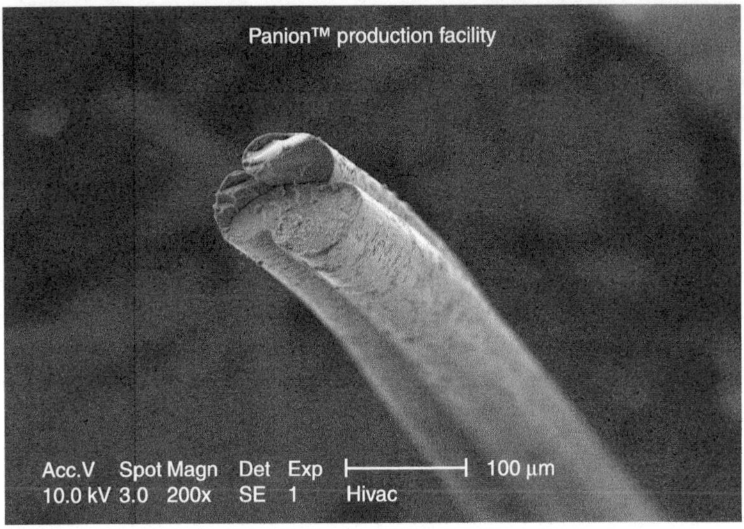

(a)

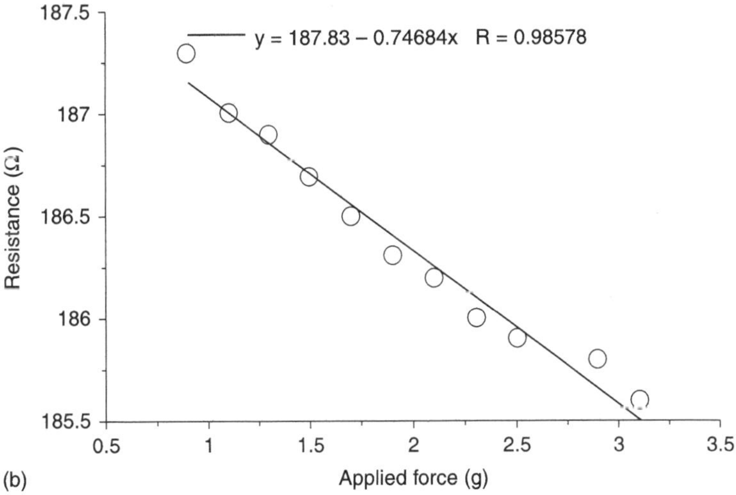

(b)

5.3 (a) PANI fiber and (b) resistance dependence against applied force (Mattes, 2004).

hopping conductivity increases. Thus, to determine the conduction type of a CPC (metallic or hopping), one solution is to observe the electrical behavior according to the temperature (Mott, 1973).

EPTMS in the textile field can be present as a coating or pure fiber of CPC. Coating textile materials (fiber, yarn, fabric, knit, etc.) with CPC is relatively easy, because the process is often appropriate to the filler's content allowing electrical conduction. Inkjet printing can be used only when the viscosity, which depends on fillers content and solvent quantity, is compatible with the process. CB is the most commonly used filler, due to its suitable price and electrical conductivity. Thus, some authors using a thermoset elastomer and CB have developed a series of items of clothing that can detect movements of particular body parts (Lorussi *et al.*, 2004; Bartalesi *et al.*, 2006). Other similar studies (My Heart, 2011) have also shown the same possibility.

Apart from clothing, this type of sensor, based on CB and a highly elastic matrix (thermoset elastomer or thermoplastic elastomer), can be used in aeronautics or for the monitoring of composite structures. In the former, the CPC sensor consists of selective coating (track) localized on a fabric (parachute canopy in Fig. 5.4(a,b)) (Cochrane *et al.*, 2007, 2010; Cochrane, 2011) and in the latter, the CPC sensor consists of coated yarns introduced during the weaving process in the preparation of the composite (Nauman *et al.*, 2011(a,b)).

The melt spinnability of low aspect ratio conductive fillers (the majority of conductive fillers) is limited. Indeed, for this type of filler, percolation concentrations (ϕ_C) are often incompatible with spinning processes in terms of the melt flow index. However, CPC based on CNT have a high aspect ratio (above 10^5) and are spinnable to produce yarns used as mechanical sensors. The electrical response of CNT-based CPC to a mechanical stress is classically explained by the increase of distance between individuals or aggregated CNT (this distance exceeds the hopping and tunneling distance) (Wood *et al.*, 2000; Bilotti *et al.*, 2010; Ciselli *et al.*, 2010).

Even if a single fiber of CPC has electromechanical behavior that can make an EPTMS, when this fiber is included in textile structures (i.e. by weaving, knitting, non-woven or braiding), the total (material + structure) electromechanical behavior becomes complex. Thus an elongation textile structure sensor composed of fiber, based on CNT, belongs to the 'potentiometer per contact' sensor family.

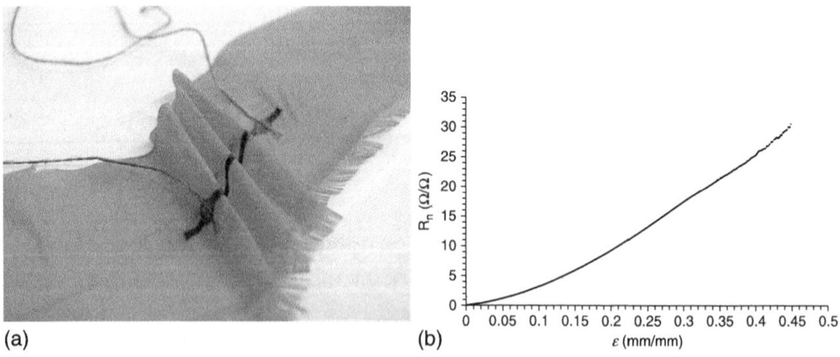

5.4 (a) Elongation sensors made of CPC track (Cochrane *et al.*, 2010) and (b) its resistive response against elongation ($k = 31$) (Cochrane *et al.*, 2011).

Textile applications of elongation resistive sensors

The relatively low cost sensors (i.e. potentiometer per contact based on commercial conductive yarn, stainless steel) have been employed in robotic, design, musical instrument, keyboard and health monitoring applications (Eng, 2009; Fashioning Technology, 2011).

Applications of elongation sensors within the human-centered field are mainly in:

- medical or paramedical, to detect and record human motions for rehabilitation or to monitor patient health (Calvert *et al.*, 2008; My Heart, 2011);
- sport, to learn good posture (Lorussi *et al.*, 2004; Bartalesi *et al.*, 2006);
- glove to recognize sign language, to monitor movement or give tactile sensibility to robots (Bartalesi *et al.*, 2006; Yoshikai *et al.*, 2009).

Other fields of application are the monitoring of textile structures such as parachutes (Cochrane *et al.*, 2007, 2010; Cochrane, 2011) or structural composites (Nauman *et al.*, 2011a,b).

5.2.2 Resistive pressure sensors

Resistive pressure sensors are less developed because, in general, capacitive sensors are preferred. Nevertheless 'all or none' textile (switch) sensors exist and are used, for example, to create deported keyboards for mp3 players. The use of conductive foam can provide 'textile' pressure sensors. In this case, as stress flattens the structure, the foam porosity decreases artificially (i.e. the density of conductive material increases) until local disappearance and the material as a whole becomes more conductive. Foam 'densification' increases the number of conduction paths and thus makes the material more conductive. The use of resilient foam (i.e. with a highly elastic behavior) provides a very good reversibility (reduced wear of the sensor) (Dunne *et al.*, 2006). Standard polyurethane (PU) foam can be impregnated by conductive material (CPC, ICP) to obtain pressure sensing composites (Brady *et al.*, 2005). Foam can easily be introduced and connected between two textiles layers.

Pressure sensors can also be based on the use of conductive yarns combined with hook and loop (Velcro™) (Perner-Wilson and Buechley, 2011) or spacer fabrics (Buechley *et al.*, 2008). In these systems, pressure sensors operate as potentiometers per contact, whereby only the efficiency of contact between the conductive material changes (changes in the overall macroscopic structure). Some tutorials (Lovell and Buechley, 2010) and craft books, such as *Fashioning Technology* (Fashioning Technology, 2011) and *Fashion Geek* (Eng, 2009), propose interesting varieties of textile-based switches.

5.3 Chemical resistive sensors

In addition to mechanical impact, environmental factors such as liquids or volatile solvents can considerably modify the electrical resistivity of the CPC or ICP. This allows for the detection of external stimulus and creates chemical sensors in textile forms. Detection mechanisms are based on the modification of a material's electrical conductivity. However, the electrical modification is different for the two families of CPC and ICP.

5.3.1 CPC chemical resistive sensors

The CPC polymer matrix has the ability to swell. If a 'good' solvent (liquid or volatile) is in contact with a CPC, it diffuses into the polymer and expansion results. Hence, the inter-fillers distance increases and the conductive network of electrical fillers is modified. An increase of electrical resistance is observed due to the disconnection of individuals or clusters fillers (Electrical Percolation Type Solvent Sensor, EPTSS). Concerning the liquid solvents, a review (Villmow *et al.*, 2011a) defined what signifies a 'good' solvent, as referred to above. In fact, the dissolution of a polymer starts with the first step where the polymer swells. The solvent is described as 'bad' if it does not lead to any swelling of the polymer. To obtain efficient sensors, a high rate of detection, reversible and reproducible responses, must be achieved. That is why the choice of the solvent/polymer system governs the selectivity and detection range. The example in Fig. 5.5 shows that the matrix choice is crucial in chemical sensing.

The tested CPC is based on a poly(ethylene-co-ethyl acrylate) filled with 37 wt.% of carbon black (EEA-CB composite). Several successive cycles of

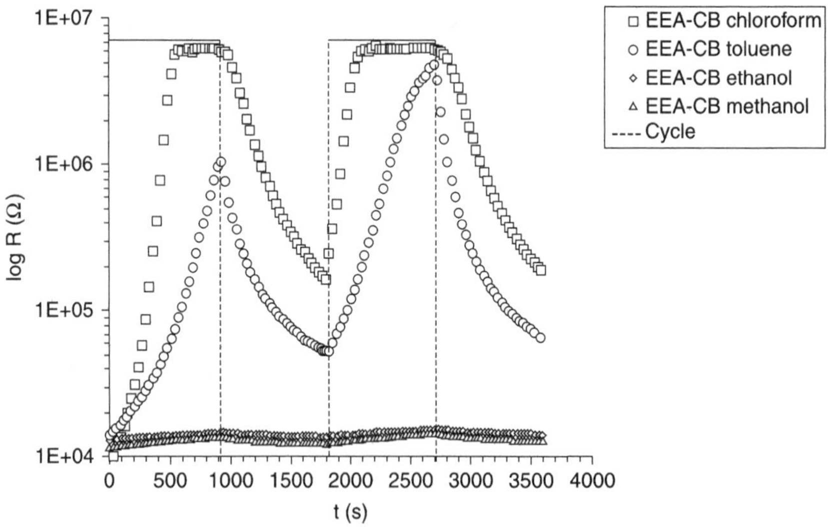

5.5 Evolution of electrical resistivity of a poly(ethylene-co-ethyl acrylate)-carbon black (EEA-CB) in the presence of solvents (Feller and Grohens, 2004).

solvent exposition (first solvent vapor, then air) was applied. The responses are entirely different in function to the nature of the used solvent. A good solvent, such as chloroform or toluene, diffuses inside the polymer and causes swelling and the disconnection of fillers. In the case of chloroform, a plateau in resistivity is reached faster than that of toluene. In fact, the latter has a lower diffusion within the polymer. However, the EEA-CB shows no sensitivity to ethanol and methanol, because both are non-solvents within the polymer matrix. Two detection mechanisms can be considered according to the affinity of the polymer with the vapor or liquid, that is, if the detected substance is a solvent or a non-solvent of the polymer.

Change of polymer matrix by solvents

As explained earlier, with the solvent detection by CPC, the polymer solubility in the considered solvent plays a major role. Indeed, the polymer swelling leads to the filler's separation by disconnection of macromolecular chains. The solubility of the polymer and diffusion of solvent through the chains determine the EPTSS sensitivity. As with temperature sensing (Section 5.3), a positive vapor coefficient (PVC), corresponding to the increase in resistance in the presence of solvent, and a negative vapor coefficient (NVC), are identifiable. Authors have presented these effects on different systems. Zhang et *al*. (2005) presents the detection of tetrahydrofuran (THF) by nanocomposite polystyrene (PS)/CNT. The sudden swelling of the polymer first causes a marked increase in electrical resistance (the PVC effect) and after a long exposure to THF, the polymer solubility causes a decrease in the viscosity, which allows exchange of the fillers and the reformation of the electrical network (the NVC effect).

Change of polymer matrix by non-solvents

In the detection of vapors by CPC involving a non-solvent, water vapor has been widely studied. Thus, hydrophobic polymers were tested. In the presence of water vapor, a drop in the electrical resistance of the composites was observed, corresponding to an increase in electrical conductivity. Two mechanisms may explain the increase of electrical conductivity (Yang *et al.*, 1996; Rubinger *et al.*, 2007; Su and Wang, 2007). The first is the absorption of water molecules by the polar groups of the polymer, whereby electrolysis of adsorbed water allows the transport of protons (H^+) between the anode and the cathode. The adsorbed water leads to a decrease in the ionic interactions between polar polymer groups and thus enhances ion mobility. The second is the adsorption of water molecules by fillers, such as CNT. The second mechanism becomes more dominant as the filler content in the composite increases. The sensitivity of these composites can be improved by the addition of salts. In the case of the composite Poly(Methyl MethAcrylate)(PMMA)/CNT (Su and Wang, 2007), the incorporation of potassium hydroxide (KOH) increases the amplitude of sensitivity due to ion mobility within the composite.

One CPC can be sensitive to several solvents. Therefore, the use of a combination of chosen CPC can improve the selectivity of the global sensor or can have some interesting applications. For example, an e-*nose* has been created by combining specific CPC. Composed of an array of CNT-based CPC transducers, vapors of nine single organic solvents can be discriminated using a classical recognition technique and principal component analysis (Castro *et al.*, 2011).

Change of conductive fillers

Different kinds of electrical conductive fillers can be used in this CPC (see Introduction). However, for chemical detection, the use of carbon fillers – carbon black (Dong *et al.*, 2004), carbon nanoparticles (Bouvree *et al.*, 2009) and CNT (Pötschke *et al.*, 2010) – is recommended due to their temperature and oxidation stability. But some studies have been realized with combinations, more or less surprising, of fillers. As an example, an EPTSS has been developed by fabricating a composite consisting of a conducting phase and an insulating phase, whose volume increases by the incorporation of a desired substance. This composite of montmorillonite with carbon shows a remarkable increase of resistivity in a humid atmosphere, due to the volume expansion of the montmorillonite phase by the incorporation of water. The resistivity increases in response to stepwise water vapor increase like a humidity sensor (Tsuyumoto and Iida, 2011). Another factor can be considered, because the fillers themselves are sensitive to the presence of a solvent. For example, by their large specific surface area, hollow geometry, nanometer size and high electronic mobility, CNTs have shown remarkable properties as chemical sensors. The porosity can absorb a significant amount of gases and liquids and some authors (Kong *et al.*, 2000; Valentini *et al.*, 2004) have observed a change of electrical properties after gas adsorption, and proposed the use of CNTs as sensors. The sensitivity of electrical fillers to vapors is smaller compared to the sensitivity of the polymer matrix.

Application on textiles

Conductive composites have been reported in several forms. Compression moulded plates (Villmow *et al.*, 2011b), or spray deposition by layer-by-layer techniques have been used to form the transducer film on a substrate (Castro *et al.*, 2009). Concerning textiles, melt-spun monofilament (Narkis *et al.*, 2000; Pötschke *et al.*, 2010) or multifilament (Devaux *et al.*, 2011; Fan *et al.*, 2011) fibers or directly in the structure of a woven fabric (Rentenberger *et al.*, 2011) have been studied. The liquid sensing properties were tested on the biphasic blend polypropylene (PP)/polycaprolactone (PCL) and a comparison between compression-moulded discs and melt-drawn filaments was achieved, indicating higher responses from the discs (Pötschke *et al.*, 2011). Also traditional textiles can be modified for use as chemical detectors by coating, inkjet printing or surface treatment. In previous studies, authors have studied the gas sensing ability of PPy coated on different polyester textile fabrics (Kincal *et al.*, 1998), or of zinc oxide (ZnO) nanorod growth on the surface of cotton fabrics (Lim *et al.*, 2010).

5.3.2 ICP chemical resistive sensors

For ICP, the variations of electrical conductivity are caused not only by polymer swelling. The perturbation of the electronic network by chemical substances changes the electrical properties of the polymer. The extent of these changes can reveal the presence of chemical groups. The presence of a substance can disrupt the backbone network, inducing an electrical conductivity decrease in a doped polymer and conversely an electrical conductivity increase in an undoped polymer. The change of conductivity is the measurement principle in conducting polymers. The nature of the polymer and its degree of doping determines the generated perturbations (Lange *et al.*, 2008). For example, if PANI is oxidized in the presence of nitrogen dioxide (NO_2) gas (electron acceptor), the result is a decrease in its electrical resistance (Xie *et al.*, 2002). However, in PANI initially doped by protonation (addition of H^+), the presence of NO_2 leads to an increase of its electrical resistance (Yan *et al.*, 2007). In general, the electron donating gases (H_2S, NH_3, N_2H_4) reduce the electrical conductivity of the conducting polymer, while the electron accepting gases (NO_2, SO_2, O_3) dope the polymer and increase its electrical conductivity. Conducting polymers can be disrupted by chemical reactions, as discussed above, but also by physical adsorption of vapor. The adsorption of benzene, toluene and hexane leads to swelling of conductive polymers with a good affinity for these solvents and increases the distance between the lateral contact macromolecular chains. The distance of electronic groups decreases the electrical conductivity by a tunneling effect within the polymer.

The phenomena that can disrupt the electrical network of conductive polymers can be described as (Li *et al.*, 2007):

- the oxidation–reduction chemical reactions between the polymer and the solvent;
- the acid-base reactions between the polymer and the solvent;
- the solubility of the polymer in the solvent;
- the polarity of the polymer can create links, dipole–dipole.

The particular case of water vapor was also studied. Mostly, the electrolyte nature of water promotes the electrical conductivity of polymers by dipole–dipole interactions. Also, proton-conducting polymers improve the mobility of protons, leading to better charge transfer (Casalbore-Miceli *et al.*, 1997).

5.3.3 Textile applications of chemical resistive sensors

Main textile applications are health and personal protective equipment (i.e. military). Numerous products have already been developed or commercialized in several sectors to detect liquid or volatile solvents. In most cases, different physical principles, not only based on resistive measurement, are employed and integrated into the textile structure. For instance, the medical field has focused on the detection of substances that can be found in human sweat. The European project *Biotex* (Biotex, 2011) aims to develop a textile sensor (as a 'patch') for biochemical analysis of body fluids. Sweat is initially absorbed by the textile and conveyed using hollow fiber with a hydrophilic nature to a small area of the textile called the 'collector'. This storage area can increase the concentration of fluid at one point. The sensor on the surface detects the presence of a specific chemical such as sodium. Other sensors detecting potassium or chloride may also be inserted in the fabric. These sensors can be used to prevent the risks associated with obesity, diabetes and high level sport.

CPC and ICP are also developed for use as chemical sensors in the textile area. The aim of another European project *INTELTEX* (Inteltex, 2011) is to develop new textile structures based on CPC for several sensors types. Some of these are solvents (acetone or acetic acid) (Rentenberger *et al.*, 2011) or water vapor (Devaux *et al.*, 2011) detection in multifilament forms or woven structures.

A patent (Qi and Mattes, 2010) describes conducting-polymer based multifunctional sensing fabrics suitable for monitoring humidity, breath, heart rate, blood (location of wounds), blood pressure, skin temperature, weight and movement, in a wearable, electronic embedded sensor system. Conducting PANI fibers have been woven, knitted, stitched or braided into a fabric comprising non-conducting fibers, enabling the fabric with monitoring and detection abilities.

5.4 Temperature resistive sensors

As for most conductive materials, temperature plays a major role on the electrical conductivity of CPC or ICP. Of course, compared to mechanical and chemical, some other mechanisms are observed to produce temperature sensors based on the two families.

5.4.1 CPC temperature resistive sensors

For CPC, in 1945, Frydman introduced the positive temperature coefficient (PTC), defining the non-linear increase in electrical resistivity as a function of temperature (Fig. 5.6). Major contributions of temperature change on CPC electrical resistivity are:

- the rearrangement of the fillers;
- the volume expansion of the matrix (difference in expansion between the matrix and fillers).

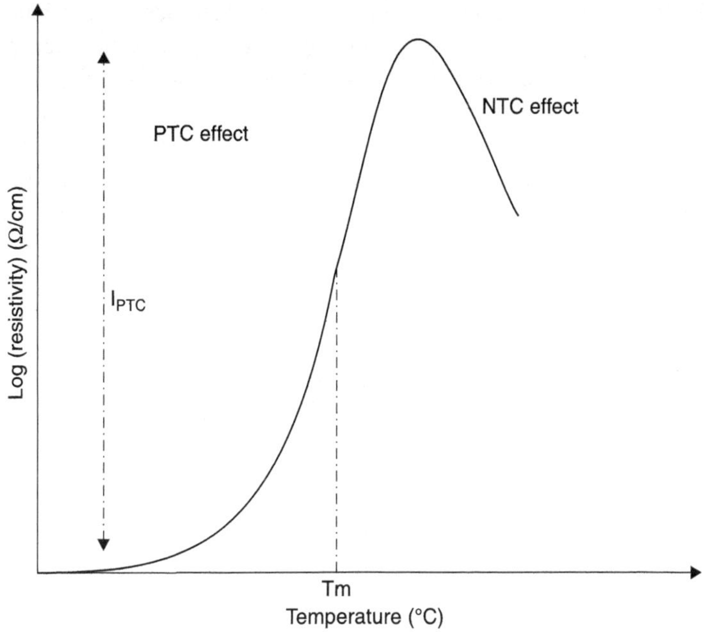

5.6 Evolution of electrical resistivity versus temperature highlighting the positive temperature coefficient (PTC) and negative temperature coefficient (NTC) effects (I_{PTC}: intensity of PTC effect, Tm: melt temperature).

Other contributions, such as thermal expansion (α, playing a role in the dimensions of the materials S and L, see Eq. 5.1) and dependence of electrical conductivity (σ, i.e. $1/\rho$) according to temperature of fillers, are negligible.

The matrix volume expansion leads not only to a reduction of the metallic conduction but also to tunneling and hopping conductions. This phenomenon is due to an increase in the distance between particles, until the current flow is cut (Feller, 2011). The disconnection of fillers conductive paths ranks CPC temperature sensors amongst electrical percolation type temperature sensors (EPTTS). The polymer phase transition temperatures (melting temperature T_m and glass transition temperature T_g) and the crystallinity degree are predominantly responsible for the PTC effect. Indeed, when increasing the temperature, the crystalline regions melt and become disorganized, causing a dilution of the particles that are localized in the amorphous phase if their size does not permit their intercalation into the crystalline phase. This leads to an increase in the distance between fillers and global resistivity (Zhao *et al.*, 2003). In Fig. 5.6, beyond the melting temperature of the polymer matrix, the NTC effect (negative temperature coefficient) appears. It corresponds to a diminution due to fillers reorganizing in the molten state, due to a molecular mobility, which in turn increases interparticle interactions.

The PTC effect is difficult to model because many factors can influence the passage conductor/insulator, such as being near the percolation threshold. The PTC is dependent on all parameters that influence the fillers percolation. As shown by Feller *et al.* (2002(a,b)) in Fig. 5.7, when short carbon fibres (SCF) are introduced, whether in a polyester matrix or poly (epoxy) (Fig. 5.7), the PTC intensity (I_{PTC}, defined in Fig. 5.6) has a greater amplitude before the percolation threshold (1.88 vol.% of SCF). Beyond an excessive fillers content (>1.88 vol.%), SCF tend to agglomerate and form nodes, which reduce the number of contact paths. The disconnection is efficient (i.e. has a significant effect on electrical resistance) if the ratio between disconnected conductive path and connected conductive path is high. It is necessary in this case to be slightly above the percolation threshold for easy disconnection.

The list of influent parameters is difficult to establish, especially as the various experimental results on different systems are sometimes contradictory. Some parameters influencing the sensing response can be:

- the influence of molar mass (Zhang *et al.*, 2005);
- the adhesion work between fillers and the polymer matrix (Park *et al.*, 2005);
- the electrical fillers migration from amorphous to crystalline phase (Zhao *et al.*, 2003);
- the cycle numbers thermal cycles (Meyer, 1974; Tang *et al.*, 1993).

Concerning the kinds of electrical fillers, carbon fillers (He, 2005; Ni, 2006; Lisunova *et al.*, 2007) are commonly used in CPC temperature sensing. Metallic fillers have also been implemented (Rybak *et al.*, 2010). The choice of conductive fillers is also

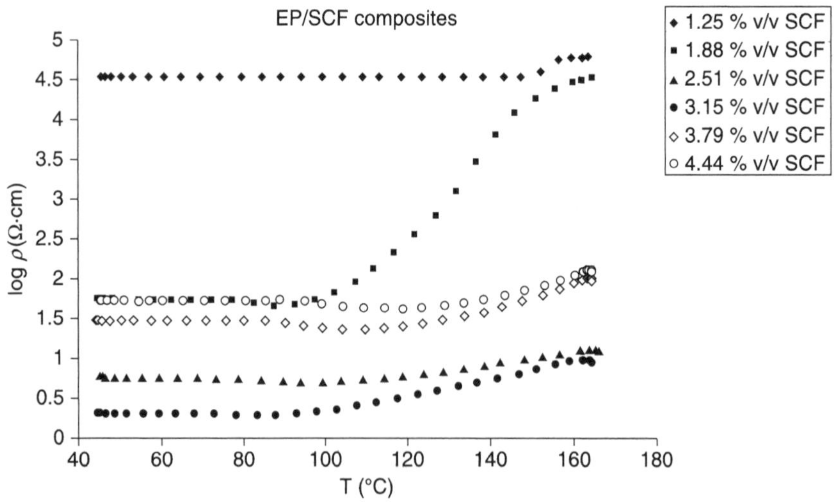

5.7 Influence of filler content on the intensity of the PTC (Feller *et al.*, 2002).

an influential factor, which modifies the temperature response. As an example, the amplitude I_{PTC} is higher in a nanocomposite based on polyethylene (PE) with CB than with CNT. This behavior is explained by the high level of contact in the CNT network, which is more stable during matrix volume expansion (Lisunova *et al.*, 2007).

5.4.2 ICP temperature resistive sensors

The ICP temperature dependence of resistance depends on the conduction regime (metallic or semi-conductor) of the polymer. With an increase in temperature, two types of behavior can be observed:

- If IPC has metallic conduction, they behave like metal and the resistivity increases in a large temperature range and a PTC effect can be observed (Lee *et al.*, 2006).
- If IPC is semi-conductor (hopping or tunneling effect) the resistivity decreases in a large temperature range, near room temperature, and an NTC effect can be observed (Bowman and Mattes, 2005).

In both cases, electrons–phonon interactions can play a key role. First, electrons–phonons interaction, due to thermal agitation, disturbs electrons movement. Consequently, for metallic type conduction, an increase of resistivity is noticed with temperature. Second, with the hopping or tunneling conduction, phonons excite loose electrons, which can move through the material (via π bond and hold, see Introduction), providing a kind of semi-conduction.

5.4.3 Textile applications of temperature resistive sensors

Regarding the effect of heat on CPC, only observations of change in conductivity as a function of temperature are studied (PTC, NTC). However, some studies are clearly oriented in the detection of a threshold temperature. According to Boiteux *et al.* (2007), the application prospects of their system (polypropylene/polyamide filled with iron particles) would be in smart materials: temperature sensors, 'circuit breakers', but also as thermistors, or as self-regulating heating elements.

In the field of textile techniques, CPC are used in temperature sensors in the *INTELTEX* project (Section 5.3.3). The elaboration of textile sensors is ensured by the incorporation of CNT in one or more polymers. The final goal of this work is integration into personal protective equipment (PPE) for firefighters; a new textile composite based on the use of innovative nanofillers enables them to be alerted to a critical elevation of the surrounding temperature (Cayla *et al.*, 2009). Concerning ICP, Mattes (2005) has developed a fiber (*PANION*) created from the conductive polymer PANI. Similarly, *SENSATEX* has developed a T-shirt (SmartShirt: Sensatex, 2011) to monitor the vital signs of the wearer, such as body temperature, by patented conductive fiber.

Some temperature sensors are already used in the textile field and show that temperature measurement is a real need. The European project PROeTEX (Proetex, 2011), amongst others, uses thermocouples to achieve a required fire fighter's coat integrating multiple physiological sensors to position, as well as environmental (temperature, presence of toxic gases). Another example is *thermochromic* material, which changes color due to changes in temperature (only temperature threshold detection in this case). For instance, in 2003, the American company Wickedglow Industries (Bodyfaders, 2011) launched a T-shirt range that changed color with heat from the body, or from contact.

5.5 Conclusion and future trends

Future advances will follow three directions:

1. improvement of existing sensors materials;
2. transfer from research to practical application; and
3. textile integrated components (electronic, sensor, actuator and energy).

The work on existing sensors materials will improve general reliability of sensors (number of supported cycle, range of detection, etc.) in all areas of application. Of course, these improvements will also result in the creation of new textile sensors to detect and quantify another solvent, detect another physical value (light, acceleration, etc.) or detect simultaneously temperature and solvent (*all in one* sensor), for instance.

The transfer from research (not only on textile materials) to textile public application by realizing more integrated textile components will see the achievement of fine fibre sensors (not only film or rod sensors) and/or woven, knitted, non-woven or braided sensors structure. This transfer to consumer application requires the development of low cost and reliable industrial production devices (spinning, printing and inkjet printing of CPC or ICP). The aim is to open sensor textile technologies to new business sectors and find application for existing textile sensor devices.

Long-term research and development will focus on the integration of electronics into textile materials (or in/on structure), in order to make fully integrated intelligent textile systems. The future integrated electronic devices are general components (resistor, capacitor e.), integrated circuits, non-textile sensors (i.e. optical sensors), actuators (mechanical, optical (screen) or chemical (molecules, drug, etc.)) and energy devices (electricity production, electrical (capacitor) or chemical (cell) storage). Thus, textile sensors plus fully integrated electronic components will be more efficient and 'invisible' to users.

5.6 References

Bartalesi R, Carbonaro N, Lorussi F, Tesconi M, Tognetti A, *et al.* (2006), 'Smart textiles: Toward a wearable motion capture system', *Proceedings of the Ninth International*

Symposium on the 3-D analysis of Human Movement, Valenciennes, France, 28–30 June.

Besançon R M (1985), *The Encyclopedia of Physics*, Third edition, New York, Van Nostrand Reinhold Co.

Bilotti E, Zhang R, Deng H, Baxendale M and Peijs T (2010), 'Fabrication and property prediction of conductive and strain sensing TPU/CNT nanocomposite fibres', *J Mater Chem*, 20(42), 9449–9455.

Biotex (2011), 'Bio-sensing textile for health management', available from: http://www.biotex-eu.com/ [Accessed 28 November 2011].

Bodyfaders (2011), 'Hyperchromic colour transforming', available from: http://www.bodyfaders.com/ [Accessed 28 November 2011].

Boiteux G, Mamunya Y P, Lebedev E V, Adamczewski A, Boullanger C, *et al.* (2007), 'From conductive polymer composites with controlled morphology to smart materials', *Synth Met*, 157, 1071–1073.

Bouvree A, Feller J F, Castro M, Grohen Y and Rinaudo M (2009), 'Conductive polymer nano-bioComposites (CPC): chitosan-carbon nanoparticle, a good candidate to design polar vapour sensors', *Sensor Actuat B-Chem*, 138, 138–147.

Bowman D and Mattes B R (2005), 'Conductive fibre prepared from ultra-high olecular weight polyaniline for smart and interactive textile application', *Synth Met*, 154, 29–32.

Brady S, Lau K T, Megill W, Wallace G G and Diamond D (2005), 'The development and characterization of conducting Polymeric-based sensing devices', *Synth Met*, 154, 25–28.

Buechley L, Eisenberg M and Catchen J (2008), 'The LilyPad Arduino: using computational textiles to investigate engagement, aesthetics, and diversity in computer science education', *CHI '08 Proceedings of the Twenty-sixth Annual SIGCHI Conference on Human Factors in Computing Systems*, pp. 423–432.

Calvert P, Duggal D, Patra P, Agrawal A and Sawhney A (2008), 'Conducting polymer and conducting composite strain sensors on textiles', *Mol Cryst Liq Cryst*, 484, 291–302.

Casalbore-Miceli G, Camaioni N, Yang M J, Zhen M, Zhan X W and D'Aprano A (1997), 'Charge transport mechanism in pressed pellets of polymer proton conductors', *Solid State Ionics*, 100, 217–224.

Castro M, Lu J, Bruzaud S, Kumar B and Feller J F (2009), 'Carbon nanotubes/poly(e-caprolactone) composite vapour sensors', *Carbon*, 47, 1930–1942.

Castro M, Kumar B, Feller J F, Haddi Z, Amari A and Bouchikhi B (2011), 'Novel e-nose for the discrimination of volatile organic biomarkers with an array of carbon nanotubes (CNT) conductive polymer nanocomposites (CPC) sensors', *Sensor Actuat B: Chem*, 159, 213–219.

Cayla A, Campagne C, Rochery M and Devaux E (2009), 'Quand les fibres deviennent capteurs', *Mattech*, 6, 348–351.

Ciselli P, Lu L, Busfield J C C and Peijs T (2010), 'Piezoresistive polymer composites based on EPDM and MWNTs for strain sensing applications', *e-Polymers*, 14, 1–13.

Cochrane C (2011), *Développement d'un système de mesure d'allongement pour voilure de parachute*, Sarrebruck, Edition Universitaire Européennes.

Cochrane C, Koncar V, Lewandowski M and Dufour C (2007), 'Design and development of a flexible strain sensor for textile structures based on a conductive polymer composite', *Sensors*, 7(4), 473–492.

Cochrane C, Lewandowski M and Koncar V (2010), 'A flexible strain sensor based on a conductive polymer composite for in situ measurement of parachute canopy deformation', *Sensors*, 10(9), 8291–8308.

Devaux E, Aubry C, Campagne C and Rochery M (2011), 'PLA/Carbon nanotubes multifilament yarns for relative humidity textile sensor', *J Eng Fiber Fabr*, 6(3), 13–24.

Dong X M, Fu R W, Zhang M Q, Zhang B and Rong M Z (2004), 'Electrical resistance response of carbon black filled amorphous polymer composite sensors to organic vapors at low vapor concentrations', *Carbon*, 42, 2551–2559.

Donnet J B, Bansal R C and Wang M J (1993), *Carbon Black*, Second edition, Chapter 3, New York, Marcel Dekker.

Dunne L E, Brady S, Tynan R, Lau K, Smyth B, *et al.* (2006), 'Garment-based body sensing using foam sensors', *AUIC '06 Proceedings of the 7th Australasian User Interface Conference*, 50, 165–171.

Dyre J C and Schroder T B (2002), 'Hopping models and AC universality', *Phys Status Solidi B*, 230(1), 5–13

Eng D (2009), *Fashion Geek: Clothes Accessories Tech.*, Cincinnati, North Light Books.

Fan Q, Qin Z, Villmow T, Pionteck J, Pötschke P, *et al.* (2011), 'Vapor sensing properties of thermoplastic polyurethane multifilament covered with carbon nanotube networks', *Sensor Actuat B: Chem*, 156, 63–70.

Fashioning Technology (2011), available from: http://www.fashioningtech.com/page/textile-sensors-1 [Accessed 28 November 2011].

Feller J F (2011) *Des Composites Polymères Conducteurs Aux éco-Composites Polymères*, Sarrebruck, Edition Universitaire Européennes.

Feller J F, Linossier I and Grohens Y (2002a), 'Conductive polymer composites: comparative study of poly(ester)-short carbon fibres and poly(epoxy)-short carbon fibres mechanical and electrical properties', *Mater Lett*, 57, 64–71.

Feller J F, Linossier I and Levesque G (2002b), 'Conductive polymers composites (CPCs): comparison of electrical properties of poly (ethylene-co-ethyl acrylate)-carbon black with poly(butyleneterephthalate)/poly(ethylene-co-ethylacrylate)-carbon black', *Polym Adv Tech*, 13, 714–724.

Feller J. F, Chauvelon P, Linossier I and Glouannec P (2003), 'Characterization of electrical and thermal properties of extruded tapes of thermoplastic conductive polymer composites (CPC)', *Polym Test*, 22, 831–837.

Fraden J (1996), *Handbook of modern sensors, physics, designs and applications*, Third edition, New York, Springer.

Frydman E (1945), 'Improvements in or relating to resistance elements having positive temperature/ resistance characteristics', *UK patent application 604695*, 8 July 1945.

Glazzard M, and Kettley S (2010), 'Knitted stretch sensors for sound output', *Proceedings of the Fourth International Conference on Tangible, Embedded, and Embodied Interaction*, Cambridge, MA, 25–27 January, pp. 391–392.

He X J, Du J H, Ying Z and Cheng H M (2005), 'Positive temperature coefficient effect in multiwalled carbon nanotube/high-density polyethylene composites', *Appl Phys Lett*, 86, 62–112.

Heinrich H G and Saari D P (1975), 'Exploratory parachute canopy stress measurements during inflation and at steady state', *Proceedings of the AIAA 5th Aerodynamic Deceleration Systems Conference*, Albuquerque, Mexico.

Inteltex (2011), 'Intelligent multi-reactive textiles', available from: http://www.inteltex.eu/ [Accessed 28 November 2011].

Kirkpatrick S (1973), 'Percolation and conduction', *Rev Mod Phys*, 45(4), 574–588.

Kim B, Koncar V and Dufour C (2006), 'Polyaniline-coated PET conductive yarns: study of electrical, mechanical, and electro-mechanical properties', *J Appl Polym Sci*, 101(3), 1252–1256.

Kim B, Koncar V, Devaux E, Dufour C and Viallier P (2004), 'Electrical and morphological properties of PP and PET conductive polymer fibers', *Synth Met*, 146(2), 167–174.

Kincal D, Kumara A, Child A D and Reynolds J R (1998), 'Synthetic conductivity switching in polypyrrole-coated textile fabrics as gas sensors', *Synth Met*, 92, 53–56.

Kong J, Franklin N R, Zhou C, Chapline M G, Peng S, *et al.* (2000), 'Nanotube Molecular Wires as Chemical Sensors', *Science*, 28, 622–625.

Lange U, Roznyatovskaya N V and Mirsky V M (2008), 'Conducting polymers in chemical sensors and arrays', *Anal Chim Acta*, 614, 1–26.

Lee K, Cho S, Park S H, Heeger A J, Lee C W and Lee S H (2006), 'Metallic transport in polyaniline', *Nature* 441, 65–68.

Li B, Santhanam S, Schultz L, Jeffries, E L M, Iovu, *et al.* (2007), 'Inkjet printed chemical sensor array based on polythiophene conductive polymers', *Sensor Actuat B: Chem*, 123, 651–660.

Lim Z H, Chia Z X, Kevin M, Wong A S W and Ho G W (2010), 'A facile approach towards ZnO nanorods conductive textile for room temperature multifunctional sensors', *Sensor Actuat B: Chem*, 151, 121–126.

Lisunova M O, Mamunya Y P, Lebovka N I and Melezhyk A V (2007), 'Percolation behaviour of ultrahigh molecular weight polyethylene/multi-walled carbon nanotubes composites', *Eur Polym J*, 43, 949–958.

Lorussi F, Rocchia W, Scilingo E P, Tognetti A and DeRossi D (2004), 'Wearable, redundant fabric-based sensor arrays for reconstruction of body segment posture', *IEEE Sens J*, 4(6), 807–818.

Lovell E and Buechley L (2010), 'An e-sewing tutorial for DIY learning', *IDC '10 Proceedings of the 9th International Conference on Interaction Design and Children*, pp. 230–233.

Mattes B R (2004), 'Electronic textile based on intrinsically conducting polymer fibre', *Workshop Regarding SMART Textiles*, Ghent, Belgium, 15 June.

Mattes B R (2005), 'Monitoring chemical vapors with protective clothing based on conducting polymer fibre', *Proceedings of the NATO Advanced Research Workshop Intelligent Textiles for Personal Protection and Safety*, Zadar, Croatia, 7–10 September.

Meyer J (1974), 'Stability of polymer composites as positive temperature coefficient resistors', *Polym Eng Sci*, 14, 706–716.

Mott N F (1973), 'Transport in disordered materials', *Phys Rev Lett*, 31(7), 466–467.

My Heart project IST 507816 (2011), available from: http://www.hitech-projects.com/euprojects/myheart [Accessed 28 November 2011].

Narkis M, Srivastava S, Tchoudakov R and Breuer O (2000), 'Sensors for liquids based on conductive immiscible polymer blends', *Synth Met*, 113, 29–34.

Nauman S, Lapeyronnie P, Cristian I, Boussu F and Koncar V (2011a), 'On-line measurement of structural deformations in composites', *IEEE Sens J*, 11(6), 1329–1336.

Nauman S, Cristian I and Koncar V (2011b), 'Simultaneous application of fibrous piezoresistive sensors for compression and traction detection in glass laminate composites', *Sensors*, 11(10), 9478–9498.

Neelakandan R and Madhusoothanan M (2010), 'Electrical resistivity studies on polyaniline coated polyester fabrics', *J Eng Fiber Fabr*, 5(3), 25–29.

Ni Q, Wu S H, Fu Y, Natsyki T and Kurashiki K (2006), 'Properties of CNT/NiTi/UPR three-phase composites', *Proceedings of Seventh International Conference on the Science and Application of Nanotubes*, Nagano, Japan, p. 322.

Park S J, Kim H C and Kim H Y (2005), 'Roles of work of adhesion between carbon blacks and thermoplastic polymers on electrical properties of composites', *J Colloid Interf Sci*, 255, 145–149.

Perner-Wilson H and Buechley L (2010), 'Making textile sensors from scratch', in: *TEI 10, Proceedings of the Fourth International Conference on Tangible, Embedded, and Embodied Interaction*, ACM New York, pp. 349–352.

Pötschke P, Andres T, Villmow T, Pegel S, Brünig H, *et al.* (2010), 'Liquid sensing properties of fibres prepared by melt spinning from poly(lactic acid) containing multi-walled carbon nanotubes', *Compos Sci Technol*, 70, 343–349.

Pötschke P, Kobashi K, Villmow T, Andres T, Paiva C M and Covas A J (2011), 'Liquid sensing properties of melt processed polypropylene/poly(e-caprolactone) blends containing multiwalled carbon nanotubes', *Compos Sci Technol*, 71, 1451–1460.

Proetex (2011), 'Advanced e-textiles for fire-fighters and civilian victims', available from: http://www.proetex.org [Accessed 28 November 2011].

Qi B and Mattes B R 2010, 'Multifunctional conducting polymer structures', *Patent application 20100170886*, 7 August.

Rentenberger R, Cayla A, Villmow T, Jehnichen D, Campagne C, *et al.* (2011), 'Multifilament fibres of poly(caprolactone)/poly(lactic acid) blends with multiwalled carbon nanotubes as sensor materials for ethyl acetate and acetone', *Sensor Actuat B: Chem*, 160(1), 22–31.

Rubinger C P L, Martins C R, De Paoli M A and Rubinger R M (2007), 'Sulfonated polystyrene polymer humidity sensor: Synthesis and characterization', *Sensor Actuat B: Chem*, 124, 42–49.

Rybak A, Boiteux G, Melis F and Seytre G (2010), 'Conductive polymer composites based on metallic nanofiller as smart materials for current limiting devices', *Compos Sci Technol*, 70, 410–416.

Sensatex (2011), Sensatex company, available from: http://www.sensatex.com/smartshirt.html [Accessed 28 November 2011].

Shyr T W, Shie J W and Jhuang Y E (2011), 'The effect of tensile hysteresis and contact resistance on the performance of strain-resistant elastic-conductive webbing', *Sensors*, 11(2), 1693–1705.

Skrifvars M, Rehnby W and Gustafsson M (2008), 'Coating of textile fabrics with conductive polymers for smart textile applications', *Ambience 08*, Borås, Sweden, 2–3 June, pp. 100–103.

Stenger-Smith J D (1998), 'Intrinsically electrically conducting polymers: synthesis, characterization and their applications', *Prog Polym Sci*, 23, 57–79.

Strazdiene E, Blazevic P, Vegys A and Dapkuniene K (2007), 'New tendencies of wearable electronics application in smart clothing', *Electro Elect Eng*, 1(73), 21–24.

Su P G and Wang C S (2007), 'In situ synthesized composite thin films of MWCNTs/PMMA doped with KOH as a resistive humidity sensors', *Sensor Actuat B: Chem*, 124, 303–308.

Tang H, Piao J, Chen X, Luo Y and Li S (1993), 'The positive temperature coefficient phenomenon of vinyl polymer/CB composites', *J Appl Polym Sci*, 48, 1795–1800.

Tsuyumoto I and Iida Y (2011), 'Percolation-type chemical sensor: electrical- and humidity-sensing properties of carbon–montmorillonite composites', *Int J Appl Ceram Tec*, 8(4), 793–799.

Valentini L, Cantalini C, Armentano I, Kenny J M, Lozzi L and Santucci S (2004), 'Highly sensitive and selective sensors based on carbon nanotubes thin films for molecular detection', *Diam Relat Mater*, 13, 1301–1305.

Villmow T, Pegel S, John A, Rentenberger R and Pötschke P (2011a), 'Liquid sensing: smart polymer/CNT composites', *Mat Today*, 14(7–8), 340–345.

Villmow T, Pegel S, Pötschke P and Heinrich G (2011b), 'Polymer/carbon nanotube composites for liquid sensing: Model for electrical response characteristics', *Polymer*, 52, 2276–2285.

Window A L (1992), 'Materials, construction, performance and characteristics', in: *Strain Gauge Technology*, Second edition, Chapter 1, New York, Springer.

Wood J R, Zhao Q, Frogley M D, Meurs E R, Prins A D, *et al.* (2000), 'Carbon nanotubes: from molecular to macroscopic sensors', *Phys Rev B*, 62, 7571–7575.

Wu J, Zhou D, Too C and Wallace G. (2005), 'Conducting polymer coated Lycra', *Synth Met*, 155, 698–701.

Xie D, Jiang Y, Pan W, Li D, Wu Z and Li Y (2002), 'Fabrication and characterization of polyaniline-based gas sensor by ultra-thin film technology', *Sensor Actuat B: Chem*, 81, 158.

Yan X B, Han Z J, Yang Y and Tay B K (2007), 'NO_2 gas sensing with polyaniline nanofibers synthesized by a facile aqueous/organic interfacial polymerization', *Sensor Actuat B: Chem*, 123, 107

Yang M J, Sun H M, Casalbore-Miceli G, Martelli A and Ridolfi G (1996), 'Poly(propargyl alcohol) doped with sulfuric acid, a new proton conductor usable for humidity sensor construction', *Synth Met*, 81, 65–69.

Yoshikai T, Fukushima H, Hayashi M and Inaba M (2009), 'Development of soft stretchable knit sensor for humanoids whole-body tactile sensibility', *IEEE-RAS International Conference on Humanoid Robots*, Paris, France, December 7–9, 624–631.

Zhao Z, Yu W, He X and Chen X (2003), 'The conduction mechanism of carbon black-filled poly(vinylidene fluoride) composite', *Mater Lett*, 57, 3082–3088.

Zhang B, Fu R W, Zhang M Q, Dong X M, Lan P L and Qiu J S (2005), 'Preparation and characterization of gas-sensitive composites from multi-walled carbon nanotubes/polystyrene', *Sensor Actuat B: Chem*, 109, 323–328.

Zhang C, Ma C A, Wang P and Sumita M (2005), 'Temperature dependence of electrical resistivity for carbon black filled ultra-high molecular weight polyethylene composites prepared by hot compaction', *Carbon*, 43, 2544–2553.

Zhang H, Tao X, Yu T, Wang S and Cheng X, (2006), 'A novel sensate 'string' for large-strain measurement at high temperature', *Meas Sci Technol*, 17, 450–458.

6
Soft capacitance fibers for touch-sensitive smart textiles

S. GORGUTSA and M. SKOROBOGATIY,
Ecole Polytechnique de Montréal, Canada

DOI: 10.1533/9780857093530.1.154

Abstract: We start with an overview of state-of-the-art tactile sensor technology. We then detail tactile sensors based on the novel soft capacitor fibers developed by our group and go on to introduce a theoretical model to describe the electrical response of a single capacitor fiber using an electrical ladder network model and show that such a fiber can function as a touch-sensitive sensor. Effects of different parameters on the sensing performance of a soft capacitor fiber are also studied. Finally we argue that capacitor fibers are ideally suited for wearable touch-sensing and as proof of this concept we present the prototype of a woven touchpad sensor featuring a one-dimensional array of capacitor fibers, and then study its performance.

Key words: flexible electronics, fibers, conducting polymers, sensors.

6.1 Introduction: overview of capacitive sensing

Touch sensing as a human interface device (HID) technology is becoming increasingly popular and ubiquitous, finding applications in smart phones, computers and responsive garments, to name a few. Various touch sensing systems have been developed based on different physical principles including resistive, capacitive, infrared, surface acoustic wave, electromagnetic, near field imaging, etc. Resistive (Cok *et al.*, 2004) and capacitive (Kalendra and Piazza, 1994) methods have been widely used in conventional touch screens of commercial products such as mobile phones, PDAs and consumer electronics devices. Resistive touch screens are composed of two material sheets that are coated with a resistive material, commonly indium tin oxide (ITO), and separated by an air gap or microdots. When a finger presses the screen, the two sheets are connected at the touch position, which changes the current flow in the screen. A sensing circuit then detects the changes and locates the touch position. A capacitive touch sensor is based on the capacitive coupling effect. A typical design involves coating the screen with a thin, transparent metallic layer, in order to form a collection of capacitors on the surface. Thus, when a user touches the surface, the disturbance caused by the finger changes the capacitance and current that flows on the display.

A significant limitation for most of these technologies is that they are only capable of detecting a single touch. Multi-touch techniques allow touch screens to recognize touches of multiple fingers or inputs of multiple persons simultaneously. The multi-touch detection mechanisms can be classified into three categories: sensor array, capacitive sensing, and vision and optical-based ones. A sensor array touch surface consists of a grid of touch sensors that work independently. When a user exerts multiple touches on the surface, the system can identify activated sensors and determine these touch positions simultaneously. An example, originally proposed by Lee *et al.* (1985), is the Fast Multiple-Touch-Sensitive Input Device (FMTSID) – one of the first multi-point touch sensor-based devices. The system consists of a sensor matrix panel, rows of sensors, A/D converter and a control CPU. The design of the sensor matrix is based on the technique of capacitance measurement between a fingertip and a metal plate. A capacitive touch method uses the capacitive coupling between two conductors to sense a touch. Typically, the touch surface contains a mesh of horizontal and vertical antennas, which function as either a transmitter or a receiver of electric signals. Examples based on this technology include Rekimoto's SmartSkin (Rekimoto *et al.*, 2002) and DiamondTouch developed by Dietz and Leigh (2001).

In addition to touch screens, touch sensors with force-sensing abilities have been studied as tactile sensors for many years (Dahiya *et al.*, 2010). Such sensors have found their applications in artificial skin for robot applications (Engel *et al.*, 2003), minimally invasive surgery (Eltaib and Hewitt, 2003), wearable computers (Hoshi and Shimoda, 2006), and mobile or desktop haptic devices (Ruspini *et al.*, 1997). Up-to-date tactile sensors have mainly focused on silicon-based sensors that use piezoresistive (Beebe *et al.*, 1995; Wolffenbuttel and Regtien, 1991) or capacitive sensing mechanism (Chu *et al.*, 1995; Gray and Fearing, 1996; Leineweber *et al.*, 2000). The silicon tactile sensors have limitations of mechanical brittleness, and hence are not capable of sustaining large deformations. Polymer-based tactile sensing approaches that use piezoelectric polymer films (Dario and De Rossi, 1985; Kolesar *et al.*, 1996; Yuji *et al.*, 2006), pressure-conductive rubber (Shimojo *et al.*, 2004), carbon fiber-based polymer composite (Park *et al.*, 2009) and conductive polymers (Kim *et al.*, 2009) have also been reported. These sensors provide good spatial resolution but the applied force range is low due to the limited thickness of a membrane.

To integrate tactile sensors into the smart-textiles applications we could, for example, use arrays of conventional rigid electric devices embedded into a textile matrix, as was done in early prototypes of 'smart apparel'. However, this approach does not provide acceptable wearing comfort or good mechanical properties. This motivated recent efforts to develop truly wearable smart textiles. So far, capacitive sensing is probably the most promising technique for the textile-based sensors, as it does not depend on the applied mechanical force (including bends and stretching) and also enables multi-touch and gesture recognition functionality. In the next section, we will discuss novel all-polymer soft capacitor fibers recently developed by our research group and their applications in tactile sensing.

6.2 Soft capacitor fibers for electronic textiles

Recently there have been several reports on capacitor fibers compatible with a textile weaving process. By adding an external inductance, such fibers make a resonant LC circuit, thus allowing the use of many highly sensitive resonant detection techniques, which are able to detect small changes in the capacitor structure. Among several proposals for a capacitor fiber, we note a multicore fiber capacitor in the form of a bundle of approximately 50 μm-sized coaxial cables connected in parallel at a micro-level (Cheng and Hong, 2001). However, such capacitors were never actually fabricated. Load-bearing composite textiles comprising a large number of simple coaxial cables have been recently reported by the Air Force Research Laboratories (Baron *et al.*, 2006) for distributed storage of electrical energy directly in the fuselage of an airplane for pulsed weapons applications. The invention envisions reduction in the aircraft payload by combining mechanical and electric functionalities in the same fiber. Finally, the group of Baughman (Dalton *et al.*, 2003) proposed carbon nanotube-based textile threads that can be adapted to build super-capacitor textiles after soaking such threads in electrolyte.

Another novel type of electronic fiber was recently developed in our laboratory – high capacitance, soft fiber from conductive polymer composites. One key advantage of our fibers is that they do not require the use of electrolytes for their operation, which is especially desirable for wearable applications. Another key advantage is that the fibers can be made fully polymeric (no metallic electrodes) and very soft for applications in wearable sensing. Because of the relatively high capacitance of the fiber (60–100 nF/m), it can also be used for energy storage applications. In terms of capacitance, our fibers take an intermediate position between the coaxial cables and super-capacitors. Thus, capacitance of a coaxial cable with comparable parameters is typically 1000 times smaller than that of our fibers.

6.2.1 Fiber materials and fiber fabrication

Soft capacitor fibers are fabricated by the fiber drawing technique, which consists of three steps. The first step involves rolling or stacking conductive and dielectric films into a multilayer preform structure. During the second step, the preform is consolidated by heating it to temperatures somewhat above the polymer glass transition temperature (T_g). Consolidation is needed as the diffusion of the polymer molecules through interfaces of adjacent layers fuses these layers together, thus reducing the effect of property mismatch of different materials in the drawing process. Finally, at the third step, consolidated preform is drawn into fibers using a fiber drawing tower. Drawing is typically performed at temperatures higher than the polymer T_g. During successful drawing, the resultant fibers generally preserve the structured profile of a preform, thus fibers with complex microstructures can be fabricated via homologous reduction (during drawing) of a macrostructure of the preform. The described drawing technique is directly analogous to the one used in manufacturing microstructured polymer optical fibers (Gao *et al.*, 2006).

Flexible multilayer capacitors discussed in this chapter generally consist of two conducting polymer layers serving as two electrodes of a capacitor, and two isolating polymer separator layers. To draw successfully, the preform materials should be compatible with each other in terms of their rheological and thermo-mechanical properties. In our first tests, we have attempted drawing a thin continuous layer of low-melting temperature metal sandwiched between two isolating polymer layers. As a metal we have used Bi_{58}/Sn_{42} alloy with a melting point of 138°C. Various polymers were tested as isolating layers. However, we found that in all cases it was difficult to preserve the laminated structure with a thin metal foil during the drawing process. Particularly, when melted, metal foil would break into wires during drawing, thus destroying the continuous electrode structure. We have rationalized this observation by noting that the visco-elasticity, ductility and interfacial tension of alloy and the surrounding polymer cannot match well at drawing temperatures. For example, at the temperature for the polymer to be drawn into fibers, the viscosity of melted metal becomes very low, thus it is easy for a thin sheet of melted metal to develop flow instability and form several larger wires to minimize surface energy associated with a polymer/metal interface.

Another potential problem during drawing of metal sheets is that when the polymer surrounding the molten metal becomes too soft, it can no longer hold the melt; as a consequence, a large drop of metal would form at the preform end, even before drawing starts, thus draining the rest of the preform from the metal. From these initial experiments we have concluded that drawing of a thin metallic sheet sandwiched between two plastic sheets is in general problematic due to a strong mismatch of the material properties during the drawing process.

After realizing the challenge of drawing metallic electrodes in the form of thin sheets, a natural option to remedy this problem was to substitute metals with thermoplastic conductive polymers as electrodes. Unfortunately, thermoplastic intrinsic conductive polymers suitable for drawing are not available commercially. The only thermoplastic conductive polymers that are currently available commercially are either carbon black (CB) filled or most recently carbon nanotube filled films. In our research we have mostly used polyethylene(PE)-based CB-filled films (BPQ series) provided by Bystat International Inc. The film, with a thickness of 91 μm, has a surface resistivity of 17 KΩ/sq. The measured volume resistivity is around 2.2 Ωm along the surface. To find the isolating materials that can be co-drawn with this conductive film, we have tried various polymer films, such as polyvinylidene fluoride (PVDF), polycarbonate (PC), polyethylene terephthalate (PET), polymethyl methacrylate (PMMA), and others. Among all the attempted polymers, we have found that the two best materials for the isolating layer are low density polyethylene (LDPE) film or a PC film.

6.2.2 Capacitor fiber designs

We have developed three distinct fiber capacitor geometries that will be discussed in this chapter. They all share the same concept, but each of them possesses its own features. The first fiber type has a cylindrical geometry with two plastic electrodes in the form of a spiraling multilayer (Fig. 6.1(a)). The central part of a fiber was either left empty with the inner plastic electrode lining up with the hollow core, a metallic electrode was introduced into the hollow core during drawing, or the core was collapsed completely thus forming a plastic central electrode. In all these fibers, the second electrode was wrapped around the fiber. The second fiber type (Fig. 6.1(b)) is also of cylindrical multilayer geometry; however, it features two hollow cores lined with two plastic electrodes. The fiber is wrapped into an isolating material, so there is no direct contact with the environment. During drawing, two metallic electrodes were introduced into the

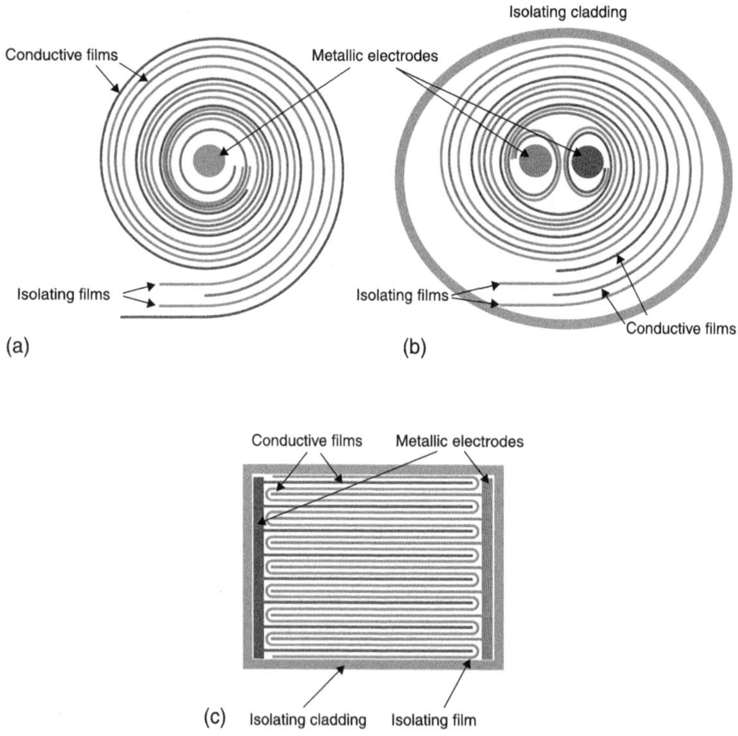

6.1 (a) Schematic of a cylindrical capacitor fiber preform featuring a spiraling multilayer comprising two conductive and two isolating films. (b) Schematic of a cylindrical capacitor fiber with two electrodes in the center. (c) Schematic of a rectangular preform prepared by encapsulating a zigzagging stack of two conductive and an isolating layer inside a rectangular PMMA tube.

fiber cores. Finally, the third fiber type features a square electrically isolating tube comprising a zigzagging stack of plastic electrodes (Fig. 6.1(c)) separated by a zigzagging dielectric layer. The metallic electrodes were integrated on the left and right sides of a tube for ease of connection.

Fibers of the first type (Fig. 6.1(a)) were fabricated by co-rolling the two conductive polymer films, which were physically and electrically separated with the two isolating LDPE films. In the resultant fiber, the inner conductive film forms one electrode inside the hollow fiber core, while another electrode is created by the other conductive film at the fiber surface. A similar fabrication strategy was used in the second fiber type (Fig. 6.1(b)), with the only exception being the positioning of two isolated fiber cores in the fiber center, while encapsulating the fiber into an isolating HDPE plastic wrap. Finally, fibers of the third type (Fig. 6.1(c)) were created by encapsulating a zigzagging stack of the electrodes and isolating layers inside a rectangular PMMA tube.

6.2.3 Capacitor fiber connection and potential applications

An important issue, when designing any smart fiber, concerns connection of such fibers either to each other or with the external electrical probes. In view of various potential applications of a capacitor fiber, we have explored several connection geometries. In Figs. 6.2(a–e), we show four complete designs for a capacitor fiber. In each figure we present both the structure of a preform before drawing, as well as the structure of a resultant fiber.

Design one is presented in Figs. 6.2(a) and (b), where we show a circular hollow core fiber with the first electrode formed by the conductive layer lining the hollow fiber core (Fig. 6.1(a)), and the second electrode formed by the other conductive layer wrapping the fiber from outside. The outside electrode is exposed for ease of access. The hollow core can be either collapsed (Fig. 6.2(b)) or left open during drawing (Fig. 6.2(a)). In general, to access the electrode inside the fiber core, we have to use a needle-like electrical probe; in fact, we have used 50 to 100 μm diameter hypodermic needles to perform electrical characterization of this fiber. One of the advantages of the hollow core fibers is that they are very soft due to lack of metallic components in their structure and, therefore, are most suitable for integration into wearable textiles. Moreover, the hollow fiber core can be filled with functional liquids, which will be in direct contact with one of the electrodes. This can be useful for various sensing applications, where physical or chemical properties of a liquid could be interrogated electrically.

Design two is presented in Fig. 6.2(c), where we show a circular hollow core fiber with one of the electrodes formed by a small 100 μm diameter copper wire, which is integrated into the fiber core directly during drawing. With a tension-adjustable reel installed on the top of a preform, the copper wire can be passed through the preform core, pulled down and embedded into the fiber center during drawing by collapsing

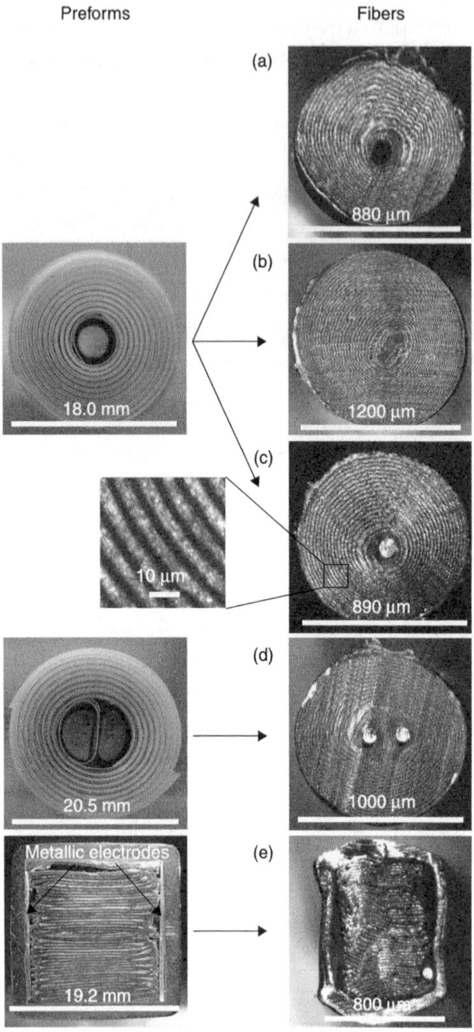

6.2 Design I: hollow core fiber with the first electrode lining the inside of a hollow core, and the second plastic electrode wrapping the fiber from outside. During drawing, fiber hollow core can be left open (a) or collapsed (b), depending on the application requirement. Design II: Hollow core fiber can be drawn with a metallic electrode in the center. Such an electrode can be a copper wire (c) in contact with the plastic electrode lining the hollow core. Design III: fiber containing two hollow cores. The cores are lined with two plastic electrodes electrically separated from each other. Fiber is drawn with two copper wires threaded through the hollow cores in the preform (d). Design IV: square fiber capacitor. Fiber features a zigzagging stack of two plastic electrodes separated with an electrically isolating PC layer (e). Also two metallic electrodes are placed in contact with plastic electrodes, and the whole multilayer is encapsulated inside a square PMMA tube.

plastic cladding around it. The second electrode is formed by the other conductive layer wrapping the fiber from outside, similar to the first design. The main advantage of this design is the ease of connection to the inner electrode, as the plastic capacitor multilayer can be easily stripped from the copper wire. This fiber has lower effective resistivity compared to the hollow core fiber, as one of the electrodes is made of a highly conductive metal. Despite the copper electrode in its structure, the fiber is still highly flexible. As the outside electrode is exposed, this fiber can be used for the detection of electromagnetic influence or as a proximity sensor.

Design three is presented in Fig. 6.2(d), where we show a circular fiber containing two hollow cores positioned in the middle of the fiber. Each core is lined with distinct conductive layers, which are forming electrodes one and two. The cores with electrodes are electrically isolated from each other. Moreover, the whole preform is then wrapped into several layers of pure LDPE plastic to isolate the capacitor layers from the environment. The preform is then drawn with two copper wires threaded through its holes. The resultant fiber features two copper electrodes and a fully encapsulated capacitor multilayer. Such fibers can be interesting for energy storage applications, due to ease of connection and electrical isolation from the environment.

Finally, design four is presented in Fig. 6.2(e), where we show a thin PMMA tube of square cross section, comprising a zigzagging multilayer of the two conductive layers separated by a single electrically-isolating PC layer. The first plastic electrode is located to the left and the second plastic electrode is located to the right of the isolating PC layer. At the left and right inner sides of the square tube, we place foils of Bi_{58}/Sn_{42} alloy in contact with the plastic conductive layers. During fiber drawing, wire-like metallic electrodes are created from the foils.

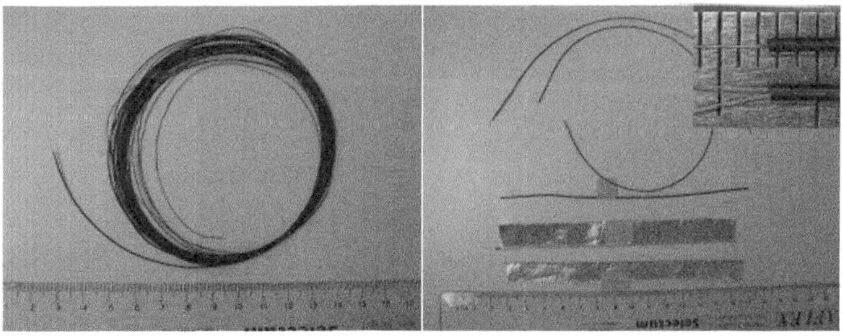

(a) (b)

6.3 (a) Capacitor fiber fabricated from the preform shown in Fig. 6.2(c). The fiber features a central 100 μm-thick copper wire, as well as an exposed conductive plastic electrode on the fiber surface. (b) To perform electrical characterization of the fibers, embedded copper wire is used as the first electrical probe, while the second electrical probe is an aluminium foil wrapped around the fiber conductive surface. The inset is an enlarged view of the fibers with single and double copper wire electrodes.

Finally, the structure of the resultant fiber is similar to the one of an encapsulated fiber with two copper electrodes.

In comparison with standard capacitors, we notice that a 10 nF ceramic capacitor measures about 600 × 300 μm and 10 μF components measure 2.0 × 1.25 mm. The fiber capacitor does not possess advantages over the standard capacitors in terms of size, but the flexibility and softness it features are essential for applications in wearable smart textiles. Encapsulation of RC series in a single fiber makes the circuit in wearable e-textiles more compact and reliable, because it may reduce the number of connection joints. Although the equivalent resistance of the capacitor is very high for a short fiber, which is limited by the properties of available conductive films, it can be reduced simply by increasing the length of the fiber, as demonstrated in the following.

6.3 Electrical characterization of the isolated capacitor fiber

To characterize electrical properties of our capacitor fibers, we used the measurement circuit presented in Fig. 6.4, where the fiber capacitor is connected to a function generator (GFG-8216A, Good Will Instrument Co., Ltd) through the reference resistor R_{ref} = 480 kΩ. The function generator provides a sinusoidal signal of tunable frequency ω = (0.3 Hz to 3 MHz). An oscilloscope (GDS-1022, Good Will Instrument Co., Ltd) measures the input voltage V_{Ch1} (ω) on channel 1 and the output voltage over the reference resistor V_{Ch2} (ω) on channel 2. A 10X probe (GTP-060A-4, Good Will Instrument Co., Ltd) was used to acquire the experimental data. The voltage produced by the function generator is fixed and in the whole frequency range of interest equals to V_{Ch1} = 2 V. In our experiments, we

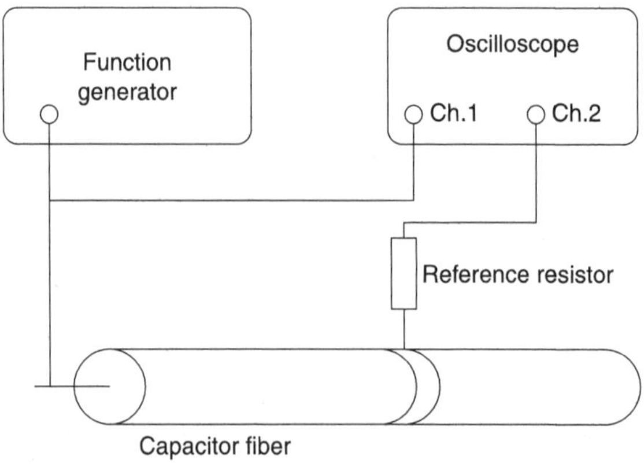

6.4 Schematic of a measurement set-up.

measured both the amplitudes and the phase difference between channels 1 and 2. Due to high resistivity of our fibers, and also to fit the experimental data at higher frequencies ($\omega > 1$ kHz), we have to take into account the effective impedances of an oscilloscope.

In this section, we present the properties of fiber capacitance and resistance as a function of various fiber geometrical parameters. Most of the presented measurements were performed on fibers featuring a single copper electrode in their cores, while the second electrode was formed by the plastic conductive layer on the fiber surface (Fig. 6.2(c)). The fiber was co-drawn with a 100 μm-thick copper wire in its center. In the preform, both conductive layers are 75 μm thick, while the two insulating layers are made of 86 μm thick LDPE films. To characterize capacitance fibers, we used embedded copper wire as the first electrical probe, while the second electrical probe was made by wrapping aluminum foil around a part or the whole of the fiber (Fig. 6.3).

6.3.1 RC ladder network model of a soft capacitor fiber fully covered with a foil probe

The high resistivity of conductive composite films endows the capacitor fiber with a distributed response. In particular, the fiber electrical properties can be well-described by an RC ladder circuit. First, we consider the case when the fiber outer electrode is fully covered with a highly conductive foil probe. In this case, the problem becomes two-dimensional (2-D) (no longitudinal currents) and the RC ladder circuit presented in Fig. 6.5 describes transverse currents in the capacitor cross section.

In Fig. 6.5, a schematic of the ladder model is presented, where R_t corresponds to the transverse resistance of a single conductive film spiraling from the fiber

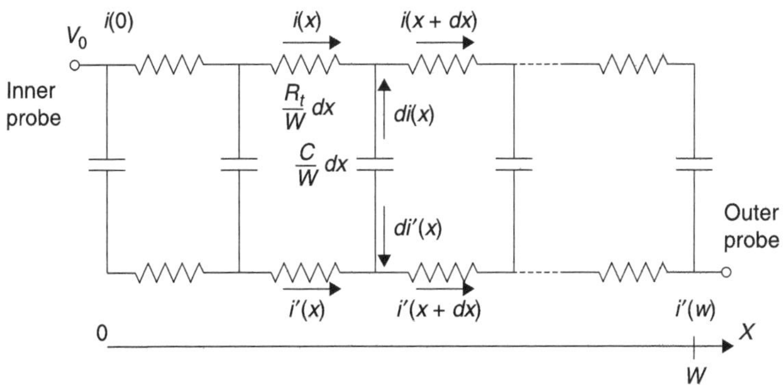

6.5 Ladder network model of the capacitor fiber fully covered with a foil probe.

core towards its surface. The value of the transverse resistance can then be approximated as:

$$R_t \approx \rho_v \frac{W}{Ld_c} \quad [6.1]$$

where ρ_v is the volume resistivity of the conductive films, L is the length of the fiber, and W and d_c denote, respectively, the width and thickness of the conductive electrodes wrapped in the fiber cross section (Fig. 6.2). To measure transverse resistance, we have to ensure that there are no longitudinal (along the fiber length) currents in the fiber. In practice, to deduce transverse resistance, we cover the outer fiber electrode (high resistance electrode) with a metallic foil, and then measure fiber AC response by applying the voltage between the inner copper electrode and the outer metallic foil.

For the longitudinal currents, fiber resistance will be:

$$R_l \approx \rho_v \frac{L}{Wd_c} \quad [6.2]$$

which for longer samples ($L > W$) is much higher than the transverse resistance. To measure the longitudinal resistance, we have to ensure that there are no transverse (perpendicular to the fiber length) currents in the fiber. In practice, it is difficult to measure the longitudinal resistance directly. In principle, if the electrode length (fiber length) is much longer than the net width of a conductive electrode wrapped in the fiber cross section, longitudinal resistance can be deduced from the AC measurement, where the high resistance outer electrode of a fiber is grounded at one end, while the inner high resistance electrode of the fiber is connected to a voltage supply at the other. Note that the low resistance copper electrode has to be removed from the fiber for this measurement and DC measurements are not possible, as in the described configuration, the fiber acts as a capacitor. Clearly, when connecting to a fiber using a continuous probe (i.e. foil) along the whole fiber length, fiber resistance will be dominated by its transverse component.

Another fundamental parameter that determines fiber performance is the fiber capacitance, denoted as C in Fig. 6.5. As thicknesses of the dielectric and conductive layers in the fiber are hundreds of times smaller than the fiber diameter, fiber capacitance can be approximated using an expression for the equivalent parallel-plate capacitor:

$$C \approx 2\varepsilon_0 \varepsilon \frac{WL}{d_i} \quad [6.3]$$

where ε is dielectric constant of the isolating films, ε_0 is permeability of the vacuum, and d_i is thickness of the rolled isolating films.

As shown in Fig. 6.5, $i(x)$ and $i'(x)$ denote the current flowing in the conductive film connected to the inner probe and outer probe, respectively. V_0 is the voltage difference between the inner probe and outer probe. We assume that the resistivity of the conductive film is a position-independent and frequency-independent parameter. Appling Kirchhoff's Voltage Law (KVL) and Kirchhoff's Current Law (KCL) to the ladder circuit, leads to the following equations:

$$\int_0^x \frac{R_t}{W} i(l)\,dl + \frac{W}{j\omega C}\frac{di'(x)}{dx} + \int_x^W \frac{R_t}{W} i'(l)\,dl = V_0 \qquad [6.4]$$

and

$$di(x) = -di'(x) \qquad [6.5]$$

with boundary conditions:

$$i(0) = i'(W) \text{ and } i'(W) = 0. \qquad [6.6]$$

Equations 6.4 to 6.6 can be solved analytically, and yield the following expressions for the effective transversal capacitance and effective transversal series resistance:

$$C_F(\omega) = -\frac{1}{\omega R_t \cdot Im\big(f(B)\big)} \qquad [6.7]$$

$$R_F(\omega) = \frac{R_t}{2} + R_t \cdot Re\big(f(B)\big) \qquad [6.8]$$

where

$$f(B) = \frac{1+\cosh(B)}{B\cdot\sinh(B)}; \quad B = \sqrt{2j\omega R_t C}. \qquad [6.9]$$

Note at low frequencies, such as $B \to 0$, Eqs. 6.7 and 6.8 reduce to the frequency-independent values as:

$$C_F = C, R_F = \frac{2}{3}R_t. \qquad [6.10]$$

6.3.2 Frequency dependent response of the capacitor fibers

As predicted by Eqs. 6.7 and 6.8, effective capacitance and effective resistance of the capacitor fibers are dependent on the operational frequency, with limiting values at low frequencies given by Eq. 6.10. In this section, we present results of experimental studies of the frequency dependent response of the capacitor fibers. To interpret correctly the measured response $V_{Ch2}(\omega)$ of the electric circuit, it is important to consider the complex impedance of the oscilloscope and the electric probe used in the characterization. This is mainly due to the relatively large resistance of our fibers. We must also be aware that conductive films used in the fiber fabrication can show significant frequency-dependence of their electrical properties. For example, it has been reported (Nakamura and Sawa, 1998) that near the percolation threshold, the resistivity of CB/polymer films decreases with increasing frequency. To find the effective circuit parameters of an oscilloscope, we first measured the response of a known resistor having a similar resistance to that of a fiber (~477 kΩ). We noted that complex impedance of the measuring circuit (oscilloscope) becomes important only at frequencies higher than 100 kHz. We then studied the frequency response of the conductive film, and found that its resistivity is frequency independent below 300 kHz. Therefore, in all the experiments that followed, we have operating frequencies lower than 100 kHz.

In Fig. 6.6, we present experimentally measured frequency dependent fiber resistance and capacitance of a fiber fully covered with a foil probe and with diameter of 0.93 mm and length of 137 mm. At low frequencies, both C_F and R_F are virtually constants, while they decrease at frequencies higher than 1 kHz. This behavior is similar to that of a standard electrolytic capacitor and can be well-explained by the RC ladder network model with a characteristic response frequency of $1/(R_tC)$~4 kHz. We can also see that Eqs. 6.7 and 6.8 provide good predictions of the experimental data, by assuming $C = 9.4$ nF and $R_t = 26$ kΩ in the model.

6.3.3 Effect of the capacitor fiber length

In order to study dependence of the fiber properties as a function of the fiber length, we have used two fiber samples of different lengths that were drawn from

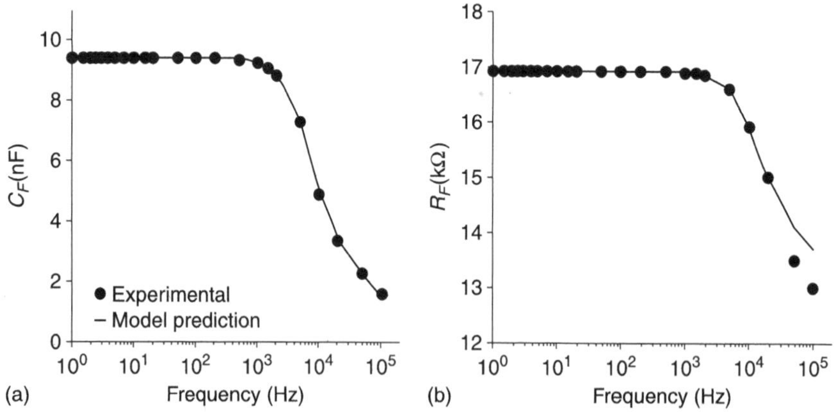

6.6 Comparison of experimental data and model predictions of frequency responses of a fiber capacitor. (a) Effective capacitance versus frequency. (b) Effective resistance versus frequency.

the same preform. Sample #1 and sample #2 had outer diameters of 920 to 980 μm and 720 to 760 μm, respectively. Both samples were drawn from the same preform at speeds around 100 mm/min at 180°C. The two samples were then cut into sections of different lengths ranging between 10 and 60 cm, and then wrapped with aluminum foil with 100% coverage ratio. The experiments were conducted at low frequencies (ω <1 kHz), so that effective capacitance and resistance can be considered as constant. In Fig. 6.7a, we present measured fiber capacitance as a function of fiber length and observe a clear linear dependence. From this data, we see that for all the fibers the capacitance per unit length is around 69 nF/m (inset in Fig. 6.7(a)), which is very close to the value of 69.5 nF/m measured for the capacitance of the fiber preform. This is easy to rationalize from Eqs. 6.3 and 6.10. As W/d_i is constant during drawing (because of the largely homologous drawing), hence C_F/L should be the same for any fiber produced from the same preform, regardless of the fiber size. The reason why our fibers can obtain large capacitance is because the value of W/d_i is much larger than that of a coaxial cable with one capacitive layer. In contrast, fiber resistance decreases inversely proportional to the fiber length. In fact, it is rather the product $R_F{\cdot}L$ that is approximately constant, as shown in the inset of Fig. 6.7(b). Equations 6.1 and 6.10 indicate that if ρ_v is constant, $R_F{\cdot}L$ should also be a constant, because W/d_c is the same for fibers drawn from the same preform. However, we also find that the thinner fiber shows a larger value of $R_F{\cdot}L$. This diameter dependency is implied in the volume resistivity ρ_v in Eq. 6.1. It is reported that the resistivity of CB/polymer

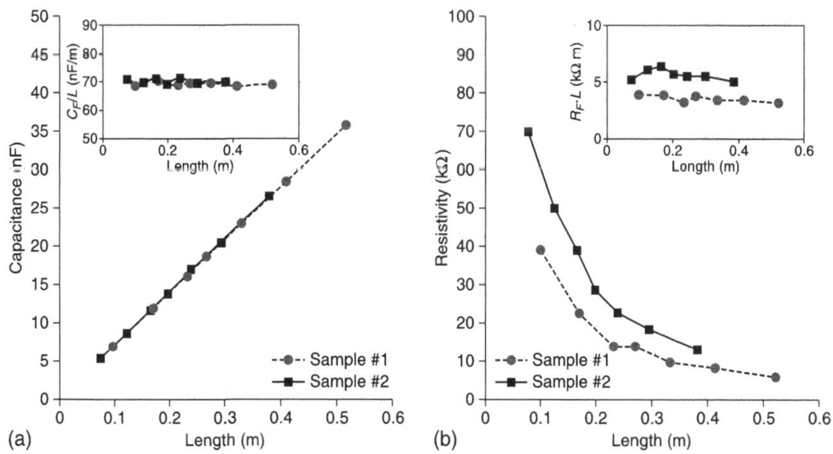

6.7 Dependence of: (a) fiber capacitance and (b) fiber resistivity on fiber length. Two data sets denoted by square and round points correspond to the two fiber samples of different diameters drawn from the same preform. Insets: dependence of the (a) fiber capacitance per unit length C_F/L and (b) fiber resistivity factor $R_F{\cdot}L$ on the fiber length.

composites increase as the material is stretched, and the value is proportional to the elongation ratio in logarithm scale (Schulte *et al.*, 1988; Feng and Chan, 2003). Thus, we observed that the thinner fiber has a larger $R_F \cdot L$ value.

6.3.4 Effect of the operation temperature on the fiber's electrical properties

The effect of the temperature of operation on the electrical properties of a capacitor fiber is presented in Fig. 6.8. Sample #1 was 135 mm long and had a diameter of 840 μm, while sample #2 was 133 mm long and had a diameter of 930 μm. To control the temperature of the two samples, they were fixed on top of a hot plate. Our measurements at low frequencies show that fiber capacitance per unit length remains almost independent of the temperature of operation, while fiber resistivity increases rapidly as the temperature rises. This result is in good correspondence with the recent reports on positive temperature coefficient (Tang *et al.*, 1997; Yu *et al.*, 1998) for the resistivity of the composites of CB and LDPE in the 0 to 100°C temperature range. The effect of thermal expansion and a consequent increase of the average distance between CB particles are thought to be the main reasons for the positive temperature coefficient of the conductive polymer composites. This interesting property promises various applications of capacitor fibers in self-controlled or self-limiting textiles responsive to temperature or heat.

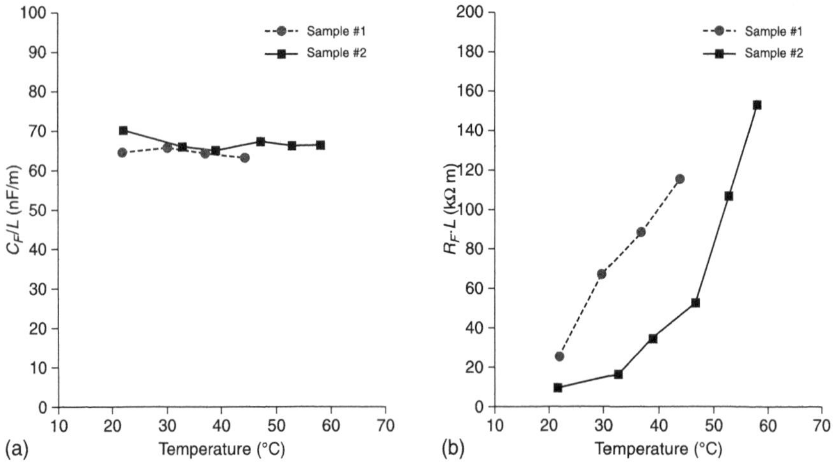

6.8 Effect of the temperature of operation on electrical properties of a capacitor fiber. (a) Capacitance per unit of length C_F/L. (b) Resistivity factor $R_F \cdot L$. Sample #1 has a diameter of 840 μm and a length of 135 mm. Sample #2 has a diameter of 930 μm and a length of 136 mm.

6.3.5 Effect of the fiber drawing parameters

Electrical performance of the capacitor fibers is equally affected by the fiber geometrical parameters and by the fiber material parameters. In this section, we show that fiber fabrication parameters, such as fiber drawing temperature and fiber drawing speed, can have a significant effect on the fiber resistivity, while also somewhat affecting the fiber capacitance. Generally, capacitor fibers presented in this work can be drawn at temperatures in the range of 170 to 185°C, with drawing speeds ranging from 100 to 300 mm/min.

Generally, we find that the fiber capacitance C_F/L is largely independent of the fiber diameter and drawing parameters, and equals to that measured directly in the fiber preform. In contrast, fiber resistivity parameter $R_F \cdot L$ is significantly affected by the drawing parameters. From Eqs. 6.1, 6.3 and 6.10, we can relate fiber capacitance per unit length and fiber resistivity parameters as:

$$\frac{R_F L}{C_F/L} \propto \frac{\rho_v}{\varepsilon_0 \varepsilon} \qquad [6.11]$$

While fiber capacitance is indeed almost independent of the fiber geometrical and processing parameters, fiber resistivity is strongly influenced by them. This can be rationalized by concluding that bulk resistivity of the CB polymer composite can change significantly during the drawing procedure from its original value in the fiber preform. To further validate our observation that drawing at lower temperatures results in higher resistivities, we performed a set of stretching experiments at room temperature on the planar conductive films. It was found that the resistivity of the conductive film increases as much as by two orders of magnitude from its original value when the films were stretched unheated to about twice their length. Similar observation has been reported in the literature for CB-filled polymers and polymer composites (Schulte *et al.*, 1988; Feng and Chan, 2003).

Conclusions of these experiments can be rationalized as follows. During the drawing process, the conductive polymer composite in the preform undergoes various processing stages including heating, melting, stretching, annealing and cooling. The stretching increases the distance between the individual CB particles in the stretching direction, thus disrupting the conductive network. During the annealing above glass transition temperature, CB particles aggregate together through Brownian motion and form a continuous network. The destruction and reconstruction of the conductive network is highly dependent on the concentration and properties of CB and processing parameters such as mixing strength, temperature and time and the annealing temperature and time, etc. (Wu *et al.*, 2000; Zhang *et al.*, 2007). If the fiber is drawn at a lower temperature and higher speed, the stronger viscous stress of the polymer matrix may disintegrate CB particles and decrease their aspect ratio, thus making the conductive network more difficult to form. The annealing conditions have a direct influence on the formation of the conductive network (Wu *et al.* 2000). The rising of the annealing temperature decreases the viscosity of the matrix polymer, thus facilitating the movement and aggregation of the CB particles. This, in turn, reduces the time needed for the CB particles to build up a conductive network. It is well-reported that the resistivity of polymer composites filled with CB (Wu *et al.*, 2000; Cao *et al.*, 2009) or carbon nanotube (Alig *et al.*, 2008), decreases as the annealing temperature and time increase. This explains our experimental observation about lower drawing speeds resulting in lower fiber resistivities; this is because a lower drawing speed leads to a longer annealing time in the furnace. In addition, a lower drawing speed corresponds to a weaker stretching and slower deformation. This not only avoids CB particles being disintegrated, but also provides them with more time to create a conductive network before cooling down to room temperature.

6.4 Capacitor fiber as a one-dimensional distributed touch sensor

In this section, we study properties of the individual soft capacitor fiber as a one-dimensional (1-D) tactile sensor integrated into the textile base. To ease the integration and connection of the fibers, a small-diameter (typically 50–150 μm) copper wire was embedded into the fiber during the drawing process by passing it inside of the fiber preform (Fig. 6.9) central hole and letting the preform collapse around the wire in the neckdown region. For the reader unfamiliar with fiber drawing procedure, we would like to explain that the neckdown region is a 'transitional' region between the preform and the resulting fiber, where the preform converges to the size of the resulting fiber. Introduction of a copper wire into the fiber drawing process actually improves the fiber drawability. In this configuration, copper wire in the fiber center serves as a low resistivity electrode attached to one of the two high resistivity conductive plastic electrodes of a fiber capacitor. The other high resistivity conductive plastic electrode of a fiber is brought to the fiber surface to ease connection. Typical fibers fabricated using the above-mentioned fabrication procedure have a sub-millimeter diameter, contain at least 30 layers, and have a typical capacitance of 100 nF/m, which is almost three orders of magnitude higher than that of a standard coaxial cable of comparable diameter.

High resistivity of the conductive polymer electrodes together with high capacitance endows the fiber with interesting electrical properties, which as we will see below, makes the fiber well suited for distributed sensing applications of touch. In a typical implementation of a touch sensor, we ground the outer fiber electrode (high Ω electrode in Fig. 6.9) on one end, while applying the AC voltage to the copper wire (low Ω electrode in Fig. 6.9) on the other end (Fig. 6.10). We then read out the voltage on the fiber outer electrode (high Ω electrode) at the same end where the copper wire is connected. In our experiments, we found that the read-out voltage is highly sensitive to the position of the touch along the fiber.

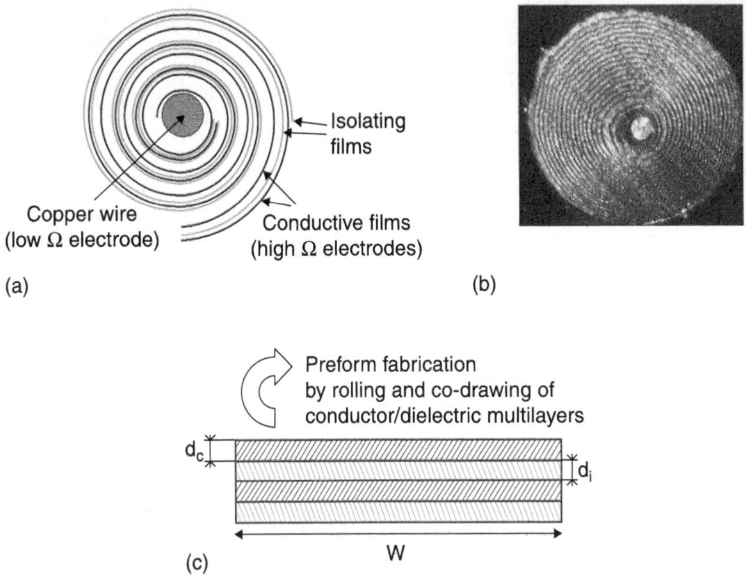

6.9 (a) Schematic of a capacitor fiber featuring a spiraling multilayer comprising two conductive and two isolating films. Black curves represent two conductive films, while grey curves represent isolating films. (b) Photo of the cross-section of a drawn capacitor fiber with a cooper wire embedded in the center. (c) Fiber preform is made by co-rolling four polymer layers (2 conductive and 2 isolating) into a Swiss roll structure. The number of resultant layers in a spiral structure is proportional to the width of the unrolled multilayer. In the drawn fiber, thicknesses of the conductive d_c and isolating d_i layers can be smaller than 10 μm, while the unwrapped width W of the layers fitting into a 1 mm diameter fiber can be in excess of 3 cm.

In the following we describe basic principles behind touch sensing using a single capacitor fiber and then consider a theoretical model that describes capacitor fiber response to the touch.

6.4.1 Electrical response of a single capacitor fiber

To characterize a single fiber, we use the general electrical layout outlined earlier. In particular, a sinusoidal signal of voltage amplitude V_0 and constant frequency (10 Hz–10 kHz) is provided by an external function generator and is applied to the copper wire electrode of a fiber (Fig. 6.10). Voltage response is acquired using a sliding contact on the outer electrode of a fiber (high Ω electrode). The sliding contact is connected to the oscilloscope as before through a 10X probe. This arrangement allows measuring the voltage distribution along the fiber length, both when the fiber is touched or not. As a human body is largely composed of conductive electrolytes covered with a layer of dielectric skin, the human body

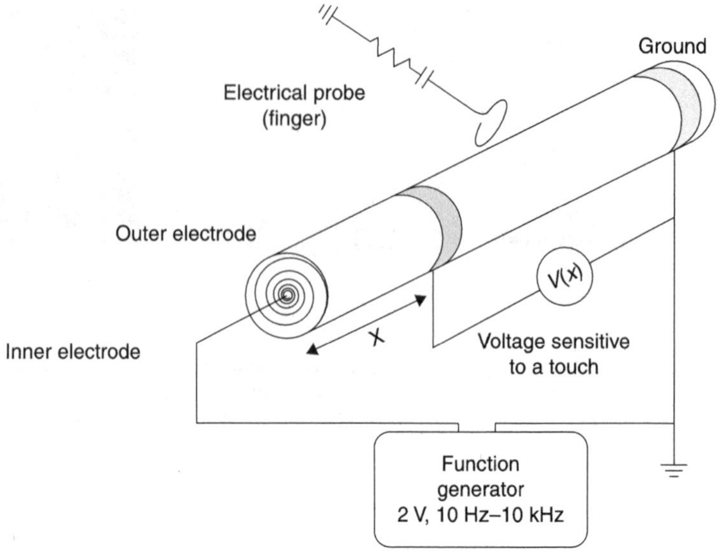

6.10 Schematic of a one-dimensional slide sensor based on a single capacitor fiber.

can be approximated with an equivalent electrical circuit comprising a resistor connected in series to a capacitor. Typical values of an effective resistance and capacitance are 1.44 kΩ and 150 pF, respectively. In our experimental studies of a single fiber response to touch, we use an equivalent human probe with these effective electrical parameters to guarantee repeatability between measurements and to simplify theoretical interpretation of the acquired data.

In Fig. 6.11, we present voltage distribution along the fiber length for an isolated fiber (Fig. 6.11(a)), and for the fiber that is touched with an equivalent human probe (Fig. 6.11(b)). Here x/L represents the normalized position of a sliding contact along the fiber length. The fiber length used in our experiments was L = 12.3 cm. $|V(x)/V_0|$ represents the voltage measured by the sliding contact along the outer electrode of a capacitor fiber. When using an equivalent human probe (Fig. 6.11(b)), we fix it at position x/L = 0.3. Different data sets in Fig. 6.11 correspond to the four different frequencies of a function generator used in our measurements (10 Hz, 100 Hz, 1 kHz, 10 kHz).

In the DC case (not shown in the figures), voltage along the outer electrode of a capacitor is constant and zero, as the outer high-resistive electrode provides a DC-connection to ground. For an isolated fiber (Fig. 6.11(a)) in the AC case, voltage distribution along the fiber length shows the same trend at all frequencies. In particular, it saturates exponentially fast from zero at the grounded end to some maximal value at the other end of the fiber.

From Fig. 6.11(b), it is clear that when operating at higher driving frequencies, the touch position can be determined from the dip in the voltage distribution along

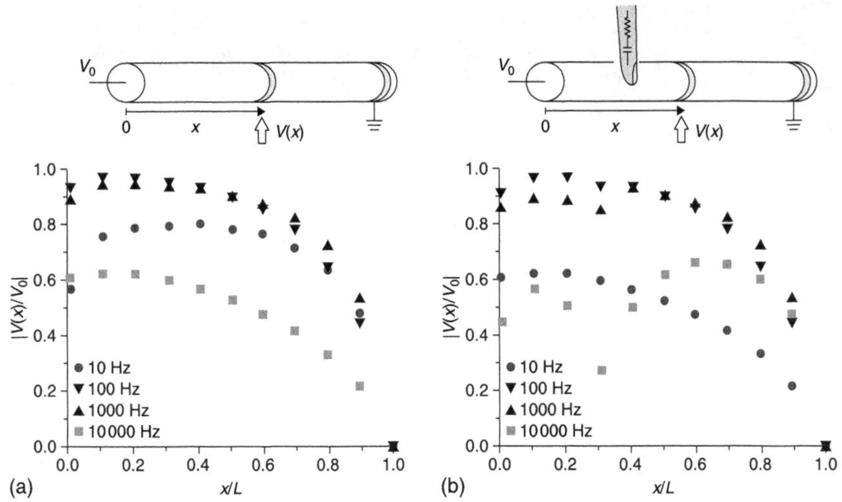

6.11 Voltage distribution along the outer fiber electrode for: (a) an isolated fiber; and (b) a fiber touched with an equivalent human probe. Four data sets correspond to the different driving frequencies of 10 Hz, 100 Hz, 1 kHz and 10 kHz. Voltage distribution along the fiber touched with a probe shows a dip in the vicinity of a touching position.

the fiber outer electrode. For practical implementation of a slide sensor, it is, however, inconvenient to require knowledge of the voltage distribution across the whole fiber length. Therefore, the important question is whether position of touch can be determined by continuously measuring the voltage at a single fixed point on the fiber outer electrode. To answer this question, in Fig. 6.12, we present the value of voltage as measured at the fiber extremity ($x = 0$) as a function of the touch position along the fiber. Now x/L represents the touch position of the equivalent human probe, while $|V(0)/V_0|$ represents the voltage measured at the fiber end opposite to the grounded end. We observe that for driving frequencies above 100 Hz, the voltage at the fiber extremity changes significantly, depending on the position of touch. In turn, this allows building a 1-D sliding sensor with a convenient acquisition procedure based on a single point measurement.

6.4.2 RC ladder network model for the capacitor fiber featuring one highly conductive, and one highly resistive, electrode

To model electrical response of the capacitor fibers, we start with the RC ladder network presented in Section 6.3 and first consider a stand-alone capacitor fiber without touching. As before (Eqs. 6.1 and 6.2), we define the conductive

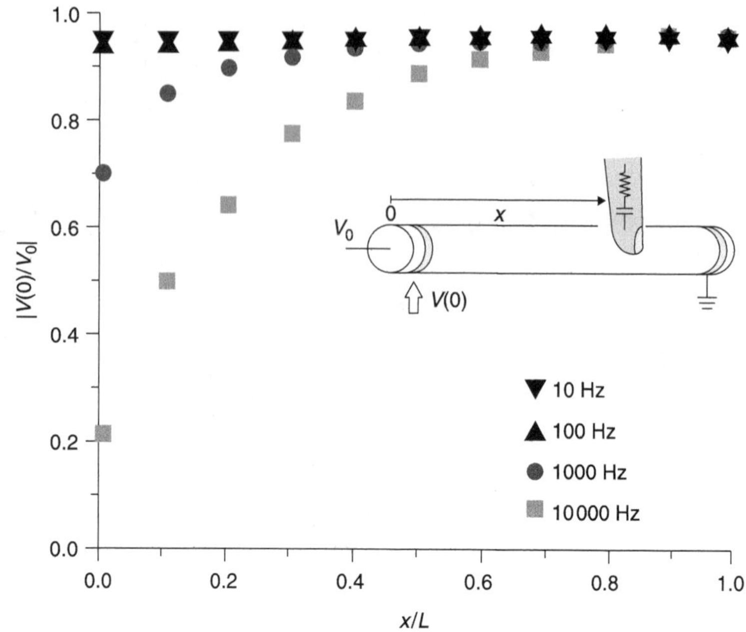

6.12 Voltage measured at the extremity of a capacitor fiber opposite to the fiber grounded end.

polymer electrode transverse resistance as $R_i \approx r_t/L$, where the electrode resistivity is $r_t = \rho_\upsilon W/d_c$. Longitudinal resistance is given by $R_l \approx r_l/L$, where $r_l = \rho_\upsilon/(Wd_c)$ is the electrode longitudinal resistivity. Finally, the total fiber capacitance is proportional to the fiber length, while the fiber capacitance per unit length is simply $C_t \approx 2\varepsilon_0\varepsilon W/d_i$ (Eq. 6.3), where d_i is the thickness of the isolating films in the fiber, ε is the dielectric constant of the isolating films, and ε_0 is permeability of the vacuum.

We note that relatively high capacitance of our fibers is due to the small thickness of the isolating films, and large net width of the conductive layers. Thus, a typical fiber features conductive and isolating layers with thicknesses smaller than 10 μm, while the net width W of the layers wrapped into a 1 mm diameter fiber can be in excess of 3 cm.

Now let us consider a capacitor fiber as a sequence of thin cross sections of length dx (Fig. 6.13(a)), each having a longitudinal resistance $dR_l = r_l dx$ for the outer (surface) electrode. We then consider electrical response of an individual cross section, while assuming that along the fiber length the individual fiber sections are connected via longitudinal resistance elements dR_l (Fig. 6.13b). Electrical response of an individual fiber cross section is modeled as an RC network, where transverse resistance elements $dR_t = r_t/dx$ are connected via capacitance $dC_t = C_t dx$ elements. In Section 6.3, we have shown an equivalent

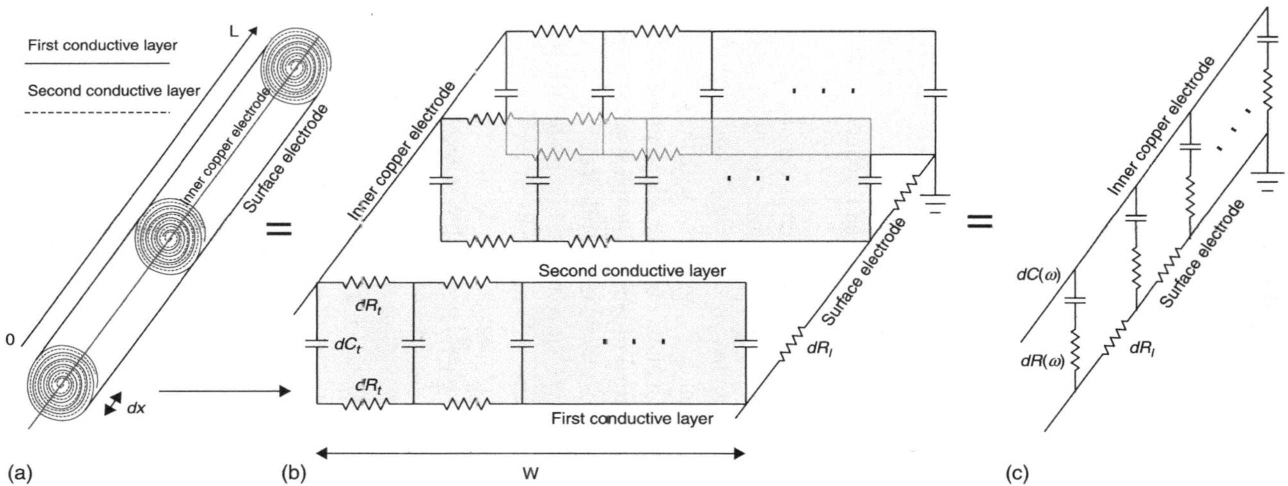

6.13 Ladder RC network model of a stand-alone capacitor fiber. (a) The fiber is modeled as a sequence of fiber cross sections of small length *dx* connected in series via longitudinal resistive elements (high resistivity outer electrode), while assuming that the inner copper electrode has a constant potential along its length. (b) Electrical response of an individual fiber cross section is modeled as an RC network, where transverse resistivity elements are connected via capacitance elements. (c) Also the scheme above corresponds to the longitudinal resistance element, which is the frequency dependent resistance and capacitance of an individual fiber cross section of length *dx*. It could be shown that an equivalent circuit that describes electrical response of an individual fiber cross section is given simply by the frequency dependent resistivity connected in series with frequency-dependent capacitance. Finally, the electrical response of a fiber is modeled as another RC network with frequency-dependent resistivity and capacitance.

circuit that describes electrical response of an individual fiber cross section of length dx is given by the frequency dependent resistance $dR(\omega)$ connected in series, with frequency dependent capacitance $dC(\omega)$, where:

$$dC(\omega) = -\frac{dx}{\omega r_t \cdot \mathrm{Im}\left(f(B)\right)}, \quad [6.12]$$

$$dR(\omega) = \left(\frac{1}{2} + \mathrm{Re}\left(f(B)\right)\right)\frac{r_t}{dx}, \quad [6.13]$$

and

$$f(B) = \frac{1+\cosh(B)}{B \cdot \sinh(B)}; \; B = \sqrt{2j\omega \cdot r_t C_t}. \quad [6.14]$$

Electrical response of a stand-alone fiber can, therefore, be modeled as another RC ladder with frequency-dependent $dR(\omega)$, $dC(\omega)$, and frequency-independent parameters dR_l (Fig. 6.13(c)).

Now that the model for a stand-alone capacitor fiber is defined, we modify it slightly in order to analyse a 1-D slide sensor. In particular, the fiber is assumed to be touched at a position x_b with a finger having effective electric parameters R_b = 1.44 kΩ, C_b = 150 pF. Moreover, to simplify comparison with the experiment, we include in our model the effective circuit of an oscilloscope probe used in our measurements. The probe is attached at a position x_p on the fiber surface (Fig. 6.14), and the effective circuit parameters of a probe and oscilloscope are R_p = 10 MΩ, C_p = 200 pF. The necessity to include effective circuit of a probe into the model comes from the realization that resistance of a standard 10X probe (10 MΩ) used in our experiments has the same order of magnitude as the transverse resistance of the short fiber segments used in our studies. For example, transverse resistance of the 10 cm-long fiber pieces typically ranges in 0.1 to 1 MΩ. Moreover, we can show that at frequencies lower than $\nu \approx 1/(2\pi R_p C_b) \approx 100\,Hz$ or higher than $\nu \approx 1/(2\pi R_b C_p) \approx 55\,kHz$, the effective impedance of a probe becomes smaller than that of a finger, therefore the probe effective circuit has to be included in the model to accurately explain experimental measurements.

In Fig. 6.14, we distinguish three parts of an RC ladder network. The first part is located to the left of the probe, where $i''(x)$ denotes the longitudinal current flowing in the polymer conductive film, while $di''(x)$ denotes the transverse current flowing in the thin section of length dx. To the right of the probe, while still before the finger touch position, the longitudinal and transverse currents in the polymer electrode are denoted as $i(x)$ and $di(x)$. Finally, to the right of the touch position, the corresponding currents are $i'(x)$ and $di'(x)$. V_0 is the voltage difference between the inner copper electrode and the outer electrode at $x = L$. We also assume that the fiber material parameters are position and frequency independent. Furthermore, we consider that the probe is attached to the left of touch position $x_p < x_b$. We now apply the KVL and KCL to the ladder circuit to

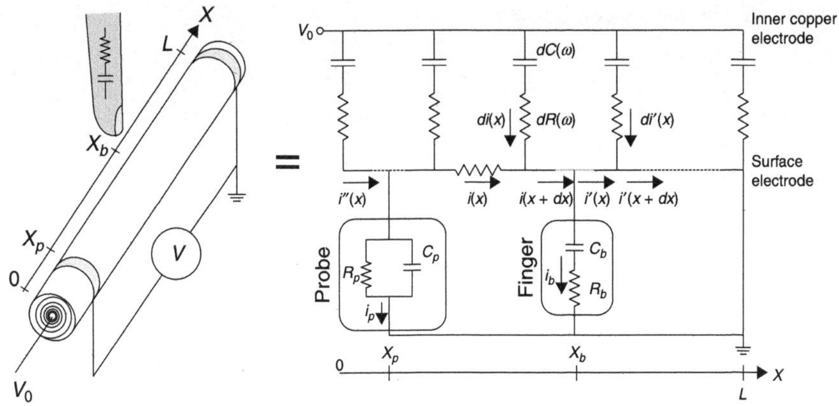

6.14 The ladder network model of a one-dimensional slide sensor. The fiber is assumed to be touched at a position x_b with a finger having effective electric parameters R_b, C_b. Moreover, to simplify comparison with experiment, we include in our model the effective circuit (with parameters R_p, C_p) of an oscilloscope probe used in our measurements; the probe is attached at a position x_p.

arrive at the following equations for any position x along the fiber. Thus, using KVL, we get:

$$
\begin{aligned}
0 < x < x_p \quad &; \quad di''(x)\left(\frac{1}{j\omega\, dC(\omega)} + dR(\omega)\right) + \int_x^{x_p} r_l\, i''(l)\, dl = V_0 - V(x_p) \\
x = x_p \quad &; \quad i_p \frac{R_p}{1 + j\omega C_p R_p} = V(x_p) \\
x_p < x < x_b \quad &; \quad di(x)\left(\frac{1}{j\omega\, dC(\omega)} + dR(\omega)\right) + \int_x^{x_b} r_l\, i(l)\, dl = V_0 - V(x_b) \\
x = x_b \quad &; \quad i_b\left(\frac{1}{j\omega\, C_b} + R_b\right) = V(x_b) \\
x_b < x < L \quad &; \quad di'(x)\left(\frac{1}{j\omega\, dC(\omega)} + dR(\omega)\right) + \int_x^{L} r_l\, i'(l)\, dl = V_0.
\end{aligned}
\qquad [6.15]
$$

Using the KCL, we get:

$$
\begin{aligned}
0 < x < x_p \quad &; \quad i''(x) + di''(x) = i''(x + dx) \\
x = x_p \quad &; \quad i''(x_p) = i(x_p) + i_p \\
x_p < x < x_b \quad &; \quad i(x) + di(x) = i(x + dx) \\
x = x_b \quad &; \quad i(x_b) = i'(x_b) + i_b \\
x_p < x < L \quad &; \quad i'(x) + di'(x) = i'(x + dx).
\end{aligned}
\qquad [6.16]
$$

Finally, the boundary conditions are:

$$i''(0) = 0. \qquad [6.17]$$

To solve these equations, we can first differentiate Eqs. 6.15 with respect to x to obtain three similar second-order differential equations with respect to i, i' or i'' in the following form:

$$\frac{d^2 i(x)}{dx^2}\left(\frac{1}{j\omega dC(\omega)/dx} + dx \cdot dR(\omega)\right) - r_l i(x) = 0. \qquad [6.18]$$

Then, in each of the three sections of the fiber, we can write a solution for the currents as:

$$i = C_1 e^{\tilde{B}x} + C_2 e^{-\tilde{B}x} \quad ; \quad i' = C_3 e^{\tilde{B}x} + C_4 e^{-\tilde{B}x} \quad ; \quad i'' = C_5 e^{\tilde{B}x} + C_6 e^{-\tilde{B}x} \qquad [6.19]$$

where

$$\tilde{B} = \sqrt{\frac{r_l}{\left(\dfrac{1}{j\omega\, dC(\omega)/dx} + dx \cdot dR(\omega)\right)}} = \sqrt{\frac{r_l}{r_t\left(\dfrac{1}{2} + f(B)\right)}} \qquad [6.20]$$

and B, $f(B)$ are defined in Eqs. 6.12 and 6.13. Insertion of Eq. 6.19 into the remaining Eqs. 6.15 to 6.17 results in a set of linear equations from which the constants, C_1–C_6, can be determined. Finally, from the known current distributions, voltage distribution in the given position x measured from the surface electrode can be easily found as:

$$V(x) = V_0 - \frac{d^2 i(x)}{dx^2}\left(\frac{1}{j\omega\, dC(\omega)/dx} + dx \cdot dR(\omega)\right) \qquad [6.21]$$

Comparison of experimental data with predictions of a theoretical model

In Fig. 6.15 we present experimental data and the theoretical RC ladder model Eqs. 6.15 to 6.21 prediction for the dependence of voltage measured at $x_p = 0$ as a function of the equivalent human probe touch position x_b. In each graph, different sets of curves correspond to distinct fibers, which are different from each other in a single parameter. Thus, in Fig. 6.15(a), we present measurements of three fibers drawn using preforms containing different number of conductive layers, and as a consequence, having different capacitance C. The preforms were drawn using the same temperature profile and drawing speed so as to guarantee similar values of the bulk resistivity of the polymer electrodes. Then, the fiber geometrical parameters, such as layer thicknesses, electrode width and

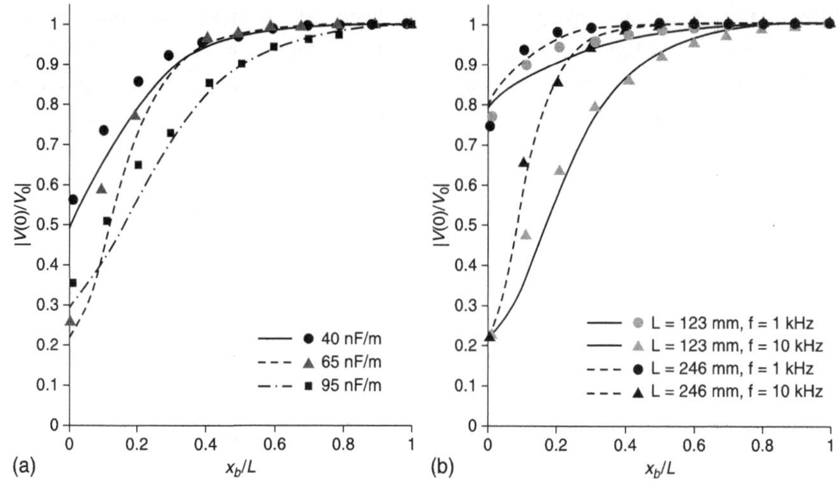

6.15 Fiber response to the touch of the equivalent human probe – comparison between predictions of the RC ladder model and experimental data. (a) Response at 10 kHz of the 3 distinct fibers of the same length and different capacitances $C_t = 40\,nF.m^{-1}$, $C_t = 65\,nF.m^{-1}$, $C_t = 95\,nF.m^{-1}$. The rest of the geometrical and electrical parameters of the fibers are similar to each other. (b) Response at 1 and 10 kHz of the two distinct fibers of the different lengths. The rest of the geometrical and electrical parameters of the fibers are identical to each other, as a shorter fiber was obtained by cutting a longer fiber in half.

fiber length, were measured using the optical microscope. Parameters of the oscilloscope effective circuit were measured independently. Finally, bulk resistance of the conductive layers in a fiber was measured, as described in Section 6.3, by wrapping the fiber outer electrode in foil and then extracting the transverse resistance and, consequently, the bulk resistance of conductive layers from the fiber AC response. In this arrangement, the currents are purely transverse and the RC ladder model was shown (Section 6.3) to give precise fits for the bulk resistivity parameter. We then use all the model parameters found in the independent measurements to predict the response of the fiber to the touch. From Fig. 6.15(a) we see that at the operating frequency of 10 kHz, the experimental curves are well described by the RC ladder model, which does not use any fitting parameter.

Similarly, in Fig. 6.15(b), we present fiber response to the touch for two identical fibers of different lengths. In these experiments we first use the fiber of length 24.6 cm and then cut it in half to 12.3 cm and repeat the measurement. The rest of the input parameters necessary for the use of a theoretical model were measured as described above. From the figure, we see that fiber response is well described by the RC ladder model, both at 1 and 10 kHz operating frequencies.

Moreover, we see that this particular fiber becomes insensitive to touch if the touch point x_b is further than 10 cm from the measuring point x_p.

The measurement of fiber response as a function of fiber length, presented in Fig. 6.15(b), brings about an important question about the maximal length of a 1-D slide sensor, and about the optimal frequency of operation. We note that the functional form of the currents (Eq. 6.19) flowing between the probe and the measurement point is exponential with the characteristic length:

$$\tilde{L} = \frac{1}{\mathrm{Re}(\tilde{B})} = \frac{W}{\sqrt{2}} \frac{1}{\mathrm{Re}\left(\left(1+2f(B)\right)^{-\frac{1}{2}}\right)} \qquad [6.22]$$

where we used $\sqrt{r_t/r_l} = W$ – the total width of the rolled electrodes. We now use asymptotic expansions of the function $f(B)$:

$$f(B) = \frac{1+\cosh(B)}{B \cdot \sinh(B)} \underset{B=\sqrt{2j\omega \cdot r_t C_t}}{=} \begin{cases} \dfrac{1}{B} & ; \quad \omega >> \dfrac{1}{r_t C_t} \\ \dfrac{2}{B^2} & ; \quad \omega << \dfrac{1}{r_t C_t} \end{cases} \qquad [6.23]$$

to obtain the following limiting values for the characteristic length of the current decay:

$$\tilde{L} = \frac{W}{\sqrt{2}} \begin{cases} 1 & ; \quad \omega >> \dfrac{1}{r_t C_t} \\ \dfrac{1}{\sqrt{\omega r_t C_t}} & ; \quad \omega << \dfrac{1}{r_t C_t} \end{cases} . \qquad [6.24]$$

Note that for the slide sensor of length L to be sensitive along its whole length, we have to require that $L \sim \tilde{L}$. From Eq. 6.24, it means that at high frequencies $\omega >> 1/(r_t C_t)$ the maximal sensor length is limited by the net width of a polymer electrode wrapped into the fiber cross section. Note that for most of our fibers, the region of high frequency is in the vicinity or above 1 kHz. Furthermore, in a typical fiber of $D = 1$ mm diameter, we can currently fit $N \sim 10$ to 50 turns of the conductive electrode, which results in the net width of a conductive electrode in the fiber $W \sim \pi DN \sim 3$ cm–15 cm. Therefore, for the operation frequencies in the vicinity or above 1 kHz, the maximal length of the capacitor fiber-based slide sensor is currently limited to several tens of centimeters.

In principle, operating at lower frequencies allows match of the fiber length and characteristic current decay length $L \sim \tilde{L}$ for any desired length of the fiber. However, this demands very low operation frequencies of $\omega \sim (W/L)^2/(r_t C_t)$, which even for a relatively short 1 m-long fiber can be as low as 1 to 10 Hz. However, operation at low frequencies is prone to strong electrical interferences and noise. Moreover, at low frequencies of ~1 Hz, the finger has very large impedance, >> 1 MΩ, which is mismatched with that of our fiber. This makes the sensor of very low sensitivity at low frequencies.

6.4.3 Interpreting electrical data from a single-fiber touch sensor

Finally, a 1-D slide sensor was characterized in a realistic setting with a student touching the sensor consecutively at different positions along the fiber length. At this time the sensor was touched with an actual finger and not an equivalent probe. The driving frequency was 1 kHz and was provided by the analogue output of an ADC card. The same card was also employed as a signal acquisition unit with an acquisition rate of 20 kHz. Such high acquisition rates were only used for the purpose of resolving the time response of a sensor and in practice significantly slower acquisition rates (<10 Hz) could be used. To deduce the touch position, we continuously measured the voltage at the endpoint of the fiber outer electrode. Figure 6.15(b) presents an example of the measured voltage as a function of time. To operate the sensor, we first record for several seconds the voltage level ('idle' level) without touching the fiber (see Fig. 6.16). We then

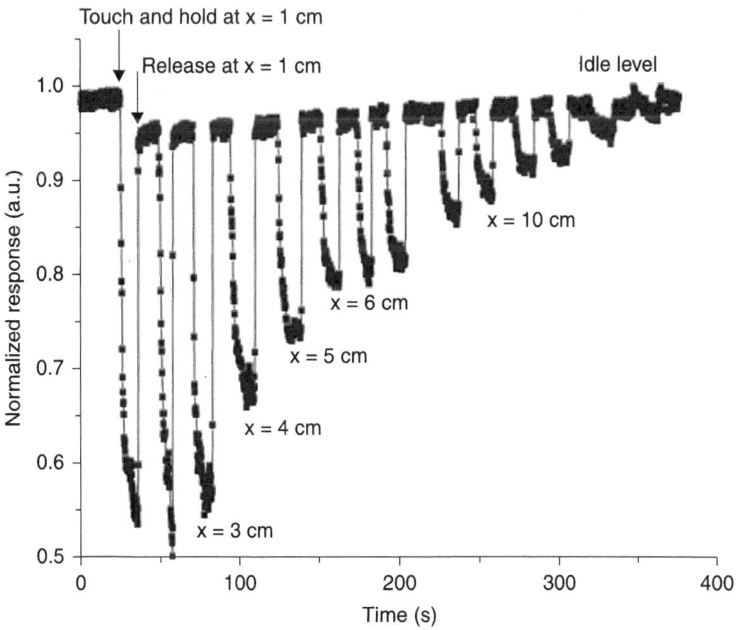

6.16 Typical time resolved response of a one-dimensional slide sensor. Dips in the measured voltage correspond to the touch and release events.

touch the sensor at $x = 1\,cm$ from the fiber end. At the moment of touch, the signal drops very fast (in a matter of 1 s) to almost its equilibrium value, and this is followed by a much slower (~10 s) relaxation to the actual equilibrium value. We believe that the fast timescale corresponds to a pure electrical response of a fiber, while the slow timescale corresponds to small pressure induced changes in the fiber electrical parameter due to fiber deformation under touch. As soon as the finger is removed, the signal returns back to the 'idle' level in a matter of 0.2 s. Next, by touching, holding and releasing the fiber at various positions along its length, we can record a complete calibration curve. Then the position of touch can be determined from the voltage level at the bottom of dips corresponding to touching events. This is, of course, the simplest implementation of a touch sensor. To avoid person-dependent calibration of the sensor, we can, for example, perform two voltage measurements from opposing sides of a fiber (by simultaneously flipping the ground from one side to the other), then use analytical models to extract the effective electrical parameters of a human figure and find the touch position.

6.4.4 Effect of the fiber length on sensitivity

To study the effect of fiber length on its sensing performance, we take two pieces of fiber, named sample #1 and sample #2, from the same batch of drawing. Sample #1, with a length of 123 mm, is a part of sample #2, with a length of 246 mm. Thus, they share the same electrical and geometric parameters. The samples have an equal diameter of around 1.1 mm. The geometric parameters, such as the thickness and width of films (d_c, d_i and W), can be measured from microscopic observations or calculated from the preform geometry and draw-down ratio. The fiber's electrical parameters were characterized independently by the method reported in Wu *et al.* (2000). The measured capacitance per unit length for both fibers equals 93 nF m^{-1}. The volume resistivity of the conductive film, calculated from transverse resistivity r_t, equals 4.5 Ωm. Figure 6.15(b) also shows a comparison of the experimental data and model predictions for samples #1 and #2 at frequencies of 1 and 10 kHz. The data below 100 Hz are not displayed, because the disturbance caused by finger touch at these frequencies is very small.

We can see that both fibers exhibit a voltage drop at the left end, which becomes more prominent as frequency goes higher. This phenomenon is precisely predicted by the model. In order to find the reasons that cause this observation, we made calculations using a model without the oscilloscope in the circuit, or simply set the value of C_p extremely small and the value of R_p extremely higher in the present model. The results show that if we neglect the effect of the oscilloscope, the voltage becomes a constant without any drop when the measurement point approaches the left end of the fiber. This implies that this voltage drop on this end of the fiber is rather caused by the decreased impedance of the oscilloscope at higher frequencies than by the fiber itself. From Fig. 6.15(b) we can also see that the longer fiber, denoted by the grey color, has a longer part in the middle with its voltage be a constant, while the short fiber, denoted by the black color, has more part of its length with a changing voltage. This can be rationalized by the distributed nature of the fiber. The edge effect becomes more dominant when the fiber length decreases. In Fig. 6.15(b), we compare the effect of frequency on the fiber performance. For both fibers, the overall voltage decreases as frequency increases, which is related to the capacitance and resistance of the fiber.

6.4.5 Effect of the operational frequency on sensitivity

The maximal voltage amplitude at the outer fiber electrode is strongly dependent on the operation frequency and reaches its maximum at frequencies between 100 Hz and 1 kHz. When the fiber is touched with an equivalent human probe, voltage distribution along the fiber outer electrode shows high sensitivity to the operation frequency (Figs 6.9(a,b)). Thus, at low driving frequencies (10 Hz, 100 Hz), voltage distribution along the fiber outer electrode is virtually unchanged by the touch. At higher frequencies (>100 Hz), a dip appears in the voltage distribution in the vicinity of the point of contact with an equivalent human probe. Finally, at high frequencies (>1 kHz), this dip becomes very pronounced and easy to detect. Overall frequency response of the capacitor fiber can be understood from the basic electric circuit theory. Thus, the fiber used in these experiments had capacitance per unit length of C_t = 93 nF/m and transverse resistance per unit length of R_t = 14.5 kΩ·m (bulk resistance per unit length of conductive layers 4.8 Ω·m). The characteristic frequency associated with a corresponding RC circuit is $v = 1/(2\pi R_t C_t) = 120\,Hz$. Electric response of a capacitor fiber is, in fact, that of a high pass filter. Namely, at frequencies well below 120 Hz, the fiber operates in a quasi-DC regime, where it simply breaks the electrical circuit. Fiber response to touch is, therefore, minimal at low frequencies. At frequencies comparable or higher than 120 Hz, the capacitor fiber acts as s distributed complex impedance with relatively low resistance. At these higher frequencies, touching the capacitor fiber can modify significantly the local current flows and voltage distributions, thus resulting in sensitivity of its various electrical parameters to touching.

6.5 Fully woven two-dimensional touch pad sensor using one-dimensional array of capacitance fibers

6.5.1 Sensor design and fabrication

Soft but highly elastic mechanical nature of fibers makes them easy to use in conventional weaving process (Fig. 6.17(b)). In our laboratory we have used a table Dobby loom (Leclerc Voyager 15¾″ inch 4 s) to weave the fibers into a wool-based textile matrix (Fig. 6.17(b)). The resulting 15 × 10 cm woven touch pad contained 15 capacitor fibers, each 12 cm long (Fig. 6.17(c)).

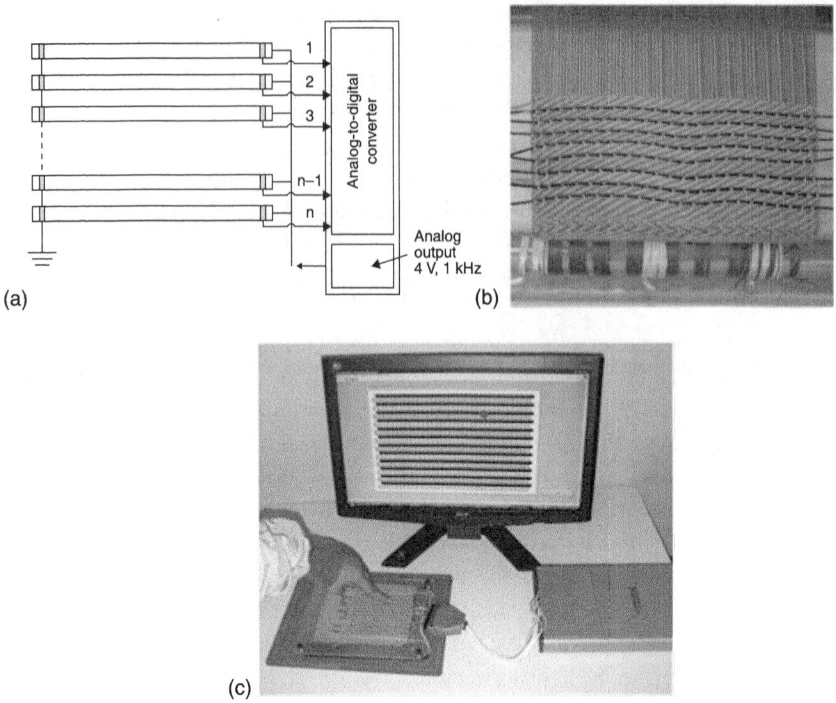

6.17 The woven touch pad sensor. (a) Schematic representation of the woven two-dimensional touch pad sensor featuring a one-dimensional array of the capacitor fibers. All the connections to and from the fibers are done using a 120 μm-diameter copper wire. Fibers in the array have the common ground and a common source; however, they are interrogated individually. Analog output of the ADC board is used as a function generator that provides a sinusoidal signal at 1 kHz with an amplitude of 4 V. (b) Weaving a two-dimensional touch pad sensor on a Dobby loom. (c) Woven touch pad connected to the ADC board, as well as the monitor image of a textile with an interpreted touch position.

In our experimental set-up, we use the same connection as described earlier: the inner (low Ω) electrodes of all the fibers (copper wires) are connected to the analogue voltage source integrated into the Analog-to-Digital Convertor (ADC) card, providing a sinusoidal signal at 1 kHz with amplitude of 4 V (Fig. 6.17(b)). The outer (high Ω) electrodes of the fibers are connected to the individual channels of the ADC card (National Instruments USB-6343 X-series DAQ), in order to measure the voltage at their endpoints. The connections are made using thin cooper wires (0.120 μm-diameter) and secured with a conductive epoxy.

We have already shown in Section 6.4 that the readout voltage measured at the end of the individual fiber is highly sensitive to the position of the touch along this fiber, which in this case acts like a tactile slide sensor. Thus, a 1-D senor array integrated into the textile allows us to build a touch pad sensor that can localize in two dimensions the position of touch (Fig. 6.17(c)). In practice, we use the ADC board plugged into a PC to acquire the signal from all the fibers and the touch position can then be monitored on the computer screen via custom LabVIEW software.

6.5.2 Cross-talk and channel calibration

Finally, we address the question of cross-talk between the fibers in the woven 2-D touch pad sensor. As discussed earlier, the sensor comprises a 1-D fiber array separated by 1 cm of textile (Fig. 6.17). In principle, false responses in our system could be induced by cross-talk between individual fibers, which may lead to interference effects between measured signals. In this case, the electrical signal at one channel may induce the signal at the other channel, which might be detected as a false touch. In Fig. 6.18, we show an example of a typical recorded voltage response as a function of time for the three neighboring fibers in the touch pad. No detectable cross-talk between the fibers was observed during recording. Dips on the graphs correspond to the individual touch events that take place at different moments in time. As can be seen in Fig. 6.18, after each touch and release, all the sensor channels successfully return back to the idle state. We notice that fibers in

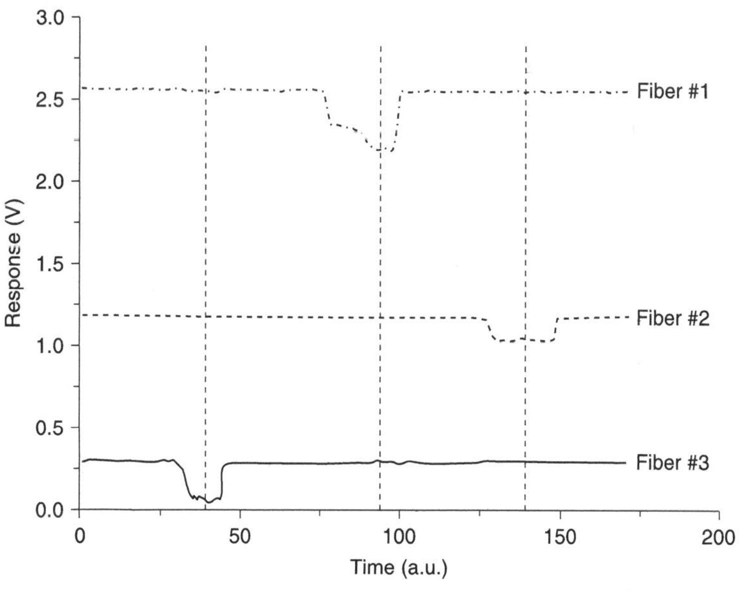

6.18 Recorded voltage response as a function of time for the three neighboring fibers in the woven touch pad. No detectable cross-talk between the fibers was observed. Dips on the graphs correspond to the individual touch events that take place at different moments in time.

Fig. 6.18 have different idle levels, which is explained by all of them having slightly different resistance and capacitance, so the real (not normalized) signal is presented mainly for demonstration purposes. Despite the fact that in the simplest interrogation configuration each individual fiber incorporated into the textile is not multi-touch sensitive (only the closest to the outer electrode touch will be detected), the absence of inter-channel cross-talk enables simultaneous interrogation of the individual slide sensors. In this sense, a 2-D touch pad has a partial multi-touch functionality.

6.6 Conclusion

In conclusion, we have described fabrication, electrical characterization and integration into tactile sensors of the novel all-polymer soft capacitor fibers. The capacitor fibers featuring relatively high capacitance and resistance were fabricated using fiber drawing technique. For ease of connection, a thin copper wire was integrated into the fiber core during the drawing procedure. Soft-capacitor fibers have a typical capacitance per unit length of 69 nF/m, and a typical resistivity parameter of 5 kΩ·m. Our measurements and theoretical modeling show that the fiber capacitance is a very stable, geometry defined parameter independent of the fiber diameter, and fiber fabrication parameters. In contrast, fiber resistivity has a very strong positive temperature coefficient, is highly sensitive to stretching and it is strongly dependent on the fiber drawing parameters.

Next, an individual capacitor fiber was demonstrated to act as a slide sensor that allowed determining the touch position along its length by measuring the fiber AC response at a single point at the fiber surface. Electrical response of such a sensor was described by the RC ladder model, with the modeling data in excellent agreement with experimental observations.

Developed capacitor fibers are soft, small diameter, lightweight and do not use liquid electrolytes, thus they are ideally suited for the integration into textile products. At the end of the chapter, we have demonstrated that by weaving a 1-D array of capacitor fibers (in parallel to each other), a fully woven 2-D touch pad sensor can be built. Performance of a touch pad sensor was then characterized and the absence of the inter-channel crosstalk was confirmed. We also note that a 2-D touch pad has a partial multi-touch functionality.

6.7 References

Alig I, Skipa T, Lellinger D and Pötschke P (2008), 'Destruction and formation of a carbon nanotube network in polymer melts: rheology and conductivity spectroscopy', *Polymer*, 49, 3524–3532.

Baron W, Blair M and Fries-Carr S (2006), 'Airframe structure-integrated capacitor', *US Patent 6981671*.

Beebe D J, Hsieh A S, Denton D D and Radwin R G (1995), 'A silicon force sensor for robotics and medicine', *Sens Actuat A*, 50, 55–65.

Cao Q, Song Y, Liu Z and Zheng Q (2009), 'Influence of annealing on rheological and conductive behaviors of high-density polyethylene/carbon black composites', *J Mater Sci*, 44, 4241–4245.

Cheng L F and Hong Y P (2001), 'Multifiber ceramic capacitor', *J Mater Sci: Mater Electron*, 12, 187–191.

Chu Z, Sarro P M and Middelhoek S (1995), 'Silicon three-axial tactile sensor', *Transducers*, 1, 656.

Cok R S, Bourdelais R R and Kaminsky C J (2004), 'Flexible resistive touch screen', *US Patent 0212599 A1*.

Dahiya R S, Metta G, Valle M and Sandini G, (2010), 'Tactile sensing: from humans to humanoids', *IEEE Trans Rob*, 26, 1–20.

Dalton A B, Collins S, Muñoz E, Razal J M, Ebron V H, *et al.* (2003), 'Super-tough carbon-nanotube fibres', *Nature*, 423, 703.

Dario P and De Rossi D (1985), 'Tactile sensors and the gripping challenge', *IEEE Spectr*, 22, 46–52.

Dietz P and Leigh D (2001), '*Diamond*Touch: a multiuser touch technology', *Proceedings of UIST, the 14th Annual ACM Symposium on User Interface Software and Technology*, pp. 219–226.

Eltaib M E H and Hewitt J R (2003), 'Tactile sensing technology for minimal access surgery: a review', *Mechatronics*, 13, 1163–1177.

Engel J, Chen J and Liu C (2003), 'Development of polyimide flexible tactile sensor skin', *J Micromech Microeng*, 13, 359–366.

Feng J and Chan C M (2003), 'Effects of strain and temperature on the electrical properties of carbon black-filled alternating copolymer of ethylene-tetrafluoroethylene composites', *Polym Eng Sci*, 43, 1064–1070.

Gao Y, Guo N, Gauvreau B, Rajabian M, Skorobogata O, *et al.* (2006), 'Consecutive solvent evaporation and co-rolling techniques for polymer multilayer hollow fiber preform fabrication', *J Mater Res*, 21, 2246–2254.

Gray B L and Fearing R S (1996), 'A surface micromachined microtactile sensor array', *Proceedings of the IEEE International Conference on Robotics and Automation*, pp. 1–6.

Gu J F, Gorgutsa S and Skorobogatiy M (2010), 'Soft capacitor fibers using conductive polymers for electronic textiles', *Smart Mater Struct*, 19, 115006.

Hoshi T and Shinoda H (2006), 'A sensitive skin based on touch-area-evaluating tactile elements', *Proceedings of the Symposium on Haptic Interfaces for Virtual Environment and Teleoperator Systems*, pp. 89–94.

Kalendra P W and Piazza W J (1994), 'Automatic calibration of a capacitive touch screen used with a fixed element flat screen display panel', *US Patent 5283559*.

Kim K, Lee K R, Kim W H, Park K B, Kim T H, *et al.* (2009), 'Polymer-based flexible tactile sensor up to 32×32 arrays integrated with interconnection terminals', *Sens Actuators A*, 156, 284–291.

Kolesar E S, Dyson C S, Reston R R, Fitch R C, Ford D G and Nelms S D (1996), 'Tactile integrated circuit sensor realized with a piezoelectric polymer', *Proceedings of the 8th IEEE International Conference on Innovative System Silicon*, pp. 372–381.

Lee S, Buxton W and Smith K C (1985), 'A multi-touch three-dimensional touch-sensitive tablet', *Proceedings of the SIGCHI Conference on Human Factors in Computing Systems*, pp. 21–25.

Leineweber M, Pelz G, Schmidt M, Kappert H and Zimmer G (2000), 'New tactile sensor chip with silicone rubber cover', *Sens Actuat A*, 84, 236.

Nakamura S and Sawa G (1998), 'Percolation phenomena and electrical conduction mechanism of carbon black-polyethylene composites', *Proceedings of the International Symposium on Electrical Insulating Materials*, pp. 1–54.

Park C S, Park J and Lee D W (2009), 'A piezoresistive tactile sensor based on carbon fibers and polymer substrates', *Microelectron Eng*, 86, 1250–1253.

Rekimoto J, *et al.* (2002), 'SmartSkin: An infrastructure for freehand manipulation on interactive surfaces', *Proceedings of the SIGCHI Conference on Human Factors in Computing Systems*, pp. 113–120.

Ruspini D C, Kolarov K and Khatib O (1997), 'Proceedings of the symposium on haptic interfaces for virtual environment and teleoperator systems', *Proceedings of the 24th Annual Conference on Computer Graphics and Interactive Techniques*, pp. 345–352.

Schulte B, Tischer W, Waldenrath W and Kaloff H (1988), 'Stretched polycarbonate films filled with carbon black', *US Patent 4791016*.

Shimojo M, Namiki A, Ishikawa M and Makino R (2004), 'A tactile sensor sheet using pressure conductive rubber with electrical-wires stitched method', *IEEE Sens J*, 4, 589–595.

Tang H, Chen X and Luo Y (1997), 'Studies on the PTC/NTC effect of carbon black filled low density polyethylene composites', *Eur Polym J*, 33, 1383–1386.

Wolffenbuttel M R and Regtien R P L (1991), 'Polysilicon bridges for the realization of tactile sensors', *Sens Actuat A*, 26, 257–264.

Wu G, Asai S, Zhang C, Miura T and Sumita M (2000), 'A delay of percolation time in carbon-black-filled conductive polymer composites', *J Appl Phys*, 88, 1480–1487.

Yu G, Zhang M Q and Zeng H M (1998), 'Carbon-black-filled polyolefine as a positive temperature coefficient material: effect of composition, processing, and filler treatment', *J Appl Polym Sci*, 70, 559–566.

Yuji J and Sonoda C (2006), 'PVDF tactile sensor for static contact force and contact temperature', *Proc IEEE Sens*, 738–741.

Zhang W, Dehghani-Sanij A A and Blackburn R S (2007), 'Carbon based conductive polymer composites', *J Mater Sci*, 42, 3408–3418.

Part II
Technologies

7

Textile fabrication technologies for embedding electronic functions into fibres, yarns and fabrics

J. EICHHOFF, A. HEHL, S. JOCKENHOEVEL and T. GRIES, RWTH Aachen University, Germany

DOI: 10.1533/9780857093530.2.191

Abstract: In this chapter textile fabrication processes along the textile production chain will be presented. Starting from fibre and yarn production processes the general principles will be explained, also outlining how functionality can be brought into the fibre during the fibre and yarn production process. The chapter then outlines the basics of fabric production technologies. Here the integration of conductive materials, sensors and components will be discussed. Finally, technologies for subsequent functionalisation, such as embroidery and printing, will be presented.

Key words: spinning, weaving, braiding, knitting, embroidery, functionalisation.

7.1 Introduction

In this chapter the basics of textile production technologies will be explained. The structure of this chapter is depicted in Fig. 7.1. Starting at the process of fibre production, the general principle of melt-spinning polymeric fibres will be explained. Natural fibre production using the example of cotton will be discussed. The next stage in the textile process chain is the production and functionalisation of yarns. Staple fibre spinning processes will be reviewed. Functionalisation technologies, such as electro-coating and winding, will be presented to illustrate the production of electrically conductive fibres and yarns. In the next part of this chapter, basic fabric production technologies are discussed. Weaving, knitting and braiding, as the most common fabric production principles, will be presented. Finally, processes for the subsequent functionalisation of textiles, such as sewing, embroidery or printing, will be outlined.

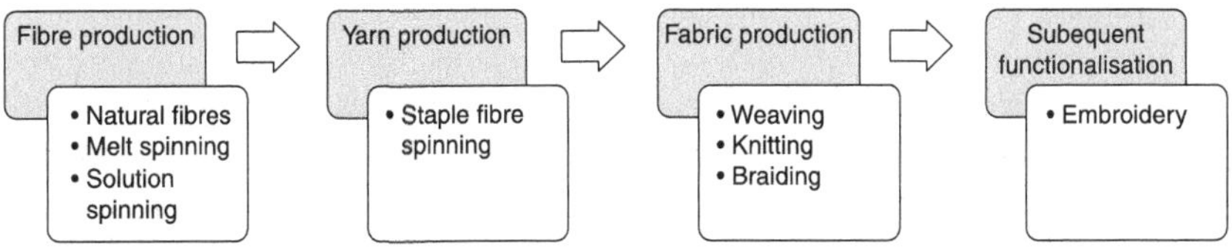

7.1 Textile fabrication processes.

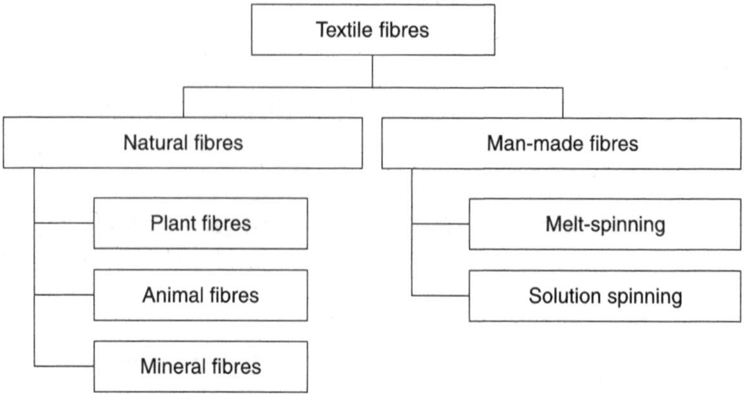

7.2 Classification of fibre materials.

The first step in the textile process chain is the production of fibres. The world of fibre materials is divided into natural and man-made fibres (Fig. 7.2). The following sections deal with natural fibres and continuous (man-made) fibres, respectively.

7.2 Fibre and yarn production processes: natural fibres

In the area of natural fibre materials, a distinction can be made between fibres originating from plants (i.e. cotton) and fibres originating from animals (i.e. wool). There are also specialty fibres like mineral fibres (asbestos), but the most common natural fibre materials used are cotton and wool.

Cotton is an agricultural plant, which is mechanically harvested. The cotton plant is a small shrub and the cotton fibres grow in the capsules containing the cotton seed (Fig. 7.3). The fibres have a typical length of 25 to 30 mm. Such short fibres are called *staple fibres*. When the cotton capsules are ripe, they pop open, exposing the white cotton fibres (Fig. 7.3). During the harvesting process, the harvester grabs the cotton plant at the trunk and strips off all cotton capsules, leaves and smaller twigs. During this process the cotton plant is destroyed, so it has to be planted newly the next season.

In subsequent processes (Wulfhorst *et al.*, 2006) the fibres are separated from the seeds and other materials like capsule fragments, leaves, twigs and dirt. In a carding machine (Fig. 7.4) the fibres are combed by a series of rotating drums and moving carding bars equipped with metal combing teeth. As a result, a soft uniform fibre band is formed. In the fibre band, called sliver, the fibres lie parallel to each other. The sliver is held together by the friction between the parallel fibres, which provides just enough tenacity for subsequent production steps.

7.3 Cotton plant with ripe cotton capsules.

7.2.1 Staple fibre spinning

During the spinning process, the transition from a loose fibre bundle to an actual yarn takes place. The bundle of parallel fibres is twisted, which gives the yarn its tenacity. The tenacity of staple fibre yarns is based on inter-fibre friction. By

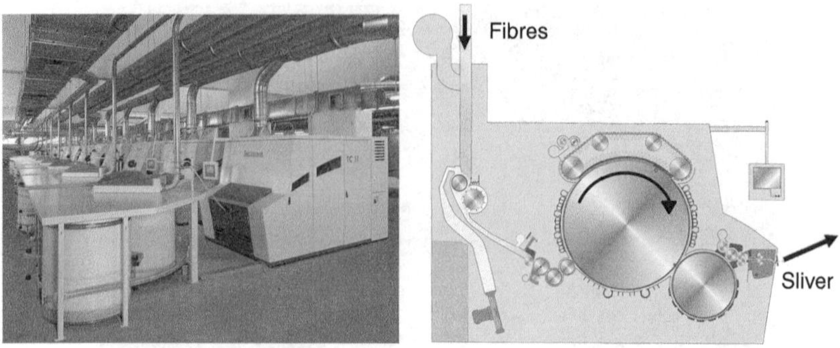

7.4 Left: Carding machine line in a factory. Right: Schematic drawing of a carding machine (images used with permission of Trützschler Spinning).

adding a twist to the fibre bundle, the axial load on the yarn is deflected into a radial force, which as a consequence induces a perpendicular (axial) reactive friction force. There are three predominant spinning technologies, which have been established in the industry: ring spinning, rotor spinning and airjet spinning. These three principles are presented in the following sections.

7.2.2 Ring spinning

Ring spinning is one of the oldest automated spinning principles. Figure 7.5 depicts a ring spinning position. The spinning machine is fed with a sliver of about 900 tex[1] (called 'flyer') to a drawframe. A drawframe consists of wheel pairs, which revolve at different speeds. The first pair is slower than the second pair, which draws the sliver apart, thus making it finer. The bobbin is actively rotated, while the spinning frame moves up and down. The flyer is guided through a stationary former and then through a traveller before being wound onto the bobbin. The traveller consists of a small metal clamp and slides on a ring with a T-shaped profile (for details, see Fig. 7.6).

During the spinning process the bobbin is actively rotated. The flyer is being wound onto the bobbin, dragging the traveller along the ring as the bobbin rotates. In this way, a twist is induced in the flyer, which continues back to the drawframe. Therefore the twist, which provides the yarn tenacity, is induced between drawframe and bobbin. The amount of twist induced depends on the difference between the bobbin rotation speed and the speed of the traveller. The spinning frame moves up and down during the process to wind the produced yarn in a way that allows rewinding in subsequent processes. Ring-spun yarns exhibit a high quality concerning evenness and tenacity. But the ring-spinning process is also slow, as only about 20 m/min are achievable.

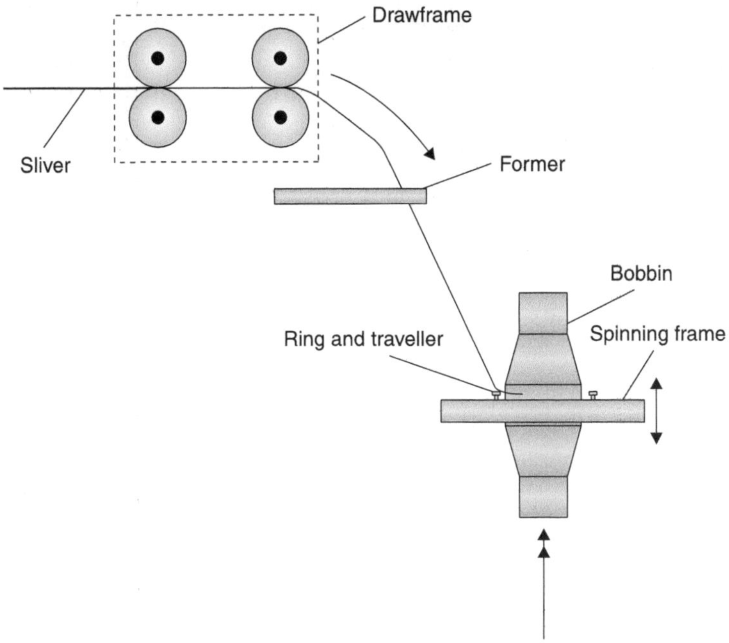

7.5 Schematic drawing of a ring-spinning position.

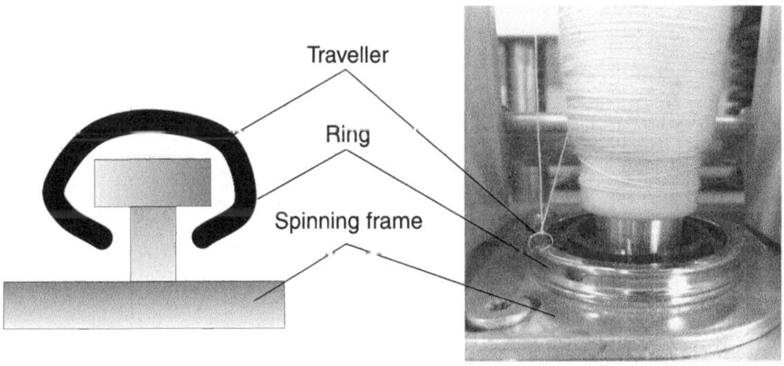

7.6 Ring and traveller.

7.2.3 Rotor spinning

To explain the rotor spinning principle, a simple analogy is applied. The rotor spinning principle works in a similar way to the production of candyfloss. Candyfloss is produced by melting sugar in a small drum rotating at high speed. Due to the centrifugal force, the molten sugar exits the small drum through the openings in the

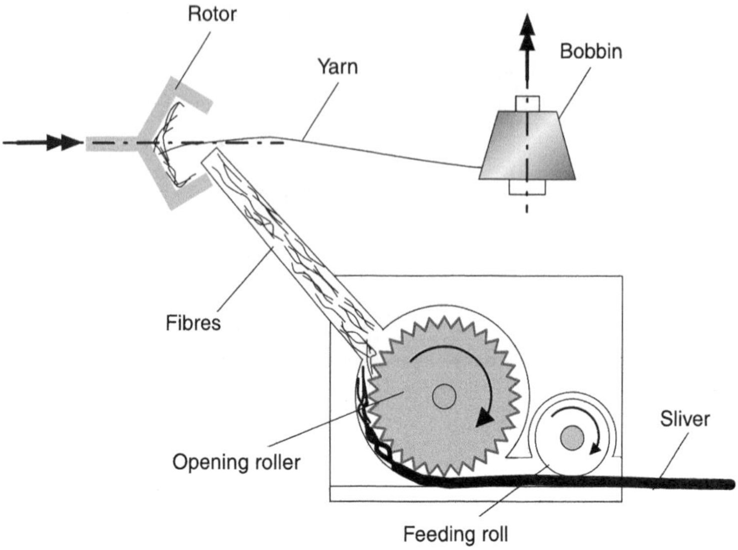

7.7 Rotor spinning principle.

drum walls. In mid-air the sugar solidifies and forms fibres. The fibres are caught in a larger drum totally surrounding the smaller drum. A wooden stick is used to catch the sugar fibres and start the 'spinning' process of the candyfloss.

Rotor spinning works in a similar fashion (Fig. 7.7). A sliver of staple fibres is fed to an opening roller, which mainly consists of a drum equipped with a dense net of sewing teeth. The opening roller rips the fibre bundle apart and blows the single fibres into a rotor, which revolves at a very high speed (up to 150,000 min^{-1}). The yarn formation is started by inserting a short piece of yarn into the rotor. The end of the yarn rotates and is twisted together with the loose fibres, which are located in the rotor. By transporting the freshly spun yarn out of the rotor, a continuous yarn formation process can be established.

The rotor spinning process eliminates the need for a drawframe, since the fibre bundle is directly split into single fibres. The production speed of the rotor spinning process is around 100 to 200 m/min, which makes this process more productive than ring spinning. On the downside, the mechanical properties of rotor-spun yarns are lower than those of ring-spun yarns.

7.2.4 Airjet spinning

The airjet spinning technology is a more recent development. Here a jet of air is used to produce the yarn (Fig. 7.8). The drawing frame delivers the fibre band, which is sucked into a conical opening and moved over the needle. An air vortex

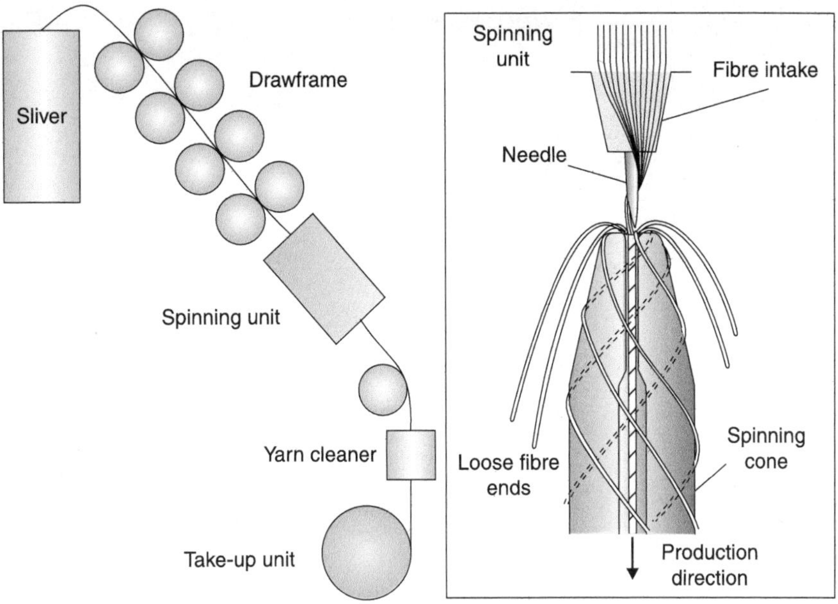

7.8 Airjet spinning process.

separates some fibre ends and bends them over the spinning cone. These bent back fibres wrap themselves around the core of parallel fibres, which gives the airspun yarn its tenacity. The airjet spinning technology is very productive, since production speeds are around 380 to 450 m/min (ring spinning: 20 m/min). On the downside, only the fibres wrapped around the parallel core contribute to the strength of the yarn. Therefore the strength of airspun yarns is lower compared to ring and rotor spun yarns.

7.3 Fibre and yarn production processes: continuous (man-made) fibres

7.3.1 Melt-spinning

Man-made fibres represent all fibre materials that have been artificially produced. There are diverse established spinning processes for producing fibres. Melt-spinning hereby accounts for the largest part of man-made fibre materials. Virtually any thermoplastic polymeric material can be melt-spun. Figure 7.9 shows the principle components of a melt-spinning plant. Polymeric granules are fed into an extruder, which melts and mixes the polymer. At the end of the extruder a homogenous, molten polymeric spinning mass is pressed into the spinneret by a spinning pump. The polymer is forced through the narrow openings of the

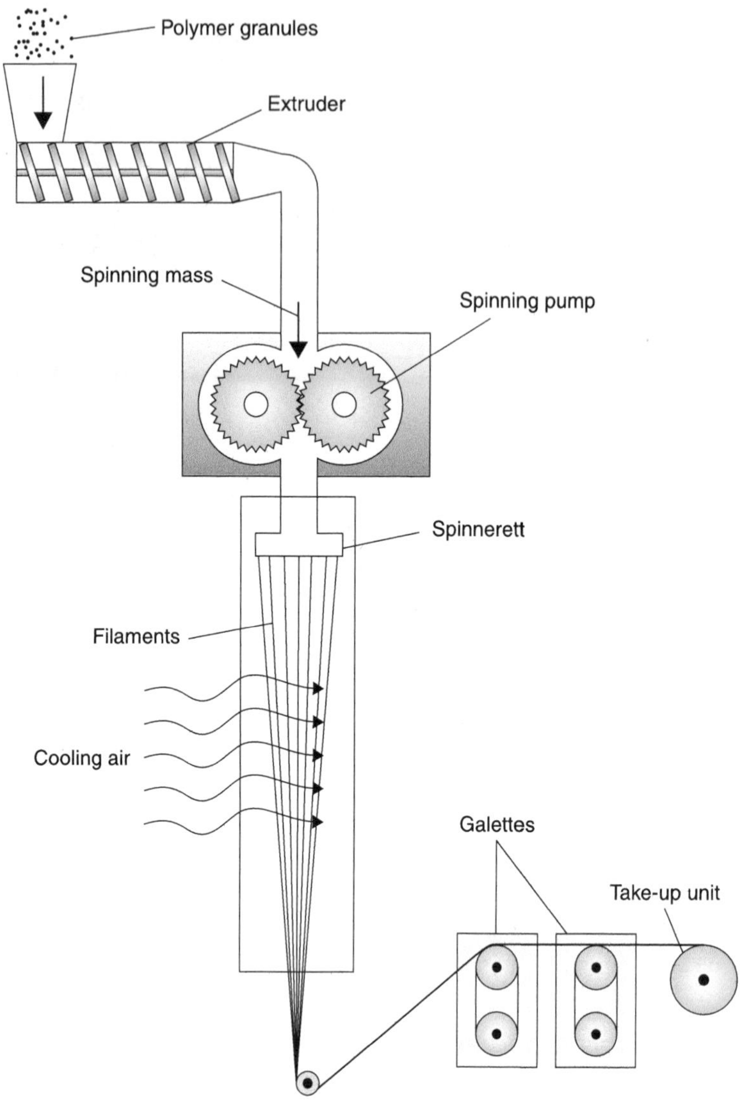

7.9 Principle components of a melt-spinning plant.

spinneret, resulting in a fine thin stream of molten polymer, called filaments. A spinneret can contain several hundred openings, resulting in several hundred filaments. In the subsequent cooling chamber, the filaments are treated with cooling air. During the cooling process, a certain amount of elongation is induced into the yarn by gravity. When the polymer solidifies, the filament bundle (yarn) is guided over several pairs of heated rotating drums, called galettes. The yarn is heated to glass temperature to soften the filaments and therefore make them more susceptible to mechanical strain. The strain, which is induced by the difference in rotation speed of the galettes, stretches the yarn. During the stretching process, the yarn becomes finer and the molecule chains of the polymer are oriented along the yarn axis, which heightens the mechanical tenacity of the yarn. Multiple galette pairs can be used to achieve the desired yarn fineness. Common materials, which can be melt-spun, are polyester (PES), polyethylen (PE), polypropylen (PP), polyamide (PA) and polyvinylidenfluoride (PVDF). Finally, the yarn is wound onto a bobbin. Filament yarns are also called 'endless fibres', due to the fact that on a bobbin each filament is one continuous piece of the yarn.

7.3.2 Solution spinning

Solution spinning follows another approach (Fig. 7.10). The polymeric material is dissolved and the spinning mass is pumped through the spinneret. Two principles of solution spinning can be applied, wet-spinning and dry-spinning.

In the *wet-spinning* process, the spinneret is located in a basin with a chemical liquid. By chemical precipitation, the polymer is spun out and the filament bundle can be drawn from the basin. Typical materials that are spun with the solution spinning method are viscose and ultra-high density polyethylene (UHDPE).

The process depicted in Fig. 7.10 is called 'wet-spinning', since the formation of the fibres takes place in the presence of a liquid chemical medium.

The *dry-spinning* process has a similar process diagram to the melt-spinning process. The polymer is dissolved and the spinning mass pumped through the spinneret. Hot air is injected, the solvent evaporates and is sucked out of the chamber. With the solvent evaporated, the polymer solidifies, forming the filaments.

7.4 Functionalisation of fibres and yarns

In the foregoing sections, the production of fibres and yarns has been discussed. The production processes represent standard textile production methods. For smart textiles, electrically conductive materials are necessary to distribute data and energy. The materials and processes affiliated with electrically conductive yarns are presented in Chapter 2 of this book. However, a very short summary of the possibilities to functionalise yarns in the yarn formation process are given.

A staple fibre spinning is a mixture of staple fibre materials, which are employed in the clothing industry to modify yarn characteristics. Yarns for knitted woollen sweaters for example, contain a certain amount of acrylic fibres to increase the wear and rubbing resistance of the yarn. In terms of smart textiles, a certain amount of electrically conductive staple fibres can be mixed into the yarn to achieve a certain level of electrical conductivity. Since adding only a percentage of electrically conductive fibres will not result in high conductivities, such yarns are often employed in protective clothing to achieve an anti-static effect. The

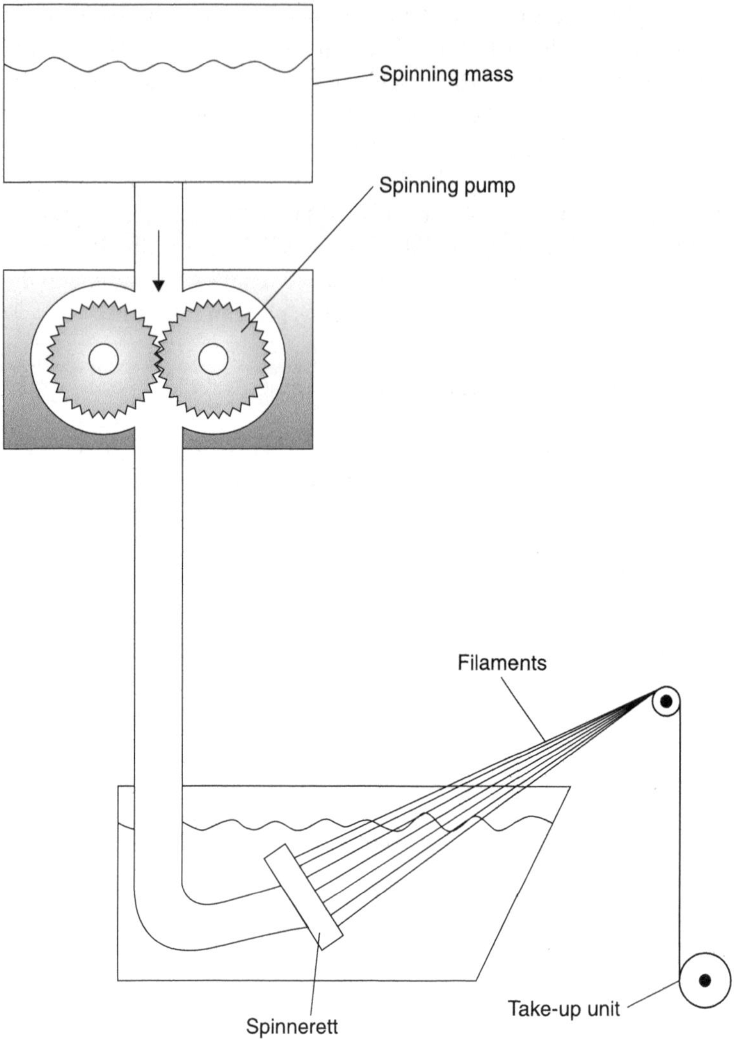

7.10 Solution spinning process.

yarns for the inner lining of fire fighter jackets for example, contain a mixture of mainly aramide fibres (heat resistance) and a low percentage of carbon fibres (which are electrically conductive). Usually about 2% of the fibre material is carbon.[2] This eliminates the risk of static sparks, which might ignite a combustible gas that the fire fighter is exposed to.

Another method to functionalise staple fibre yarns is **winding and twisting**. By twisting a fine metal wire around a staple or multifilament fibre core yarn, an electrical conductivity can be achieved (Fig. 7.11). While the metal wire acts as an

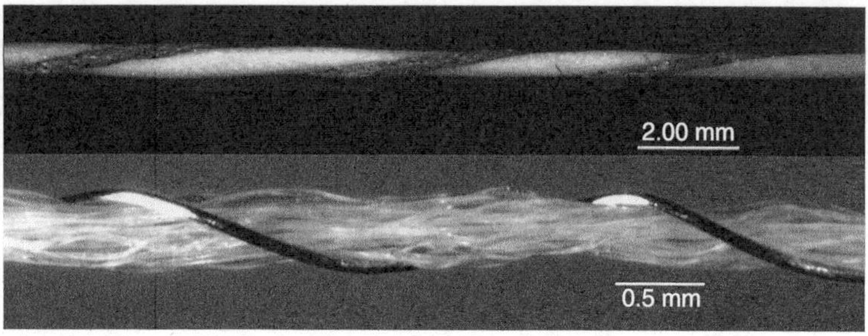

7.11 Electrically conductive material wound around a core yarn. Top: silver-coated Polyamide yarn. Bottom: silver wire.

electrical conductor, the staple fibre yarn is the load carrier. Flexibility of the complete yarn is given by the spiral-shaped path of the wire. When the core yarn is stretched, the angle of the wire helix becomes flatter, providing structural elongation without stressing the wire material.

In man-made fibres, electrical conductivity can either be achieved by spinning **intrinsically conductive polymers** or using **additives** such as **carbon nanotubes** or **carbon black particles**. An example of a polymer that can be produced in an electrically conductive form is PVDF. Walter *et al.* (2011) produced a piezoelectric sensor fibre made of PVDF, which was embedded in a fibre-reinforced sample part. To exploit the piezoelectric effect in a fibre (– generation of an electrical potential as a reaction to mechanical deformation), the PVDF fibres were melt-spun, drawn, implemented into a fibre-reinforced sample and then polarised between two electrodes under the influence of heat. The sample part was then subjected to an alternating mechanical load and the potential of the piezo fibres electrically measured. It was shown that the piezoelectric effect in fibres allows the measurement of mechanical deformation, thus enabling the monitoring of a technical sample part.

By adding particles to the spinning mass, polymers can be made electrically conductive, but a dilemma arises with the inclusion of particles. The more particles that are added, the better the electrical conductivity will be. However, the mechanical properties of the yarn will decrease as more particles are mixed into the molten polymer. The maximum conductivity, which can be achieved by these processes, is 1.000 S/cm (Siemens [S]: $1\,S = \Omega^{-1}$).

A common method for achieving electrical conductivity is **coating the yarns with a fine metal layer** (Schwarz *et al.*, 2009). By employing an electrochemical process, a fine layer of silver, copper or gold can be deposited on the surface of the fibres, giving the material a good surface conductivity (Fig. 7.12). The advantage of this process is that very good electrical resistance is achievable (up to 50 Ω/m for plied yarns[3]). However, often the metallic layer on such yarns is very susceptible to mechanical friction and rubbing. The metallic coating is damaged

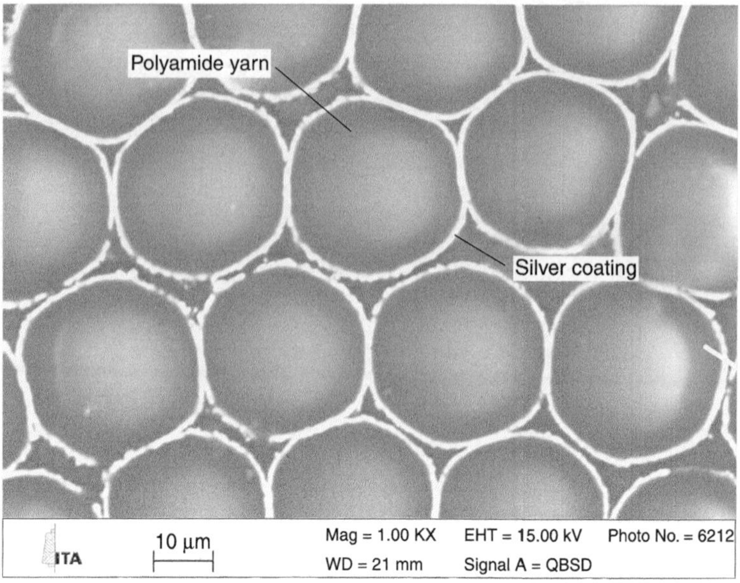

7.12 Silver-coated polyamide fibres.

during the textile production process (weaving, knitting etc.). Subsequent washing of the product induces even more rubbing, creating cracks in the metal surface and rubbing the metallic material off partially. Also, the price for such yarns is comparatively high and therefore not fit for the mass market (100–1.000 €/kg).

The 'holy grail' of electrically conductive fibres would be a material with:

- a high electrical conductivity;
- a high mechanical tenacity;
- wash resistance;
- is comparatively cheap and available in large quantities.

In current research projects, electrically conductive materials are being researched and good progress has been made, broadening the range of possibilities for smart-textiles developers. Future research will intensify this trend.

7.5 Fabric production: weaving

In the fabric production process, the step from linear textile systems (yarns) to two-dimensional (2-D) textiles (fabric) is made. The three most common production methods for fabrics are weaving, knitting and braiding. This section discusses weaving. Weaving is one of the oldest fabric production technologies, dating back thousands of years. In the weaving process, two yarn systems are used, warp and weft thread (Fig. 7.13).

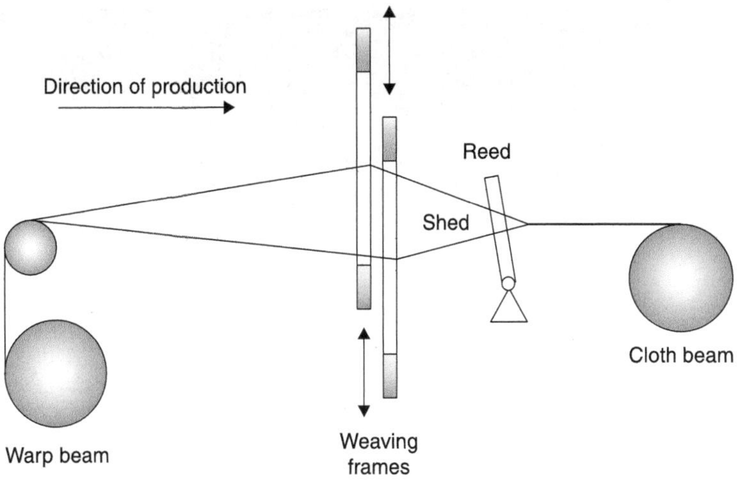

7.13 The basic principle of weaving.

The warp threads are oriented in the direction of production, and the weft threads are oriented perpendicular to the direction of production. The warp threads are stored on a warp beam. Depending on the width of the fabric, several hundred to thousands of warp threads are spooled onto a warp beam. From the warp beam the warp threads are guided through several weaving frames. The weaving frames can be moved up and down, pulling a part of the warp threads with them. In the space between the warp threads (called the shed), the weft thread can be inserted. After the weft thread has been inserted into the shed, the reed dashes the weft thread against the fabric border. After that, the weaving frames move in opposite directions, binding the weft thread into the fabric and another weft thread insertion cycle can begin. The produced fabric is rolled up on the cloth beam.

There are several methods for inserting the weft thread into the fabric, the most common being *shuttle*, *rapier* and *airjet* technology. The **shuttle** weaving process is the oldest weft insertion technology. A shuttle consists of a small frame, which holds a bobbin with the weft material. During the weft insertion phase, the complete amount of thread is transported through the shed. This produces a closed rim on the side of the fabric (Fig. 7.14).

With the **rapier technology**, the weft thread is inserted into the shed using two grippers (Fig. 7.15). The rapier/gripper on the left side of the machine clamps the weft thread and pulls the thread from an external bobbin into the shed. In the middle of the fabric, the two rapiers meet. The thread passes from one gripper to the other and the rapiers retract. The reed dashes the weft thread against the fabric border and the weft thread is cut. The weaving frames change position and the gripper takes hold of the bobbin-side weft thread end, which has been cut off.

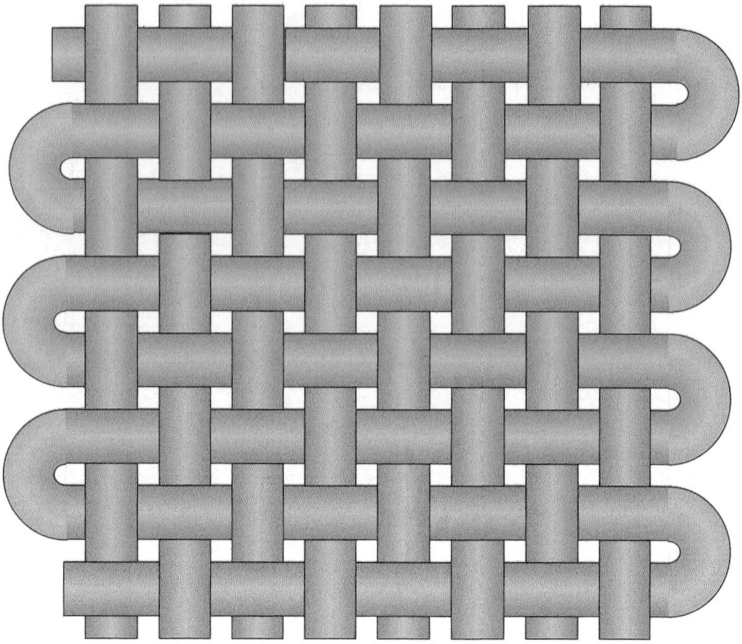

7.14 Closed fabric rim.

Since the weft thread is cut after each insertion cycle, the border of the fabric exhibits loose thread ends – the weft is not one continuous thread.

Another possibility for inserting the weft thread is **airjet technology** (Fig. 7.16). Compressed air is used to shoot the weft thread through the shed. A series of air nozzles are placed along the weft path, which provide an air cushion on which the weft thread travels towards the other end of the fabric. With airjet technology, the weft thread is also cut, producing the same fabric edge type as the rapier technology. Since no mechanical parts have to be moved to insert the weft thread, airjet weaving machines have a high productivity. Up to 900 weft insertions per minute are feasible.

7.15 Left: Rapier head with gripper. Right: Rapier loom weft insertion (image used with permission of Lindauer Dornier).

7.16 Airjet weaving loom (image used with permission of Lindauer Dornier).

7.5.1 Weave patterns

The weave pattern is determined by the configuration of the weaving frames. The most common standard weave patterns are plain weave (Fig. 7.17, left), twill weave (Fig. 7.17, middle) and atlas weave (Fig. 7.17, right).

At the plain weave, the warp thread passes alternately over and under a weft thread. With the twill weave, the warp thread passes over two and under one weft thread alternately. Also the twill weave exhibits an offset of one weft thread (Fig. 7.17, middle). At the atlas weave, the warp threads pass over three weft threads and then below one weft thread. This creates a unique pattern for each weave. More sophisticated weave patterns are possible by using more weaving frames.

When weaving frames are employed, a certain set of warp threads is tied to a weaving frame. Only the complete set of warp threads can be moved up or down,

7.17 Typical weave patterns. Left: Plain weave. Middle: Twill weave. Right: Atlas weave.

7.18 Jacquard weaving loom (image used with permission of Lindauer Dornier).

thus limiting the weavable patterns. To achieve complex patterns, **Jacquard technology** can be used. With Jacquard technology, each warp thread can be controlled individually by a set of harness cords. Figure 7.18 depicts a weaving loom with a Jacquard attachment. A frame resides over the weaving machine. Above the warp threads, a Jacquard attachment is installed. Each warp can be individually pulled up or let down. This enables weaving of complex designs.

7.5.2 Functionalising woven fabrics

The individual characteristics of each weaving pattern and weft insertion principle offer a variety of possibilities towards the development of smart textiles. In most applications, conductive yarns are used in the weft, since weft yarns can be exchanged easily when another product is to be manufactured on the machine. In general, functional yarns can also be used in the warp. But this is more complicated and cost-intensive, since a special warp beam must be produced with an exact spacing of the warp threads. Also the warp yarns are exposed to more process steps, machine elements and therefore friction and tension. The warp yarns must be spooled onto the warp beam, then guided through the weaving frames and bent up and down during the shed action. This exposes the warp yarns to an increased amount of stress, possibly damaging the electrically conductive coating or breaking conductive filaments, which can cause short circuits in the woven product.

This implies also that there are challenges when weaving electrically conductive yarn material. In order to achieve high productivity, it is desirable to weave at high speeds. The weaving process must be controlled in such a way that mechanical stress on the yarns and abrasion are kept to a minimum. It has to be kept in mind that weaving is inseparably connected to the production of dust. Hundreds of yarns are in motion, rubbing against each other and producing fine fibre fragments, which float in the air; this is mainly the case when staple fibres are used for weaving, since the loose fibre ends sticking out of a yarn body can break or be pulled out of the yarn. Electrically conductive loose fibres, which float in the air, are a danger to the electronics in the machinery, since the fine fibres can pass through tiny gaps and short circuit the control boards of the weaving machine. The electrical parts of the weaving machines can, of course, be encapsulated against these fibres. The problem of short circuiting caused by conductive fibres is severe when, for example, carbon fibre fabrics for fibre-reinforced parts are produced. Machinery processing carbon fibres must be specially encapsulated, which is cost-intensive. Therefore it is desirable to use as few conductive fibre elements as possible, which can be achieved by employing functional fibres only in the weft.

Using *shuttle weaving* technology and an electrically conductive yarn in the weft, it is possible to weave a fabric in which the conductive weft thread is going through the complete textile (Fig. 7.14). Such a textile could be used, for example, as a large area heater. Only the ends of the yarns would have to be in contact to an electrical current, which makes the contacting process easy. Fabrics woven with shuttle technology could also be employed as a sensor textile in applications where the detection of an event on a large area is required (e.g. deformation or damage of the fabric).

In fabrics woven with **rapier** or **airjet technology**, each weft thread is cut. The highest degree of design flexibility is achieved by using a combination of Jacquard technology and a three-layered woven fabric. A three-layered fabric is created by employing multiple weaving frames, which create two sheds. Using this set-up and electrically conductive warp and weft threads, it is possible to establish an electrical contact at defined positions in the fabric (Fig. 7.19). The top and bottom

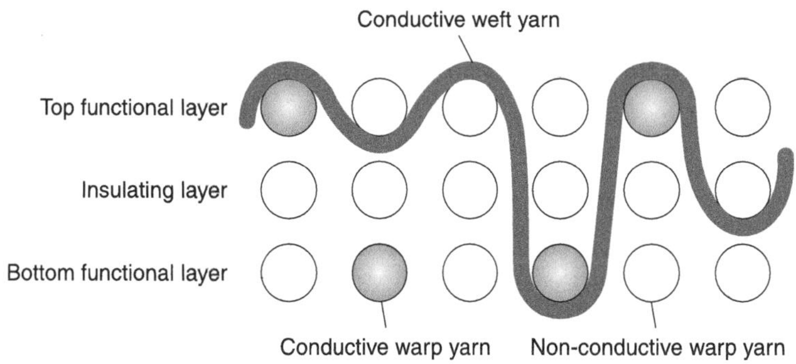

7.19 Textile circuit board realised by employing a three-layered Jacquard woven fabric.

warp thread layer contain electrically conductive yarns. The middle layer electrically separates the two outer layers. By employing Jacquard weaving technology, each warp thread can be controlled individually. Complex patterns can be woven into the fabric, which control the contact points between the conductive yarns. This method allows the production of textile circuit boards with exactly defined electrical paths and interconnections.

7.6 Fabric production: knitting

The term knitwear includes two main textile techniques, weft and warp knitting (Spencer, 2001; Weber and Weber, 2008) (Table 7.1). After weaving, it is the most common method of manufacturing textile fabrics. Because of the interlooped structure of the knitted fabric, the properties are completely different to woven fabrics. The difference in weft and warp knitting originates in the way the needles move during the production and in the way the yarn is supplied. Weft knitting is a one fibre technique, which means that only one fibre is needed to build the stitches. The needles are moved separately, whereas the warp knitting needles are moved simultaneously. Therefore, all needles need the fibre material at the same time. For this reason, the yarn is supplied with the help of warp beams. The most important knitwear fabrics are circular knitted, warp knitted, flat-knitted fabrics and fully-fashioned fabrics.

The knitted textile structure evolves from loops that are intermeshed row after row. The needle hook is responsible for the formation of a new loop with the supplied yarn. During the upward movement of the needle in order to catch the

Table 7.1 Classification of weft and warp knitting

	Weft knitting	Warp knitting
Needle arrangement		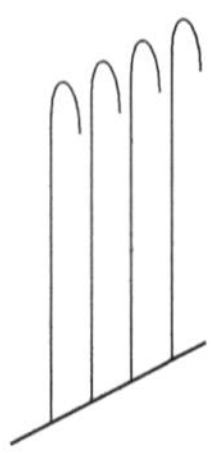
Needle movement	Separately	Simultaneously
Yarn supply	Single yarn	Depending on machine
Techniques	Flat knitting 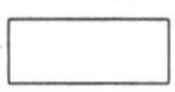	Fully-fashioned knitting machine • Single yarn supply 
	Circular knitting 	Flat warp knitting machine • Warp beam supply 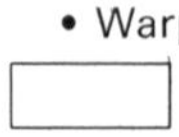

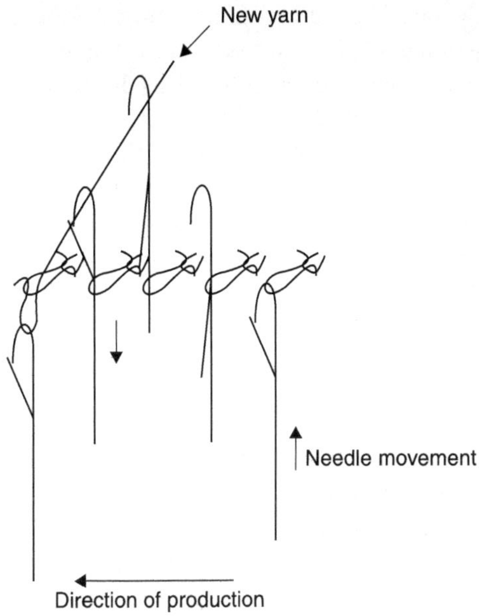

7.20 Needle movement during loop formation.

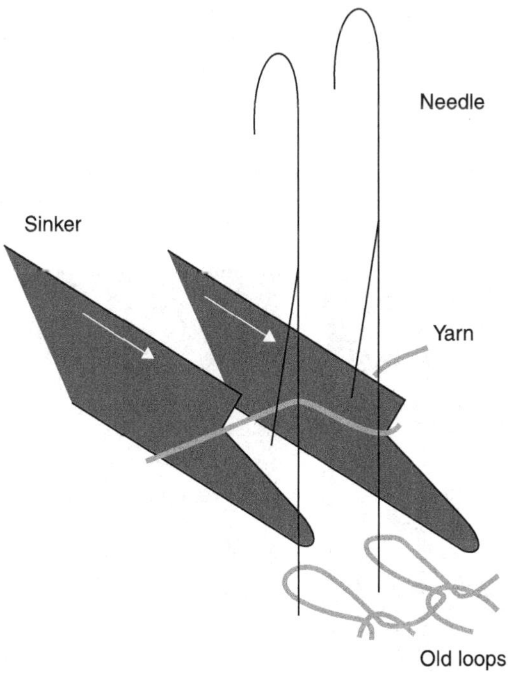

7.21 Sinker.

yarn to build a new loop, the old loop slides down the needle (Fig. 7.20). This causes the opening of the needle. The needle hook is now open to catch the yarn. The newly built loop is drawn through the old loop from the previous knitting circle. During this movement, the needle is closed. Now the old loop can be released as the new loop remains in the needle hook.

The sinker is also important for the production of knitwear (Fig. 7.21). It is a thin metal plate, which can have different shapes. Each sinker is positioned between two needles and its main purpose is to help build the loop. Furthermore, it holds the loops that were formed in the previous circle down when the needle moves upwards and downwards to build the new loops.

7.6.1 Circular knitting

Circular knitting machines always produce a tube-shaped fabric. They exist in different sizes or diameters, depending on the field of application. The needles and sinkers in these machines are arranged in a circle. The machines can be divided into two different types, depending on the number of sets of needles:

- *single set*: plain;
- *two sets*: rib, interlock, spacer fabric.

Both single set und double set machines also exist as Jacquard machines, which are needed for special designs. In these machines, the movement of each needle can be controlled from each cam. Common products that are produced with circular knitted fabric are T-shirts. For production, nearly every material can be used. The form varies from filament to staple fibre yarn. For special purposes, also monofilaments and wires are used.

Machines that possess just one set of needles are only able to produce plain-knitted structures (Fig. 7.22). In these structures, one side of the fabric shows right

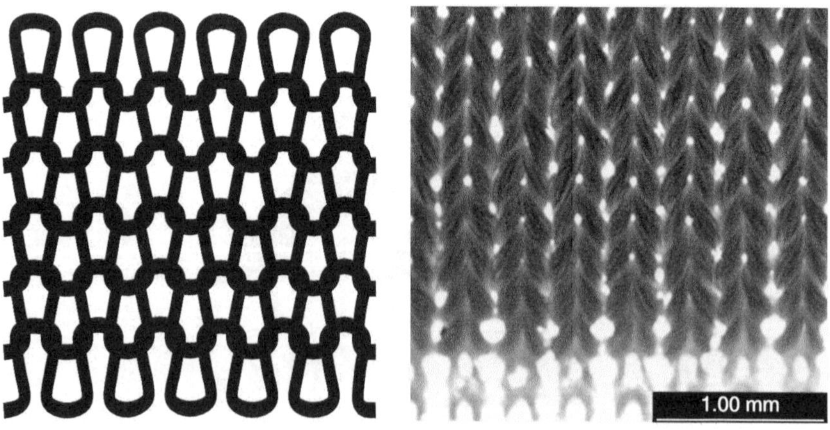

7.22 Plain-knitted structure.

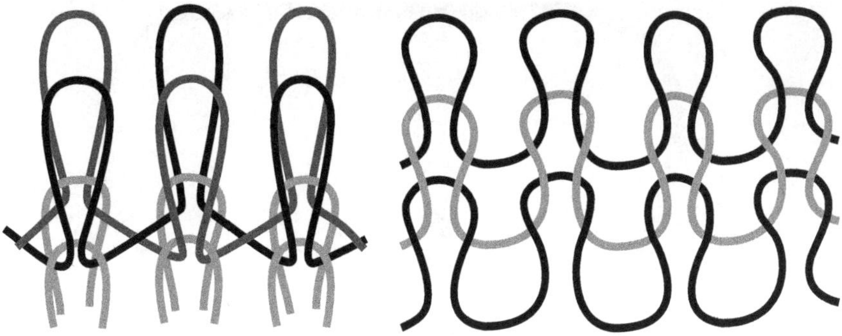

7.23 Left: Interlock structure. Right: rib structure.

loops and the other side rib loops. The following picture shows the loop structure of a plain knitted fabric.

The interlock structure was derived from the rib structure (Fig. 7.23). For the production of this kind of fabric, two needle sets are necessary and the needles need to be arranged in a different way. The loops are formed in two different directions (Fig. 7.24). The result is a fabric with smooth surfaces on both sides. This is due to the right loop structure on each side. The rib structure shows rib loops on both sides of the fabric. These fabrics can be produced using loop- or needle transfer.

7.6.2 Flat knitting

Machines for the production of flat-knitted fabrics are the most versatile of the weft-knitting machines. Flat-knitting machines can be divided into two groups:

1. hand-propelled and hand-manipulated models;
2. automated, electronically-controlled, power driven machines.

Normally, the machine for flat knitting has two stationary beds that are arranged in an inverted V formation. These beds possess tracks in which the needles can be moved. The fabrics produced by a flat-knitting machine are mainly coarse and intensely patterned. An advantage of flat-knitted products is that vertical and horizontal stationary threads can be integrated into the fabric. In this case, the fabric serves to fix these threads. Fabrics produced this way can be used for technical textiles. Common products produced on conventional flat knitting machines are outer-wear, such as jumpers that consist of staple fibre yarns.

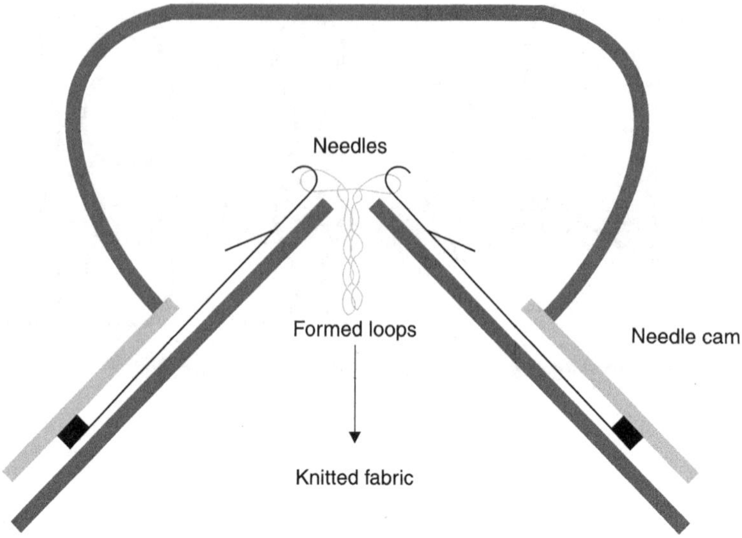

7.24 Loop formation.

7.6.3 Warp knitting

Dependent on the angle of take-off of the fabric from the needle and dependent on the pattern possibilities, it is possible to divide warp-knitting machines into two different types:

1. warp-knitting machines;
2. Raschel warp-knitting machines.

Raschel knitting machines often possess more guide bars. A common warp-knitted structure is shown in Fig. 7.25. As mentioned before, the needles in warp knitting machines are moved simultaneously. The structure that one single thread forms is shown in grey.

Mainly filament yarn is used for these fabrics, due to the high yarn tension during production. The yarn breaks more often if staple fibres are used, which is also possible. In comparison with filament yarns, the machine efficiency is therefore lower, because broken fibres need to be repaired.

7.7 Fabric production: braiding

Note: The information in this section can be found in Eichhoff (2012). Braiding is a fabric production method, which requires at least three yarns. By intertwining the yarns alternately following a certain algorithm, the braid is formed. According

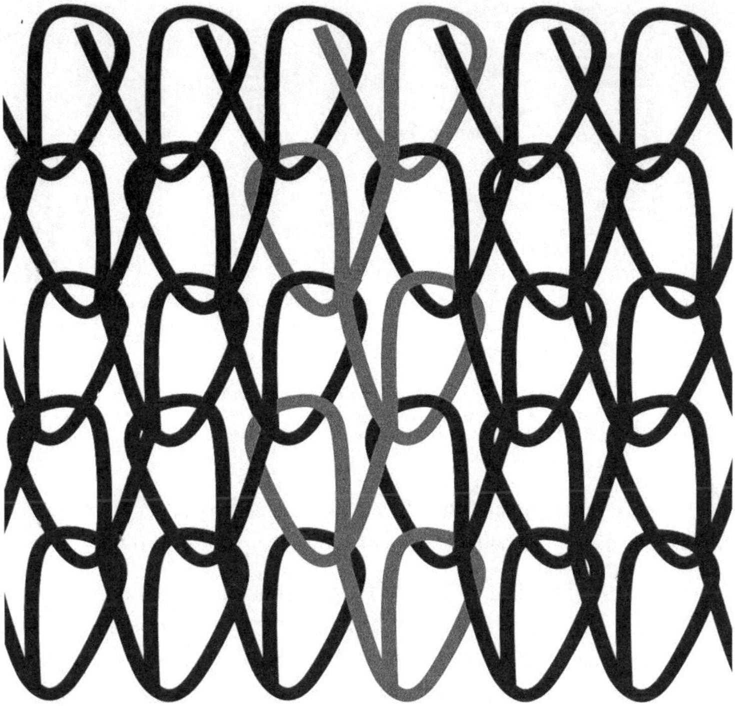

7.25 Warp-knitting structure.

to DIN 60.000, braids can be defined as 'areal- or voluminous bodies with a regular thread density and a closed product surface, whose braid- (bobbin-) yarns are intertwined diagonally towards the product edge.' In contrast, the threads in a woven fabric are intertwined in a perpendicular fashion towards the product edge. Industrial braids are produced on braiding machines. The main elements of a braiding machine are depicted in Fig. 7.26.

The yarn is stored on bobbins, which reside on braiding carriers. The braiding carriers store the yarn material, control the yarn tension, release yarn material in a controlled manner during the braiding process and move the bobbins as dictated by the braiding machine mechanism.

Horngears hold the braiding carriers and transport them from one horngear to the next. The braiding carriers are grouped in two sets, with opposite movement directions (clockwise and counter-clockwise). The layout of the horngears dictates the form of the braid. Two different horngear layouts are depicted in Fig. 7.27.

> [A tubular] braid consists of two systems of yarns, alternately passing over and under each other, causing a zig-zag pattern on the surface of the fabric. One

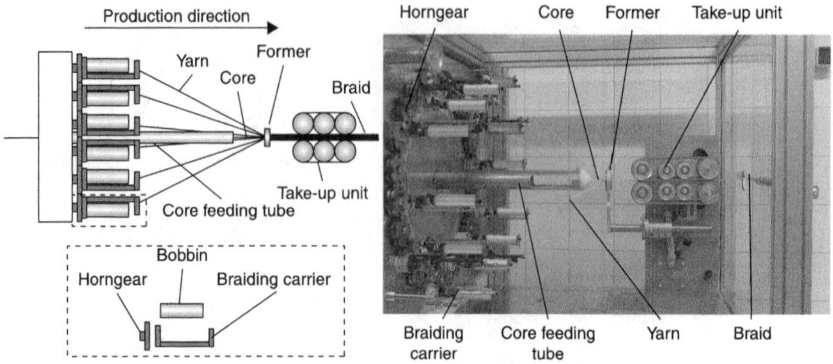

7.26 Main elements of a braiding machine.

system of yarns moves helically clockwise with respect to the fabric axis while the other moves helically counter-clockwise (Ko *et al.*, 1989).

For tubular braids (Fig. 7.27, left) the horngears are arranged in a concentric fashion. The horngears have alternating directions of rotation, turning subsequently clockwise and counter-clockwise. This way a braiding carrier takes an (approximated) sinusoidal path around the centre of the braid (Fig. 7.28).

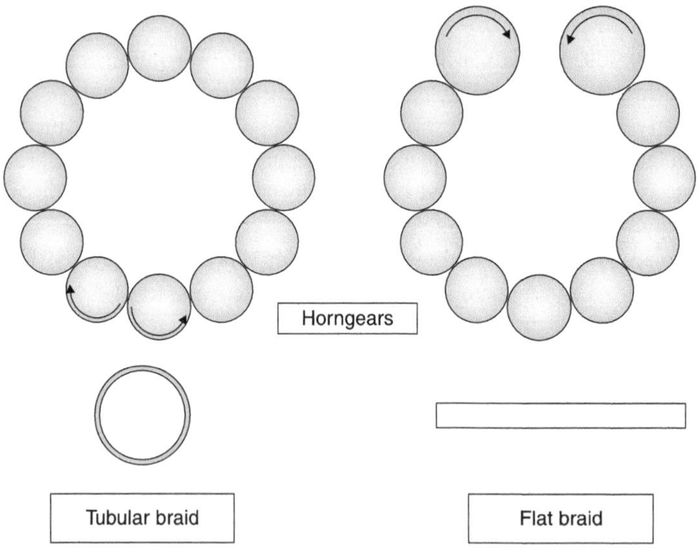

7.27 Two layouts of horngears on a braiding machine. Left: circular (tubular) braid. Right: flat braid.

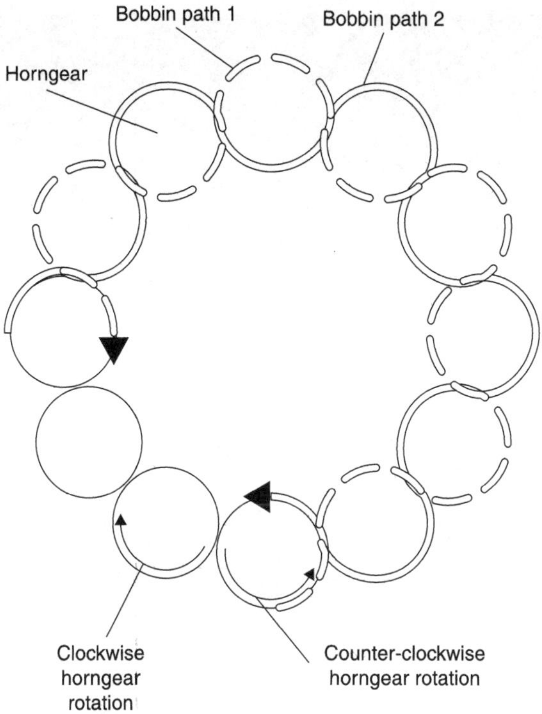

7.28 Path of braiding carriers.

The two sets of braiding carriers constantly change inward and outward position, intertwining the yarns and forming the braid.

Flat braids can be produced by arranging the horngears in an open pattern (Fig. 7.27, right). The bobbins travel along the horngears until they reach one of the final horngears. There they complete a full turn on the horngear and travel back. The resulting textile fabric is a non-tubular, flat braid. The braid is guided through a former. At the point of braiding (also called the fell) the yarns intertwine, forming the braid. A take-up unit controls the movement of the braid and transports the finished product away from the former to provide space for the new fabric.

Braids can have a biaxial or a triaxial structure (Fig. 7.29). Biaxial braids have two sets of yarns, their structure as described above. Triaxial braids have one additional, (axial) thread system. In addition to the two helical yarn systems in the textile, a third yarn system is introduced. This technology is used, for example, in fibre-reinforced composites. Reinforcement yarns can take up the load only along the axis of the yarn. The axial yarns are introduced into the braid using stationary feeding tubes.

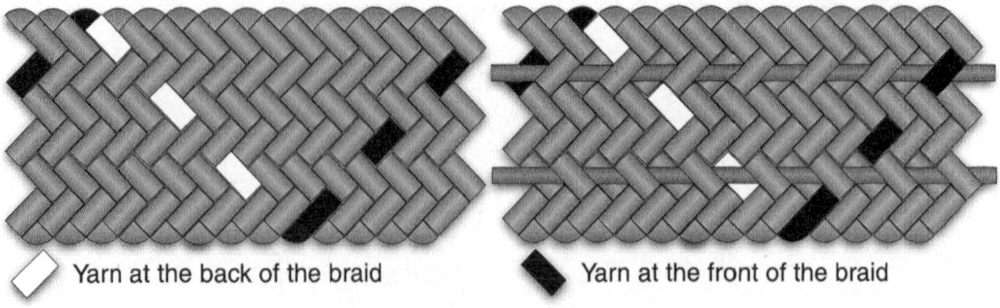

7.29 Left: Biaxial braid. Right: Triaxial braid.

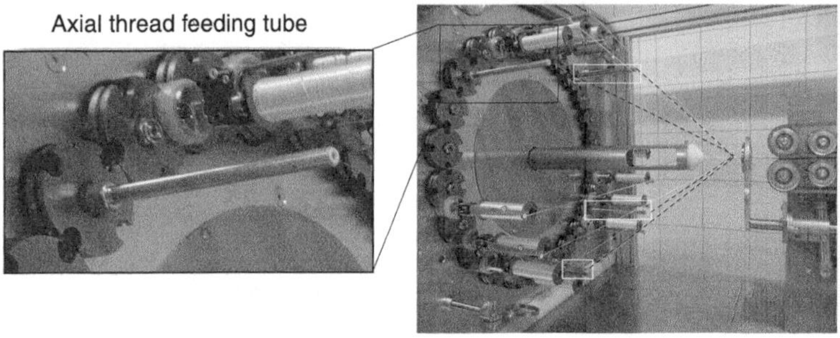

7.30 Introduction of a third yarn system on a braiding machine. raiding with 12 textile yarns in helical direction and 4 metal wires in axial direction.

In Fig. 7.30, the machine set-up for introducing axial threads into a braided structure is depicted. In the example shown, 12 braiding carriers are used to form the braid. On four horngears, an axial yarn feeding tube is installed. The feeding tube is screwed onto the axis of the horngear and does not rotate. Through this tube, an additional thread can be fed into the braiding process. The bobbins for the axial threads are placed outside the machine. In the example depicted in Fig. 7.30, four copper wires are fed into the braided structure (dotted black lines).

7.7.1 Functionalising braids

By using electrically conductive yarns in braided structures, sensors and data leads can be realised. Two examples will be given in this section. In the **Smart Rope** project, which was conducted at the Institut für Textiltechnik of RWTH Aachen University, a yarn-based monitoring system for braided ropes has been developed. An electrically conductive yarn was braided into a textile fibre rope.

When the rope is exposed to mechanical stress, the sensor fibre is also elongated and changes its electrical resistance. This change can be measured and provide information on the status of the rope. The Smart Rope system has been constructed in different application scenarios. The sensor system has been integrated into the shroud lines of a parachute (Fig. 7.31). When the chute opens and the jumper is slowed down, a high impulse is forced upon the shroud lines to stop the jumper from free fall. After each jump, the shroud lines have to be inspected for damage. With smart-textile sensors, the inspection time could be reduced from two hours to ten minutes.

The sensor threads have also been integrated into personal safety lines, elevator ropes and mooring lines for ships and oil rigs, to assess the condition of the rope in order to provide additional security.

In the **Profitex** project (www.project-profitex.eu), a data-transmitting security rope for fire fighters has been developed. The security rope has been braided with integrated data leads. The electrical leads were used as braiding yarn in a biaxial braid. By using four copper wires as braiding yarns in the functional layer, individual data leads could be realized. Since the copper wires run in the same set of bobbins, they never interlace. The copper wires run in four parallel helixes. Figure 7.32 shows an example of such a braided data cable. Four layers are braided over a core rope. The functional layer includes the electrical copper wire leads. The insulating layer separates the functional layer from the electromagnetic interference (EMI) shielding layer. The EMI shielding layer protects the signal-carrying lines from being influenced by external electromagnetic waves and is a necessary element in any high-frequency signal cable. The braided data cable is covered by a mantle layer, which protects it from mechanical wear.

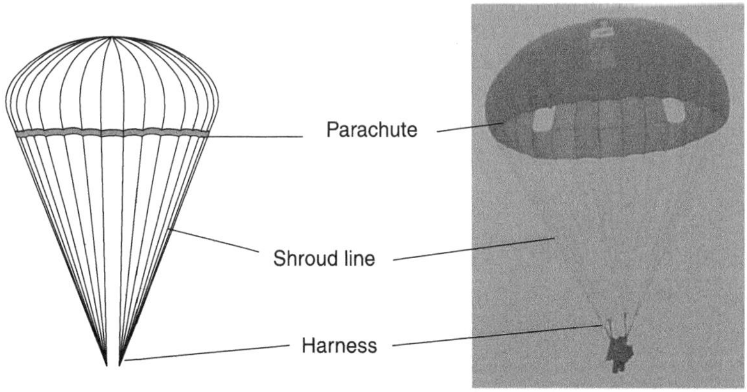

7.31 Sensorised shroud lines in a parachute.

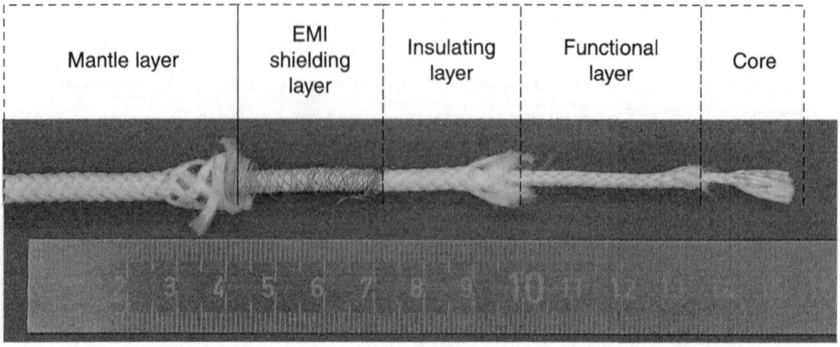

7.32 Braided data cable with four electrical leads in the functional layer.

7.8 Embroidery

Embroidery is a method that can be used to apply yarn materials to a textile substrate in a defined pattern (Gries and Klopp, 2007). In this section, two embroidery methods are presented. At first the 'classic' embroidery method will be discussed, which employs a two-thread system. After that, a variant of the embroidery process will be presented – the tailored fibre placement (TFP) method. This method introduces a third yarn system, which enables the exact placement of sensor yarns, tubes or stiff materials like metal wire, etc.

7.8.1 Standard embroidery (two-thread system)

Figure 7.33 depicts an embroidery machine. The upper thread (1) is stored on a conical bobbin. It is guided through two yarn brakes (2,3) and through the yarn tensioning lever (4). The lever moves up and down to compensate for the movement of the needle (5) during the embroidery operation. The basic fabric (7) is held under tension in the embroidery frame. A computer unit (6) controls the braiding operation.

During the embroidery process, the frame with the basic fabric is moved in X- and Y-direction to create the pattern. The needle punches through the fabric, interlacing the upper thread with the bottom thread by means of a rotating gripper, which is located below the basic fabric. The bottom thread capsule is depicted in Fig. 7.34.

The resulting stitch can be observed in Fig. 7.35. A symmetrical stitch is created, which has the same appearance on the upper side as on the bottom side of the fabric. With the embroidery process, complex patterns can be realised. Conductive yarns can be used to realise circuits and sensors (Tao, 2001, 2005;

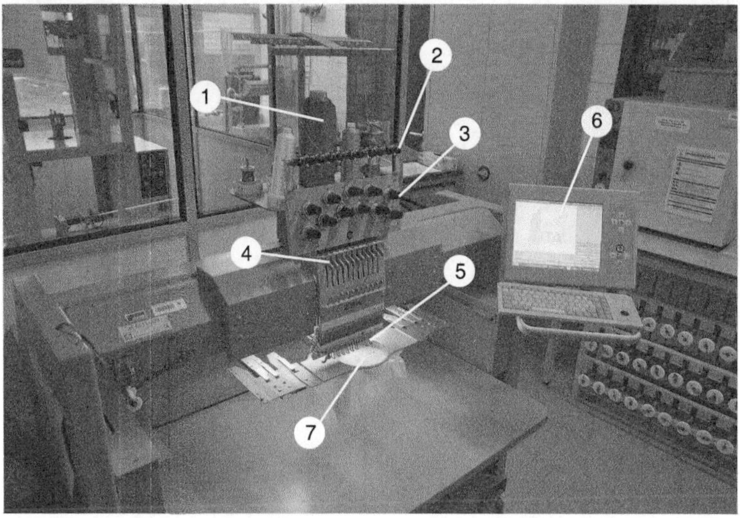

7.33 Embroidery machine elements: 1-bobbin; 2,3-yarn brakes; 4-yarn tensioning level; 5-needles; 6-computer unit; 7-asic fabric.

7.34 Capsule containing the bottom thread spool.

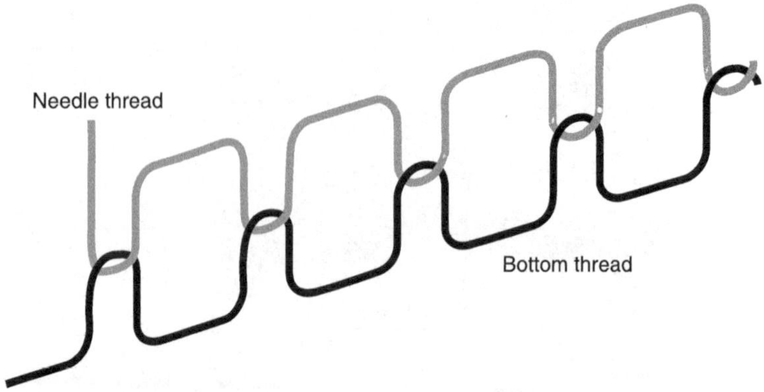

7.35 Double lock stitch.

Linz, 2005, 2006, 2007). Since the process is computer-controlled, it is very accurate and reproducible.

Figure 7.36 shows an example of embroidered circuits. A thin electrical cable has been used as bottom thread. The needle thread is PES. By adjusting the yarn tension in such a way that the bottom thread receives a higher tension than the needle thread, the bottom thread is placed straight onto the textile substrate. The needle thread does not exert enough force to pull the (stiff) bottom thread into a loop, therefore only small needle thread loops fixing the electrical cable are visible.

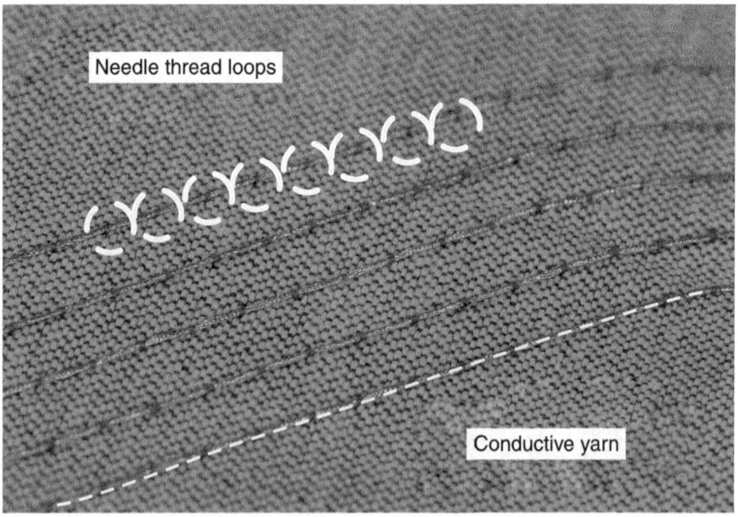

7.36 Embroidered conductive lines.

7.8.2 Tailored fibre placement (three-thread system)

TFP was originally invented to apply high tenacity rovings like glass or carbon fibre material onto a textile substrate to create 'preforms' for fibre-reinforced polymers. Preforms are dry textiles, which have the contour and dimensions of the technical part that will be produced. By infusing the preform with a polymeric resin, the fibre-reinforced part is created. The TFP method employs a three-yarn-system. In addition to the top yarn and the bottom yarn, a functional yarn is introduced. Figure 7.37 depicts the principle components of the TFP process.

The roving is placed on the textile substrate (basic fabric). The needle punches left and right of the roving into the basic fabric, creating a zigzag stitch across the roving. The top and bottom sewing threads act only to fix the roving. As the roving is never penetrated by the needle, no damage is induced in the yarn. The accuracy of the computer-controlled embroidery process allows a very close spacing of the roving material, therefore a closed surfaced can be created using this process.

Two examples of applications using the TFP method can be observed in Fig. 7.38. In the left image, the production process using the TFP method is depicted. In this image, the embroidery head places carbon fibres on a textile substrate. The pattern could be used as a preform for fibre-reinforced parts or as a smart textile, representing a resistor heating. The right image shows a scanning electron microscopy (SEM) image of an electrical cable, which has been applied using the TFP method (Fig. 7.39). The zigzag stitch of the bottom thread is clearly visible in this image.

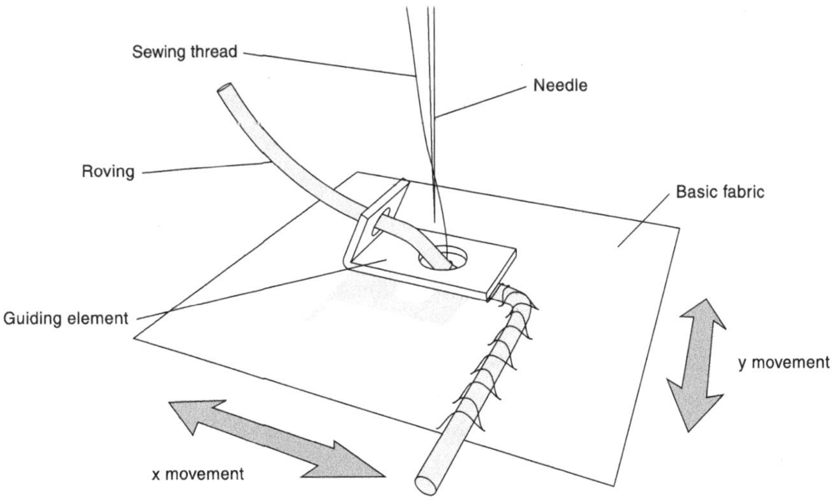

7.37 The tailored fibre placement (TFP) Method.

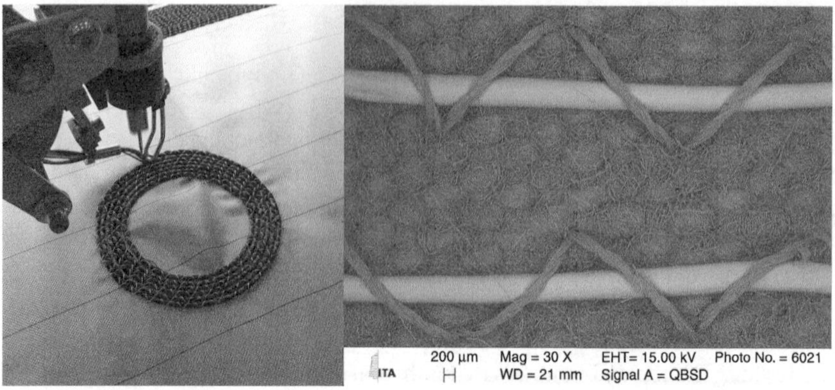

7.38 Left: TFP embroidery with carbon fibres. Right: electrical wire applied to a textile substrate using the TFP method.

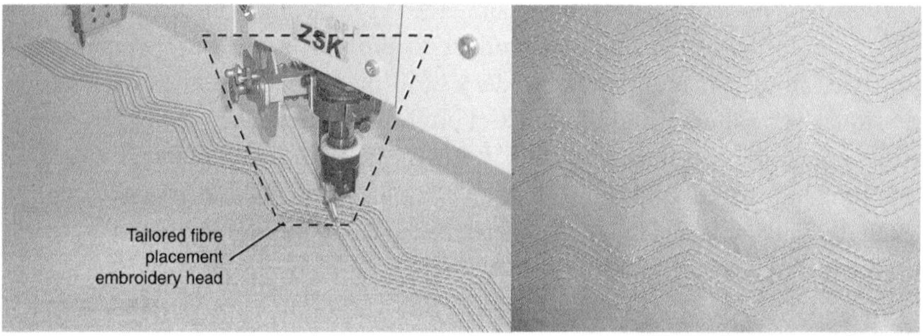

7.39 Application of electrical wire using the TFP method. Left: TFP embroidery process. Right: embroidered data lines.

7.8.3 Functionalising textiles using embroidery

'Regular' embroidery (two-thread system) can be used to apply yarns and fine metal wire onto a textile. The TFP method was initially employed to place high tenacity fibres on a basic fabric to create preforms. TFP allows the exact placement of fibre and fibre-like materials onto a textile substrate. Electrically conductive yarns and metal wire can be embroidered onto the fabric. Also, the use of tubes is possible. With this method, liquid-powered heating or cooling applications can be realised (Fig. 7.40).

Applying electrically conductive yarns offers a variety of braiding possibilities (Fig. 7.41):

- textile electrodes for body function measurement (Tao, 2001, 2005; Linz, 2006, 2007);

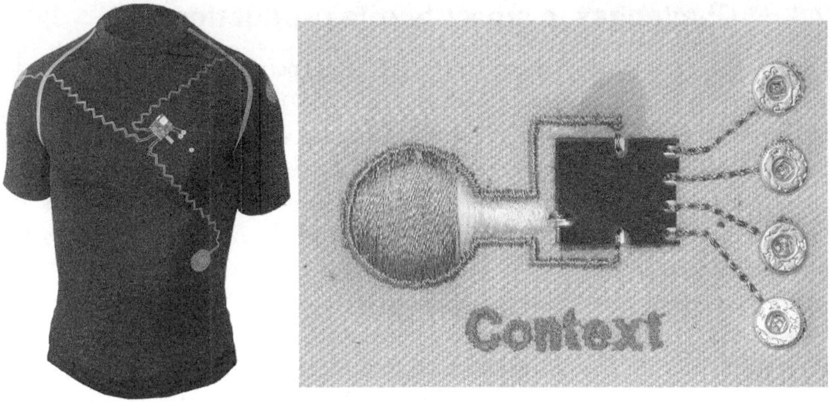

7.40 Applications making use of embroidered contacts. Left: an ECG shirt with embroidered electrodes for measuring an electrocardiogram (Linz *et al.*, 2006). Right: sensor for measuring an electro-myogram (EMG) (Linz *et al.*, 2007).

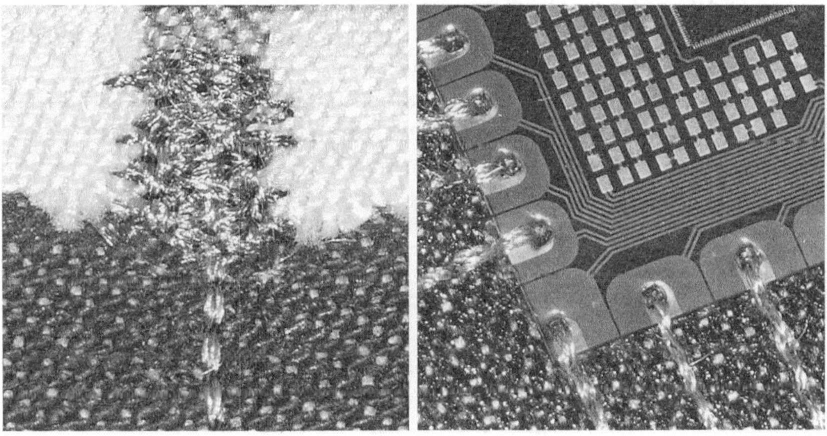

7.41 Machine embroidered electrical contacts to woven keypad (left) and to electronics substrate (right) (Linz *et al.*, 2005).

- circuits (by laying the fibres in the pattern of the electrical circuits) (Linz, 2005, 2006);
- heating applications[4] (Kochman and Gurevich, 2001);
- strain and deformation sensors (resistive or capacitive sensors);
- moisture sensors (either capacitive or short-circuit based);
- capacitive sensors (embroidering conductive pads into a three-dimensional (3-D) fabric).

7.9 Challenges in smart-textile production

Textiles and electronic devices are two very different partners and marrying them to a smart textile is a complex task (Schwarz *et al.*, 2010). In the world of electronics, an accuracy of production in the nanometer-scale is possible. Chips and circuits can be placed with an incredibly repeatable accuracy on the substrate. The most complex circuits can be realised in a small space. The substrates on which electronics are built are mostly stiff-printed circuit boards (PCB). Complex electronic devices can be manufactured at very low cost (cell phones with colour displays, built-in radios and a multitude of other functions are available for around 25 €).

However, textiles are flexible and the precision of manufacturing ranges in some tenths of a millimetre. Textiles produce dust and during their use some fibres will break and be set free. Marrying these two very different partners is a delicate task. When approaching smart-textile development, we should not target substituting a certain application built on regular PCB electronic circuitry with textiles. Referring to the above example, transferring cell phone circuitry onto a fabric is most probably not the way we should take. Regular PCB electronics are too advanced for smart textiles to be substituted.

Smart textiles must be regarded as a new medium, a new substrate for alternative applications. The strength of smart textiles lies in the flexibility and versatility of fabrics. Using the fabrics as conductors, sensors and actors can provide applications and products, which are not feasible with regular PCB electronics. Example products, to which these criteria match, are illuminated curtains (Philips: Luminous Textiles;[5] SensingTex: Illuminating Textile[6]), sensorised floor mats to detect movement in a building (Future Shape: SensFloor[7]) or monitoring fibre-reinforced parts with textile sensing systems.[8]

PCB electronics will not vanish. The 'real intelligence' will still be located in the evaluation and control units attached to smart textiles. Smart textiles offer a wide variety of applications and open exciting new markets in the future. A symbiotic co-existence of PCB electronics and textile-based electronics has been demonstrated in various research projects. The main challenges which impede the breakthrough of smart textiles are, as outlined earlier:

- robust, cheap and highly conductive yarn materials;
- automated production systems (placing electronics, contacting, encapsulating);
- robust interfaces (stiff PCB electronics ⇔ flexible fabrics).

By designing Smart-Textile solutions, which take into account the behaviour of textiles and the methods of textile production processes, integral products can be developed that are durable, effective and producible at low cost.

7.10 Notes

1 tex is the unit of measure for the fineness of a yarn. 1 tex = 1 g / 1.000 m; 1 dtex = 1 g / 10.000 m

[2] Source: Texport Fire fighting Jacket 'Fire Breaker Action Tough'. Outer shell: Nomex tough fibres with 75% m-Aramide (Nomex), 23%p-Aramide (Kevlar) and 2% P140 (Carbon), www.texport.at

[3] '*plying*' means that two or more yarns of the same type are twisted together to achieve a thicker yarn with a higher tenacity

[4] Heizteufel, Berlin (Germany), www.heizteufel.de

[5] http://www.lighting.philips.com/main/application_areas/luminous-textile/index.wpd

[6] http://www.sensingtex.com/index.php/en/products/illuminating-textile

[7] http://www.future-shape.com/de/technologies/11/sensfloor

[8] Walter *et al.* (2011) (monitoring FRP parts), (Messervey, 2010; Measures, 2001 (reinforced concrete)

7.11 References

Eichhoff J (2012), *Data-transmitting Textile Fibre Ropes for Firefighting Applications*, Diss, Aachen, Techn. Hochsch. Dissertation 2012. Aachen, Shaker.

Gries T and Klopp K (Eds.) (2007), *Füge- und Oberflächentechnologien für Textilien: Verfahren und Anwendungen*. Berlin, Heidelberg: Springer.

Ko F, Pastore C and Head A (1989), *Handbook of Industrial Braiding*. Covington, KY, Atkins & Pearce.

Kochman A and Gurevich A (2001), 'Soft electrical textile heater and method of assembly', *US Patent US 6,229,123*.

Linz T, Kallmayer C, Aschenbrenner R and Reichl H (2005), 'New interconnection technologies for the integration of electronics on textile substrates,' *Ambience 2005*, Tampere, Finland, September.

Linz T, Kallmayer C, Aschenbrenner R and Reichl H (2006), 'Fully integrated EKG shirt based on embroidered electrical interconnections with conductive yarn and miniaturized flexible electronics,' *Proceedings of the IEEE BSN International Workshop on Wearable and Implantable Body Sensor Networks*, Cambridge, MA, April.

Linz T, Gourmelon L and Langereis G (2007), 'Contactless EMG sensors embroidered onto textile,' *Proceedings of the IEEE International Workshop on Wearable and Implantable Body Sensor Networks*, Aachen, Germany.

Measures R M (2001), *Structural Health Monitoring with Fiber Optic Technology*. San Diego, CA, Academic Press.

Messervey T (2010), 'Improved Structural Health Monitoring,' *SPIE Newsroom*, 10 March, available from: http://spie.org/x39282.xml [accessed 21 December 2012]

Schwarz A, Hakuzimana J, Westbroek P and Van Langenhove L (2009), 'How to equip para-aramide yarns with electro-conductive properties', *Proceedings of the Sixth International Workshop on Wearable and Implantable Body Sensor Networks, BSN 2009*, Berkeley, CA, 3–5 June. Piscataway, NJ, IEEE, pp. 278–281.

Schwarz A, Van Langenhove L, Guermonprez P and Deguillemont D (2010), 'A roadmap on smart textiles', *Textile Prog*, 42(2), 99–180.

Spencer J (2001), *Knitting Technology: A Comprehensive Handbook and Practical Guide*, Third Edition. Abington UK, CRC Press.

Tao X (2001), *Smart Fibres, Fabrics and Clothing*. Cambridge, UK, Woodhead Publishing Co.

Tao X (Ed.) (2005), *Wearable Electronics and Photonics*. Cambridge, UK, Woodhead Publishing Co.

Walter S, Steinmann W, Seide G and Gries T (2011), 'Piezoelectric sensor fibers based on meltspun poly(vinylidene fluoride) and electrically conductive polymer nanocomposites', The American Association of Textile Chemists and Colorists (AATCC); The Fiber Society; National Textile Center (Hrsg): *Symposium on New Frontiers in Fiber Materials Science*, 11–13 October, Charleston, SC. Research Triangle Park, NC, AATCC, pp. 296–313.

Weber K P and Weber M (2008), *Wirkerei und Strickerei. Technologische und bindungstechnische Grundlagen*. Deutscher Fachverlag.

Wulfhorst B, Gries T and Veit D (2006), *Textile Technology*. Munich, Hanser.

8

Fabrication technologies for the integration of thin-film electronics into smart textiles

C. ZYSSET, T. KINKELDEI, N. MÜNZENRIEDER and G. TRÖSTER, ETH Zurich, Switzerland and
K. CHERENACK, Philips Research Eindhoven, The Netherlands

DOI: 10.1533/9780857093530.2.227

Abstract: Smart textiles are a key component for sensing physiological parameters close to the human body, supporting and assisting people in daily living. In addition, smart textiles offer potential in technical textiles: they comprise car and airplane interiors where lighting and emergency signaling are potential applications, in bandages for wound monitoring or automatically adjusting furniture in homes. The chapter starts with an introduction outlining the state-of-the-art smart-textile production. Then the processes for the fabrication of electronics on flexible plastic stripes, mechanical requirements due to the integration into woven textiles and the integration method of the flexible stripes into woven textiles are described. The feasibility of the technology is demonstrated with three prototypes: one prototype measures humidity and temperature, the second one has an integrated bus system and the last one is a textile integrated active matrix with light emitting diodes. At the end of the chapter, potentials of the technology are described to overcome issues related to smart textiles production, such as user acceptance, washability or large-scale production.

Key words: smart textiles, flexible plastic substrate, thin-film device.

8.1 Introduction

Integrating sensors or general electronics into textiles enables ubiquitous sensing to support people in their daily life. The sensing tasks involve physiological signals as well as environmental parameters for assistance in healthcare and rehabilitation, sports and high-risk professions such as fire fighting (van Langenhove and Herteleer, 2004; Paradiso *et al.*, 2005; Mattmann *et al.*, 2007). Textiles, which are usually worn close to the body, provide a suitable platform to integrate all kinds of electronics to measure physiological signals and environmental parameters. However, such textiles with integrated electronics will only succeed if they are accepted by the users requiring an unobtrusive and seamless integration of the electronics in the textiles. Moreover, smart textiles enable other fields of applications: one is technical textiles comprising for example carpets, curtains or car interiors (Savio and Ludwig, 2007). As for body worn textiles, unobtrusive integration of the electronics is required to maintain the look and the feeling of textiles. Another field is textiles for medical applications such as bandages.

A common approach to fabricate smart textiles is based on printed circuit boards (PCBs)[1] attached to a textile. The PCBs serve as substrates for microprocessors, sensors and data storage units (Park *et al.*, 2002; Harms *et al.*, 2008; Locher and Tröster, 2007a; Martin *et al.*, 2009). As an example, the Smash shirt (Fig. 8.1), developed for upper body position monitoring, consists of a main PCB containing a microprocessor and plugs for acceleration sensor units. The sensor units are distributed over the whole shirt and connected with the main PCB using cables to store the acquired data on a single device (Harms *et al.*, 2008). Attaching PCBs to textiles renders parts of the textiles rigid, which reduces the comfort of the wearer. In Linz *et al.* (2007), flexible PCBs are applied to reduce the rigidity of the textile.

To establish interconnections between individual PCBs on textiles, several methods are applied: one approach is to integrate insulated copper wires into textiles during weaving (Locher and Tröster, 2007b; Martin, 2009). Other approaches are screen-printing of silver paste on textiles (Locher and Tröster, 2007b), gluing insulated copper wires onto textiles (Harms *et al.*, 2008), encapsulate meandering copper lines into polymer and attach the whole structure to textiles (Gonzalez, 2008) or use embroidery (Linz, 2007).

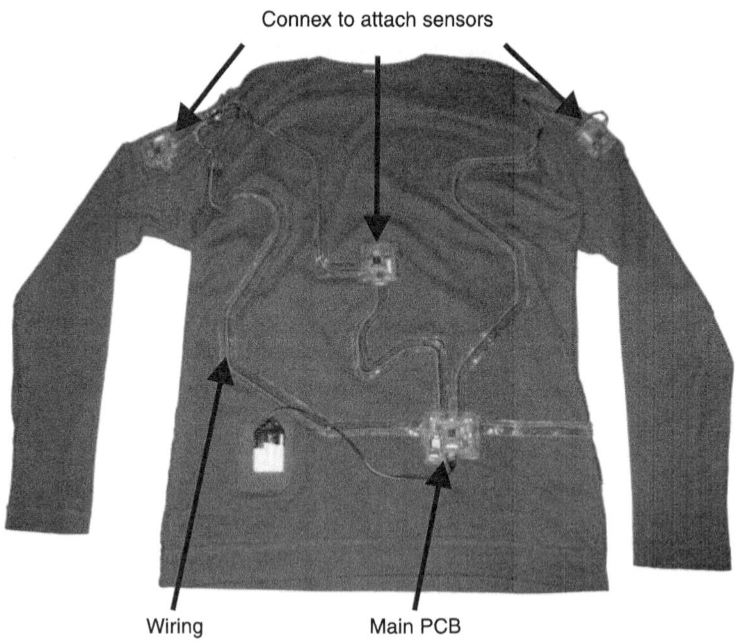

8.1 Smash shirt, with main PCB for data storage, and wiring to connect to the connex, where the sensors are plugged in.

A different approach to fabricate smart textiles is to integrate electronic devices such as resistors or transistors into single threads (Bonderover and Wagner, 2004; Abouraddy *et al.*, 2007; Hamedi *et al.*, 2007). This approach of integrating electronics into textiles is unobtrusive, but the circuits achieved so far consist of a maximum of two transistors. The aim of this section is to describe a different approach to integrating electronics into woven textiles using flexible plastic stripes, which comprise electronics in the weft direction. Conductive textile threads in the warp direction are used to interconnect the flexible stripes. In Section 8.2, the basic ideas of the technology are described, as well as the required process steps of fabricating the electronics and merging them with woven textiles. In Section 8.3, the feasibility of the technology is presented with three prototypes. Section 8.4 addresses aspects of the mechanical stability of plastic stripes with electronics woven into textiles. Sections 8.5 and 8.6 give an outlook on possible contributions of the technology to the field of smart textiles and resources to find further information.

8.2 Merging flexible electronics and smart textiles

As described in the Introduction, a common approach to fabricate smart textiles is attaching rigid PCBs to textiles and establish interconnects among the PCBs by using, for example, copper wires. The attached PCBs offer the possibility to integrate sensing systems with processing and storage capabilities on textiles, but they also introduce rigid parts in the resulting textiles, which reduce the comfort of the wearer.

The goal of merging flexible electronics and textiles is to develop a technology that combines the advantages from both the electronics and textile world. The electronics integrated into textiles should be capable of sensing and transmitting the recorded signals while the textile remains bendable and drapable. In addition, the technology should rely on existing processes to be compatible with standard machinery and production facilities from both the electronics and the textile fabrication chains.

The technology presented in this section makes use of the inherent x-y grid structure of weft and warp threads of a woven textile. In the weft direction, flexible plastic stripes, called e-stripes, are integrated during the weaving process into textiles. E-stripes serve as substrates for various kinds of electronics such as thin-film devices,[2] surface mount devices (SMDs) and integrated circuits (ICs). Furthermore, the e-stripes provide interconnect lines between the electronics and contact pads on the e-stripe. In the warp direction, conductive textile threads are used to establish connections among the contact pads on individual e-stripes to generate electronic systems integrated into woven textiles.

8.2.1 Technology principles

The x-y grid structure of warp and weft threads in woven textiles is beneficial for electronics integration, because electronic bus systems and matrices are arranged in a similar grid structure. Figure 8.2 shows an exemplary bus system, a matrix and schematically the structure of a woven textile. Bus and matrix systems are used to control sensors and actuators in an efficient way by sharing interconnect lines.

In a bus system (Fig. 8.2(a)), a number of devices, here labeled as 'Device n' to 'Device n+2', are interconnected with shared lines for communication and power supply. In Fig. 8.2(a), four bus lines are shown. However, depending on the communication scheme among the individual bus devices, this number can vary. Common bus systems are I^2C and SPI. Both bus systems require four shared lines, of which two are used for power supply and the other two for communication. Another example is the 1-Wire protocol, which requires two lines, one for common ground and one for communication and power supply.

A matrix system (Fig. 8.2(b)) consists of n columns and m rows. At each crossing point of a column line and a row line, an electronic device is located, which can be individually selected by sending an electrical signal on the specific row and column lines. As an example, when row number j and column number i are enabled, Device(ij) is selected. A prominent application of matrix systems are flat panel displays, where each individual pixel is controlled by selecting specific row and column lines (Crawford, 2005).

To integrate electronics into textiles, the inherent arrangement of warp and weft threads is used (Fig. 8.2(c)). The warp direction is used to integrate conductive threads, which serve as bus lines or as column lines in a matrix system. The weft direction of woven textiles is used to integrate flexible plastic stripes during the

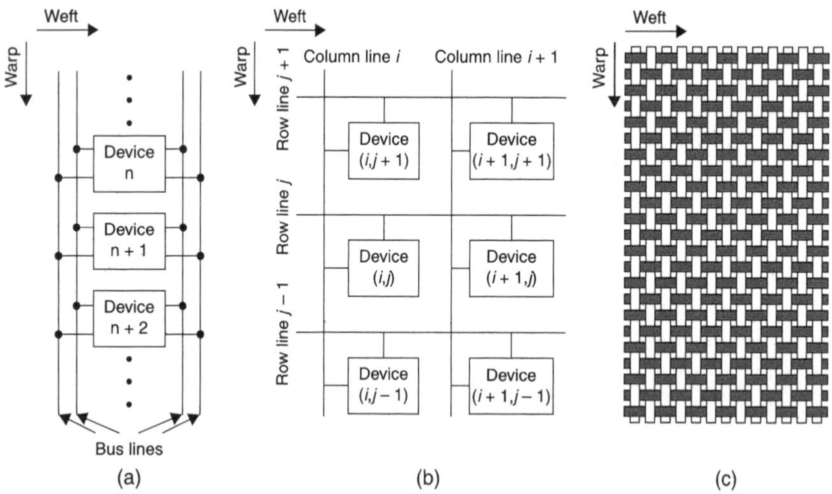

8.2 (a) Bus system with n devices connected to four shared bus lines. (b) Example of a matrix arrangement with two columns and three rows. (c) Arrangement of warp and weft threads in a woven textile, showing the similarities to the electronic bus and matrix system.

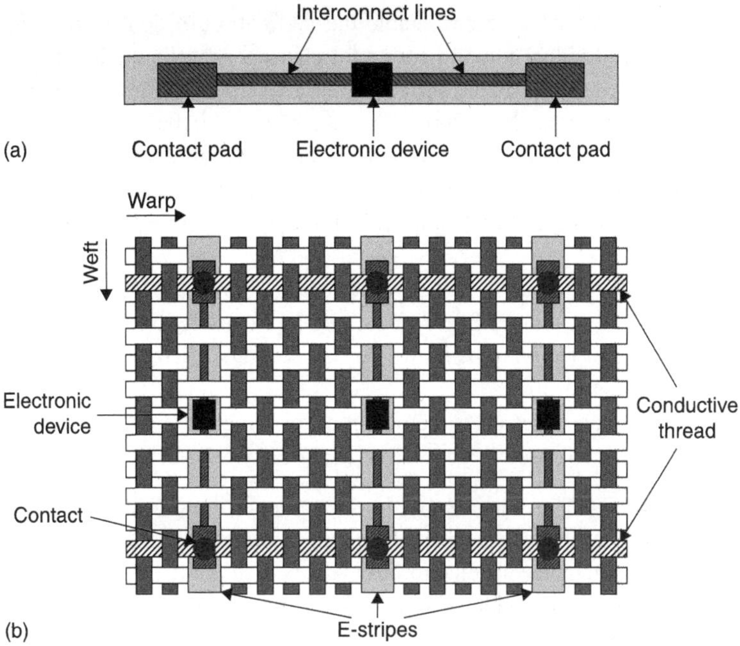

8.3 (a) Schematic of an e-stripe with an electronic device, interconnect lines and contact pads for conductive warp threads. (b) Schematic of the approach to integrate electronics into woven textiles. In weft direction, e-stripes are integrated and in warp direction, conductive threads are woven and contacted to the e-stripes. The white threads in warp and dark grey threads in weft are non-conductive textile threads.

weaving process. The e-stripes, providing a basis for electronics, are processed using clean-room techniques before weaving. The e-stripes serve as carriers for ICs, thin-film sensors, thin-film transistors (TFTs), interconnect lines and contact pads for the conductive threads in the warp direction. Figure 8.3(a) shows a schematic of an e-stripe containing an electronic device, interconnect lines and contact pads for the conductive threads. E-stripes in weft and conductive threads in warp are interspaced with standard non-conductive textile threads. After weaving, an electrical contact is established between the contact pads on the e-stripes and the conductive threads. In addition, the electrical contact is mechanically reinforced. Figure 8.3(b) shows the principle of the technology.

The integration of e-stripes based on flexible plastic substrates in the weft direction offers several advantages, besides the ability to carry electronics:

- Flexible stripes enable the reduction of rigid areas in smart textiles compared to PCBs.

- Metal layers on e-stripes enable the fabrication of contact pads and wiring among individual components on the e-stripe, as well as a flexible distribution of the electronics on the surface by selective patterning of the metal layer on the e-stripe (Crawford, 2005; Cherenack *et al.*, 2010a).
- Flexible substrates and hence e-stripes are a suitable platform for a variety of devices and circuits. They can range from thin-film devices, like transistors or sensors, to bare dies (semiconductor chips without packaging) and small SMDs.
- Due to the integration in the weft direction, the location of the e-stripes in the textile can be chosen freely, compared to integration in warp direction where a change of the warp beam would be necessary to change the location of the electronics.
- Flexible substrates offer the potential for large-scale production by applying printing or roll-to-toll techniques[3] (Service, 1997).

All these advantages converge in two essential attributes of this technology: first, the location of the electronics on the e-stripes is variable as well as the location of the e-stripe and hence the electronics in the woven textile. Second, the electronics can be chosen from a wide range of devices, such as SMDs, thin-film devices and bare die chips.

8.2.2 Stripe fabrication

Fabricating e-stripes, which serve as substrates for sensors, interconnect lines and contact pads, comprise the following steps: cleaning of the substrate surface, deposition and structuring of material layers, cutting the substrate into stripes and, if required, attaching SMDs or bare die ICs. As a base material for the e-stripes, a 50 μm thick polyimide foil, in particular Kapton E (from DuPont), is used. Kapton was chosen as the substrate due to its comparatively low coefficient of thermal expansion (CTE) of 16 ppm per °C, a high glass transition temperature (T_g) of 355 °C, which allows process temperatures of up to ca. 300 °C, and its stability against solvents, which are used in further processing steps (Kapton, 2011).

Before material layers are deposited onto the Kapton foil, a thorough cleaning is necessary to ensure a stable adhesion of the subsequent material layers. A stable adhesion of the material layers on the flexible substrate is needed, because woven e-stripes are exposed to bending radii as small as 165 μm (Fig. 8.4(a)). A bending radius of 165 μm of a 50 μm thick substrate imposes a strain of ca. 12% on the material layer on the substrate (Suo *et al.*, 1999). In the case of deposited material layers made of gold, copper or titanium, a strain of 12% causes cracks in the metal layers (Fig. 8.4(b)). The cracks can emerge, of course, from brittle material layers, but also from improper adhesion caused by dust particles or other remnants between the plastic substrate and the deposited material layers. Dust particles and other remnants cause tiny areas where the two layers do not adhere and the strain

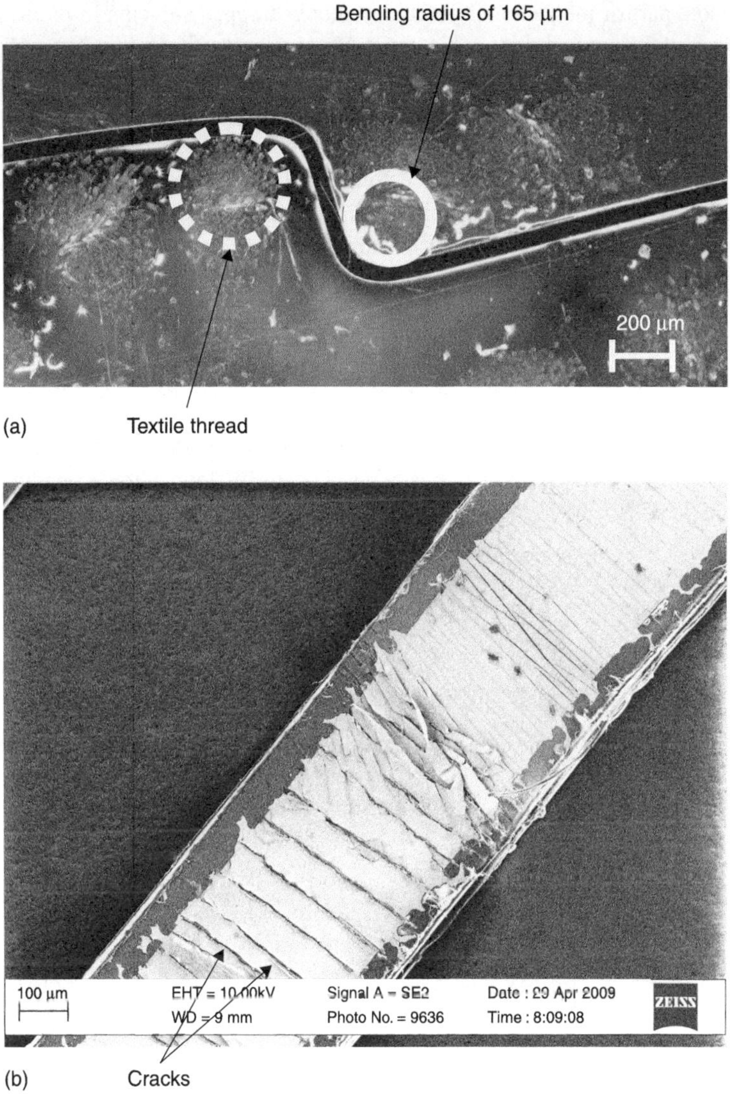

8.4 (a) Cross section of a woven textile with an integrated e-stripe. The e-stripe is exposed to bending radii as small as 165 μm corresponding to strains of up to 12%. (b) SEM picture of a cracked metal layer after weaving. The cracks occur because no cleaning step was applied before layer deposition.

is localized. Therefore, a cleaning step is required to prevent cracking (Cherenack *et al.*, 2010b).

During the cleaning process, dust particles and other remnants on the surface of the substrate that reduce the adhesion are removed. In addition, heating, which is also a part of the cleaning process, removes trapped solvents in the Kapton foil. First, the Kapton foil is sonicated in acetone and isopropanol (IPA) for 10 min. Sonicating is a process where the liquid, in the present case the acetone and IPA, are actuated with ultrasound. The mechanically actuated liquid promotes the physical removal of dust particles and remnants on the surface of the plastic substrate. To remove the trapped solvents and other small molecules, the plastic substrate is stored at 200 °C for 24 h in a vacuum oven. Furthermore, the substrate is then exposed to oxygen plasma for 3 min to activate the substrate surface and further promote the adhesion of subsequent material layers (Lin and Liu, 2008).

The effectiveness of the cleaning process was tested by fabricating 1 mm wide stripes with a 500 μm copper layer on top. The stripes were bent between two movable plates. While the bending radius was decreased, the resistance of the copper layer was measured. The results of these measurements are shown in Fig. 8.5, with the resistance normalized to the resistance of the copper layer without applied bending radius. The cleaning method allows bending radii as small as 120 μm, which is smaller than the measured bending radius of 165 μm in woven textiles. At radii smaller than 120 μm, the copper layer starts to crack (Cherenack, 2010a).

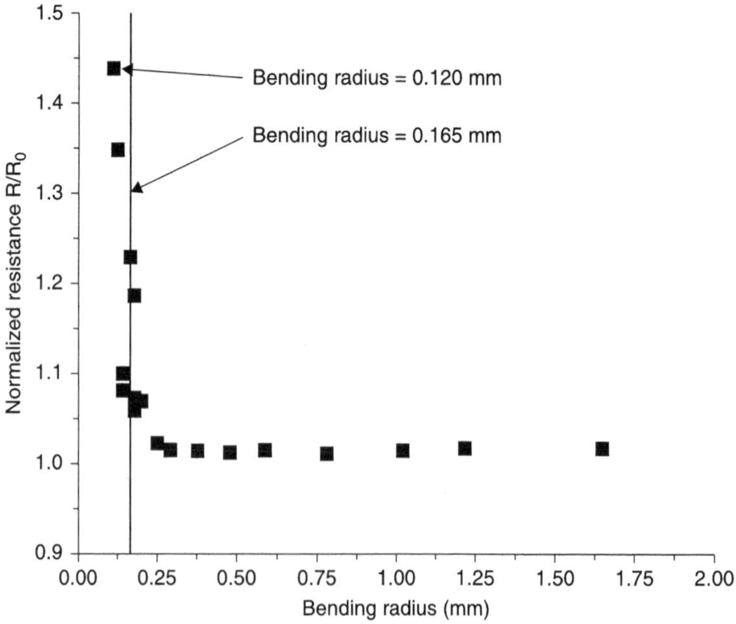

8.5 Bending radius versus normalized resistance of a 500 μm copper layer on a 50 μm thick Kapton substrate. At radii smaller than 120 μm, the copper layer starts to crack. As a vertical line, the smallest measured bending radius of 165 μm in woven textiles is indicated.

After cleaning, material layers such as metals, insulators, organic materials and semiconductors are deposited and structured on the flexible substrate. For the material deposition, several methods are applied:

- *Electron beam evaporation*: The workpiece is placed in a vacuum chamber. Within this chamber, an electron beam (e-beam) is used to heat up a target material such as gold, titanium etc. The electron beam causes atoms of the target material to evaporate. The vapor is distributed in the whole vacuum chamber and precipitate on the workpiece as well as other parts in the vacuum chamber in line of sight.
- *Sputtering*: To deposit material on a substrate with sputtering, the workpiece is placed in a vacuum chamber. Plasma is guided on the target material to evaporate it. The material in the vapor phase transits from the gaseous state back to the solid state and covers the workpiece and the rest of the chamber. Sputtering is used to fabricate oxide semiconductors, such as indium-gallium-zinc-oxide (IGZO), which are used for (flexible) TFTs (Yabuta *et al.*, 2006, Münzenrieder *et al.*, 2011).
- *Spin-coating*: To spin a coating onto a substrate, the coating material is placed in dissolved form on the substrate. By spinning the substrate, the dissolved material starts to expand and cover the whole surface of the substrate, resulting in a uniformly thick material layer. After spin-coating, the material is irradiated or heated to remove the solvents, leaving a layer of the desired material.
- *Atomic layer deposition (ALD)*: The workpiece is placed in a vacuum chamber. Normally two reactive gases are subsequently led into the chamber. Both gases react on the surface and result in a single atomic layer of the desired material. By repeating this process, several atomic layers are deposited.

To fabricate interconnect lines, contact pads, sensors, TFTs, etc. the deposited material layers on the flexible substrate need to be structured. Structuring the material layers is achieved by using the following techniques:

- *Etching*: Figure 8.6(a) shows an etching process. First, a material layer is deposited on the substrate. In a second step, a photo-resist is spin-coated onto the material layer. The photo-resist is structured by using a mask, which covers parts of the photo-resist and exposing it to UV-light and dipping it in a developer. Then, the substrate is exposed to an etchant, which removes the parts of the material layer not protected by the photo-resist. Finally, the remaining photo-resist is removed using an appropriate solvent.
- *Lift-off*: Figure 8.6(b) shows a lift-off process. In the first step, photo-resist is spin-coated onto the substrate and structured in the same way as for etching. Second, a material layer is deposited on the whole workpiece. In the third step, the workpiece is dipped into a solvent bath, which dissolves the photo-resist. The material layer, which is on top of the photo-resist, is removed with the photo-resist, leaving behind the desired structure on the substrate.

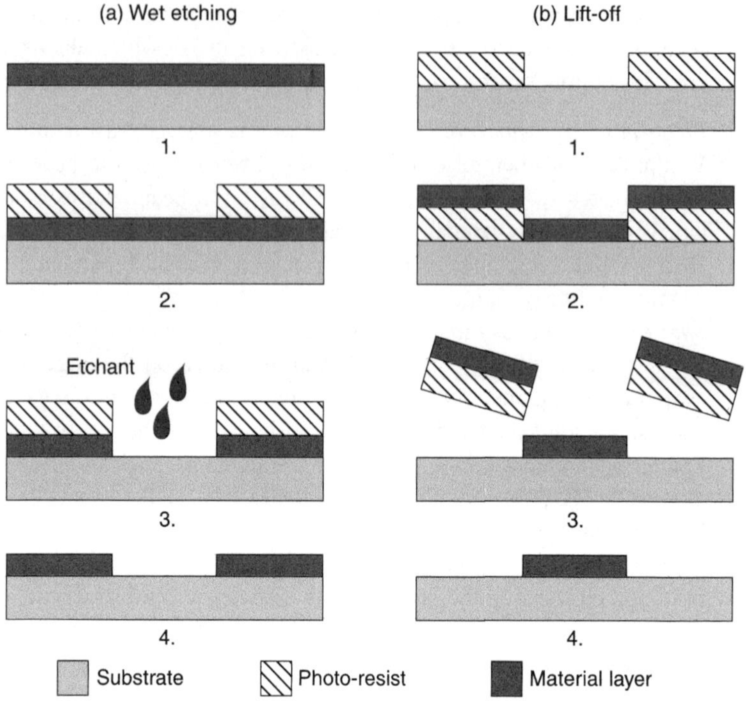

8.6 (a) Shows the process steps for wet etching to structure a material layer. In (b) the main steps of a lift-off process are depicted.

The deposition and structuring methods described here are the ones that are used to fabricate e-stripes. However, they are only a selection among a large variety of available processes (Menz *et al.*, 2001).

After depositing and structuring material layers, the substrate is separated into individual stripes. For cutting the substrate, a wafer saw with a diamond blade is used. This method allows the fabrication of stripes as narrow as 50 μm, with a deviation from the cutting path of less than 7 μm (Kinkeldei *et al.*, 2009). When bare die ICs or SMDs are required on the e-stripes, they are attached by soldering or gluing with a conductive adhesive. The devices are protected using glob-top (epoxy resin), which is cured for 1 h at 100 °C. In Fig. 8.7, an example of a separated e-stripe with contact pads for conductive threads, interconnect lines and two digital temperature sensors mounted on the e-stripe is depicted. The stripe is 2 mm wide and 4.5 cm long.

8.2.3 Weaving

The stripes are woven into a 4.5 cm wide textile band using a commercial weaving machine of type NFREQ 42 from Müller Frick. Prior to weaving, conductive threads are integrated into the warp beam. The conductive threads in warp

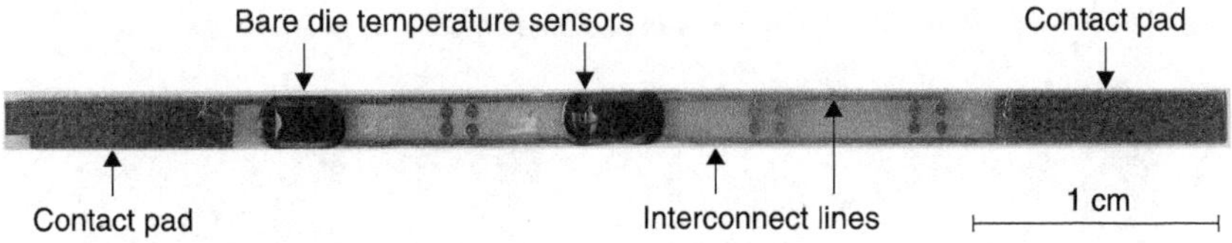

8.7 Exemplary e-stripe with contact areas, interconnect lines and bare die integrated circuit digital temperature sensors.

direction serve, for example, as bus lines (for a bus system) or as column lines (in a matrix system). For the demonstrators, which are shown later in this section, the conductive threads are knotted into the warp beam. To ensure that the conductive warp threads cross the contact pads on the e-stripes, it is necessary that the conductive threads are part of the same heddle frame.

Inserting an e-stripe into the textile involves several steps, which are specifically tailored for the demonstrators:

1. The weaving speed of the machine is lowered.
2. The insertion of weft threads is blocked by stopping the weft thread feed to open space in the textile for the e-stripe. The number of blocked weft threads is equal to the weaving density (threads/mm) multiplied with the width of the e-stripe (mm). For the used weaving machine, the weaving density is set to 2.5 weft threads per mm and the exemplary stripe from Fig. 8.4 has a width of 2 mm, resulting in 5 omitted weft threads for this particular e-stripe.
3. To insert the e-stripe into the textile, the weaving machine is stopped and the e-stripe is inserted manually into the weaving machine. During the insertion, the contact pads on the e-stripe have to be aligned against the conductive threads in the weft direction to enable contacting after weaving.
4. The weaving machine is run in slow mode for at least two weaving cycles, to ensure that the textile is woven correctly and the e-stripe is not damaged. Then the weaving speed is increased.

The described steps are repeated for each e-stripe.

This weaving process is tailored to the used band weaving machine and the fabrication of the demonstrators. However, the process can be generalized for large-scale weaving machines and has to comprise the following steps: first, the machine has to open space for the e-stripes. Second, the stripes have to be inserted into the textile and aligned against the conductive warp threads. Finally, non-conductive textile threads have to be woven until the next e-stripe is inserted.

8.2.4 Contacts

After weaving, the contacts between the contact pads on the e-stripes and the conductive threads in the warp direction have to be established. For the

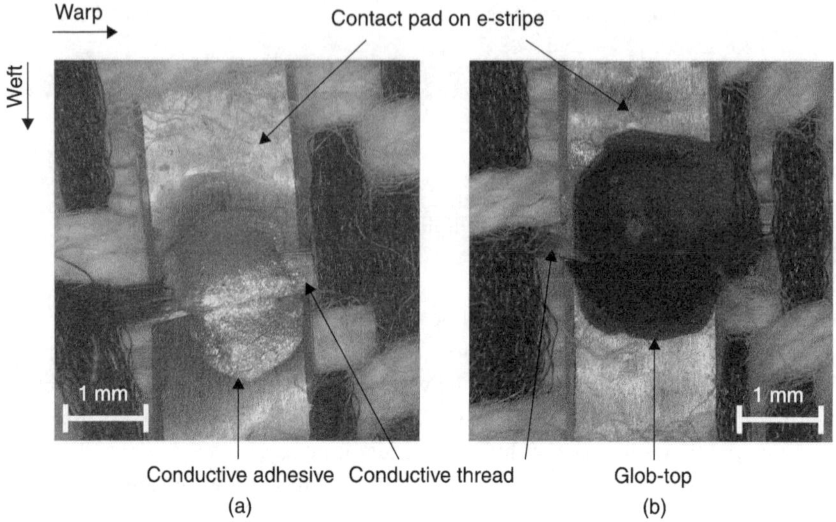

8.8 (a) Electrical contact with conductive adhesive between conductive thread in warp direction and contact pad on e-stripe in weft direction. (b) The electrical contact covered with glob-top for mechanical reinforcement.

demonstrators, this is done by using a silver-filled epoxy adhesive (Epo-Tek H20E), which is cured for 24 h at 60 °C to establish an electrical contact. To mechanically reinforce the contact, glob-top (Epo-Tek 7139) is used and cured for 1 h at 100 °C. Figure 8.8(a) shows an exemplary electrical contact with conductive adhesive and Fig. 8.8(b) shows the contact covered with glob-top. As for weaving, the process of establishing the contacts can be generalized. First, an electrical contact has to be established using, for example, soldering or gluing with a conductive adhesive and second, the contact should be mechanically stabilized by using epoxy.

8.3 Demonstrators

The feasibility of e-stripe technology is demonstrated with three prototypes: the first, which is able to sense temperature and humidity, consists solely of thin-film devices and does not require conductive threads in the warp direction. The second incorporates a bus structure in the woven textile and uses bare die ICs as bus devices. The last prototype is a matrix structure, which combines thin-film devices and SMDs.

8.3.1 Temperature and humidity sensitive textile[4]

The integration of thin-film devices into a woven textile is demonstrated with temperature and humidity sensors on e-stripes. The sensors are designed to fit on a 1 mm wide stripe. To measure the temperature, a resistance temperature device (RTD) was used. The sensing principle is based on the fact that metals change their resistance, depending on their temperature characterized by the temperature coefficient α^5. The humidity sensor is based on the conductive polymer PEDOT:PSS (Baytron P AI 4083), which absorbs water molecules leading to a change in the resistance of the polymer. The RTD shows a meander shape with a line width of 20 μm and a spacing of 20 μm. The total length of the RTD is 22 mm. For the humidity sensor, finger electrodes are used with a length of 1 mm, a spacing of 20 μm and 40 fingers. The fingers are covered with the conductive polymer PEDOT:PSS. Three interconnect lines are used to contact the sensors. These interconnects are routed to both ends of the e-stripe, to enable contact after weaving using standard copper wires. A schematic of the sensor part of the e-stripe is shown in Fig. 8.9.

To fabricate the sensors, a 7.6 × 7.6 cm^2 Kapton E substrate is used. After the cleaning process (Section 8.2.3), to achieve sufficient adhesion of the material layers on the substrate, 5 nm of titanium and 100 nm of gold are deposited using e-beam evaporation. The titanium/gold layer is structured to form the meander of the RTD, the finger electrodes for the humidity sensor and interconnect lines for contacting the sensors using a lift-off technique. In the next step, PEPOT:PSS is spin-coated on the structured electrodes with 1200 rpm for 1 min. The PEDOT:PSS film is dried for 10 min at 180 °C on a hotplate. The resulting thickness of the polymer film is 80 nm.

The substrate is then cut into 1 mm wide stripes using a wafer saw and woven into a textile (Sections 8.2.3 and 8.2.4). The resulting textile and a photo of the sensors are shown in Fig. 8.10. The functionality of the textile integrated sensors was tested in a climate chamber at temperatures between 30 and 60 °C and humidity between 30%RH and 80%RH. Figure 8.11 shows the measured

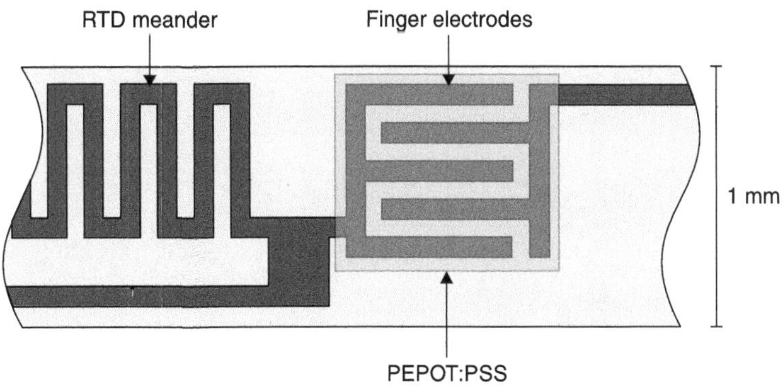

8.9 Schematic of the sensor part of the e-stripe with the finger electrodes and the PEDOT:PSS for the humidity sensor and the meander-shaped resistive temperature device.

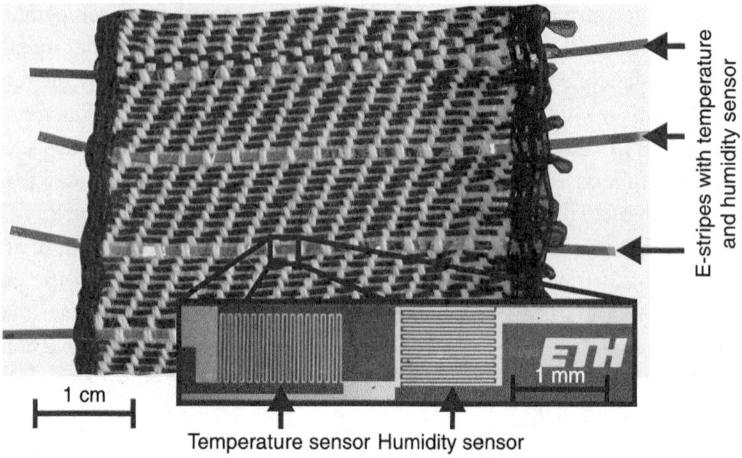

8.10 Prototype of temperature and humidity sensitive textile. The inset displays the sensor area of the integrated e-stripe with the meander-shaped temperature sensor and the PEDOT:PSS humidity sensor.

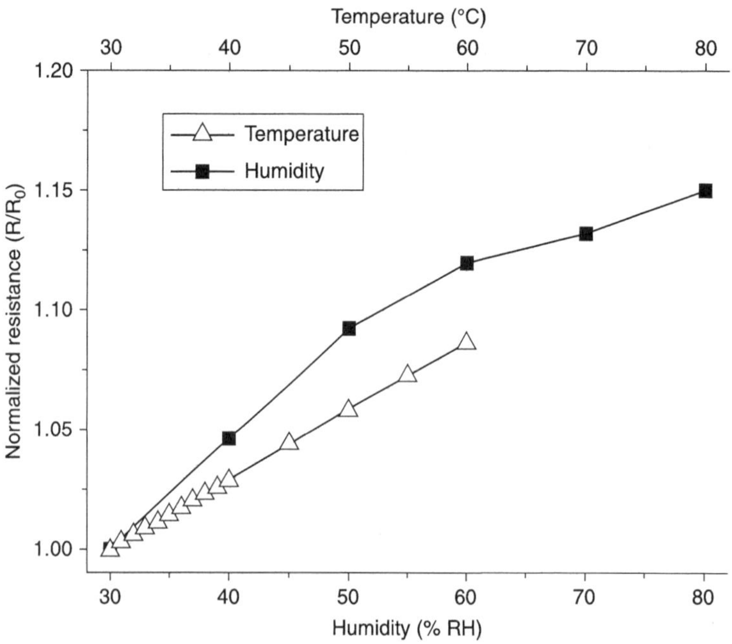

8.11 Sensor responses for the temperature and humidity sensitive textile, obtained in a controlled environment of 30 to 60 °C with a constant humidity of 50 %RH for the temperature sensor characterization. For the humidity sensor characterization, the humidity was varied between 30 and 80%RH at a constant temperature of 30 °C.

resistance values for both sensors. The resistances are given normalized to the initial resistance values at 30 °C and 30%RH, respectively. The resistance values for the RTD show a linear behavior with a temperature coefficient α of 0.0028 °C^{-1}. The PEDOT:PSS based humidity sensor exhibits a linear behavior below 50%RH. Above 50%RH, a non-linearity is observed, which is caused by a water layer in the sensor material reducing the sensitivity of the sensor and hence decreasing the resistance change (Kus and Okur, 2009; Kinkeldei, 2011a).

8.3.2 Textile integrated bus system[6]

To address a multitude of devices and in particular sensors in a textile, a bus structure (Fig. 8.2(a)) is applicable. For the integration of a bus structure into a woven textile, the weft direction is used to integrate e-stripes with the bus device, interconnect lines and contact pads. In the warp direction, conductive threads are integrated as bus lines. The integration of a bus structure is demonstrated with 5 e-stripes, which contain 1 to 3 bare die temperature sensors per e-stripe, resulting in a total of 10 sensors in the textile. The temperature sensors include the 1-Wire bus protocol (DS18S20, Maxim-IC (Maxim)), allowing sensor read-out and power supply with only two bus lines. One line is used for ground, the other for power and data communication.

The e-stripes are fabricated on Pyralux, which is a 25 μm Kapton foil sandwiched between two 18 μm copper layers (Pyralux, 2011). From a base substrate with a size of 4.5 cm × 6 cm, the bottom copper layer is completely etched away while the top copper layer is structured to form interconnect lines, contacts to attach the digital temperature sensors and contact pads of 2 mm width and 9 mm length on both ends of the stripe for contacting the conductive threads. The distance between two adjacent temperature sensors is 6 mm, to ensure that the temperature sensors on the woven e-stripes are not covered with a warp thread. Five contact areas for the temperature sensors are structured to choose the location of the sensors on the e-stripe. The substrate is then separated into 2 mm wide and 4.5 cm long stripes using a wafer saw. The digital temperature sensors are attached by flip-chip soldering.[7] The temperature sensors are encapsulated with a glob-top cover for protection. A single stripe with two temperature sensors is shown in Fig. 8.11.

Before the e-stripes are woven into the textile, two conductive threads (Bekinox VN 14/1x) are inserted into the warp beam with a distance of 3.6 cm. These two conductive threads are used to interconnect the contact pads at both ends of the e-stripes. The weaving process is then performed (Section 8.2.4) on a commercial band weaving machine of type NFREQ 42 from Müller Frick.

Figure 8.12(a) shows the woven prototype with five e-stripes in the weft direction. The conductive threads in the warp direction are not visible, because they are integrated in the textile. In Fig. 8.12(b), a schematic of the demonstrator is shown. By connecting the two conductive threads in the warp direction with an

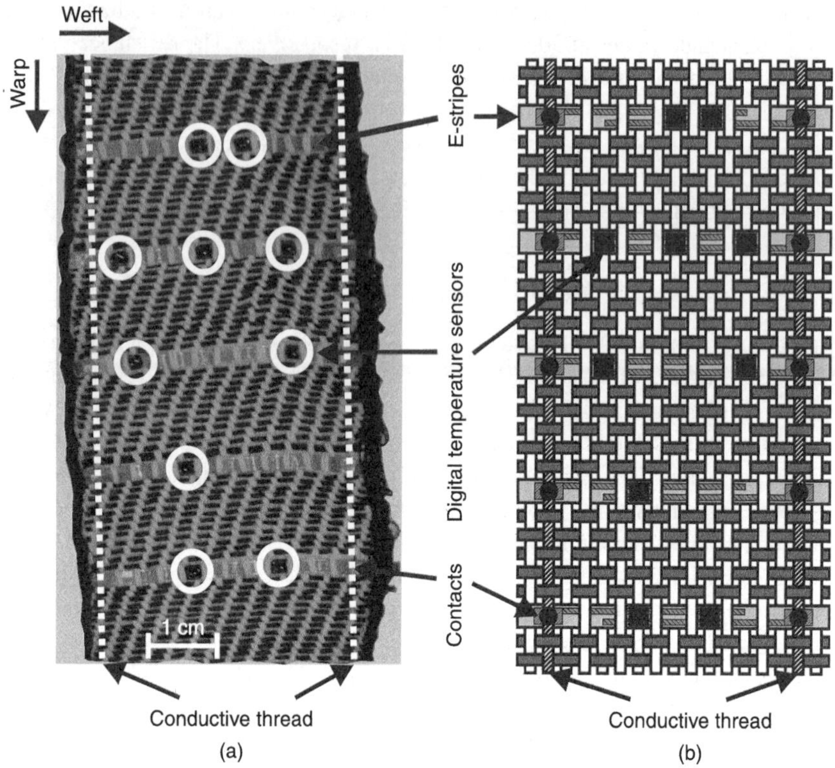

8.12 (a) Prototype with integrated bus system. The textile contains ten digital temperature sensors (encircled white) distributed on five e-stripes. The conductive threads in warp direction serving as bus lines are not visible due to the integration. (b) Schematic of the textile integrated bus system.

appropriate measurement unit, it is possible to read out the temperature values of all 10 integrated sensors.

The ability to bend the textile was investigated using the demonstrator with the integrated temperature sensors. First, the textile was wrapped around cylindrical rods in warp and weft directions. The textile was bent to a radius of 0.75 mm without a failure of the electrical system in the textile. However, the areas of around 2 mm × 2 mm, where the digital temperature sensors are located, were locally stiff and did not bend during the experiment.

The bending rigidity of the textile was evaluated by using a Shirley Stiffness Tester (Schwartz, 2008). The influence of the integrated e-stripes compared to the same textile without e-stripes was evaluated. For the textile without integrated e-stripes, a bending rigidity G of 13.45 g cm^2/cm and for the demonstrator a bending rigidity G of 17.45 g cm^2/cm was measured. This 30% increase in the bending rigidity is accounted to the e-stripes, which are less flexible than the textile threads (Zysset *et al.*, 2010a).

8.3.3 Textile integrated matrix

As proposed in Section 8.2.1, matrix systems integrated into woven textiles can be used to address a multitude of sensors and actuators in the textile. Therefore, a demonstrator was fabricated with three flexible stripes in the weft direction. Each e-stripe contains three SMD light emitting diodes (LEDs, type ROHM SML-311UTT86K), where each LED is controlled with a TFT, allowing turning the LED on and off. In the warp direction, three conductive threads are used to control the gates of the TFTs. With this arrangement of e-stripes and conductive threads, a 3 by 3 LED matrix with individually controllable LEDs is achieved.

In Section 8.2.3, strains of up to 12% on the material layers of woven e-stripes are reported. In comparison, TFTs based on oxide semiconductors start to crack at strains larger than 1% (Cherenack, 2010b). Oxide semiconductors for the TFTs are used, because they can be fabricated below 300 °C, which is necessary for processing on flexible substrates. In addition, oxide semiconductor TFTs provide a larger electron mobility of around $10\,cm^2V^{-1}s^{-1}$ compared to, for example, amorphous silicon TFTs ($\sim 1\,cm^2V^{-1}s^{-1}$) (Kattamis *et al.*, 2007), enabling faster switching and higher current density. To avoid cracks in the transistors, the TFTs are protected with the SMD LEDs, which are rigid and placed on top of the TFTs. This renders parts of the e-stripe rigid, which is an approach often used in flexible electronics (Graz *et al.*, 2011).

To enable the integration of TFTs as part of a textile integrated LED matrix, e-stripes are designed (Fig. 8.13(a)). On the left side of the stripe we see the power contact pad, which is used to select the row (Fig. 8.2(b)). The ground contact pad is located on the right side. From both pads, interconnect lines run along the stripe. Between the two interconnect lines, the LEDs and TFTs are located in a way that the LED fully covers the TFT and therefore provides mechanical rigidity to protect the TFT. In addition, between the two interconnect lines, the gate control pads are placed, which are later connected with the conductive threads in the warp direction. The distance between two adjacent LEDs on the e-stripe amounts to 1.5 cm; the e-stripe is 2 mm wide and 5 cm long. The LEDs have a footprint of 1.6 mm × 0.8 mm. Figure 8.13(b) indicates which parts of the e-stripe remain flexible after fabrication and which are rigid for protection of the TFTs.

The e-stripes are fabricated on a 7.6 cm × 7.6 cm Kapton E (Kapton, 2011) base substrate. First, the flexible foil is cleaned using the process described in Section 8.2.3. After cleaning, the gate contacts for the TFTs are patterned with a lift-off step: first, a photo-resist is spin-coated and patterned and 5 nm chromium and 30 nm platinum are deposited using e-beam evaporation. Then the lift-off is performed. In the next step, a 25 nm aluminum oxide (Al_2O_3) layer serving as gate dielectric is deposited at 150 °C using ALD. Next, the IGZO semiconductor is

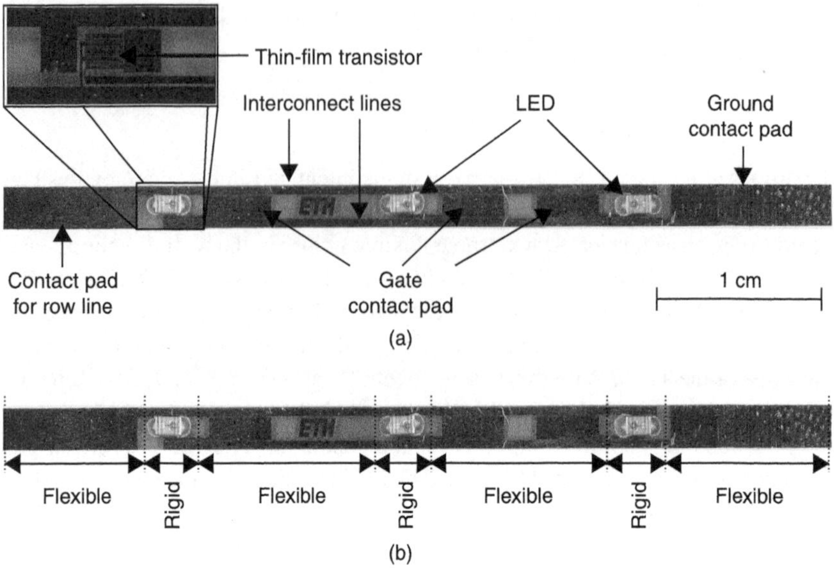

8.13 (a) Flexible stripe with contact pads for row line, ground and gate contacts of the thin-film transistors, interconnect lines and attached light-emitting diodes. The inset shows a thin-film transistor which is located beneath a light-emitting diode. (b) Indication of which parts of the stripe are flexible and which are rigid.

magnetron sputtered. Both layers are wet etched to form the transistor stack. The drain and source contacts, as well as the interconnect lines, the ground and supply pad and the gate contact pads, are formed using a lift-off process to structure a 10 nm titanium and 100 nm gold layer, which is e-beam evaporated. Finally, the whole stripe is covered with an insulating 25 nm Al_2O_3 layer, which is only opened at the contact pads. Figure 8.14 shows a schematic cross section of the TFT.

The e-stripes are woven on a commercial band weaving machine of type NFREQ 42 from Müller Frick, using the weaving process described in Section 8.2.4. In the warp direction, four conductive threads were integrated. Three of these conductive threads serve as column lines to control the gate contacts of the TFTs. The fourth conductive thread acts as a common ground line for all three e-stripes. Figure 8.15 shows a photo of the prototype with the integrated 3 by 3 active LED matrix, where the four LEDs in the corners and the LED at the center are turned on.

By selectively controlling (through addressing the power interconnect lines on the e-stripe and the conductive threads) each of the nine integrated LEDs, it is possible to generate patterns on the textile. This is achieved by operating the LEDs in pulsed mode with a frequency of 142 Hz, guaranteeing a continuous light signal for the human eye.

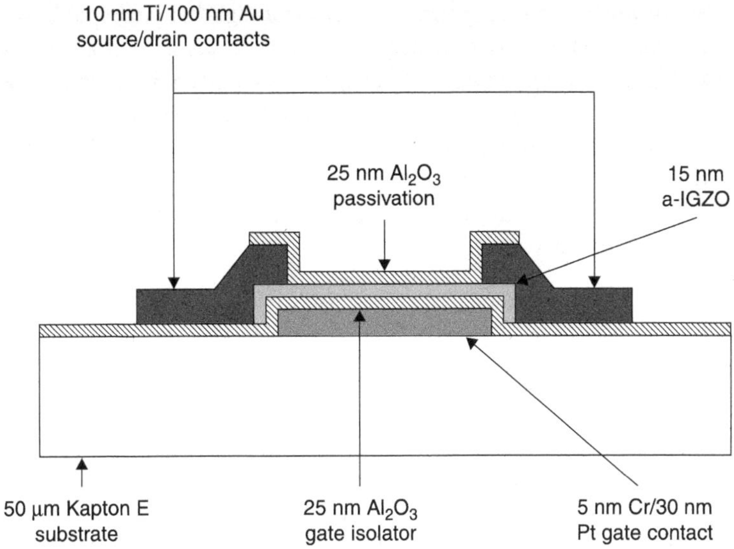

8.14 Cross section of the thin-film transistor which is used to control the LEDs.

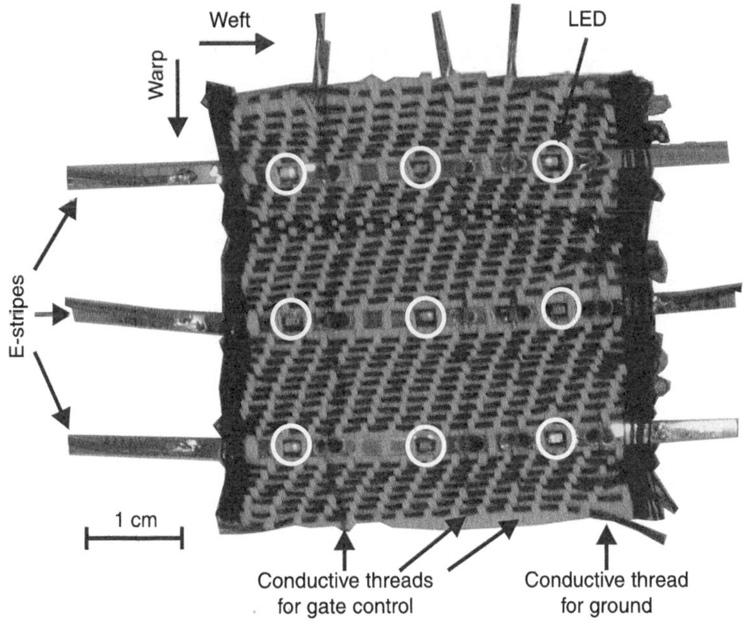

8.15 Prototype of textile integrated active matrix. The conductive threads in warp direction for thin-film transistor gate control and ground line are not visible due to textile integration. The LEDs are encircled for better visibility.

Besides being able to generate patterns through individual addressing of the LEDs, the transistor performance is investigated before and after weaving. As described in Münzenrieder *et al.* (2011), applying strain to a transistor leads to changes in the electron mobility and the switching (threshold) voltage. However, the measurements for woven TFTs protected against strain with a rigid device, indicate that the transistors are not affected by weaving. The two parameters, electron mobility and threshold voltage, remain at constant values of $9.6\,cm^2V^{-1}s^{-1}$ and 1.7 V, respectively.

8.4 Mechanical reliability of contacts

For the demonstrators shown in Sections 8.3.2 and 8.3.3, electrical contacts between conductive threads in the warp direction and contact pads on the e-stripes are crucial for establishing the desired electronic systems in woven textiles. The fabrication of the contacts with conductive adhesive and glob-top is described in Section 8.2.5. However, the destruction of such a contact can lead to partial failure of the electronics in the textile. Since potential applications of smart textiles are clothes worn by people, the smart textiles and the contacts are exposed to external forces. Therefore, the mechanical reliability of the contacts between conductive threads and contact pads on the e-stripes is investigated. Because of the shape of the contacts (they extend out of plane of the textile), they are exposed to shear forces that are exerted from objects moving, for example, a finger or a hand, over the textile surface.

Figure 8.16 shows a cross section of a contact between a conductive thread and a contact pad on a woven e-stripe. The figure shows the textile threads, the woven e-stripe and the contact to the conductive thread. In addition, the direction of the applied shear force is indicated. Maximum shear forces of 20 N were applied to the contacts and measured with a Correx force gauge. The tests were made at 20 contacts and did neither lead to detachment of the contacts nor to electrical failure of the contacts. Shear forces of 20 N are well above the 8 N of force, which a single fingertip can exert (Ledermann and Taylor, 1972). Furthermore, in a user study with seven participants, a shirt with the prototype from Section 8.3.2 integrated was worn for periods of more than 20 min and part of the time beneath a fire fighter jacket without sensor failure (Zysset *et al.*, 2010b).

However, 20 N shear forces do affect the whole system of conductive threads and e-stripes. The force applied to the contacts is transferred via the contact itself to the Kapton substrate, leading to an elongation of the e-stripe. The 20 N are equal to a stress 400 MPa, which is calculated using the cross-sectional area of the e-stripe of 2 mm × 25 μm. Using the Young's modulus of Kapton of 5.4 GPa (Kapton, 2011), the stress of 400 MPa is equal to a strain of 7.4%, which is smaller than the 12% exerted by weaving.

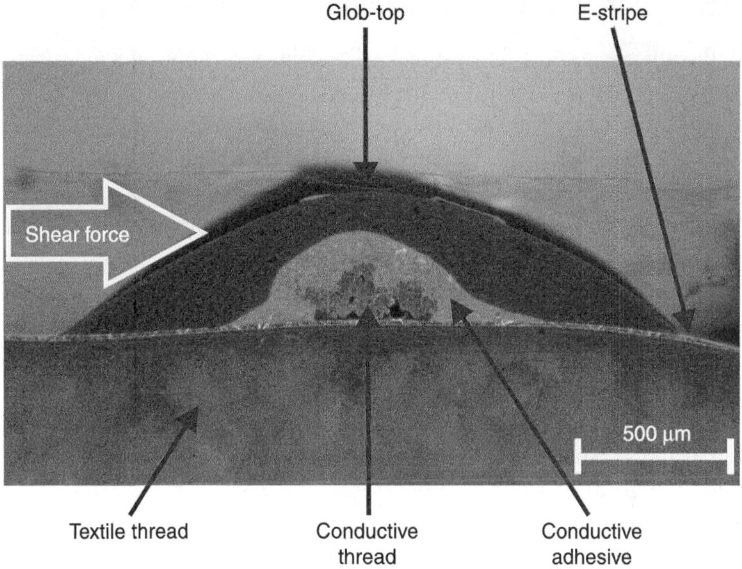

8.16 Cross section of a contact between a conductive thread and a contact pad on a woven e-stripe. The direction of the shear force applied to the contact is indicated by the white arrow.

8.5 Conclusion and future trends

This chapter presented an approach to fabricate smart textiles where the electronics are integrated during the weaving process. Following this approach enables the fabrication of smart textiles, which remain drapable (Section 8.3.2) compared to common approaches, where rigid parts are mounted on textile substrates. Despite the many approaches of smart-textile fabrication investigated and developed since the mid-1990s, smart textiles account for only a small percentage of the worldwide textile market (Tröster, 2011). To increase the market share of smart textiles, several challenges are faced and need to be addressed. For at least some of the challenges, the approach presented here could contribute to overcome them and turn the challenge into a potential.

The following points address several challenges, which could be addressed with regards to the smart-textile approach presented in this chapter:

- large-scale production
- washability
- user acceptance.

8.5.1 Large-scale production

Although electronics fabrication and textile fabrication are industries that can provide products at an affordable price level due to mass production, these two sectors of industry have not yet found a common ground to provide smart textiles. This is mainly due to the fact that smart textiles are not produced on a large scale but rather manually, which makes them expensive. The approach presented in Section 8.2 offers potential for scaling the technology up to smart-textile production on industrial weaving machines. In addition, electronics fabrication is based on existing processes, such as material deposition and structuring, which allow the fabrication of long e-stripes by using roll-to-roll techniques for material deposition under vacuum (Chaug *et al.*, 2004) and structuring (Noda *et al.*, 2003). Electronic devices such as ICs can be applied by automated pick-and-place packaging processes on the flexible substrates. Finally, experiments on a Dornier rapier loom have shown that it is possible to integrate long flexible stripes into woven textiles.

8.5.2 Washability

A key issue for smart textiles is washability. Early commercial smart textiles, such as the ICD+ textile coat developed by Philips and Levi (2011), required the wearer to remove all electronic components prior to washing. More recent smart-textile demonstrators, such as the Eleksen textile keyboard (Colye *et al.*, 2007), rely on waterproof packaging to protect the electronics from damage during washing.

Using flexible substrates offers potential to increase the washability of smart textiles. E-stripes with LEDs were woven into a textile and washed at 30 °C for 47 min per washing cycle. The textile was washed five times and after each cycle the LED brightness was determined. After five cycles, the e-stripes were still functional and did not show a change in LED brightness. However, increasing the washing temperature reduces the brightness and ultimately leads to failure of the device (Zysset *et al.*, 2010a).

To increase the washability, electronics on the e-stripe could be protected by sandwiching the devices between two flexible foils. Such sandwich structures were demonstrated in Kinkeldei *et al.* (2011(b)), with the purpose of reducing the strain induced by bending the electronics. In addition, improved flexible foils, which are currently investigated for organic LED fabrication with a low permeability for water and other small molecules (McCormick, 2011), could be used to protect the electronic devices.

8.5.3 User acceptance

To increase the user acceptance of smart textiles, different issues have to be addressed: in case the smart-textile measures physiological parameters, the measurements have to be reliable, the smart textiles need to be affordable, handling has to be simple to address not only technology affected people, the added electronic functionality should be integrated unobtrusively in terms of appearance and also haptic perception of the textile, etc. (Schaar and Ziefle, 2011).

Progress in electronics fabrication, in particular higher integration density of solid state electronics according to the ITRS roadmap (ITRS, 2010), the advent and majoring of new technologies such as organic electronics (Katz and Huang, 2009), printed electronics (Berggren *et al.*, 2007) and novel materials, will decrease the size of the e-stripes and increase the integration density of sensors, actuators, data processing units, etc. on the e-stripes. From all these trends in research and development, integration of electronics during the weaving process and smart-textile development in general, will benefit and offer potential to increase the user acceptance.

By overcoming the challenges of washability, large-scale production and user acceptance, smart textiles will increase their market potential. However, until smart textiles are a common item in our daily lives, they will first enter into the market of technical textiles as car or airplane interiors as well as in medical applications where monitoring of vital parameters on the human body is necessary (Tröster, 2011).

8.6 Sources of further information and advice

Further information on the topic of microelectronics fabrication and devices can be found in Sze (1982) and Menz *et al.* (2001). For thin-film electronics, Wong (2010) provides information on flexible substrates, amorphous silicon, organic materials and thin-film sensor devices. To get more information on organic electronics, Klauk (2006) addresses a multitude of topics related to materials and fabrication.

Additional material on mainly electronic aspects of smart textiles is provided by the *Proceedings of the International Symposium on Wearable Computers*, as well as the IEEE database[8] Talk2MyShirt[9] is a blog reporting on various activities in smart-textile development. The web portals ProETX[10] and SmartTextiles[11] list several companies involved in smart-textile development. On the Systex[12] web page, information on research and possible partners for smart-textile development in industry and academia are listed.

8.7 Notes

1 A printed circuit board (PCB) provides mechanical support and electrical interconnects for electronic components such as resistors, capacitors, sensors, microprocessors, etc. Structured copper layers on a non-conductive substrate connect the electronic devices, which are soldered onto the PCB.

2 Thin-film devices are made of individual layers with thicknesses, ranging from nanometers to micrometers.

[3] Roll-to-roll techniques refer to processes, where a plastic substrate or a metal foil is processed by re-reeling the substrate from one roll to a second one. During the re-reeling, material layers are deposited and structured to form electronics. For the deposition of the materials, several methods such as printing under ambient atmosphere or sputtering under vacuum are applicable.

[4] This section is based on Kinkeldei *et al.* (2011a).

[5] The formula for a RTD is $R = R_0(1 + \alpha(T - T_0))$, with R the resistance, R_0 the resistance at the known temperature T_0 and α the temperature coefficient.

[6] This section is based on Zysset *et al.* (2010a).

[7] Flip-chip bonding is a technique often used in micro-fabrication to interconnect a bare die to external circuitry with solder bumps or conductive adhesive. The solder bumps were deposited on the pads on the bare die on the top side. Similarly, the conductive adhesive is deposited onto the pads on the bare die while facing upwards. To attach the bare die to the external circuitry, it is flipped over and landed on the external circuitry. By heating up the whole system, the contacts made either of solder or conductive adhesive are established. Therefore, the process is called flip-chip bonding (Tummala, 2001).

[8] http://ieeexplore.ieee.org/Xplore/dynhome.jsp

[9] http://www.talk2myshirt.com/blog/

[10] http://www.proetex.org

[11] http://www.smarttextiles.net

[12] http://www.systex.org

8.8 References

Abouraddy A, Bayindir M, Benoit G, Hart S, Kuriki K, *et al.* (2007), 'Towards multimaterial multifunctional fibres that see, hear, sense and communicate', *Nat Mat*, 6(5), 336–347.

Berggren M, Nilsson D and Robinson N D (2007), 'Organic materials for printed electronics', *Nat Mat*, 6, 3–5.

Bonderover E and Wagner S (2004), 'A woven inverter circuit for e-textile applications', *IEEE Electron Dev Lett*, 25(5), 295–297.

Chaug Y S, Haubrich J E, Sereda M and Liang R C (2004), 'Roll-to-roll processes for the manufacturing of patterned conductive electrodes on flexible substrates', *MRS Proc*, 814(1), 19.6.

Cherenack K, Kinkeldei T, Zysset C and Tröster G (2010a), 'Woven thin-film metal interconnects', *IEEE Electron Dev Lett*, 31(7), 740–742.

Cherenack K, Münzenrieder N and Tröster G (2010b), 'Impact of mechanical bending on ZnO and IGZO thin-film transistors', *IEEE Electron Dev Lett*, 31(11), 1254–1256.

Coyle S, Wu Y, Lau K T, De Rossi D, Wallace G and Diamond D (2007), 'Smart nanotextiles: a review of materials and applications', *MRS Bull*, 32, 434–442.

Crawford G P (2005), *Flexible Flat Panel Displays*. Chichester, UK, Wiley & Sons.

Gonzalez M, Axisa F, Bulcke M, Brosteaux D, Vandevelde B and Vanfleteren J (2008), 'Design of metal interconnects for stretchable electronic circuits', *Microelect Reliab*, 48(6), 825–832.

Graz I M, Cotton D P J, Robinson A and Lacour S P (2011), 'Silicone substrate with in situ strain relief for stretchable thin-film transistors', *Appl Phys Lett*, 98(12), 124101.

Hamedi M, Forchheimer R and Inganäs O (2007), 'Towards woven logic from organic electronic fibers', *Nat Mat*, 6(5), 357–362.

Harms H, Amft O, Tröster G and Roggen D (2008), 'Smash: distributed sensing and processing garment for the classification of upper body postures', *Proceedings of the ICST 3rd International Conference on Body Area Networks*, 22.

ITRS Roadmap updates (2010), available from: http://www.itrs.net/links/2010itrs/home2010.htm [Accessed 14 November 2011].

Kapton E Datasheets: DuPont (2011), available from: http://www2. .dupont.com/Kapton/en_US/assets/downloads/pdf/E_datasheet.pdf [Accessed 20 September 2011].

Kattamis A Z, Cherenack K H, Hekmatshoar B, Cheng I C, Gleskova H, *et al.* (2007), 'Effect of SiN_x gate dielectric deposition power and temperature on a-Si:H TFT stability', *IEEE Electron Dev Lett*, 28(7), 606–608.

Katz H E and Huang J (2009), 'Thin-film organic electronic devices', *Ann Rev Mat Res*, 39, 71–92.

Kinkeldei T, Zysset C, Cherenack K and Tröster G (2009) 'Development and evaluation of temperature sensors for textile integration', *Proceedings of the IEEE Sens*, pp. 1–4.

Kinkeldei T, Zysset C, Cherenack K and Tröster G (2011a), 'A textile integrated sensor system for monitoring humidity and temperature', *Proceedings of the International Conference on Solid-State Sensors, Actuators and Microsystems*, pp. 1156–1159.

Kinkeldei T, Münzenrieder N, Zysset C, Cherenack K and Tröster G (2011b), 'Encapsulation for flexible electronic devices', *IEEE Elect Dev*, 99, 1–3.

Klauk H (2006), *Organic Electronics: Materials, Manufacturing, and Applications*. Weinheim, Wiley-VCH.

Kus M and Okur S (2009), 'Electrical characterization of PEDOT: PSS beyond humidity saturation', *Sens Actuat B*, 143, 177–181.

Ledermann S and Taylor M (1972), 'Fingertip force, surface geometry and the perception of roughness by active touch', *Attention, Perception & Psychophysics*, 12(5), 401–408.

Lin Y S and Liu H M (2008), 'Enhanced adhesion of plasma-sputtered copper films on polyimide substrates by oxygen glow discharge for microelectronics', *Thin Solid Films*, 516(8), 1773–1780.

Linz T, Gourmelon L and Langereis G (2007), 'Contactless EMG sensors embroidered onto textile', *4th International Workshop on Wearable and Implantable Body Sensor Networks (BSN 2007)*, pp. 29–34.

Locher I and Tröster G (2007a), 'Fundamental building blocks for circuits on textiles', *IEEE Trans Adv Pack*, 30(3), 541–550.

Locher I and Tröster G (2007b), 'Screen-printed textile transmission lines', *Text Res J*, 77(11), 837.

Martin T, Jones M, Chong J, Quirk M, Baumann K and Passauer L (2009), 'Design and implementation of an electronic textile jumpsuit', *Proceedings of the 2009 International Symposium on Wearable Computers*, pp. 157–158.

Mattmann C, Amft O, Harms H, Tröster G and Clemens F (2007), 'Recognizing upper body postures using textile strain sensors', *Proceedings of the 2007 International Symposium on Wearable Computers*, p. 36.

Maxim: DS18S20 1-Wire Temperature Sensor, Datasheet Maxim-IC (2011), available from: http://datasheets.maxim-ic.com/en/ds/DS18S20.pdf [Accessed 20 September 2011].

McCormick F (2011), 'Barrier films and device encapsulation', Presentation at OLED Materials for Lighting and Displays, available from http://www.ccrhq.org/oled-materials-lighting-and-display [Accessed 14 November 2011].

Menz W, Mohr J and Paul O (2001), *Microsystem Technology*. Weinheim, Wiley & Sons.

Münzenrieder N, Cherenack K and Tröster G (2011), 'The effects of mechanical bending and illumination on the performance of flexible IGZO TFTs', *IEEE Trans Electron Dev*, 58, 2041–2048.

Noda K, Sato H, Itaya H and Yamada M (2003), 'Characterization of Sn-Doped In_2O_3 film on roll-to-roll flexible plastic substrate prepared by DC magnetron sputtering', *Japan J Appl Phys*, 42, 217–222.

Paradiso R, Loriga G, Taccini N, Gemignani A and Ghelarducci B (2005), 'WEALTHY, a wearable health-care system: new frontier on e-textile', *J Telecomm Info Tech*, 4, 105–113.

Park S, Mackenzie K and Jayaraman S (2002) 'The wearable motherboard: a framework for personalized mobile information processing (PIMP)', *Proceedings of the 39th Annual Design Automation Conference*, ACM, p. 174.

Philips and Levi, ICD+ dress (2011), available from http://www.p4c.philips.com/cgi-bin/dcbint/cpindex.pl?slg=en&scy=es&ctn=levi/icd/jacket [Accessed 14 November 2011].

Pyralux: DuPont website (2011), available from http://www2.dupont.com/Pyralux/en_US/products/laminate/AP/pyralux_ap.html [Accessed 13 November 2011].

Savio D and Ludwig T (2007), 'Smart carpet: a footstep tracking interface', *21st International Conference on Advanced Information Networking and Applications Workshops*, 2, 754–760.

Schaar A K and Ziefle M (2011), 'Smart clothing: perceived benefits vs perceived fears', *5th International Conference on Pervasive Computing Technologies for Healthcare (Pervasive Health)*, pp. 601–608.

Schwartz P (2008), *Structures and Mechanics of Textile Fiber Assemblies*. Cambridge, UK, Woodhead Publishing in Textiles.

Service R (1997), 'Patterning electronics on the cheap', *Science*, 278, 383–384.

Suo Z, Ma E Y, Gleskova H and Wagern S (1999), 'Mechanics of rollable and foldable film-on-foil electronics', *Appl Phys Lett*, 74(8), 1177–1179.

Sze S M (1982), *Physics of Semiconductor Devices*. Hoboken, Wiley & Sons.

Tröster G (2011), 'Smart Clothes: the unfulfilled pledge?', *IEEE Perv Comp*, 2(2), 87–89.

Tummala R R (2001), *Fundamentals of Microsystems Packaging*. New York, McGraw-Hill.

van Langenhove L and Herteleer C (2004), 'Smart clothing: a new life', *Sci Tech*, 16(1/2), 63–72.

Wong W S (2010), *Flexible Electronics: Materials and Applications*. New York, Springer.

Yabuta H, Sano M, Abe K, Aiba T, Den T, *et al.* (2006), 'High-mobility thin-film transistor with amorphous InGaZnO channel fabricated by room temperature rf-magnetron sputtering', *Appl Phys Lett*, 89, 112–123.

Zysset C, Cherenack K, Kinkeldei T and Tröster G (2010a), 'Weaving integrated circuits into textiles', *Proceedings of the 2007 International Symposium on Wearable Computers*, pp. 1–8.

Zysset C, Kinkeldei T, Cherenack K and Tröster G (2010b), 'Woven electronic textiles: an enabling technology for healthcare monitoring in clothing', *5th International Workshop on Ubiquitous Health and Wellness (UbiHealth 2010)*.

9

Organic and large-area electronic (OLAE) technologies for smart textiles

F. ELLINGER and C. CARTA, Technische Universität Dresden, Germany, A. HÜBLER and G. SCHMIDT, Technische Universität Chemnitz, Germany, J. ZAPF, Siemens, Germany, G. TRÖSTER, ETH Zürich, Switzerland, A. TALO, Enfucell, Finland, D. KOZAKIS, Data Control Systems, Greece, D. VASSILIADIS, Exoduss, Greece, R. PARADISO, Smartex, Italy, M. KREBS, Varta, Germany, M. SCHARBER, Konarka, Germany and M. TUOMIKOSKI, VTT, Finland

DOI: 10.1533/9780857093530.2.253

Abstract: This chapter provides an overview of the state-of-the-art in the area of multifunctional, ultra-lightweight, ultra-thin and bendable organic and large-area electronic (OLAE) systems. This chapter introduces three possible applications, all based on a fully printed, textile compatible, sound module: textile integrated audio module including broadcast radio and solar supply; active receiver tag for wireless streaming of acoustic data and advertizing; and security tag system with acoustic alarm, motion and/or temperature sensors. There are reviews of the OLAE components and systems required for these applications. For the realization of these components and systems, the advantages of several flexible OLAE technologies need to be combined, while keeping cost at a minimum. Interface and packaging issues are discussed for full system integration on a common flexible foil enabling bending radii down to 1 cm. By combining these components, a variety of novel multifunctional OLAE systems can be enabled: the resulting market potential is discussed at the end of the chapter.

Key words: multifunctional organic and large-area electronics (OLAE) systems, disposable and rechargeable printed batteries, printed solar cells, printed loudspeaker, printed sensors, roll-to-roll (R2R) printed organic field effect transistors (OFET), printed analog electronics, Indium Gallium Zinc Oxide (IGZO) thin-film transistor (TFT) technology, flexible radio receiver, textile integration.

9.1 Introduction

In the recent past, remarkable progress has been made in flexible organic large-area electronics (OLAE) technologies and components. However, the integration of multiple OLAE components into *completely flexible multifunctional* systems, suitable to textile integration, is much less mature. By *completely flexible* systems, we mean the full integration of all required system components (main and auxiliary devices) into *flexible* OLAE (printed, thin-film, organic) technologies, on a common substrate or foil carrier, which maintains the advantages of bendability, ultra-thinness and ultra-low weight. The focus of the EU-funded IP project FLEXIBILITY (FLEXIBILITY Project) is the design of such *completely flexible multifunctional* systems, involving components with functionalities from multidisciplinary areas. Project partners share expertize in a diverse set of disciplines, such as audio/acoustics, circuit design, wireless communications, sensors, touch screens, solar and supply power engineering. Based on these heterogeneous components, a variety of specific *multifunctional* systems can be designed to create novel application scenarios and attractive synergic markets for the components involved. For the realization of these systems, the advantages of several flexible OLAE technologies need to be combined, while still minimizing costs. Among others, roll-to-roll (R2R) printing offers ultra-low costs per unit area (2 €/m^2 on paper, 8 €/m^2 on foil) for components requiring large areas (e.g. loudspeaker and audio amplifiers), three-dimensional (3-D) integration, as well as the integration of heterogeneous devices on one single substrate. Compact (down to 10 μm gate length), super-fast (>200 MHz transit frequency, mobilities >10 cm^2/Vs), low-loss IGZO thin-film technology can be employed to enable RF operation and wireless communications. Since the cost per unit area of this thin-film technology is higher, it is only attractive for circuits requiring very compact dimensions.

9.1.1 Applications

To further validate the capabilities of designed components and their suitability to multifunctional system integration, three specific demonstrators (1) to (3) are proposed, all showing promising market potential (FLEXIBILITY Project). Further combinations of functionalities are also possible and may be considered.

Audio-tag system

As illustrated in Fig. 9.1, a printed multipurpose sound module integrated on one single substrate, which consists of an audio speaker, amplifier and a pre-distortion circuit, is developed. Compared to modules using conventional speakers and amplifiers, much lower weight, lower costs and a very thin form-factor are possible.

A thin-film amplitude modulation/frequency modulation (AM/FM) broadcast radio is designed and employed as a signal source. An IGZO thin-film technology features transistors with transit frequencies above 200 MHz, making active circuits with high frequency operation possible. A miniaturized touch screen fully integrated into thin-film technology (including controller: integrated circuit (IC)) serves as a user interface. Two options fully integrated into flexible technologies are available and compared for the direct current (DC) power supply: a solar-powered regenerative DC supply module and a disposable battery, which does not contain any hazardous materials. The audio system is optimized for wearable textile applications and is integrated into a jacket or on the front of a duffle or messenger bag.

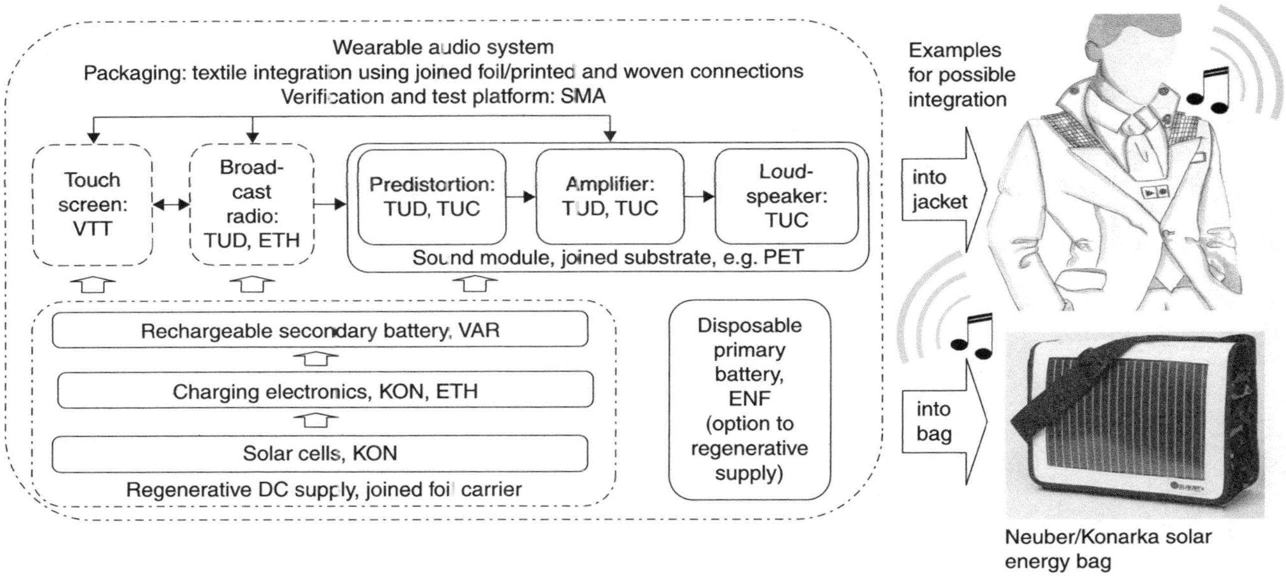

9.1 Wearable audio system completely based on ultra-thin, flexible technologies.

Wireless audio data streaming tag system

As depicted in Fig. 9.2, a flexible wireless receiver tag for wireless streaming of audio data is developed. A coverage range of up to 50 m is targeted. Since, in the envisioned scenario, only one shared transmitter is required, this component can be implemented with conventional electronics: in fact, it does not necessarily require low cost, thin form factor and bendability. The tags are required in large numbers and will be realized in flexible, low-cost technology. Based on the printed sound module and the printed motion sensor, product information or advertisements can be reproduced acoustically, for instance, in a supermarket after a customer touches a product. The streaming can be associated with selected products, product groups or products of a specific company.

Sensor security tag systems

Figure 9.3 shows the architecture of the fully printed sensor security system and application examples. An analog printed oscillator is designed and combined with the printed sound module to realize acoustic alert functionality. Two options are possible:

1. *Combination with printed infrared motion sensor*: Alert is, for example, activated if a theft is attempted (e.g. in a museum) or an item should not be touched (e.g. in a dangerous working environment).
2. *Combination with printed temperature sensor*: An acoustic alarm is activated if the temperature or its average are too high, e.g. in power electronics, food or pharmaceutical products. This helps to detect wrong operation or storage.

Stand-alone components and sub-systems

Most of the developed components integrated into the above three systems are also exploited as stand-alone products, thereby increasing the impact and the

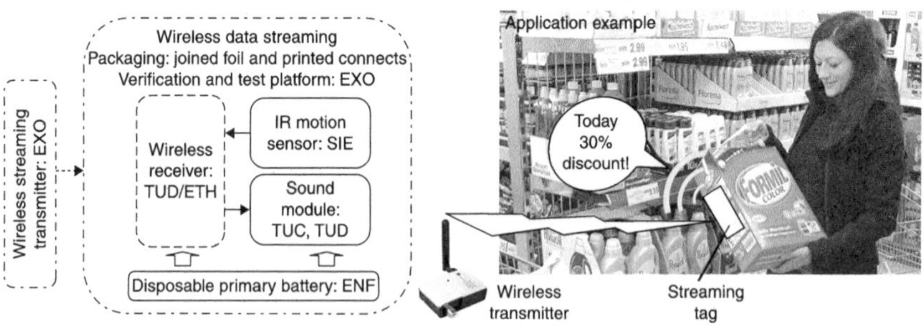

9.2 Simplified architecture of system enabling wireless streaming of acoustic data completely based on ultra-thin flexible technologies, and application examples.

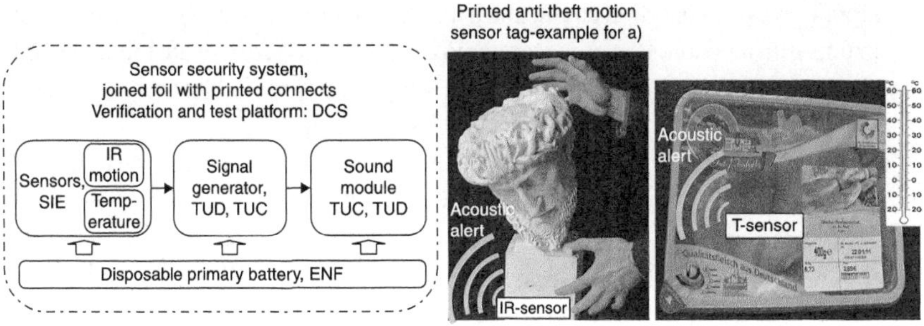

9.3 Simplified architecture and application examples for completely printed security sensor tags.

probability for overall market success. For example, the sound module (or even just the loudspeaker) can also be used for low-cost toy applications, wearable security alert systems and car entertainment. Implemented in the ceiling of a car, the sound module can be expected to enable an impressive natural space sound experience, thanks to the large speaker area. A further example is the solar-powered DC supply module, which can be used for a variety of applications with best efficiency in outdoor scenarios. Depending on the environment (e.g. sufficient area available, relation between active and sleep mode), the solar powered DC supply module can also be an excellent supply option for the systems (2) and (3), and others. Depending on the desired complexity of the user interface, the touch screen may also be employed for (2) and (3).

Progress beyond state-of-the-art

The integration of systems, such as those depicted above, poses relevant challenges and requires significant advances of the current state-of-the-art. The diverse domains, where such innovation is required, can be segmented into the following types:

- *Multifunctional systems* completely integrated in flexible technologies. A corresponding combination of functionalities and heterogeneous devices has not been demonstrated yet on flexible technologies, e.g.:
  - audio module with broadcast radio and solar supply optimized for textile integration;
  - active receiver tag for wireless streaming of acoustic data and advertizing;
  - security tag system with motion or temperature sensors and sound alarm.
- *Novel and advanced components* with functionalities and/or performances not reported before in flexible technology, e.g. a printed loudspeaker implemented on a polyethylene terephthalate (PET) or paper substrate together with an audio

amplifier, a pre-distortion circuit and a tuneable audio signal generator. Moreover, in thin-film technology, a broadcast radio and a wireless receiver are implemented. A fully flexible DC power module including rechargeable batteries, solar cells and charging electronics is designed; which yields up to 24 V.

- *Improvement of technologies* with focus on circuit- and system-level integration. Examples for R2R printed technology: novel 3-D stacking approach and integration of heterogeneous audio devices on same substrate. Examples for thin-film technology: high mobility of $>10\,cm^2/Vs$ and transit frequency of >200 MHz at very low supply voltages down to 3 V supply and bending radii <1 cm.

9.2 Flexible technologies for textile integration

In Table 9.1, the components required for the envisioned practical applications are mapped with respect to the three specific demonstrator platforms. The key demands, required areas and the technology chosen for integration are listed. Since many of the components can be used in several demonstrators and systems, the impact of these components is high. Some of the components can be used as options to enable a variety of system functionalities. Homogeneous and heterogeneous packaging approaches are used to assemble the complex demonstrator systems. As shown in Fig. 9.4, this multifunctional approach addresses the full value chain with business opportunities on all levels. This section reviews the state-of-the-art and the required developments for all the components and technologies required for such multifunctional integrated systems.

9.2.1 Loudspeakers

After the discovery of piezoelectricity in the polymer polyvinylidene fluoride (PVDF) by Kawai in 1969, different applications for this type of material have been explored. In particular the possibility to process the polymer in liquid form allows a very flexible use. Today, this group of materials mainly attracts interest as electro-active polymers (EAP), especially in robotics as artificial muscles. In 1996, Paradiso demonstrated, in a study at MIT Boston, how balloons can communicate acoustically with the help of PVDF microphones and loudspeakers. Since 2001, the South Korean P&I Corporation produces loudspeakers based on commercial PVDF foils, which are treated with an ion-assisted reaction in order to apply electrodes by using the conductive polymer PEDOT:PSS (Lee *et al.*, 2003(a,b)). Coating of the PVDF foils was realized with different methods (e.g. screen printing). Metals or metal alloys could be used as electrodes; however, such layers must be sputtered. This process stiffens the foil, which causes problems with excitation, resulting in good acoustic properties only in high-frequency ranges. The first products based on this principle are available from the Korean companies Mirae Plasma and Fils-Film Speaker.

Table 9.1 Components and demonstrators developed in FLEXIBILITY

	Components	Demonstrator tag systems				Key demands	Area**	Technology choice
	Functionality	Wearable audio	Wireless streaming	Sensor security				
				Temperature	Motion			
	Application example	Smart clothes or fun bag	Disposable advertisements	Disposable food	Anti-theft motion sensing			
	Life time	3 years	1 year	1 year	2 years			
Components based on flexible technologies (printed, thin-film, organic)								
1	Audio loudspeaker	Needed	Needed	Needed	Needed	Audio	L	Printed
2	Audio amplifier	Needed	Needed	Needed	Needed	Audio	M	Printed
3	Audio pre-distortion	Needed	Needed	Optional	Optional	Audio	M	Printed
4	Audio signal generator	Not required	Not required	Needed	Needed	Audio	M	Printed
5	Motion sensor	Not required	Not required	Not required	Needed	IR-sensor	M/C	Printed
6	Temperature sensor	Not required	Not required	Needed	Not required	R = f(T)	M/C	Printed
7	Broadcast radio receiver	Needed	Not required	Not used	Not required	RF-MHz	C	Thin-film
8	Wireless stream receiver	Optional	Needed	Optional	Optional	RF-K/MHz	C	Thin-film
9	Touch-screen	Needed	Optional	Optional	Optional	OLED	M	Print/organic
10	Disposable primary battery	Optional	Needed	Needed	Needed	Potential	L/M	Printed
11	Recharg. secondary battery	Needed	Optional	Optional	Optional	Potential	M	Printed
12	Charging electronics	Needed	Optional	Optional	Optional	Low-loss	C/M	Thin-film
13	Solar cells	Needed	Optional	Optional	Optional	OPV	L/M	Print/organic

(*Continued*)

Table 9.1 Continued

		Demonstrator tag systems				Key demands	Area**	Technology choice
Functionality		Wearable audio	Wireless streaming	Sensor security				
				Temperature	Motion			
Application example		Smart clothes or fun bag	Disposable advertisements	Disposable food	Anti-theft motion sensing			
Life time		3 years	1 year	1 year	2 years			
System and sub-system integration and packaging approaches								
14	Printed interconnects	Needed	Needed	Needed	Needed	Mainly homogeneous technologies		
15	Joined substrate, e.g. PET	Needed	Needed	Needed	Needed	As above		
16	Conductive tape	Needed	Needed	Needed	Needed	Homogeneous and heterogeneous technologies		
17	Textile integration	Needed	Not required	Not required	Not required	As above		
Auxiliary devices								
18	Wireless streaming transmitter	Not required	Needed	Needed	Needed	One for many tags ⇒ standard electronics		
19	Broadcast radio transmitter	On air	Not used	Needed	Needed	Available		

Rough estimation of edge-dimensions: **) large (L): 5–100 cm, medium (M): cm range, compact (C): mm range

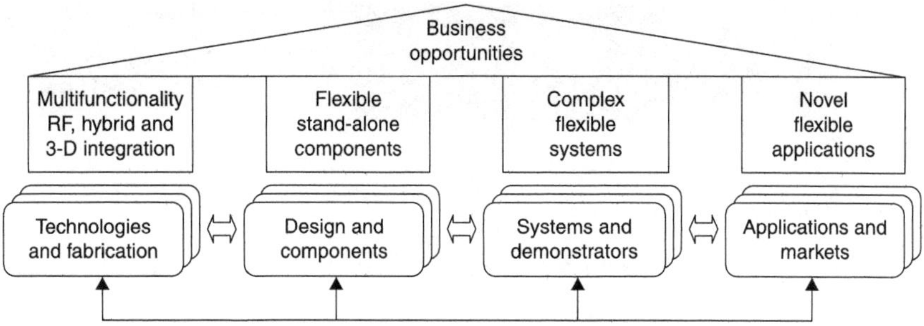

9.4 Full value chain addressed in FLEXIBILITY including the associated business opportunities.

It is also possible to radiate sound based on selective electromagnetic excitation. The company NXT plc, Cambridge, UK, uses this principle in distributed mode loudspeakers (DML) (Borwick, 2001). A modified version uses a piezoelectric compound as a vibration stimulator of plastic plates (DMA technology) (Boyd, 2005). The group of T. Sugimoto, from NHK Research Labs in Tokyo (Sugimoto *et al*., 2009), was able to demonstrate that a PVDF layer on foil is well-suited as a loudspeaker, because the distortion is much smaller compared to a single PVDF layer. In this work, a polythiophene was used for the electrodes, so that it was possible to produce a flexible and transparent loudspeaker. Despite the comparatively low conductivity of the polymer electrodes, even large-area loudspeakers with this set-up were functioning. As an alternative to systems based on PVDF and conventional polymeric (PEDOT:PSS) and metallic conductors, Yu from the University of Minnesota (Yu, 2006, 2007) coated PVDF foil with carbon nanotubes (CNTs). The main focus of this research work was to use the system for active noise control. Xiao *et al*. (2008) from Tsinghua University Beijing, China, have demonstrated that thin, transparent loudspeakers can be produced with CNTs by using the thermo-acoustic effect. The material, which is made entirely of CNTs, is produced in a complex process by growth of CNTs on silicon wafers.

The company Warwick Audio Technologies from Coventry, UK (http://www.warwickaudiotech.com/content.php?menu_id=3&page_id=23) uses a mechanical approach: an especially developed, compressible foil vibrates with the help of an electrical signal.

Flexible flat loudspeakers become increasingly attractive for smart objects, when they can be integrated with additional components such as amplifiers and signal generators on one single substrate. In the FLEXIBILITY Project, for the first time, printed loudspeakers are integrated together with printed circuits (amplifier, pre-distortion and signal generator) on one common PET or paper substrate. Nothing comparable has been reported in literature.

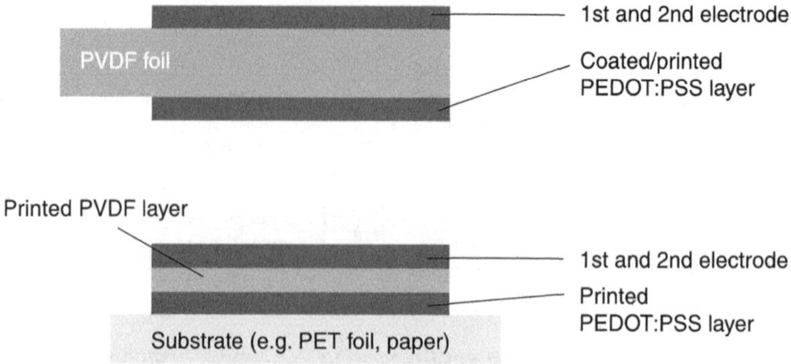

9.5 Top: Printed PVDF loudspeaker. Bottom: TUC design with printed PVDF layer on PET substrate (optionally paper substrate is possible).

The active polyvinyliden fluoride (PVDF) zone can be selectively printed, for example on a polyethylene terephthalate (PET) substrate. A layer design with an electrode first, then the PVDF, and then a second electrode printed on top (Fig. 9.5), can be realized. This allows a selective, freely-positioned design of the active loudspeakers on a foil, which is beneficial for the integration of other circuits. Such solutions are not possible with other known alternative technologies. As shown in Fig. 9.6, TUC succeeded in 2010 to produce fully-printed loudspeakers on PET-foils (Hübler *et al.*, 2012).

Before any optimization, the prototypes were capable of a frequency response of 1.5 to 4 kHz at voltages of around 50 V and sound pressure levels of 70 dB at 1 m distance. The impedance of the fabricated samples is in the order of 5 kΩ; at

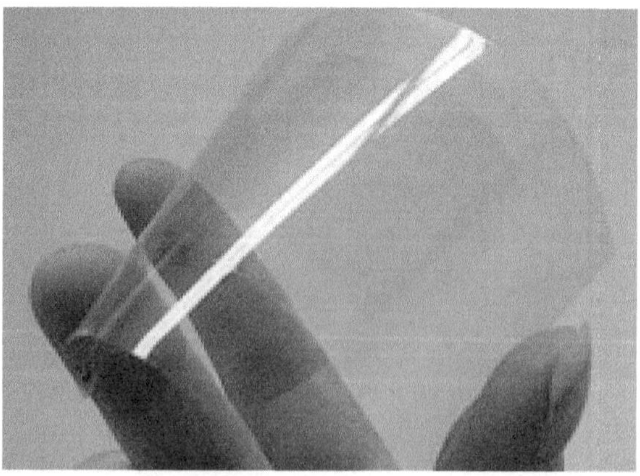

9.6 Printed TUC loudspeaker on PET foil with transparent printed PVDF and small bluish PEDOT:PPS layer.

present, with 10 mA, the current consumption is relatively high. By systematic optimization, the following parameters are expected to be achievable:

- sound power: >80 dB at 1 m;
- maximum required ac voltage amplitude below 20 V;
- impedance >5 kΩ, to minimize the current consumption.

At least two speaker versions are required: first, a compact version for security alerts with an area of more than 5 cm^2 and a bandwidth of 1 to 2 kHz; second, an audio version with an area of around 200 cm^2 and a bandwidth of 200 to 6 kHz.

The design offers several degrees of freedom, since the following parameters can be optimized: geometry (especially layer thickness and area), choice of copolymers, formulation of functional layers, polarization and piezoelectric coefficients. The integration of the speaker on paper is currently being investigated, as it reduces the costs from around 8 to 2 €/m^2 at the expense of the reliability and lifetime.

9.2.2 Technologies for electronic circuits

Mainly two technologies are available for the circuit design: R2R printing of TUC and indium gallium zinc oxide (IGZO) thin-film technology of ETH Zurich. The first has the major advantage that the full sound module, which covers a large area, can be integrated on one single substrate at the lowest possible costs. Compact circuit dimensions are enabled by 3-D integration techniques. However, the transistors fabricated in printed technology are too slow for operation at radio frequencies (RFs). The latter is possible with ETH IGZO thin-film technology, which enables the realization of a broadcast radio, wireless communication and low-loss control circuits (e.g. for DC charging), while still providing mechanical flexibility. This thin-film technology is more expensive per area than R2R printing. However, the required areas of the thin-film circuits are by orders of magnitude smaller, thus relaxing the cost issues.

Printed technology

In the past, various research groups worldwide have contributed to the development of printed electronics. While some groups focused on the development of new or improved materials, others designed simple electronic devices made of organic functional materials. Many different materials, organic and inorganic, including nano-tubes, were used for the realization of antennas, rectifiers, ring oscillators and logic circuits, as printing technologies, gravure printing and inkjet printing were used. In addition to a high number of publications on OLEDs and sensors, some works about gravure printed diodes with organic semiconductors (Lilja *et al.*, 2009) and all printed low-voltage organic transistors (Tobjörk *et al.*, 2008) were reported. A combination of reverse gravure printing and inkjet printing was used. The Imperial College in London, UK, investigated gravure printing of the semiconductor poly (3-hexylthiophene (P3HT) and the insulator poly (methyl methacrylate (PMMA), as well as the inkjet printing of silver inks used as a gate electrode (Voigt *et al.*, 2009). The Welsh Centre for Printing and Coating, UK, demonstrated the feasibility of the production of thin conductive tracks by means of flexography. These tracks can be used as highly conductive electrodes for OLEDs (Deganello *et al.*, 2010). Further important research institutes for printed electronics are VTT, IMEC and the Tampere University.

The technology under development at TUC offer features suitable for the system described in Section 9.1. Relevant examples of the latest results are available in the literature, for example:

1. The process conditions and windows to achieve maximum homogeneous functional layers have been investigated and defined (Kempa *et al.*, 2005; Reuter *et al.*, 2007; Hambsch *et al.*, 2010).
2. TUC found and explored a complementary semiconductor material semiconductor (CMOS) system (p-type pentacene and n-type: BASF Sepiolid/ Polyera ActivInk N1500; useful in many digital and analog circuits of various complexity), which can be processed as a solution.
3. Printed organic devices have been integrated into a standard computer aided design (CAD) system. Circuit data was converted into a suitable format for the pre-press necessary for low-cost mass-printing. First circuits (e.g. logic gates and amplifiers stages) have been simulated and fabricated (Bartzsch *et al.*, 2007).
4. Fully printed memory devices were developed at TUC (Deshmukh *et al.*, 2009; Reuter *et al.*, 2010).
5. The first multilayer 3-D circuit design, so that the initially poor integration density of printed circuits can be improved by orders of magnitude (Hübler *et al.*, 2010).

The results in (1) to (5) are promising starting points, but further improvements of printed technology are the object of current research, in order to enable applications such as those in Section 9.1. Among these objectives are, for example, supply voltages of the devices and circuits below 24 V, high yield, and further development of the 3-D integration techniques to decrease the required circuit area and to improve the circuit complexity per unit area.

At present, the processing of printed technology is well advanced. However, from the circuit-design point of view, no satisfying design kits, automated tools and full embedding into standard CAD tools exist. Hence, circuit design has not been efficient, thereby limiting the possible complexity and performance of printed circuits. This is especially the case for analog circuits, where more than just standard cells are required. The situation is similar to that of 50 years ago, when the design of CMOS-based ICs began. Hence, it is an important requirement to develop a design kit to allow efficient circuit design and the maximum exploitation of printing technology. This design kit must include models, layout templates of the most important devices and precise design and layout rules.

The three mass printing processes – offset printing, gravure and flexography – have specific features, which are favorable for different layer systems and are being optimized in FLEXIBILITY. At TUC, they are being developed in parallel and are chosen according to the formulation of the printed material, the required layer thickness, target lateral resolution, etc. A special demand on all printing methods is smoothness and homogeneity, as well as reproducibility of layers. Stable results can be achieved with a FET channel length of 50 μm and more, even with several thousand printed transistors. At the moment, for optimum operation, these metal oxide semiconductor field effect transistors (MOSFETs) require supply voltages in the order of 50 V. By optimizing the technology, materials and geometries, the required supply voltages are predicted to be reduced to below 24 V. A further important goal for audio applications is to achieve transistor transit frequencies of well above 10 kHz. The designed printed components, such as the loudspeaker, audio signal generator and audio amplifier plus pre-distortion, are fabricated by TUC printing technology. Relevance will be assigned to those functional materials that must have optimized conducting, semi-conducting or dielectric properties. Standard systems have been established, which can be adjusted to process conditions and are being developed further to match the needs of the circuit designers. It is important to harmonize runability, film formation features, adhesion and boundary surfaces with the contact layers and electrical functions. Three-dimensional integration and folding techniques will be used, in order to reduce the required circuit area by a factor of at least three compared to conventional two-dimensional (2-D) integration. This is beneficial for both analog (very large devices) and digital (very high device count) circuits. TUC has already designed a folded 3-D ring oscillator (Fig. 9.7).

Thin-film technology

Since the introduction of thin-film technology, the application of thin-film transistors (TFTs) have concentrated around digital driver circuits for displays. Hence, many circuits are based on ring oscillators to characterize the inverter gate delay. In Suresh *et al.* (2010), a delay of 48 ns per stage has been reported, claiming the fastest all-transparent oxide semiconductor circuit to date; however, with a supply voltage of 25 V, the circuit was fabricated on glass. In Mourey *et al.* (2009), ZnO TFTs fabricated on flexible substrate (Kapton) achieving an inverter delay of 60 ns per stage at 18 V, were reported.

Only a few works have been published concerning a deep understanding of the analog behavior and modeling of TFTs. However, in Tarn *et al.* (2010), an operation amplifier (OpAmp) and digital-to-analog converter has been reported, both fabricated using amorphous silicon TFTs. The OpAmp achieves a gain of

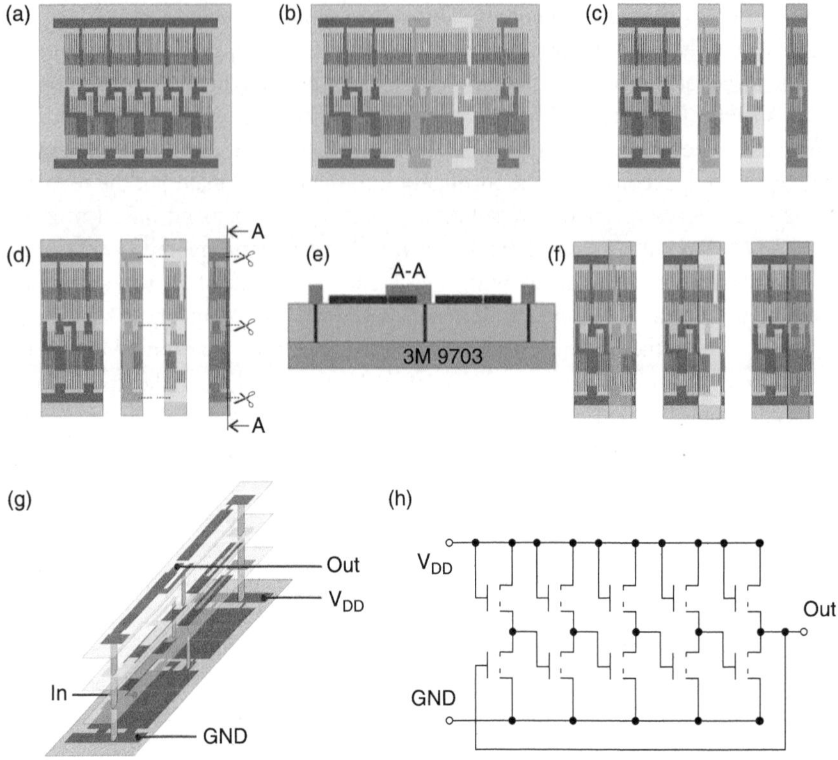

9.7 Printed three-dimensional integration approach of TUC illustrated by ring oscillator with five stages. Some layers (e.g. semiconductor and dielectrics are omitted for clarity): (a) layout; (b) stacking scheme, gray: substrate, dark gray: source/drain electrodes, other colors: gate electrodes; (c) inverter stages 3, 4 and 5 are separated from stages 1 and 2 and from each other; (d) small cuts are made at the contacts; (e) cuts are filled with PEDOT:PSS and anisotropic conductive two-side adhesive tape is attached to backside; (f) inverter stages are subsequently stuck to stage 2; (g) exploded scheme; (h) circuit schematics.

42.5 dB and a unity gain frequency of 30 kHz at a supply voltage of 25 V, which would already be sufficient for audio signals. Based on nanocrystalline ZnO TFTs fabricated on Si substrates, a record performance was achieved with a 1.2 μm AU-based gate. The devices yield a mobility of 68 cm^2/Vs, and a transit frequency of 2.45 GHz at a supply voltage above 10 V (Bayraktaroglu *et al.*, 2009). The state-of-the-art thin-film devices and analog circuits are considerably constrained in their performance, for example by rigid substrates (Bay, 2009), high supply voltages, for example more than 15 V (Mourey *et al.*, 2009), or restricted operation speed (Tarn, 2010). Moreover, to the authors' knowledge, no extensive analog design kits are available. The reliability, bendability and strainability of thin-film technology, including IGZO TFT, was demonstrated by ETH in Woo *et al.* (2010) and Cheremacl *et al.* (2010(b,c)).

In order to enable the applications of Section 9.1, the research activities of FLEXIBILITY are currently targeting the following goals:

- development of an analog design kit, including scalable models and automated layout templates for the key active and passive devices, as well as basic analog circuits such as amplifiers and oscillators;
- design of fully integrated high-Q passives, e.g. for filters required for wireless communications;
- operation voltage of 3–5 V, relaxing the requirements for batteries;
- combining of self-aligned lithography with lift-off patterning make mobilities >10 cm^2/Vs, TFTs with a f_T >200 MHz and threshold voltages of only 250 mV possible on IGZO, even at 3 V;
- the robustness of the thin-film circuits will be demonstrated by bending tests with radii below 1 cm;
- exclusive use of flexible substrates.

In the past, TFTs were mainly applied for driver circuits in displays. For such applications, there was no major need for extensive design environments featuring an analog design kit. However, as for the printed technology, a good design kit is instrumental to the design of both analog and digital circuits with complexities and performances required in the target applications. This design kit includes models and automated layout templates of the most important devices: transistors, inductors, varactors and contact pads. Compact thin-film inductances and varactors, with quality factor higher than 10, have to be designed for the filters required in the wireless radios.

Figure 9.8 and Fig. 9.9 summarize some information about the thin-film technology of ETH Zurich, which is being used as a starting point for its development. At a technology level, advances of the state-of-the-art are still needed. By combining self-aligned lithography techniques and lift-off processes, the transit frequency of the IGZO TFTs will exceed carrier saturation mobility of 10 cm^2/Vs and transit frequencies of 200 MHz at supply voltages below 5 V. At

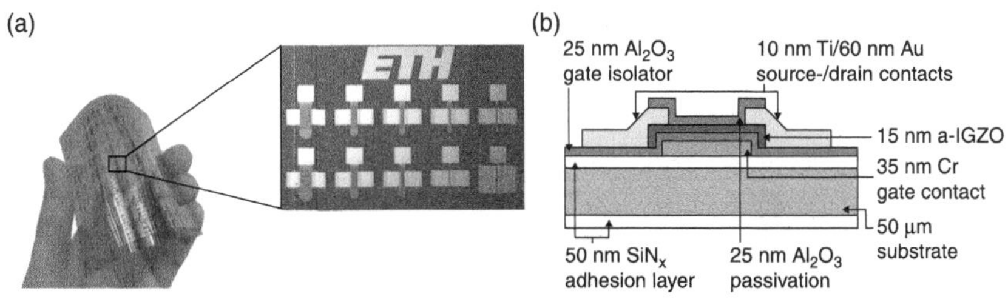

9.8 Bendable IGZO thin-film transistor technology at ETH Zurich: (a) photo of samples; (b) architecture.

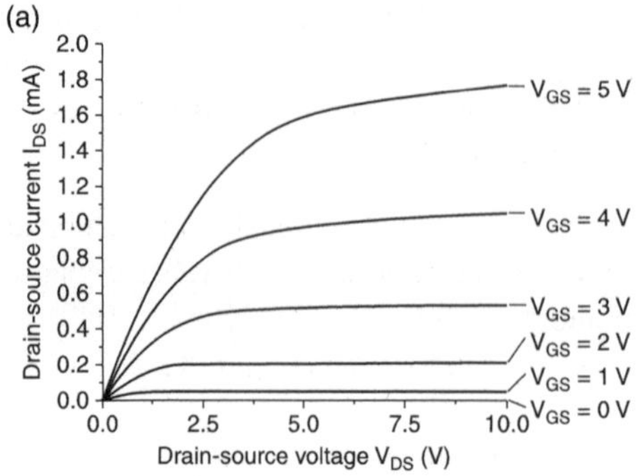

(b)

Performance parameter:	
Linear mobility μ_{LIN} (cm^2/Vs)	13.1
Saturation mobility μ_{SAT} (cm^2/Vs)	12.8
Transit frequency (MHz)	>200
Threshold voltage (mV)	250
Sub-threshold slope (mV/decade)	130
On/off ratio	$2 \cdot 10^8$

9.9 Measured IGZO thin-film field-effect transistor performance.

this low supply voltage, the transit frequency is a factor of two higher compared to existing solutions on flexible substrates. Several features make this technology attractive for flexible systems. The low-temperature deposition technology with T <150 °C enables the use of many different organic substrates. Moreover, the low threshold voltage amounting to $V_{th} = 0.25$ V and the small associated saturation voltage, allow low-loss operation down to a supply of 3 V. The process flow also guarantees a low threshold voltage shift necessary for reliable analog circuits. The micro-patterning of relevant layers and careful tuning of the layer coatings will guarantee that the fabricated devices (TFTs, resistors capacitors, inductors and varactors) can handle bending radii below 1 cm, resulting in a tensile and compressive strain not achieved so far. For testing, the bending machine shown in Fig. 9.10 is used. The bendability, reliability and robustness of TFTs have been investigated in Cherenack *et al.* (2010(b)). Appropriate process control guarantees a moderate device performance degradation, even if woven using commercial weaving machines (Cherenack *et al.*, 2010(a)). Also, the interconnect lines can survive the bending stress and strain if applying a suited layer structure (Cherenack *et al.*, 2010(c)). To increase the robustness and the bending properties, it has been demonstrated that micro-pattering of the critical device layers improve the vulnerability of aging micro-cracking (Woo *et al.*, 2010).

9.10 Bending machine for testing.

9.2.3 Motion and temperature sensors

Infrared motion sensor

At present, the operation of many motion and acceleration sensors relies on piezo-electronic (Orengo *et al.*, 2010) or Micro-Electro-Mechanical System (MEMS) (Tan *et al.*, 2008) principles. To allow a fully printed realization, the use of near-infrared photosensitive devices, which are integrated together with optical high-pass filters, is being considered. Hence, wavelengths beyond 1.1 μm are required. Such a device can detect motion via human thermal radiation. For printing, crystalline semiconductors cannot be used, but organic synthetic materials need to be employed. However, these organic photodiodes are not able to detect light above 1 μm (Natali *et al.*, 2003; Yao *et al.*, 2007).

A good approach is the use of colloidal quantum dots, which are suitable as infrared absorber for photo-detection (McD, 2005; Konstantatos *et al.*, 2006; Koleilat *et al.*, 2008). A hybrid hetero-junction photodiode containing these quantum dots is able to detect infrared up to 1.8 μm (Rauch *et al.*, 2009). This wavelength is of particular interest, since water has a low absorption in this range. Such quantum-dot-based near-infrared photodiodes are investigated and manufactured in the FLEXIBILITY Project. This fully printed motion sensor is designed by SIE and based on photosensitive devices, which are able to detect light in the near-infrared area. Therefore the detection of humans by thermal radiation is possible – even at night.

Temperature sensor

A temperature sensitive printed structure is reported in Gao *et al.* (2010). It features a 1 bit memory, which saves the information if a tag has been exposed to excessive temperature since the last reading. The temperature ranges are from 30 to 90°C and 80 to 120°C. In Kinkeldei *et al.* (2009), platinum resistance sensors optimized for textile integration are presented. VTT has developed indicators for monitoring cold chain and oxygen, which can tell whether a food package has already been opened (VTT, 2010).

A fully printed temperature sensor tailored for the FLEXIBILITY systems is currently being developed. The temperature dependence of resistors in a Wheatstone bridge is exploited. Most of the available inks for the production of resistors are optimized for a low thermal coefficient to be less sensitive regarding temperature changes, but opposite characteristics are required here. Hence, an ink suitable for temperature sensors must be searched for (e.g. using platinum, silver, copper, palladium, etc.). The values of the four required resistors will depend strongly on the lateral dimensions of the printed line. So a robust und optimized design is necessary to avoid expenses in trimming. A temperature detection range of 10 to 70°C is targeted.

9.2.4 Touch screen

A good overview of the current state-of-the-art in the area of touch screens can be found in DisplaySearch (2009). At present, there is a trend to increase the resolution and possible display area (e.g. for e-readers and phones). Most of the touch screens and displays (including those using organic technology) need CMOS electronics (e.g. for the controller IC). In FLEXIBILITY, full integration of the complete touch screen using, for example Super Yellow polymer material (Becker *et al.*, 2000), is being considered, together with the possibility to integrate the controller IC into organic electronics using mass-fabrication devices. Moreover, the compact touch screen is being optimized for minimum power consumption rather than high resolution, so as to ease the supply via solar energy.

VTT is designing a miniaturized touch screen based on printed OLEDs and printed sensor structures. The use of Super Yellow polymer material is considered. The touch screen module can have a contact or non-contact low resolution user interface, with various functions such as on/off switching, displaying sound volume, and controlling functions for volume and radio channels. The touch screen module will be compact in terms of module thickness and dimensions, cost-effective from the material and processing point of view, and fully integrated into ultra-thin printed technology. The technological challenge is the full integration of all components into a thin-film platform, which is manufactured by conventional printing processes. The possibility to use printed transistors even for the controller IC, which up to date is integrated in non-flexible CMOS, is studied. With the additional goal of minimizing the power consumption, the targets are:

- area of a few cm^2;
- operational time >1000 h;
- brightness: 500 cd/m^2;
- efficiency: 8 cd/A;
- DC power @ 4V: 26 mW/cm^2;
- DC power supply.

9.2.5 Direct current power supply

Disposable battery

Both primary (single use) and secondary (rechargeable) printed batteries have already been demonstrated (www.enfucell.com; www.tgdaily.com/hardware-features/43086-ultra-thin-batteries-can-be-printed-out; www.enas.fraunhofer.de/Images/datasheet_500_pf_bat_tcm438-141336.pdf). The one most frequently used to date is based on lithium, due to its high energy density. It may be formed as a gel to allow bendability. Oak Ridge National Research Laboratory in the US has licensed over 100 companies to make this kind of devices. Advantages of lithium-based batteries are their ability for recharging and cell voltages up to 3.7 V (e.g. graphite-lithium cobalt oxides). Moreover, batteries based on a zinc manganese dioxide cathode yielding a cell voltage of 1.5 V are popular. By connecting several batteries in series, voltages of up to 6 V were demonstrated (Tgdaily, 2009). Research institutes such as the Fraunhofer ENAS (www.enas.fraunhofer.de/Images/datasheet_500_pf_bat_tcm438-141336.pdf) and companies such as ENF (*Enfucell, 2010*), Prelonic (www.prelonic.com) and Power Paper, and licensees such as Graphic Solutions, are currently producing these. They have potentially the lowest cost and are regarded as the most environmentally friendly. However, these kinds of batteries cannot be recharged.

The disposable battery designed by ENF is based on eco-friendly zinc manganese dioxide chemistry with zinc chloride as electrolyte, contains no hazardous materials and is fully printed with traditional printing methods, such as screen printing, and is laminated into a layered structure. No expensive equipment is needed, since fabrication can take place in production lines that are modified from traditional reel-to-reel equipment. In this field, ENF owns several key patents (Zhang, 2008; Zhang and Hongyang, 2008; Zhang and Li, 2008, 2010). In this material composition, a voltage of 1.5 V and a capacity of 90 mAh can be reached per basic cell with maximum area of 60 × 72 mm. Since no metal case is required, the battery has a weight of 2.9 g, a thickness of 0.6 mm and is bendable. These batteries are for single use only, since they cannot be recharged.

Rechargeable battery

For the rechargeable battery – developed by VAR – graphite-lithium cobalt oxides and nickel metal hybrids are investigated. Ongoing activities aim at showing that structures of 100 μm are printable and that the cells can be connected in series using low-cost techniques. Both batteries are optimized for the requirements of the components and systems designed in the FLEXIBILITY project. Good tradeoffs between voltages, capacity, area and bendability have to be found. There is one significant challenge: Flexible electronics need high voltages, for example 24 V, for the printed technology. Hence, to allow fully printed active systems, the batteries have to generate up to 24 V to allow efficient operation. Such high supply voltages have not been reported to date: 6 V (Tgdaily, 09) is the highest reported so far. An improvement of the state-of-the-art by a factor of four is needed. Innovative principles (Fig. 9.11) have to be used to realize compact and low cost batteries with such high supply voltages. Most of the printed batteries on the market have shelf times of less than 1.5 years. This shelf time needs to be improved as well.

The rechargeable (secondary) batteries developed by VAR will be based on materials allowing recharging. One material composition considered is graphite-lithium cobalt oxide (organic electrolyte), offering around 3.7 V per basic cell and a high capacity per area of 5 mAh/cm^2. However, lithium is not 100% environmentally friendly. A further option not requiring lithium is to use a nickel metal hybrid (alkaline electrolyte) technology yielding 1.2 V per cell. Both approaches are being investigated. The following techniques will be studied and compared: stacked or co-planar electrodes, linear gap or inter-digitized electrode pairs, serial or parallel

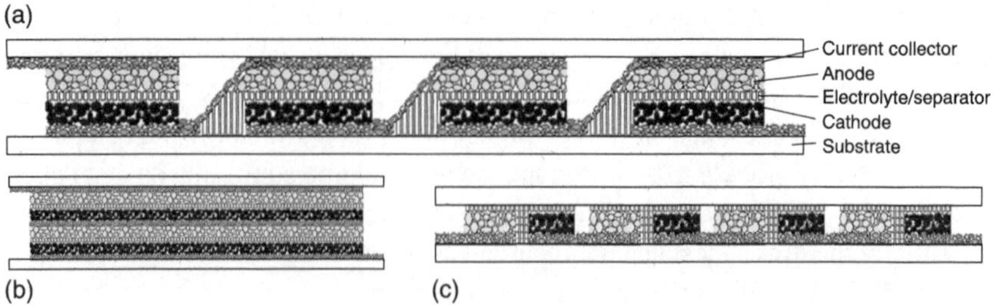

9.11 Printed battery integration approaches to get added output voltages based on basic cells: (a) series connection; (b) vertical stacking; (c) coplanar approach beside each other.

printed constructions, etc. Also new recipes for solid electrolytes are being considered. Printing of fully polymer cells is aimed for. The requirements for solid electrolytes are good ionic conductivity combined with electronic insulation, ease of printing, but flexibility after 'drying' (rubber-type) to solid. Also, one major challenge is achieving up to 24 V. Considering an effective voltage of 3 V per cell for the case of a graphite-lithium cobalt oxide battery, 8 combined cells are required. Assuming that the resolution of screen printing is in the order of 100 μm, it takes 300 μm to print one cell. That means to get a voltage of 24, a lithium-based printed battery with 8 cells has a minimum length of around 2.4 mm. The thickness of the battery is around 0.36 mm, while the width of the battery depends on the required capacity. Ongoing research aims at showing that structures of around 100 μm are printable and that the cells can be connected in series using low-cost techniques. A number of recharges of at least 500 is being targeted.

Flexible organic solar cells

Flexible organic solar cells are still in the initial phase of commercialization. Today a small number of products are available, including the Sunbags distributed by Neubers (www.energy-sunbags.de/de.html). The first Organic Photovoltaic (OPV) products have been developed for the consumer electronics market.

In the EU project, FACESS, organic solar cells are combined with thin-film batteries. Here, silicon-based transistor circuits are required for the control of the charge operation. In FLEXIBILITY, approaches are being studied to also integrate the DC charging electronics in flexible technology. In this regard, two options are investigated: first, ETH thin-film technology (favorable in terms of efficiency) and second, TUC printed technology (advantageous in terms of low cost integration into the full system). Unique to the FLEXIBILITY project is also the combination with fully printed/thin film audio and wireless receiver circuits and the goal of providing supply voltages up to 24 V.

Most of the components and all systems in the target demonstrators involve active circuits. Hence, the DC power supply is a component of crucial importance;

it also has a strong environmental impact. Two approaches are well suited with respect to environmental friendliness: a compact disposable (primary) battery, which does not contain any hazardous materials and a secondary battery, which can be recharged by solar power. Supply voltages down to 3 V are sufficient for the ETH thin-film technology. However, for the R2R printing technology, the DC supply units should provide voltages of up to 24 V. To reach this goal, several techniques are being investigated, which allow an accumulated output voltage based on connection of cells. Proposals for corresponding integration techniques are illustrated in Fig. 9.11. The capacity can be traded off with output voltage and area.

The OPV modules are being optimized to feed the rechargeable batteries. In this regard, the specific parameters, such as the maximum possible area, form-factor, costs, required capacity, light conditions, etc., will be taken into account. The major application is the supply of the wearable audio jacket (or bag). Maximum solar energy is available outdoors, where the jacket is used most of the time. Compared to standard solar cell modules, the following additional challenges have to be addressed: a high voltage of up to 24 V has to be achieved in relatively small areas. This means that a high number of around 40 solar cell strips, each providing around 0.6 V, have to be placed and connected on a relatively small module area. Hence, the actively useable cell area and the geometric fill factor decreases, because a relatively large area is required for the connections between the cells. Compared to standard solar cell modules (e.g. typical solar cell modules for energy farms), which are not so area limited, the possible efficiency is smaller. The OPV module is based on the KONARKA OPV technology providing a record cell efficiency of up to 8.3%. Considering the further losses based on the boundaries given by the target applications, such as a high voltage per module area ratio, a module efficiency of at least 5% is targeted for the modules. Life times of at least 5 years are targeted, which are sufficient for the application needs (compare also Table 9.1).

9.3 Circuit design

It is the aim of the FLEXIBILITY project to develop all required tag circuits and components in flexible technologies. That means that no additional silicon ICs are required. In this sense, FLEXIBILTY makes a significant contribution towards multifunctional fully-integrated OLAE systems. One circuit development cycle involves the simulation, the design of layout with design rule considerations, the fabrication based on the technologies, the characterization, and redesigns.

9.3.1 Printed signal generator

Signal generator circuits, fully fabricated by means of mass printing technologies, have already been demonstrated (Bartzsch *et al.*, 2007; Hübler *et al.*, 2007). These circuits are based on unipolar transistor technology and yield a relatively low oscillation frequency in the range of 10 Hz, since they were not optimized for use in the typical audio spectrum. Printed ring oscillators using gravure and screen printing were reported by CEA Liten, France (Verilhac *et al.*, 2009). Here a frequency of the 5-stage oscillator, as high as 300 Hz at a supply voltage of 50 V, has been reported. The maximum oscillation frequency in ring oscillators is bound by the accumulated gate delays of several stages.

The circuit design for the intended applications of Section 9.1 (FLEXIBILITY Project) targets the following key features:

- higher oscillation frequencies of 1 kHz up to 2 kHz;
- much lower supply voltage of <24 V;
- frequency control capabilities.

The signal generator can be integrated with the speaker and amplifier on one single substrate.

The schematics and the measured output signal of a ring oscillator circuit, developed in 2010 by TUC, are shown in Fig. 9.12. For the target applications, a tuneable signal between around 1 and 2 kHz should be generated, which is subsequently amplified by the audio amplifier and reproduced by the printed loudspeaker. All printed components (speaker, generator and amplifier) will be integrated on one single substrate (PET or paper). The most challenging goal is to increase the oscillation frequency to values suited for sound alerts and which are close to the efficiency optimum of the printed speaker, while at the same time enabling frequency tuning and decreasing the required supply voltage down to 24 V. The following strategies are taken into account:

1. integration of switched reactive elements or varactors for frequency tuning (Ellinger, 2007);
2. complementary circuits for high gain (Gray *et al.*, 2009);
3. inductive peaking to achieve higher gain, enhanced signal stability and reduced phase noise. Active inductors (equivalent circuits) are considered to reduce the circuit area compared to passive inductors (Knochenhauer *et al.*, 2011);

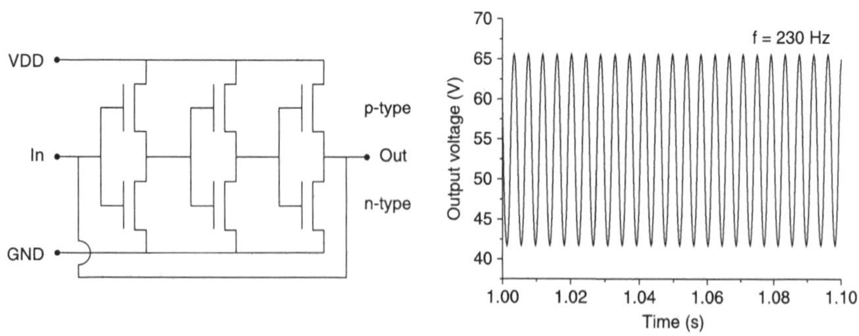

9.12 Printed ring oscillator designed by TUC [Kem10*].

4. investigation of alternative oscillator concepts beyond ring oscillators: e.g. negative resistance topologies with single stage amplifier to avoid accumulated gate delay inherent in ring oscillators; viable examples are negative resistance topologies with a single-stage feedback amplifier and enhancement of the printing resolution (Schmidt *et al.*, 2010) to improve the maximum frequency of oscillation of the transistors and to reduce capacitive parasitics (Ellinger, 2007);
5. enhancement of printing resolution to improve the maximum frequency of oscillation of the transistors and to reduce the gate delay (Schmidt *et al.*, 2010).

9.3.2 Printed audio amplifier

A large variety of audio amplifiers using several technologies exist on the market. Regarding the architectures of such amplifiers, the reader is referred to Self (2009). They are mainly based on standard silicon electronics. To the authors' knowledge, no audio amplifier integrated into printed technology has been reported to date. Preliminary modeling and simulation work carried out at TUD shows promising results. Challenges are the combination of large bandwidth, an output signal voltage of up to 20 V, and a good efficiency at supply voltages of up to 24 V.

It is expected that the research activities of the FLEXIBILITY project will yield the first printed audio amplifier. Several circuit concepts (common source, common gate, source follower, Darlington, feedback, etc.) and combination of these are being investigated. Further innovations not reported before include a printed pre-amplifier stage, which compensates for the sound pressure decrease of the printed loudspeakers towards high and low frequencies, thereby enabling larger overall bandwidth and better sound quality. The printed speakers show a level drop towards low and high frequencies, which can be compensated – to a certain extent – by approximated inverse characteristics with integrated RC filters. Such an audio amplifier can be integrated with the loudspeaker on one single PET or paper substrate. In order to enable the target application, the printed audio amplifier must be designed to drive the printed loudspeaker, so that a well audible sound pressure of >80 dB (at 1 m distance) can be reached. These circuits are currently being designed by TUD and fabricated by TUC.

9.3.3 Flexible broadcast radio receiver

An all-printed flexible AM radio operating at 640 kHz has been reported in Noh *et al.* (2009). Since this circuit operates only with passive components, the signal to noise ratio and the output signal quality are poor. It requires an external, non-integrated high-Q resonator, which is bulky and inflexible. Within FLEXIBILITY, thin-film technology featuring transistors with transit frequencies above 200 MHz is available. Hence, much higher performance, low noise amplification, higher operation frequencies and design techniques for full integration are possible.

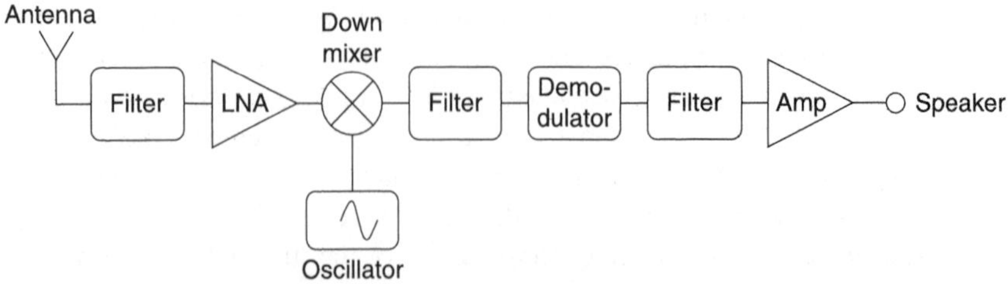

9.13 Simplified architecture for thin-film radio receiver.

Moreover, higher operation frequencies are feasible, paving the way to more advanced radio standards. Operation at long wave (LW) (148.5–283.5 kHz), or even Ultrakurzwelle (UKW) at 63 to 75 MHz or 87.5 to 108 MHz, should be possible. However, since the UKW frequency is in the range of the transit frequency, this will still be challenging. This is believed to be the first fully integrated thin-film broadcast radio developed.

This thin-film radio is compatible with use in wearable, textile applications. A first possible and simplified architecture of the radio receiver is shown in Fig. 9.13 (Rohde and Whitaker, 2000). Depending on the architecture, the circuit includes high-Q filters, a low noise amplifier, a mixer, a demodulator and an audio driver. To allow small dimensions and a high circuit yield, the device count should be below 50. For the antenna, a low-cost wire may be used, which can be easily integrated into textiles.

9.3.4 Flexible wireless data receiver tag

To date, the wireless transmission of data in flexible (organic or thin-film) technology is mainly associated with Radio Frequency Identification (RFID) tags. IMEC has, for example, demonstrated an organic thin-film RFID tag with 128 bits and a data rate of 1.5 kb/s operating at 13.56 MHz and at 24 V supply (Myny *et al.*, 2009(a,b)). This RFID system works on the basis of magnetic fields generated by the reader. Hence, the communication distance is limited to a few cm. A communication distance of up to 3 m has been demonstrated, based on a passive backscattering modulator working at 868 MHz (Pardo *et al.*, 2007). For the latter, high performance silicon technology is required for the tag. Moreover, the Austrian Company, IDENTEC Solutions, has developed an active RFID tag using high-performance silicon electronics. It operates at ultra-high frequency (UHF) and Industrial Scientific and Medical (ISM) band and enables localization up to 500 m.

In the FLEXIBILITY project, a complete, active, wireless RF receiver is being developed, which is fully integrated into thin-film technology. Ranges of up to 50 m and data rates exceeding 10 kb/s are targeted. This system is tailored for the wireless streaming system supporting audio/acoustic data. As one possible option, the operation at the amateur radio band around 25 MHz or any low frequency ISM bands will be considered. An analog modulation is used to simplify the circuit architecture. The receiver architecture has similarities with Fig. 9.13; however, with different requirements, for example, regarding the operation frequencies, which can be more relaxed. The receiver can be implemented in the IGZO technology of ETH.

9.4 Textile integration

An overview of the state-of-the-art in the area of e-textile connections can be found in Post *et al.* (2000) and Linz *et al.* (2008). ETH has shown that their thin-film technology is an excellent candidate for textile integration (Cherenack *et al.*, 2010(a–c); Woo *et al.*, 2010). Most conductive yarns are not as conductive as traditional electronic wires. Steel fibers, such as Bekinox®, have a sufficiently high conductivity, but are less flexible. Silver coated yarns are a good alternative to steel fibers, because they have a higher conductivity and are easier to process. However, their biocompatibility is worse. Good examples for silver coated technologies are Xstatic and Shieldex. Enameled copper wires, such as those produced by Elektrisola, also show high conductivity and good insulation values, but are difficult to process due to their rigidity, and new possibilities must be investigated in this field. In the EU project ProeTEX, each interconnection has been realized with elastic conductive yarns, specially designed to be processed with the machine needles; yarn conductivity is due to a monofilament of stain steel twisted around an elastic core (Lycra®), covered by polyester to grant insulation. Unfortunately, insulation was not optimal and improvements are still necessary.

The state-of-the-art in textile integration regarding prominent EU projects, such as the MyHeart project (www.research.philips.com/technologies/myheart/index.html), Proetex (www.proetex.org), Vivometrics and Stella (www.stella-project.eu), is illustrated in Fig. 9.14. In these projects, standard electronic components have been integrated. These components have a limited flexibility. The FLEXIBILITY project aims at improving the state-of-the-art by textile integration and packaging using fully flexible devices, which means that no standard electronic components are used. The integration of components on different heterogeneous flexible technologies with different properties, contact materials, etc, will be a challenge. Here, the focus is on a wearable audio system. Advanced flexible woven solutions with low loss of signal and power supply connections are studied.

Especially for the textile integration and connection, many specifications must be considered, such as flexibility, conductivity, insulation, water permeability and user comfort. Most conductive yarns are not as conductive as traditional wires. FLEXIBILITY aims at developing low-loss connections with optimum tradeoffs regarding other parameters. The following materials and technologies will be considered: steel fibers (e.g. Bekinox), silver coated yarns (e.g. Xstatic, Shieldex), enameled copper wires (e.g. Elektrisola), elastic stitched conductive yarns (stain steel or metal filaments twisted around an elastic Lycra™ core covered by polyester or traditional yarns), conductive substrates and paths fully integrated in

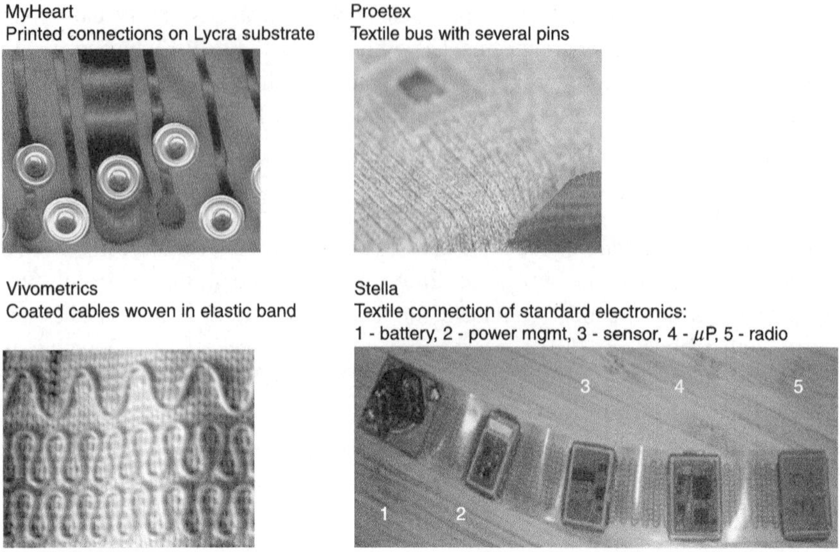

9.14 State-of-the-art for textile integration, reference to EU projects.

fabric, fabric coated with conductive layer, and elastic conductive paths printed onto fabric substrate. Further interesting techniques to be investigated are 3-D integration, multilayering, sandwiching, stitching and gluing.

The global methodology used for textile integration is illustrated in Fig. 9.15. For textile integration, the following phases are planned: a first phase involves the selection of the conductive components, in term of fibers, combination of fibers into yarns, implementation of woven or coated fabric bus. The final structure of the basic connection has to be designed, according to the insulation requirements; the use of an insulation layer or a multilayer architecture could be considered for shielding purposes. The basic connection has to be integrated with the electronics module and/or the electrical connector on a textile substrate. This is the second phase that pertains the module integration and the selection of the processes. As reported in Fig. 9.15, the integration in a single module of connection, connector and substrate, can be done through assembling, coating and printing technology, on the basis of components realized with traditional textile processes in the form of woven or knitted fabric. Once the single components and the methodology have been selected, leading to the integration of the system, all the key inputs for the design of the system are known. According to the system design, the integration of the different modules and the electronics components will be based on a further combination of the selected elements in a garment or in an accessory; or the assembling of shaped panels where all the elements have already been integrated. Overviews of reported work on this topic by SMA and ETH can be found in Paradiso *et al.* (2004, 2005(a,b), 2007(a)) Cherenack *et al.* (2010(c)) and Woo *et al.* (2010), respectively.

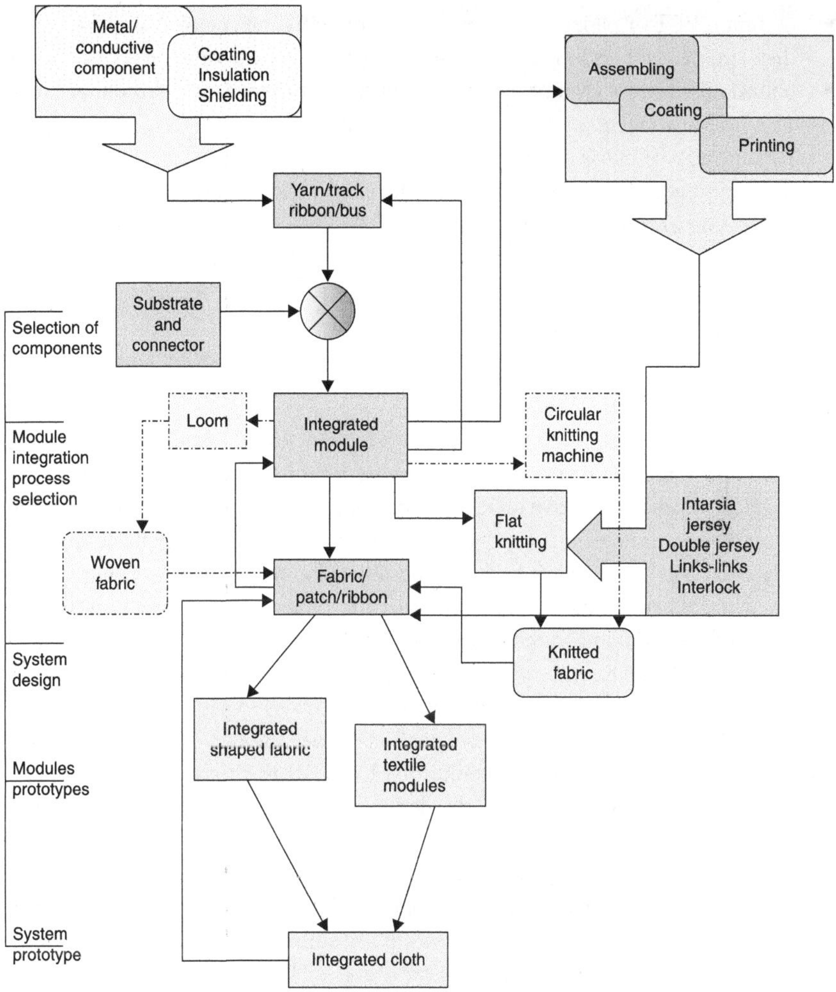

9.15 Methodology for textile integration.

9.5 Packaging integration and service life issues

The heterogeneous integration, packaging and interconnection tasks will require the joint efforts of all partners and will be coordinated by SIE. This is especially the case for the multifunctional demonstrator systems, which are based on the components and packaging technologies listed in Table 9.1. The homogeneous and heterogeneous component technologies are integrated using packaging technologies such as:

- common PET or paper substrates (e.g. to integrate the speaker, amplifier, predistortion circuit and the signal generator on one single substrate);
- common substrate (foil or paper) with printed or dispensed interconnects and vias produced mechanically or by laser beam;
- anisotropic conductive tapes;
- textile integration with woven interconnects. The low-temperature processing deposition of thin-film technology with T <150°C enables the use of a variety of organic substrates, simplifying integration with other technologies. ETH has shown that their thin-film technology can also be used for woven interconnections (Cherenack *et al.*, 2010(a));
- 3-D integration techniques will be considered where possible and beneficial, which enable a minimization of the required circuit area.

Typically, OLAE electronics and packaging have a lower life time compared to standard electronics and packaging. Hence, life time issues are very important for OLAE electronics and so have to be taken into account during the development and in the marketing strategies. The required life times for the targeted applications have already been listed in Table 9.1. The disposable food temperature sensing system and the wireless streaming systems require a life time of around one year. Higher life times of around two and three years are probably required for anti-theft motion sensing system and for the wearable audio system, respectively. Regarding all components, the lowest life time of around two years is expected for the primary printed batteries. However, they are typically replaced anyway. Most of individual components designed within the frame of FLEXIBILITY have the potential for higher life times than required for the defined demonstrators. This is, of course, beneficial for stand-alone products, for example, for the solar cells and rechargeable batteries. For these two components, a life time above five years is expected. For the circuits in IGZO thin-film technology, a life time of more than ten years is estimated.

The packaging has a strong impact on the life time. The packaging strategies are optimized considering influence parameters, such as temperature (which has non-negligible impact above 80°C), ultraviolet light (UV), bending stress, etc. Life time enhancement techniques will be applied, such as optimized layer set-up, protection coating layers, printed varnish against UV radiation, etc. Based on system evaluations, the optimum tradeoff between life time and performance can be evaluated and taken into account in future developments.

9.6 References

Alastalo A, *et al.* (2010), 'Printable WORM and FRAM memories and their applications', *LOPE-C 2010 Conference*, Frankfurt, Germany, 31 May–2 June.

Bartzsch M, Kempa H, Otto M, Hübler A and Zielke D (2007), 'Device and circuit simulation of printed polymer electronics', *Org Electron*, 8(4), 431–438.

Bayraktaroglu B, Leedy K and Neidhard R (2009), 'High-frequency ZnO thin-film transistors on Si substrates', *Electron Dev Lett, IEEE*, 30(9), 946–948.

Becker H, Spreitzer H, Kreuder W, Kluge E, Schenk H, *et al.* (2000), 'Soluble PPVs with enhanced performance: a mechanistic approach', *Adv Mat*, 12(1), 42–48.

Borwick J (2001), *Loudspeaker and Headphone Handbook*. Focal Press.

Cherenack K, Zysset C, Kinkeldei T, Münzenrieder N and Tröster G (2010a), 'Woven electronic fibers with sensing and display functions for smart textiles', *Adv Mat*, XX, 1–5.

Cherenack K, Münzenrieder N and Tröster G (2010b), 'Impact of mechanical bending on ZNO and IGZO thin-film transistors', *IEEE Electron Dev Lett*, 31(11), 1254–1256.

Cherenack K, Kinkeldei T, Münzenrieder N, Zysset C and Tröster G (2010c), 'Woven thin-film metal interconnects', *IEEE Electron Dev Lett*, 31(7), 740–742.

Deganello D, Cherry J A, Gethin D T and Claypole T C (2010), 'Patterning of micro-scale conductive networks using reel-to-reel flexographic printing', *Thin Solid Films*, 518, 6113–6116.

Deshmukh K D, Reuter K, Kempa H, West J E and Katz H E (2009), 'Tuning of threshold voltage in organic field-effect transistors with hydrophobic gate dielectric using mono-energetic low-energy electron beams and triode corona', *Appl Phys Lett*, 95, 113307.

DisplaySearch (2009), Touch panel market analysis report.

Ellinger F (2007), *Radio Frequency Integrated Circuits and Technologies*. Springer, March.

FLEXIBILITY Integrating project (flexible multifunctional bendable integrated light-weight ultra-thin systems), EU Contract Number FP7–287568; available from: http://www.flexibility-fp7.eu

Gao J, Siden J and Nilsson H E (2010), 'Printed temperature sensors for passive RFID tags', *Proceedings of Symposium of Progress in Electromagnetics Research*, Xi'an, China, 22–26 March.

Gray P R, Hurst P J, Lewis S H and Meyer R G (2009), *Analysis and Design of Analog Integrated Circuits*. Wiley.

Hambsch M, Reuter K, Stanel M, Kempa H, Fügmann U, *et al.* (2010), 'Uniformity of fully gravure printed organic field effect transistors', *Mat Sci Eng B*, 170, 93–98.

Hübler A C, Doetz F, Kempa H, Katz, H E, Bartzsch M, *et al.* (2007), 'Ring oscillator fabricated completely by means of mass-printing technologies', *Org Electron*, 8(5), 480–486.

Hübler A C, Schmidt G, Kempa H, Reuter K, Hambsch M and Bellmann M (2010), 'Three-dimensional integrated circuit using printed electronics', *Org Electron*, Elsevier.

Hübler A C, Bellmann M, Schmidt G C, Zimmermann S, Gerlach A and Haentjes C (2012), 'Fully mass printed loudspeakers on paper', *Adv Mat*,

Kawai H (1969), *Jpn J Appl Phys*, 8, 975.

Kempa H, Reuter K, Bartzsch M, Hahn U, Hübler A C *et al.* (2005), 'Stability study of all-polymer field-effect transistors; Polytronic 2005', *Proceedings of the 5th International Conference on Polymers and Adhesives in Microelectronics and Photonics 2005*, pp. 67–71.

Kinkeldei T, Zysset C, Cherenack K and Tröster G. (2009), 'Development and evaluation of temperature sensors for textile integration', *IEEE Sens*, 1580–1583.

Knochenhauer C, Sedighi B and Ellinger F (2011), 'A comparative analysis of peaking methods for output stages of broadband amplifiers', *IEEE Trans Circ Syst I*, 58(11), 2581–2589.

Koleilat G *et al.* (2008), 'Efficient, stable infrared photovoltaics based on solution-cast colloidal quantum dots', *ACS Nano*, 2, 833–840.

Konstantatos G *et al.* (2006), 'Ultrasensitive solution-cast quantum dot photodetectors', *Nature*, 442, 180–183.

Lee C S *et al.* (2003a), *J Mat Res*, 18(12), 2904.

Lee C S *et al.* (2003b), *Synth Met*, 139, 457.

Lilja K E, Bäcklund T G, Lupo D, Hassinen T and Joutsenoja T (2009), 'Gravure printed organic rectifying diodes operating at high frequencies', *Org Electron*, 10, 1011–1014.

Linz T, Vieroth R, Dils C, Koch M, Braun T, *et al.* (2008), 'Embroidered interconnections and encapsulation for electronics in textiles for wearable electronics applications', *Adv Sci Tech*, 60, 85–94.

Mourey D A, Zhao D A, and Jackson T N (2009), 'ZnO thin film transistors and circuits on glass and polyimide by low-temperature PEALD', *Electron Devices Meeting (IEDM)*, pp. 1–4.

Myny K, Steudel S, Vicca P, Beenhakkers M J, van Aerle N A J M, *et al.* (2009a), 'Plastic circuits and tags for 13.56 MHz radio-frequency communication', *Solid-State Elect*, 53, 1220–1226.

Myny K, *et al.* (2009b), 'A 128b organic RFID transponder chip, including Manchester encoding and ALOHA anti-collision protocol, operating with a data rate of 1529b/s', in: *Solid-State Circuits Conference: Digest of Technical Papers*. ISSCC, IEEE International.

Natali D, Sampietro M, Arca M and Devillanova F A (2003), 'Wavelength-selective organic photodetectors for near-infrared applications based on novel neutral ditholenes', *Synth Met*, 137, 1489–1490.

Noh J, Cho G, Lim N, Kim J and Subramanian V (2009), 'All printed flexible AM radio circuit', *Large-area, Organic and Printed Electronics Convention.*

Orengo G, Saggio G, Bocchetti S and Giannini F (2010), 'Advanced characterization of piezoresistive sensors for human body movement tracking', *IEEE International Symposium on Circuits and Systems*, pp. 1181–1184.

Paradiso R (1996), *The Interactive Balloon: Sensing, Actuation and Behavior in a Common Object*. MIT Media Laboratory, Cambridge, MA.

Paradiso R (2007), 'Knitted textile for the monitoring of vital signals', *US Patent Application 20070083096.*

Paradiso R, Loriga G and Taccini N (2004), 'A wearable system for vital signs monitoring: wearable e-Health systems for personalised health management: state of the art and future challenges', *Stud Health Tech Info*, 108, 253–259.

Paradiso R, Loriga G, Taccini N, Gemignani A and Ghelarducci B (2005a), 'WEALTHY: a wearable health care system: new frontier on e-Textile', *J Telecom Info Tech*, 4, 105–113.

Paradiso R, Loriga G and Taccini N (2005b), 'A wearable health care system based on knitted integrated sensors', *IEEE Trans Tech Biomed*, 9(3), 337–345.

Pardo D, Vaz A, Gill S, Gómez J, Ubarretxena A, *et al.* (2007), 'Member IEEE, Roc Berenguer, design criteria for full passive long range UHF RFID sensor for human body temperature monitoring', *IEEE International Conference on RFID.*

Post E R, Orth M, Gershenfeld P N and Russo R (2000), 'E-broidery: design and fabrication of textile-based', 39(3 and 4).

Rauch T, Böberl M, Tedde S F, Fürst J, Kovalenko M V, *et al.* (2009), 'Near-infrared imaging with quantum-dot-sensitized organic photodiodes', *Nat Photonics*, 3, 332–336.

Reuter K, Kempa H, Brandt N, Bartzsch M and Hübler A C (2007), 'Influence of process parameters on the electrical properties of offset printed conductive polymer layers', *Prog Org Coat*, 58, 312.

Reuter K, Kempa H, Deshmukh K D, Katz H E and Hübler A C (2010), 'Full-swing mass-printed organic inverters using a charged perflourinated electret fabricated by means of mass-printing technologies', *Org Electron*, 11, 95.

Rohde U and Whitaker J (2000), *Communications Receiver: DPS, Software Radios and Design*, Third Edition. McGraw-Hill.

Schmidt G C, Bellmann M, Meier B, Hambsch M, Reuter K, *et al.* (2010), 'Modified mass-printing technique for the realization of source/drain electrodes with high resolution', *Org. Electron*, 11, 1683–1687.

Self D (2009), *Audio Power Amplifier Design Handbook*. Focal Press.

Sugimoto T, *et al.* (2009), 'Loudspeakers for flexible displays', *Acoust Sci Tech*, 30(2), 151–153.

Suresh A, Wellenius P, Baliga V, Luo H, Lunardi, L M, Muth J F, *et al.* (2010), 'Fast all-transparent integrated circuits based on indium gallium zinc oxide thin-film transistors', *Electron Dev Lett, IEEE*, April, 317–319.

Tan D T, Roy S, Nguyen P T and Huu T H (2008), 'Streamlining the design of MEMS devices: an acceleration sensor', *IEEE Circ Syst Mag*, 8(1), 8–27.

Tarn Y C, Ku P C, Hsieh H H and Lu L H (2010), 'An amorphous-silicon operational amplifier and its application to a 4-bit digital-to analog converter', *IEEE J Solid-State Circ*, 45(5), May, 1028–1035.

Tobjörk D, Kaihovirta N J, Mäkelä T, Pettersson F S and Österbacka R (2008), 'All-printed low-voltage organic transistors', *Org Electron*, 9, 931–935.

Verilhac J M, Benwadih M, Seiler A L, Jacob S, Bory C, *et al.* (2009), 'Step toward robust and reliable amorphous polymer field-effect transistors and logic functions made by use of roll to roll compatible printing processes', *Org Electron*,

Voigt M M, Guite A, Chung D Y, Khan R U A, Campbell A J, *et al.* (2009), 'Polymer field-effect transistors fabricated by the sequential gravure printing of polythiophene, two insulator layers, and a metal ink gate', *Adv Funct Mat*, 19, 1–8.

VTT report (2010), Research, development and commercialisation activities in printed intelligence.

Woo N, Cherenack K, Tröster G and Spolenak R (2010), 'Designing micro-patterned Ti films that survive up to 10% applied tensile strain', *Appl Phys A*, 100, 281–285.

Xiao L, *et al.* (2008), 'Flexible, stretchable, transparent carbon nanotube thin-film loudspeaker', *Nano Lett*, 8(12), 4539–4545.

Yao Y, *et al.* (2007), 'Plastic near-infrared photodetector utilizing low band gap polymer', *Adv Mater*, 19, 3979–3983.

Yu R (2006), *Sens Actuat A*,132.

Yu R (2007), *Trans Cont Syst Tech*, 15, 4

Zhang X (2008), 'Thin battery and method of manufacturing a thin battery', WO2008096033 (A1), August.

Zhang X and Li S (2008), 'A battery and a method of manufacturing a battery', WO2010026286 (A1), August.

Zhang X and Li S (2010), 'Anode and a method of manufacturing an anode', WO2010026285 (A1), March.

Zhang X, Talo A and Ning H (2008), 'Thin battery with longer life time', EP2039652 (A1), August.

9.7 Appendix: abbreviations and acronyms

AM	Amplitude modulation
CAD	Computer aided design
CMOS	Complementary metal oxide semiconductor

CNT	Carbon nano tube
DC	Direct current
EAP	Electro-active polymer
FET	Field effect transistor
FM	Frequency modulation
IC	Integrated circuit
IGZO	Indium gallium zinc oxide
ISM	Industrial scientific medical
MEMS	Micro electromechanical system
MOSFET	Metal oxide semiconductor field effect transistor
OLAE	Organic large area electronics
OPV	Organic photo voltaic
P3HT	Poly (3-Hexylthiophene)
PET	Polyethylene terephthalate
PMMA	Poly (methyl methacrylate)
PVDF	Polyvinylidene fluoride
R2R	Roll to roll
RF	Radio frequency
RFID	Radio frequency identification
TFT	Thin-film transistor
UHF	Ultra high frequency

10
Joining technologies for smart textiles

I. LOCHER, SEFAR AG, Switzerland

DOI: 10.1533/9780857093530.2.285

Abstract: This chapter reviews important building blocks for the use of electronic systems on textiles. A major focus is on electrical and textile joining technologies and their properties. Furthermore, different means of electrical traces are described that are suitable for integration in textiles. Some challenges in technology and in the manufacturing processes for electronic systems on textiles are pointed out. Finally, the chapter provides an assessment of technological trends, as well as future applications.

Key words: electronic systems on textiles, joining technologies, connection technologies, e-fabrics.

10.1 Introduction

The first approaches to the integration of electronics into garments were made more than 14 years ago in 1998, with the wearable motherboard from GeorgiaTech. Whereas the first research demonstrators were rather bulky, using the garment as a supporting structure for large components and strands of wires (Post *et al.*, 2000; Jones *et al.*, 2002; Park *et al.*, 2002), recent approaches (Cherenack *et al.*, 2010) (BMBF TexoLED) (Bonderover and Wagner, 2004; Diaz, 2011) try to use the yarn itself as a functional element, making them more like conventional textiles. Some of the early approaches became redundant because electronics have shrunk so much that they can now be integrated into small devices such as today's mobile phones. The very first smart fabric products on the market, such as the MP3-Jacket, were from companies such as Infineon.

In 2012, industry and applied research institutes are utilizing various approaches in their products and demonstrators. For example, Interactive-Wear, Clothing+ and Ohmatex use textile ribbon cables as connection elements, while sealed conventional circuit technology provides the functional elements in their products. The garment itself is the carrier of the functional infrastructure. These products range from LED signal jackets to fire fighter uniforms with integrated temperature sensors. Companies such as Sefar AG, Future-Shape GmbH and NEL Ltd use the fabric itself as a circuit board to place and interconnect electronic components. Institutes, such as Fraunhofer IZM and TITV Greiz, also utilize these and other approaches in their research. When discussing smart fabrics, we might immediately think of apparel with smart functionality. However, it is interesting to see that apparel made up only about 40% of the global textile market in 2010. The rest was made up of technical and interior (home) textiles. For example, Sefar and Future-Shape supply these markets.

10.2 Components of electronic systems in textiles

A principal goal in the smart-textile world is to combine the electronic and the textile world, while maintaining the functionality of electronics and the flexibility and the robustness of textiles. At present, this goal can only be achieved by making compromises in both areas. An example of an electronic system used in textiles is shown in Fig. 10.1. The textile substrate – a woven fabric – consists of PET monofilaments and insulated copper wires, used as a conductive element. It is formed as a mesh of cross running, electrically-floating copper wires. The copper wires are connected by the removal of the wire insulation and the addition of conductive paste. Some wires must also be cut, in order to avoid electrical shorts. Figure 10.1 shows an assembled electrical circuit using various component package generations, ranging from size 0402 to size 1206. They are assembled using standard processes in the electronics industry, such as laser drilling, stencil printing and pick-and-place techniques.

Another approach for the use of electronics in textiles is depicted in Fig. 10.2 (Linz *et al.*, 2005). The textile substrate does not embed any conductive elements. Instead, a flexible substrate film carrying the electronic circuit is mounted on the textile and conductive traces are sewn onto the textile substrate connecting the film pads. Here, the conductive traces are Ag-plated, multi-filament threads produced by Statex GmbH. Special attention must be given to the process of crossing the threads, as they are not insulated. Similar approaches are employed by TITV and Forster Rohner AG. The placement of the substrate film and sewing for connection of film pads with conductive traces are particularly important to these approaches.

The two examples given are very different approaches, but which achieve similar goals. However, in both processes, a reliable joining method is essential.

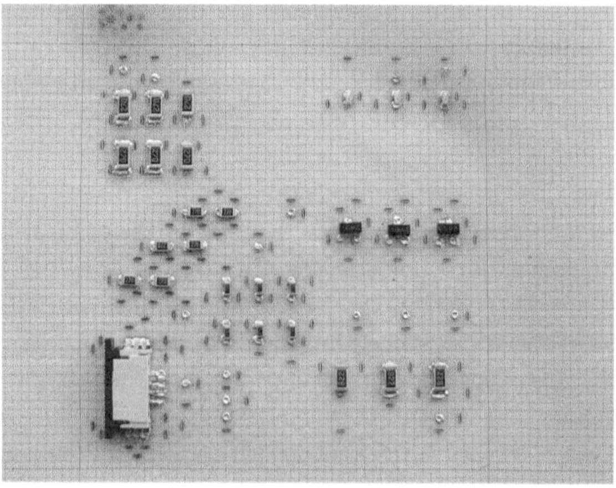

10.1 Electronic system on textile (Sefar AG).

10.2 Electronic system on textile (Fraunhofer IZM).

Such electronic systems in textiles are considered to be an intermediate step towards the full integration of electronics into a yarn. The following section reviews the various joining methods that are utilized in the two worlds of electronics and textiles, and it presents a selection of different conductive threads as well. Each joining method and its properties are briefly explained and analyzed. Some of these methods can be used to construct electronic systems in textiles using the textile as an electrical substrate.

10.3 Conductive threads as electrical traces

An ideal electrical trace in a circuit will not distort or attenuate the carrying signal. However, in reality, no electrical trace is perfect, particularly not when it is integrated into a fabric. Trace configuration should thus be optimized according to the type of signal being carried. For instance, a power signal should be sent over a low-ohmic trace, in order to avoid ohmic losses and elevated temperatures in the trace. However, a radio frequency (RF) data signal (typically of a frequency above 1 GHz) should be transmitted through a trace configuration with constant line impedance, using a substrate (e.g. based on Teflon™ or ceramics) with low dielectric losses, or the signal will face severe distortion. Line impedance is defined by the dimensions of the traces and the substrate, as well as the material of which they are composed. Furthermore, the properties of the substrate will correspond to those of the signal trace; it can be a rigid board or a flexible film. However, experience shows that low ohmic traces are more important in fabrics than constant line impedance, since RF signal transport in fabrics rarely occurs in practical applications.

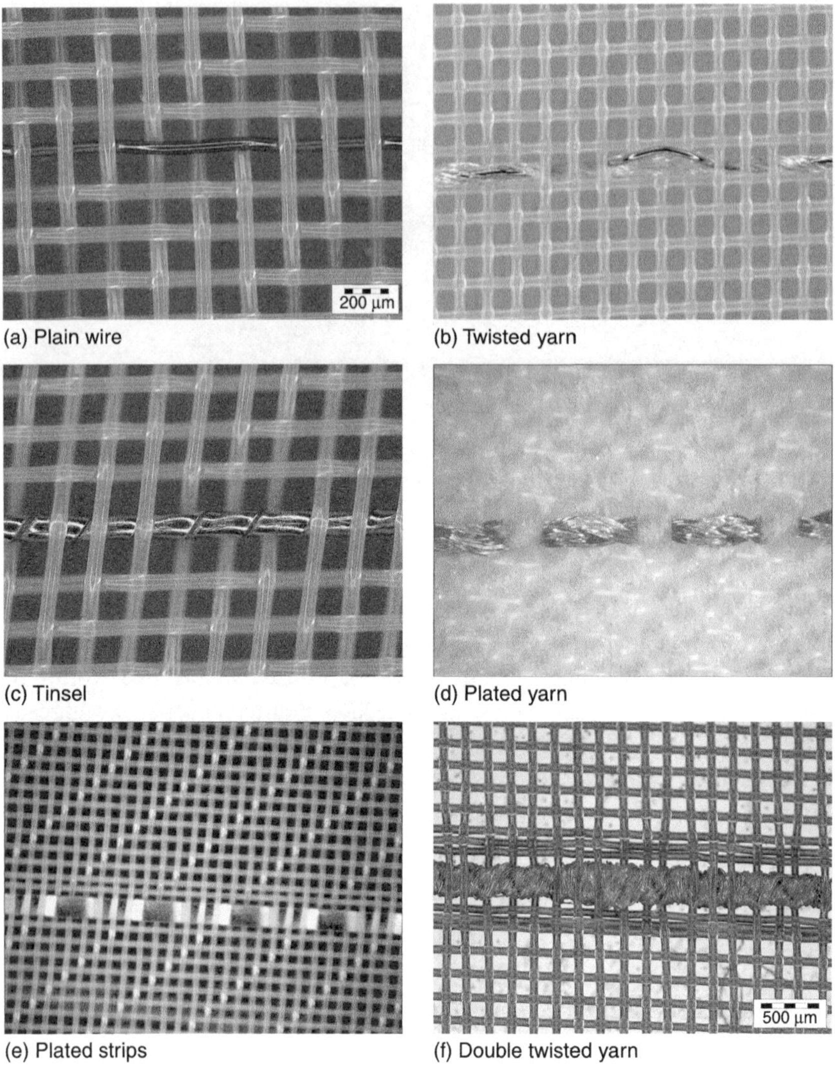

10.3 (a–f) Selection of conductive threads (Sefar AG).

The selection of conductive threads listed below, and in Fig. 10.3, is thus **not** among those that would induce constant line impedance:

- (a) *Plain wire*: Available in various metals and alloys, such as copper, copper alloys, stainless steel, nickel and tungsten. Wires possess the highest conductivity at a given thread diameter. They are sized at a minimum diameter of 10 microns and are connected using standard industrial methods. Companies such as Elektrisola offer wires with an additional, thin insulation. Wires have a well-defined position within a woven fabric, as can be seen in Fig. 10.3(a), but their mechanical strength and bending fatigue are weak in comparison to polymer threads.

- (b) *Twisted yarn*: A metal wire is twisted together with a polymer yarn, with the wire twisted into the shape of a spring. The number of twists is an important parameter to enable a defined electrical resistance and a mechanical strength higher than that of a plain wire. However, the position of the wire within the fabric is difficult to know exactly due to the spring shape.
- (c) *Tinsel*: This consists of a non-conductive, multi-filament core with a wrap of metal film. Again, the metal contour resembles a spring, while the mechanical strength of tinsel is even higher than that of twisted yarn. Tinsel is only available uncoated.
- (d) *Plated yarn*: A polymer thread, of PA for example, is plated with a layer of silver or a similar metal. The layer thickness is usually in the range of 0.1 to 1 microns. It is recommended to use multi-filament threads, since there might be cracks in the metal layer of single fibrils. A thick thread or several threads in parallel are also necessary, in order to achieve similar resistance values as those achieved with metal wires. These types of threads are usually not solderable or weldable. However, in terms of mechanical strength, they are arguably the best option of this selection, although there might be an increase in resistance over time due to natural wear. Suppliers of these types of threads are Statex GmbH, imbut GmbH, Tersuisse and Amberstrand.
- (e) *Plated strips*: These are similar to plated yarn, and consist of a polymer film with plated film of a metal such as aluminum or silver. They are also known as Lurex™ yarn.
- (f) *Double twisted yarn*: This has the same core construction as twisted yarn, but has an additional protective layer wrapped around the twisted yarn (sold by W. Zimmermann GmbH).
- *Stranded wires*: Several pure wires are twisted together to form a tiny rope, with excellent conductivity and mechanical strength. However, a strand's handle is usually very different to a textile handle.
- *Coax*: There are tiny coax cables down at 0.2 mm diameter at 50 Ω line impedance available, e.g. from Mitsubishi. However, they are rather stiff.

10.4 Introduction to joining technologies for electronics

Hundreds of reliable soldering joints are necessary on a printed circuit board (PCB), in order to make a mobile phone work and a vast number of welding joints and a huge number of connectors are needed in order to make a car run. A single bad joint can cause your car to break down. Electrical connections are crucial ingredients for every electrical system. The study of connectors can almost be considered an individual scientific field in its own right. Connections interconnect various components to merge into an entire, functioning electrical system. They are intended not to distort the electrical signals running through them and to have a continuity and reliability over time. An ideal connection has no contact resistance, constant characteristic impedance equal to its feed lines, infinite bandwidth and no attenuation; it must also fulfill certain mechanical requirements.

In electronics, there are many individual electrical joining technologies that are designed to cater to specific connector requirements and applications (Mroczkowski, 1998) (Table 10.1). Joining technologies such as soldering and welding are appropriate for permanent connectors, while others such as jacks/receptacles (connectors) are more suited to temporary connections. Some are designed for only a few insertion cycles and some are designed for several hundred cycles; some are intended for high electrical currents or high voltages and some are designed for RF (i.e. frequency with defined characteristic impedance). Some are even made to be waterproof (see IP codes according to IEC 60529 for further information about watertight properties). Connections are usually sealed in a solid encapsulation, in order to satisfy the given requirements.

In textiles, connections mainly serve a mechanical purpose, for example holding a garment permanently together and closing a jacket temporarily, but they also need to be resilient, so as to survive natural wear and cleaning. Table 10.2 lists several textile joining technologies, along with their mechanical and electrical properties.

Combining all the desired properties of electrical and textile connections in a garment becomes tricky when some requirements are contradictory to others. For example, the electrical connection should have defined characteristic impedance and robustness to withstand cleaning, but it should also be flexible to maintain the textile handle.

10.5 Overview of existing jointing technologies in the electronics and in the textile world

10.5.1 Soldering

Soldering is a process whereby two metal conductors join together using a filler metal (solder) at temperatures below 450°C. The filler metal melts establishes an intermetallic alloy at the contact surface, in between the filler and the conductor. The soldering temperature is anything below the melting temperature of the conductors. Solder connections facilitate defined, permanent low-ohmic electrical connections, although these should not be subjected to too much mechanical stress (3N/mm^2 is about the maximum strain per solder connection).

The solder plays an important role. Usually, it works with the filler metal and the flux forming a paste (small filler metal balls embedded in liquid flux) or forming a wire (filler metal wire with flux core). The flux must become activated before the filler metal starts melting, since it has to clean the contact surface and to break the oxide layer that usually covers metal surfaces. The residues of the flux must also be removed, depending on type of flux. There are clean (aggressive) and non-clean (less aggressive) fluxes.

Table 10.1 Overview of electrical joining technologies

Joining technology	Permanent	Joining partners	Electrical properties	Mechanical properties	Corrosion behavior	Protection	Remarks
Soldering, brazing	Yes	Metal + metal, solder	Low resistance, constant	Weak, no load allowed	Good	E.g. in encapsulation	Thermal process, many different solders available, not every metal is solderable
Conductive adhesive bonding	Yes	Metal + metal, adhesive	Low resistance, constant	Weak, no load allowed	Okay, absorbs water	E.g. in encapsulation	Expensive, elevated temperature or UV for curing
Welding	Yes	Metal + metal	Low resistance, constant	Weak, no load allowed	Good	E.g. in encapsulation	Various processes (resistance, US, laser welding, etc.), high temperature exposure, not every metal combination is weldable
Cold-welding, Crimping	Yes	Metal + metal	Low resistance, constant	Good	Good	Not needed	E.g. press-fit on PCBs, for connector pins
Connector	No	Metal jack + metal receptacle	Low to medium resistance, might not stay constant	Very good	Good, when proper type	Connector case	Big variety of connectors available

Table 10.2 Overview of textile joining technologies

Joining technology	Permanent	Joining partners	Electrical properties	Mechanical properties	Corrosion behavior	Protection	Remarks
Welding	Yes	Fabric + fabric (+ joining tape)	No electrical conductivity	High strength	Good	Not needed	No electrical connection, e.g. with heat, RF, US, Laser
Press snap / Spring snap button / Eyelets	No	Metal cap/socket + metal stud/post	Medium resistance, might not be constant	Very good	Okay, depending on type	Not needed	It can be a compromise to serve the two worlds
Zipper	No	Interlocking metal teeth	No electrical connection among interlocking teeth	Very good	Okay	Not needed	Air gap in-between teeth
Velcro	No	Conductively coated hooks and loops	High resistance no constant resistance	Very good	Okay, but degrading	Not needed	Suitable for anti-static applications
Sewing	Yes	Conductive yarn + conductive fabric	Medium resistance, no constant resistance	Very good	Okay, degrading	Not needed	Conductive yarn can be electrical trace as well

Not every metal is easy to solder; titanium, metal and solder must fit together and inertial gas atmosphere improves solder results. Nowadays, lead-free solders according to RoHS (Restriction of Hazardous Substances Directive, Directive 2002/95/EC) are commonly applied in the PCB industry. There are only a few exceptions where lead-based soldering is still allowed, such as in aerospace. A common RoHS-compatible, low-cost solder is the SAC (Sn-Ag-Cu), which has a eutectic melting point of 217°C. It is widely used for assembly of PCBs in the electronics industry.

Solder as a paste is usually employed in automated assembly processes in industry. The paste is screen or stencil printed, or dispensed on the surface of the PCB; the dispensing paste having a lower viscosity than the printing paste. The electronic components are then placed on the PCB in such a way as to ensure that the components' pads are imbedded in the paste. Finally, heat needs to be applied, according to the temperature profile given in the datasheet, in order to melt and establish the solder joint. Solder quality cannot be monitored during the process. Its quality is checked by visual inspection (using random samples) or by destruction (using cross sections). There are many different paste soldering processes (reflow, wave, laser, induction, flame, vapor phase, hot bar, laser, ultrasonic, to name a few).

The textile industry needs solders that melt at much lower temperatures than SAC, so that the fabric is not destroyed by the effect of high temperatures. Hwang (1996) gives an overview of the many different solder alloys: indium-based alloys are usually brittle, expensive and incompatible to SAC solders, while lead-based alloys no longer comply with RoHS guidelines. Indeed, Bi-Sn and Bi-Sn-Ag pastes seem to be the only feasible solders for the textile world, melting at 138 and 140°C, respectively. These solders are somewhat brittle, but they are compatible to SAC. Figure 10.4 shows a cross section of a woven fabric, whose imbedded tinsel thread is soldered to the copper surface of a PCB.

10.5.2 Brazing

Brazing is a similar process to soldering, but works at temperatures above 450°C, and uses specific brazing solders and fluxes. Brazing is most notably used for joining of copper pipes. Since this technology works at high temperatures though, it has little use in the textile world.

10.5.3 Thermocompression bonding (TCB)

Resistance welding is the method used to join two plain metal work pieces together by running an electrical current through them. The necessary welding heat is generated by the electrical resistance of the metals, by the contact resistance in between them and by the electrical current. No filler metal and no flux are needed. TCB is a special type of resistance welding that can be used to join

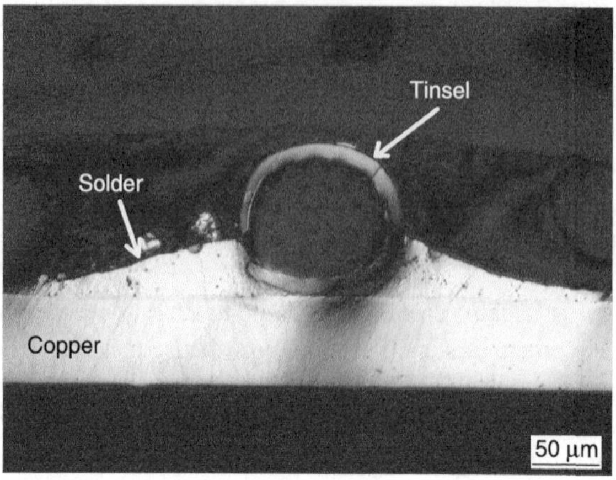

10.4 Woven tinsel thread soldered to a PCB (cross section, Sefar AG).

insulated metal surfaces. The TCB electrode consists of two legs that are connected together at the tip. The electrical current thus runs only through the electrode and generates the welding heat directly at the tip. The generated heat destroys any insulation layers of the metals instantly.

In contrast to resistance welding – where the electrical current generates heat exactly at the welding location – the heat generated from a TCB must flow from the electrode tip to the welding location at the interface of the two metals. As a result, more heat is transferred to the top than to the bottom of the two metals. This, along with the necessary melting point, limits the choice of the joining metals. The process window for TCB is therefore smaller than for resistance welding. An overview to some metal combinations is given in AVIO (2011).

TCB has three main process parameters: welding power (P), welding time (T) and pressure of the tip (P). These parameters are usually kept within a given window, to ensure quality. The off-the-shelf welding equipment controls the process by an adjustable pulse in the current (I), voltage (V) or power (P). The control mode depends on the material configuration, and it is important to differentiate between diffusion bonding and fusion welding. Diffusion bonding (solid-state bonding) occurs at 70 to 80% of the melting points of the work piece metals. Such bonds possess very good shear strength and tenacity, but less desirable peel strength. Fusion welding occurs above the melting points of the work pieces. These bonds have very good shear strength, tenacity and peel strength (Zhou, 2008).

Figure 10.5 illustrates the TCB process schematically. The electrode is placed with defined pressure on top of the work pieces; in this example, a wire is pressed onto a copper pad. The weld is then released with carefully controlled impulse shape, defined energy and time. A solid work surface is crucial for a high-quality

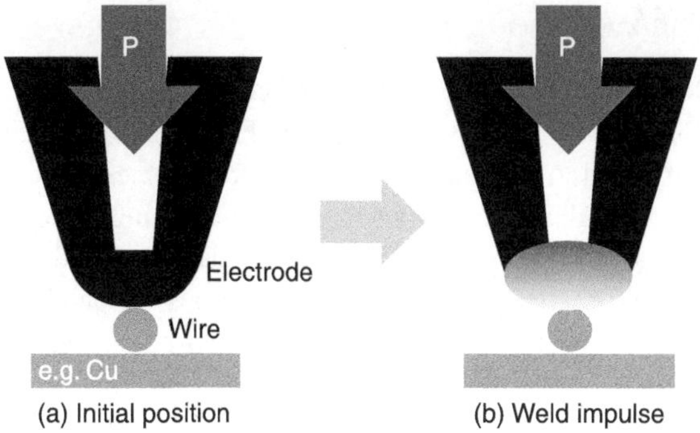

10.5 (a, b) TCB weld process.

weld. The strength of the weld can only be tested with certainty by destruction of the weld, for example by using a peel test. A weld is generated within a few milliseconds to several seconds, depending on the dimensions and materials of the two work pieces. Resistance and thermo-compression welding are typically employed for electrical coils, electric motors and light bulbs. An entire welding cycle takes about 3 to 10 seconds in an automated process set-up. Major companies, currently producing welding control units, welding heads and electrodes, include Miyachi, Avio, Kombitec, Lingl and Resistronic. Figure 10.6 shows the welding spot of a wire on a PCB surface. The shape of the welding electrode can clearly be identified. The wire itself is directly woven into the PET fabric and the PET fabric and the wire insulation around it melt and evaporate during the welding process, thereby welding the wire to the PCB.

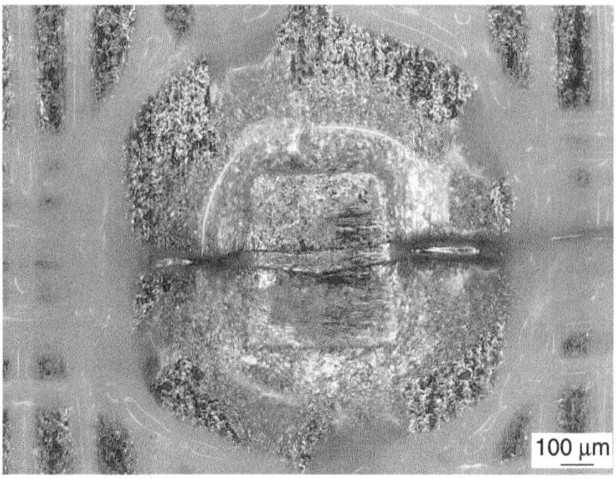

10.6 Weld of a woven wire onto a PCB (Sefar AG).

10.5.4 Ultrasonic welding

The ultrasonic welding process utilizes ultrasonic acoustic vibrations to generate heat and pressure to weld two work pieces together. Frequencies between 20 and 70 kHz are commonly used to join plastics, whereas frequencies of 20 or 40 kHz are employed for metals. Plastics, moreover, are joined using fusion welding, whereas metal joints are formed with solid-state welds. The weld process for plastics delivers the weld energy via longitudinal vibrations, perpendicular to the weld surface. The metal weld process delivers the weld energy via transverse vibrations, parallel to the weld surface. US welding is the most widely used industrial welding process (Ensminger and Bond, 2012); it is, for example, used to connect cables to surfaces and to compact copper cable strands. Its most prevalent producers include Rinco Ultrasonics, Schunk Sonosystems and Telsonic.

10.5.5 Radiofrequency (RF) welding

Fabrics can be welded using RF energy and an RF absorbing polymer tape. The tape, placed in between the two fabric pieces, is activated by RF. It then melts and bonds the two fabric pieces together. This process is usually employed when the fabric materials themselves are not weldable.

10.5.6 Bonding (using adhesives)

There are various types of electrically conductive adhesives currently in use. They can be categorized into isotropic conductive adhesives (ICA), anisotropic conductive adhesives (ACA) and non-conductive adhesives (NCA). ACAs are conductive in only one axis and NCAs are not conductive at all. They are typically used in applications with even and well-defined surfaces, such as flip chips. Usually, heat and pressure must be applied for curing. Li (2010) gives a good introduction to the aforementioned three categories.

The project 'MST4IT' funded by Germans BMBF (#16SV3425), 2008–2011, has led to some notable investigations into electrically conductive hot melts. There are also various conductive adhesives available for shielding applications, such as copper tapes with conductive acrylic adhesive (e.g. supplied by 3M) or conductive silicone paste by Chomerics. Their resistance is high in comparison to ICAs and they are designed for low currents, meaning they are prevalent in shielding and anti-static applications.

ICAs are typically two-component epoxy resins, and are employed in solar cells, LEDs and RF circuits on ceramic substrates. They are filled with silver flakes at a filling grade of 80 to 90%, making them somewhat expensive. However, they can be ordered frozen and mixed to the appropriate ratio. Usual curing parameters for epoxies is at 150°C for durations of a few minutes to several hours. The higher the temperature and the longer the time, the better the eventual conductivity of the bond (its conductivity is usually similar to solder) and the cross-linking of the polymer. However, it will also be more brittle than adhesives cured at a lower temperature and for a shorter time. The bond strength is almost as high as that of solder (Gilleo, 2004). Pot life can vary from a few minutes to several days. There are also some epoxies that can be cured using UV exposure.

Typical suppliers of ICA epoxies include EpoTek, Emerson and Cummings, and Henkel. Some epoxies, such as the EpoTek H20E, have already been used in the electronics industry for more than 30 years. Other epoxies are more suitable for laboratory experiments, such as the EpoTek E4110. There are a vast number of epoxies, and in order to obtain the optimum results in an application, it is advisable to consult the supplier. Some epoxy pastes are more suitable for dispensing processes and some are preferred for stencil and screen-printing processes. The uncured epoxy usually has a thixotropic consistency (shear stress to a thixotropic paste lowers the paste's viscosity). Unlike solder, ICAs are not able to break oxide layers, since they do not contain any flux. The component pads and circuit pads should thus be treated accordingly in advance to achieve best results. Figure 10.7 shows electronic components connected to a fabric with copper wires in warp and weft using conductive epoxy. The cuts in the copper wire avoiding shorts can be clearly seen.

10.5.7 Crimping

This technology is generally used to connect stranded wires of a cable to their connector pin terminals. The copper strand is positioned in a cavity at the back of the crimp terminal. Pressure is then applied in such a way that the back becomes wrapped in strands, establishing a cold-weld connection. Crimps have a high mechanical strength, and in contrast to soldering and adhesive bonding, crimping quality can be monitored (semi-)automatically by logging the crimping force curve. Notable crimp suppliers include Tyco, Nicomatic, Harting and Molex.

10.7 Electronic components bonded to a woven fabric substrate (Sefar AG).

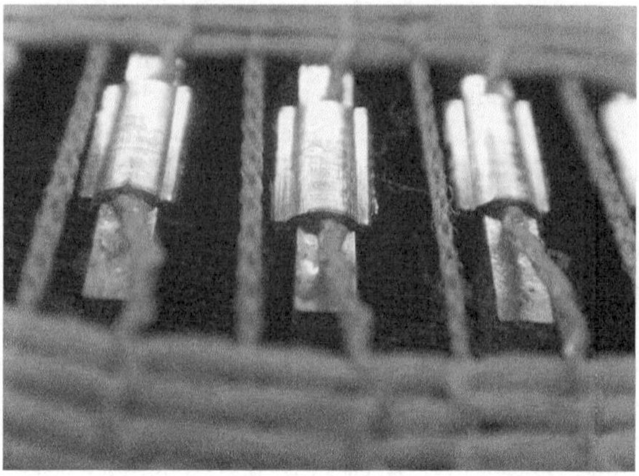

10.8 Crimping of conductive yarn (Fraunhofer IZM).

A similar, but less robust cold-welding connection is utilized for ribbon cable connections. Here, the copper strands are pressed in between adjacent tiny blades of the connector at the back of the terminal. Figure 10.8 shows crimping technology for conductive twisted yarn developed by Fraunhofer IZM.

10.5.8 Sewing and embroidery

This technology is used to join two pieces of fabric together using two threads. Sewing principle and stitch formation are described in Wulfhorst (1998). There are many types of stitches, such as lockstitch, overlock, and overlock safety. Electrical conductivity can be introduced using a conductive yarn, such as conductively plated yarn and twisted polymer/wire yarn. In contrast to soldering, welding and other techniques, a sewed connection does not establish a well-defined electrical connection. The contact is created simply by surface contact and yarn tension. As a result, electrical contact resistance is undefined and varies over time. Furthermore, moisture and liquid can settle in between, forming a galvanic element if different surface materials are present, leaving it exposed to contact corrosion and oxidation. However, it is a low-cost connection with high mechanical strength, durability and, to a large extent, a capacity to be washed. Programmable sewing machines, which in principle allow the creation of an electrical circuit layout, are available from Laesser, Oerlikon, Saurer and Tajima.

10.5.9 Velcro™

Electrical connections are possible with plated, conductive Velcro™. However, the electrical contact resistance shows high variation in each open/close cycle. This is primarily due to the fact that an undefined number of hooks, or loops, are active in each cycle, but also that the conductive plating generally starts peeling off after a couple of cycles, leading to a decaying contact quality over time. Therefore, a large area must be utilized to achieve a sufficiently low contact resistance. Companies, such as LessEMF and BlockEMF, supply conductive Velcro™. Since the technology has originally been developed for shielding applications, it is again not suitable for high currents. The question of how to connect the Velcro™ itself remains unanswered.

10.5.10 Zippers

We might think that metallic zippers would be a good approach to obtain low-ohmic connections. However, a closer look at the zipper mechanism exposes the fact that most zipper teeth do not touch each other. Instead, there is an air gap in between the teeth. However, even without this problem, the question of how to connect the zipper itself would still be problematical.

10.5.11 Press snap, spring snap buttons and eyelets

Metallic buttons/eyelets consist of corrosion resistant alloys, and enable a medium ohmic and a largely consistent electrical connection. The button itself can be electrically connected using standard welding processes (US, resistance), sewing or simply by clamping a conductive yarn in between the cap and socket of the button. The buttons are generally washable and are manufactured by companies such as YKK STOCKO Fasteners GmbH.

10.6 Summary to the joining technology overview

It has already been established that no joining technology satisfies the requirements of both worlds (namely by having good electrical properties while being robust and flexible). The industry therefore largely employs two, straightforward approaches. The company WarmX uses metallic buttons to connect their battery to their heating garment (Fig. 10.9). Mechanically, it is a very robust connection. If the buttons' resistance slightly increases over time, heating power will slightly decrease; however, this will be largely unnoticeable. Interactive-Wear, however, borrows connection technologies of the electronic world (e.g. soldering) and seals them completely in epoxy resin or in low-pressure injection hot melts to increase robustness. As can be seen in Fig. 10.10, the two white textile cable strips are connected to the black heating pad, before the connections are completely sealed in black hot melt.

In summary, the following recommendations should be acknowledged. If permanent connections are desired, soldering and welding are recommended, along with strong sealing. They possess low and constant resistances over time and processes are well understood. Furthermore, solder and welding joints can be made by hand and in a controlled industrial process. Crimping is an alternative, if only few connections are necessary or process temperature must be kept at room temperature. Buttons and electrical connectors are suggested for removable connections. Buttons

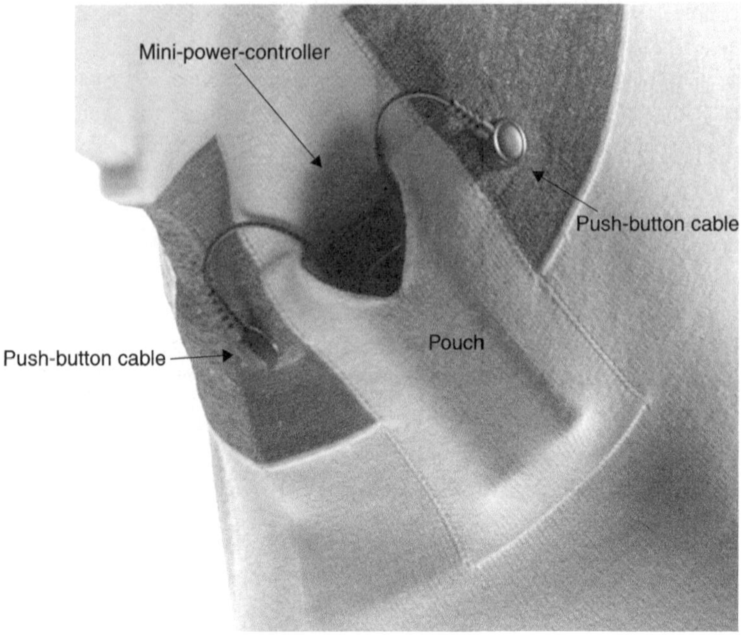

10.9 Metallic buttons to connect electrical power (WarmX GmbH).

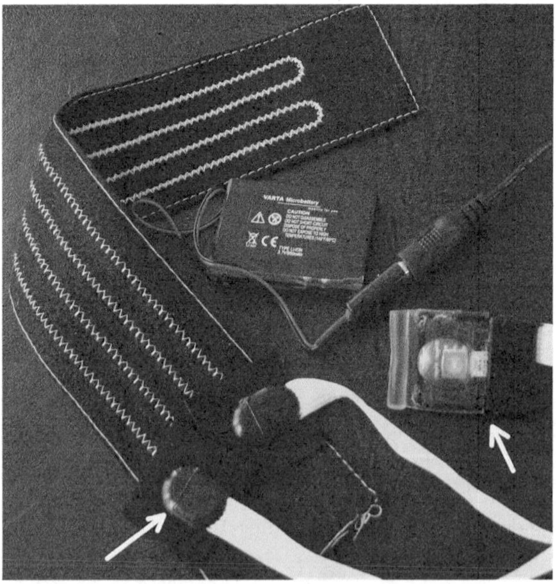

10.10 Electrical connection sealed with hotmelts (Interactive-Wear AG).

are preferred if user acceptance in the garment field is important and electrical requirements are low. Electrical connectors are recommended if the number of electrical contacts and electrical requirements are high.

10.7 Protection of electrical connections

Most electrical connections must be protected from mechanical stresses and from moisture, since they cannot withstand any mechanical impacts. In the electronics world, PCBs and their solder connections are hidden – for instance, in housings – and cable connections are consolidated with strain reliefs and sealing. Small PCBs can be entirely sealed in epoxy resin (as with solid-state relays) or in low-pressure injection hot melts (connectors) as well. A bare die on a PCB is usually sealed with a special UV curable resin, creating a so-called globtop. Furthermore, flexible circuits (using films as a substrate) are physically partitioned into two areas, a flexible area with electrical traces only and a rigid area with electronic components and connections, meaning movements apply to the flexible area alone.

The sealing of electrical connections is more challenging for electronic systems on textile, since protection should not affect textile handle and flexibility, and electrical connections should, in principle, also stay flexible. There is no generic solution to this problem, since the application determines the manner of protection. If a system is tiny enough to fit into a button, it most probably will not affect wearing comfort when solid and placed in the right location. Sometimes, an additional PVA cover or the like is wrapped around the sealed block to soften it.

Gemperle *et al.* (1998) have studied the impact of solid objects worn on the body. In larger areas, sealing materials such as silicones and polyurethanes (PU) can be used for mechanical protection. The company Stretchable Circuits (http://www.stretchable-circuits.com) has utilized PU films in their 'Klight' project, in order to seal their LEDs. Sefar laminates PU film tapes for protection of welding spots, as can be seen in Fig. 10.11.

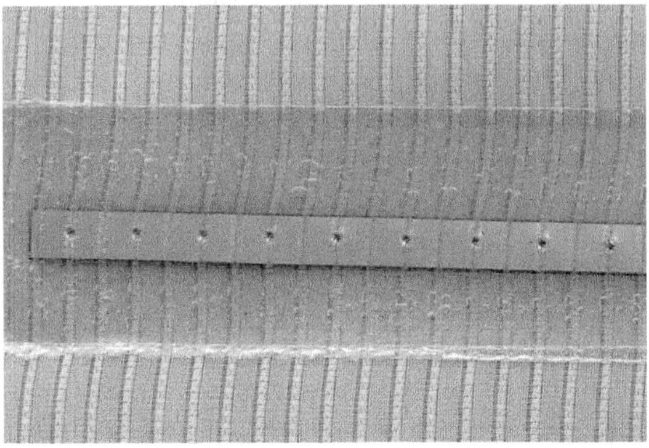

10.11 Welding spots between woven wires and FPC are protected by PU film (Sefar AG).

No matter what sealing method has been chosen, special attention must be paid to the transition area from hard to soft, that is, from seal to fabric. This area is prone to breaking of conductive traces, since the laws of physics dictate that the highest stresses and smallest bending radii will occur at this location.

10.8 Challenges for electronic systems on textiles

So far, all known approaches borrow technologies from both the electronics and the textiles industries, for the implementation of electronic systems on textiles. From the textile perspective, this is a necessary compromise, as electronic components and local encapsulations limit the flexibility and washability of fabric products, and add significant weight, especially batteries and large chips. Furthermore, the definition of robustness differs between the two industries. A slightly damaged shirt, for example one with a broken button, can still be worn. In contrast, an electronic device with a broken button may no longer function and it must be replaced. Redundancy in the imbedded circuits might improve reliability, but it also increases cost.

From the perspective of electronics, fabrics are undefined substrates with uneven surfaces and skew. They do not have a defined shape because of their low slippage resistance. After a fabric substrate has been moved, it always has a slightly different shape. This fact affects the use of automated process equipment. Fabrics are also a poor base material for electromagnetic compatibility (EMC). They can carry high static charges, especially in dry air, and damage imbedded electronics when moved (i.e. by electrostatic discharge (ESD)). Such discharges can even occur at the manufacturing stage. Moisture can reduce these discharges, but it could induce delayed corrosion.

Fabrics usually do not provide ground planes for shielding and for defining line impedances. Noise and other disturbances can easily affect the circuitry, making it unreliable. Special design rules are needed in fabrics as well. In general, electrical traces in fabrics have higher resistances than traces on PCBs given by dimensions and conductive materials. Current load capacity and the corresponding voltage thus drop along trace lengths, which must be considered in a design, or traces will heat up and digital components will no longer work due to illegal voltage levels. In addition, the circuit could start to fail due to high-ohmic, disturbance-conducting ground lines. Using several traces in parallel can lower electrical resistance, but these occupy more space. Electrical insulation is generally difficult to achieve in fabrics. Therefore, voltages should be kept below 42V, since this low level is not harmful to humans, and since ISO standards (e.g. IEC 60335) have no requirements for insulation below 42 V.

Stricter thermal limitations must also be considered in textile designs as opposed to PCB designs. Textile substrates conduct heat less effectively than copper loaded PCBs (e.g. FR4 and PI boards), while fabrics made of polyamide and polyester cannot withstand temperatures as high as PI boards. Furthermore, rigid PCB circuits can easily be cooled by fans. Large electronic components such as microprocessors with large number of pins are unsuitable from both perspectives. They limit flexibility of fabrics and are difficult to connect through fabrics.

To summarize, a good way of maintaining textile handling while integrating functionality is to mount dedicated components, such as LEDs and sensors, sparsely around the fabric. Dense circuits on textiles known from PCB technology will be impractical and unreliable. The structure of fabrics and its threads introduce certain granularity in dimension that reduces the realizable precision of circuit routings.

10.9 Challenges for automated processes in electronic systems on textiles

There are several approaches to the mass production of electronic systems on textiles. Following the business models of the mainstream apparel industry, production mainly takes place in low-income countries. The smart fabrics companies Clothing+ and Texsys have adopted this model. Another approach is to design the manufacturing process with proprietary equipment towards a specific product. The processes can be efficiently adapted to the product in question. However, a new product requires new process engineering and development. Textrace provides equipment for fabrication of textile RFID tags (RF identification), for instance. A third approach tries to adapt equipment of PCB industry and thereby establish a universal fabrication process for various products. However, this flexibility leads to inevitably high costs for equipment development. Sefar, TITV and IZM have started investigations into this approach.

In technical terms, there are numerous approaches around and myriad possible combinations of processes and materials. For example, electrical traces can be sewn, woven, printed, plated, etc. on fabrics. No single technology currently stands out as superior to the others. Depending on the product, one technology might be more suitable than another.

Several tendencies can be identified in future manufacturing chains for electronic systems on textiles. PCB industry relies on the precision of their substrates; after having etched the PCB, the stencil for solder paste printing must exactly match the etched circuit layout. Other processes, such as the component pick-and-place, use alignment marks (so-called fiducials) as optical orientation, in order to compensate for dimensional tolerances. In this manner, a process chain for systems on textiles will begin to rely more heavily on optical recognition and processes such as stencil printing will be less applicable. Instead, dispensing processes are applied that allow position adjustments by optical orientation.

Process temperatures must also be reduced compared to PCB industry. This means that low-temperature solders or conductive adhesives must be applied, since fabric substrates usually shrink or melt at elevated temperatures. One issue that remains unsolved is the testing of systems on textiles during manufacturing. The use of flying probes and needle adapters is most probably not feasible, because of the textile nature of the substrates. Nowadays, moreover, tin whiskers (Tu, 2007) can grow over time in electrical circuits, eventually leading to shorts and failure of the circuits. Similar effects might occur in textile circuits by fraying and fibril breaks. The processing of fabric substrates will certainly become more expensive than the processing of PCB substrates, since additional and slower processes will be involved, as mentioned above. However, fabric as substrate is about ten times cheaper than FPCs based on polyimide. Eventually, electronic systems on textiles might be in a similar price range as standard circuit boards.

10.10 Future trends

In the early stages of smart textiles, many so-called 'Smart Shirts' were proudly displayed, most of them handmade. The impression was given that every function can be integrated in fabrics. However, technology and manufacturing processes are still yet to reach that stage. At least some specific process steps must yet be implemented to bridge the gap between the textile and the electronics world. Happily, the smart-textile community has matured over the past few years and certain universities and companies are about to establish some of these specialized manufacturing steps.

Apart from specific processes, there are still various unsolved issues regarding materials. Many companies are currently trying to improve conductivity of polymers and coatings, for example, by using carbon nanotubes (CNTs). The goal is to combine the robustness and flexibility of polymers and the conductivity of metals. Of the various polymers, only a few can be spun to threads that eventually compose a fabric. The emerging molded injection device (MID) technology is a promising portfolio of methods to apply robust and highly conductive structures on polymers. Improvements in conductive elastic adhesives based on silicones or PUs would, furthermore, help to make connections mechanically more robust, although these adhesives are still expensive. The implementation of functionality at the fibre level of a fabric is, in short, still seen as the most elegant development possible, but also as one of the most futuristic.

Future miniaturization of electronics, as well as research in plastic electronics, will also help to further integrate the two worlds of electronics and textiles. Nevertheless, even in this harmoniously integrated world, standardization will be necessary to a certain degree, in order to enable modularization and to establish basic system components. That, again, is crucial for an accelerated progress in innovation in the world of smart textiles.

There are already smart fabric products in the fitness and outdoor business on the market. The most notable include the Polar belt and the sports bra by adidas for monitoring heart rate. Furthermore, the first smart fabric products in the field of healthcare are about to emerge, which will fulfill such functions as the monitoring and managing of vital parameters. Philips is very active in this field. The automotive industry is also interested in smart fabrics, since they allow new freedom of design and they imply potential to save weight. Some combinations of electrical systems and fabrics, such as heated seats, have already been around for a while, but have never been truly acknowledged as smart fabrics. However, the ideal application is still yet to be identified. We are convinced that a small, very simple application with limited functionality will eventually facilitate the successful launch of smart fabrics as new industrial branch.

10.11 References

AVIO (2011), 'Micro resistance welder series', available from: http://www.avio.co.jp/products/assem/pdf/mrw.pdf

Bonderover E and Wagner S (2004), 'A woven inverter circuit for e-textile applications', *IEEE Electron Dev Lett*, 2, 295–297.

Cherenack K, Kinkeldei T, Zysset C and Tröster G (2010), 'Woven thin-film metal interconnects', *Electron Dev Lett*, 31(7), 740–742.

Diaz T (2011), *Electronically Active Intelligent Yarn*. London, Smart Fabrics.

Ensminger D and Bond L (2012), *Ultrasonics: Fundamentals, Technologies and Applications*. Boca Raton, FL, CRC Press.

Gemperle F, Kasabach C, Stivoric J, Bauer M and Martin R (1998), 'Design for wearability', *Proceedings of 2nd IEEE International Symposium on Wearable Computers*, pp. 116–122.

Gilleo K (2004), *Area Array Packaging Materials: Adhesives, Pastes, and Lead-Free*. New York, McGraw-Hill Professional.

Hwang J S (1996), *Modern Solder Technology for Competitive Electronics Manufacturing*. Electronic Packaging and Interconnection Series. New York, McGraw-Hill Professional.

Jones M, Martin T and Naked Z (2002), 'A service backplane for e-textile applications', *Workshop on Modeling, Analysis, and Middle-ware Support for Electronic Textiles (MAMSET)*, pp. 15–22.

Li Y, Lu D, and Wong C P (2010), *Electrical Conductive Adhesives with Nanotechnologies*, Boston, Springer.

Linz T, Kallmayer C, Aschenbrenner R and Reichl H (2005), 'Embroidering electrical interconnects with conductive yarn for the integration of flexible electronic modules into fabric', *Proceedings of the 9th International Symposium on Wearable Computers*, pp. 86–89.

Mroczkowski R S (1998), *Electronic Connector Handbook: Theory and Applications*. New York, McGraw-Hill Professional.

Park S, Mackenzie K and Jayaraman S (2002), 'The wearable motherboard: a framework for personalized mobile information processing (PMIP)', *Proceedings of the 39th ACM/IEEE Design Automation Conference (DAC)*, pp. 170–174.

Post E R, Orth M, Gershenfeld P R and Russo N (2000), 'E-broidery: design and fabrication of textile-based computing', *IBM Syst J*, 39(3 and 4): 840–860.

TexoLED (2009) 'Textile-integrated and textile-based LEDs and OLEDs', (BMBF-MST 16SV3450), 3rd Public Status Meeting (mst-textil), Berlin, 9–10 July.

Tu K N (2007), *Solder Joint Technology: Materials, Properties, and Reliability*. New York, Springer.

Wulfhorst B (1998), *Textile Fertigungsverfahren*. Munich, Hanser Verlag.

Zhou Y N (2008), *Microjoining and Nanojoining*. Cambridge, UK, Woodhead Publishing Ltd.

11

Kinetic, thermoelectric and solar energy harvesting technologies for smart textiles

S. P. BEEBY, Z. CAO and A. ALMUSSALLAM,
University of Southampton, UK

DOI: 10.1533/9780857093530.2.306

Abstract: This chapter concerns the harvesting of ambient energy and its conversion into electrical energy for use in powering autonomous electronic systems. The chapter describes the fundamentals of energy harvesting from kinetic, thermal and solar energy sources and the application of each technique in fabrics. True integration of the technology is currently being investigated with photovoltaic (PV) fibres and printed low temperature PV materials having been demonstrated. In all cases the technology is in its infancy and it is extremely challenging to integrate this functionality in smart fabrics in an acceptable and practical way. Functionalising the surface of fabrics using novel low temperature materials deposited by standard printing processes appears the most cost-effective and practical approach.

Key words: Energy harvesting, piezoelectric fabrics, thermoelectric fabrics, solar fabrics.

11.1 Introduction

Energy harvesting (also known as energy scavenging) is the conversion of ambient energy present in the environment into electrical energy for use in powering autonomous electronic devices or circuits. Whilst it is identical in principle to large-scale renewable energy generation such as wind turbines, the amount of energy produced is much smaller, being typically tens of micro watts to a few watts. Energy harvesting can be used to provide an alternative to, or at least to augment, batteries. Whilst batteries are low cost, they contain a finite amount of energy and require periodic replacement or recharging. The disposal of batteries is also an environmental concern and, as autonomous systems proliferate, this concern can only grow. Batteries are also a limitation in smart fabrics applications, since conventional batteries are rigid bulky items that must be removed before washing. This chapter provides an introduction to energy harvesting from three sources (kinetic energy, thermal energy and solar or light energy) and discusses their implementation in fabrics. This is a relatively new topic and research in this field is in its infancy. However, it has the potential to integrate the ability to power smart textiles directly from the fabric itself.

Energy harvesting to date has focused on developing discrete devices that can generate electrical energy from kinetic energy (i.e. vibrations or movement) (Beeby *et al.*, 2006), thermal gradients (Rowe *et al.*, 1989) or incident light

(Hande *et al.*, 2007). The performance of energy harvesting is fundamentally linked to the amount and nature of the source energy present in the environment. When designing an energy harvesting solution, knowledge of the application constraints and the details of the energy source have to be known in advance. In conventional kinetic energy harvesting, bespoke devices are typically designed for each application and there is rarely a simple solution that addresses a multitude of applications.

This chapter covers kinetic, thermal and solar energy harvesting and describes the fundamental principles by which they function. More detailed information on energy harvesting together with practical examples of applications can be found in the book *Energy Harvesting for Autonomous Systems* (Beeby and White, 2010). The application of the technologies to smart fabrics is discussed along with the fabrication processes that would enable integration of the technologies within fabrics. These points are illustrated with examples of existing research that described the current state-of-the-art. Finally, the future potential for energy harvesting in smart fabrics applications is considered along with the technical challenges that need to be addressed.

11.2 Energy sources and storage: key issues

Electrical energy can be generated from other energy sources by means of a suitable transduction process. In conventional terms, large-scale power generation takes chemical energy in the form of gas or coal, for example, and converts this into thermal energy in the form of steam, which drives turbines that are connected to electromechanical generators that generate the electrical energy. However, wind turbines use rotating blades driven by the kinetic energy contained within the flow of air to drive a generator.

The energy source for most energy harvesting devices are kinetic energy in the form of forces, displacements or vibrations, thermal energy where a temperature gradient exists or solar energy in the form of incident light. The method of converting, or transducing, these energy sources will depend upon the application and the exact nature of the energy. In smart fabrics, kinetic energy sources could take the form of the motion of a person, which would cause a garment to move or a carpet to be compressed as the individual walked across it (Starner, 1996). Thermal energy could be generated from a person or from other heat sources such as pipes or in buildings. Solar energy exists wherever there is light, with sunlight containing more energy than indoor electrical lighting.

Since energy harvesting energy sources are often variable in their occurrence, the energy generated is also likely to vary. This can be problematic if the energy harvester is used to power the electronics directly, since there may be periods when no energy is available. To overcome this, the energy from the harvester is often stored in a super-capacitor or battery, which provides an energy reservoir that acts as a buffer between the harvester and the electronics. If an energy harvester is being used to recharge a portable device, then the battery exists with the device itself. If the electronic system is also built into the textile itself, then this may mean a small battery or super-capacitor must also be included to provide this storage. The optimum solution will depend on the nature of the input energy source and the requirements of the electronics, both of which depend upon the application.

11.3 Fabrication processes

The three types of energy harvesting discussed here all require the use of functional materials to perform the energy conversion. For example, thermoelectric devices require n- and p-type semiconductors. In order for energy harvesting technologies to be compatible with textiles, these materials have to be available in a flexible form, either as a fibre or coating. Active fibres can be woven into the textile following the weave used for that particular fabric. The woven active fibre fabrication approach will constrain the location of the active material to the direction of the threads in the particular weave pattern. Also, electrical connection must be made to the active fibres in order to collect the electrical power generated. Coating processes offer more flexibility, since the textile can be used as a substrate and the active materials and electrodes can be deposited onto the fabric surface in any pattern or orientation required. However, the mechanical properties of the films and their thickness might have an effect on the feel of the fabric, and so flexible, stretchable materials are required. The deposition process has to be compatible with the textiles and the constraints they impose. Maximum temperatures, for example, will be limited to around 150°C, depending upon the type of fabric.

Printing approaches, such as screen and inkjet printing, are an attractive deposition process, since they are already widely used for the patterning of textiles (Miles, 2010). Printing has also been used in the electronics industry for many years in the fabrication of circuit boards and devices and many functional materials have been developed that can be deposited in this manner. This background makes printing techniques very attractive for the functionalisation of the textile. Screen printing uses a patterned mesh on top of which the paste to be printed is smeared. A rubber squeegee is then drawn across the screen, which forces the ink through the openings in the screen onto the surface of the textile. In this way, the pattern on the screen is replicated on the textile. A wide range of pastes can be printed in this way and, in the case of functional inks, the relatively high viscosities of the pastes enables them to be heavily loaded with active ingredients.

Inkjet printing is a digital process that selectively deposits individual droplets of ink onto the surface of the textiles using print heads that enable very fine resolution of images (e.g. 1440 dpi). The print head does not come into contact with the textile and this approach can take the pattern directly from the computer and print it onto the fabric without requiring a screen to be made up in advance.

However, inkjet inks have much lower viscosity than their screen printed equivalent and therefore functional inks typically have much lower concentrations of active particles. These active particles must also be very small in size (< 200 nm) to avoid blocking the print head.

Other deposition approaches can be used, such as dispenser printing and plasma coating processes. In dispenser printing, the ink is placed in a syringe and squeezed out as required. It is similar to inkjet printing in that no screen is required and material can be deposited one drop at a time. It also enables higher viscosity inks to be printed, enabling inks with higher loadings of active materials to be digitally printed onto the substrate. However, this higher viscosity does reduce the speed of the printing process compared to inkjet. Plasma coating processes are used in the textile industry to alter the surface properties of the fabric (e.g. make it hydrophobic) or to deposit material all over the textile. If functional materials are to be deposited in this manner, it is likely they would need to be removed using some form of subsequent patterning and removal process, which adds considerably to the complexity. Plasmas can be used to coat individual yarns, which could then be woven into the fabric.

11.4 Kinetic energy harvesting for smart textiles

The conversion of kinetic energy into electrical energy is possible using the piezoelectric effect, which was discovered by J. and P. Curie in 1880. They found that if certain materials were subjected to mechanical strain, they became electrically polarised and the degree of polarisation was proportional to the applied strain. Conversely, these materials deformed when they were exposed to an electric field. These are known as the direct and indirect piezoelectric effects, respectively.

11.4.1 Piezoelectric theory

The piezoelectric behaviour occurs due to the individual polarised domains that exist within the atomic structure of the material. In the direct case, an external stress causes these domains to rotate, which results in a charge at the surface of the piezoelectric material. In the indirect case, an electric field causes the domain alignment to move, resulting in a deformation of the material.

Piezoelectric materials are widely available in many forms, including single crystal (e.g. quartz), which is naturally piezoelectric. Piezoceramics (e.g. lead zirconate titanate or PZT) are man-made materials that initially contain randomly aligned domains. These materials are polarised by an applied electric field at elevated temperature, which aligns the domains in a common direction. Thin film forms, such as sputtered zinc oxide (ZnO), must be carefully deposited to exhibit the piezoelectric effect. Screen printable thick-films based upon piezoceramic powders also require polarisation, and polymeric materials such as Polyvinylidenefluoride (PVDF) are also available.

Piezoelectric materials are anisotropic, which means the properties of the material differ depending upon the direction of forces and orientation of the polarisation and electrodes. The anisotropic piezoelectric properties of the ceramic are defined by a system of symbols and notation. This is related to the orientation of the material, the direction of measurements and the applied stresses/forces. The basis for this is shown in Fig. 11.1. The direction of polarisation will depend upon the placement of the electrodes on the surface of the material. In the example shown, electrodes would be placed on the top and bottom surface to polarise the material in the 3, or z, direction.

A series of constants used in conjunction with the axes notation is used to define the level of piezoelectric activity. For example, the piezoelectric strain constant, d, is defined as the strain developed (ε) for a given electric field (E) (d = e/E, units m/V) or the short circuit charge density (Q) for a given applied stress (S) (d = Q/S, units C/N). Piezoelectric generators that rely on a compressive strain applied perpendicular to the electrodes (i.e. parallel with the direction of polarisation direction) exploit the d_{33} coefficient of the material. Those that apply a transverse strain parallel to the electrodes utilise the d_{31} coefficient. In the case of piezoelectric films or piezoelectric elements bonded onto substrates, the piezoelectric elements are typically coupled in the transverse direction. Such an arrangement provides mechanical amplification of the applied stresses, but the d_{31} properties are typically less than the d_{33} coefficient.

Another important constant is the electro-mechanical coupling coefficients, k, which describes the efficiency with which the mechanical energy is converted within the material into electrical energy. This is defined in Eq. 11.1 below where W_i^e is the electrical energy stored across the i-axis and W_j^m is the mechanical input energy across the j-axis:

$$k_{ij}^2 = \frac{W_i^e}{W_j^m} \qquad [11.1]$$

The efficiency of energy conversion, η, for a piezoelectric element clamped to a substrate and cyclically compressed at its resonant frequency is given in Eq. 11.9, where Q is the quality factor of the generator (Richards *et al.*, 2004). This relationship suggests efficiency is improved by increasing k and Q, which provides a useful guideline when choosing materials and designing generators:

$$\eta = \frac{\frac{1}{2}\frac{k^2}{1-k^2}}{\left(\frac{1}{Q}+\frac{1}{2}\frac{k^2}{1-k^2}\right)} \qquad [11.2]$$

Other relevant piezoelectric constants include the permittivity of the material, ε, which is defined as the dielectric displacement per unit electric field and compliance, s, which is the strain produced per unit of stress. Lastly, the

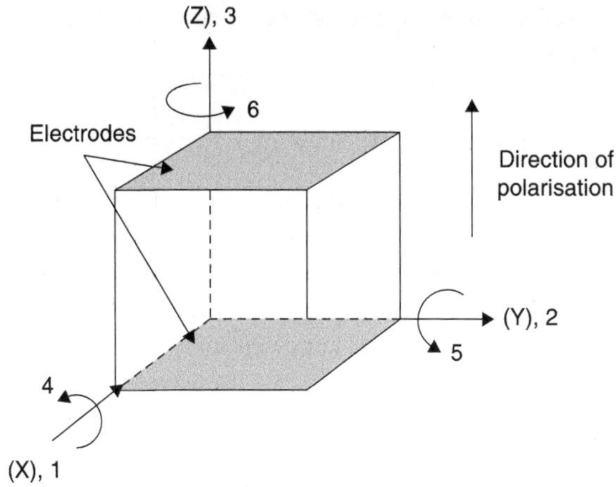

11.1 Notation of axis.

piezoelectric voltage constant, *g*, is defined as the electric field generated per unit of mechanical stress, or the strain developed for an applied charge density. These constants are anisotropic and are further defined using the system of subscripts described above. The piezoelectric constants for common materials, soft and hard lead zirconate titanate piezoceramics (PZT-5H and PZT-5A), barium titanate ($BaTiO_3$) and polyvinylidene fluoride (PVDF), are given in Table 11.1.

Piezoelectric properties vary with age, stress and temperature. The change in the properties of the piezoceramic with time is known as the ageing rate and is dependent on the construction methods, the material type and the applied stress. The changes in the material tend to be logarithmic with time, thus the material properties stabilise with age. The ageing process is accelerated by the amount of stress applied to the ceramic. Excessive stresses will depolarise materials and soft piezoceramic compositions, such as PZT-5H, are more susceptible to stress induced changes in the piezoelectric properties than the harder compositions such as PZT-5A. The application of temperatures beyond the materials Curie

Table 11.1 Coefficients of common piezoelectric materials

Property	PZT-5H	PZT-5A	$BaTiO_3$	PVDF
d_{33} (10^{-12} C/N)	593	374	149	−33
d_{31} (10^{-12} C/N)	−274	−171	78	23
g_{33} (10^{-3} Vm/N)	19.7	24.8	14.1	−330
g_{31} (10^{-3} Vm/N)	−9.1	−11.4	5	216
k_{33}	0.75	0.71	48	15
k_{31}	0.39	0.31	21	12
Relative permittivity ($\varepsilon/\varepsilon_o$)	3400	1700	1700	12

temperature will also reduce piezoelectric activity. The application of stress can also lower this Curie temperature.

11.4.2 Examples of piezoelectric energy harvesters

Piezoelectric energy harvesters are associated with many kinetic energy applications such as ambient vibration, wind, impact force and recent body motion. Piezoelectric vibration energy harvesters are typically based on a cantilever beam structure with an inertial mass located at the beam tip. The piezoelectric material is located on one or both sides of the cantilever, forming a unimorph or bimorph structure, respectively. As the environmental vibrations shake the harvester, the cantilever tip moves up and down and the piezoelectric material on the beam is strained, resulting in the generation of charge. An early example of this is shown in Fig. 11.2 (Roundy, 2004). In this case, a piezoceramic PZT-5H material was placed on both sides of a 100 μm thick brass shim with a tip mass providing a resonant frequency of 120 Hz. An optimised design with a target volume of 1 cm^3 produced a maximum power of 320 μW at 120 Hz resonance frequencies and 2.5 ms^{-2} acceleration.

The type of structure shown in Fig. 11.2 is not suitable for integration into smart fabrics, although smaller, microscale versions could in principle be used. The fabrication technologies used in Micro-Electro-Mechanical Systems (MEMS) have the potential to fabricate low cost, small size (e.g. 5 × 5 mm) devices that would be much less obstructive when embedded in a fabric. An example of such a device was again based on a cantilever micromachined from silicon, with thin-film piezoelectric materials deposited on the top surface (Marzencki *et al.*, 2008). The device has a volume of 5 mm^3 and produces around 1 μW, but requires high frequency vibrations above 800 Hz and large acceleration, which will limit practical applications.

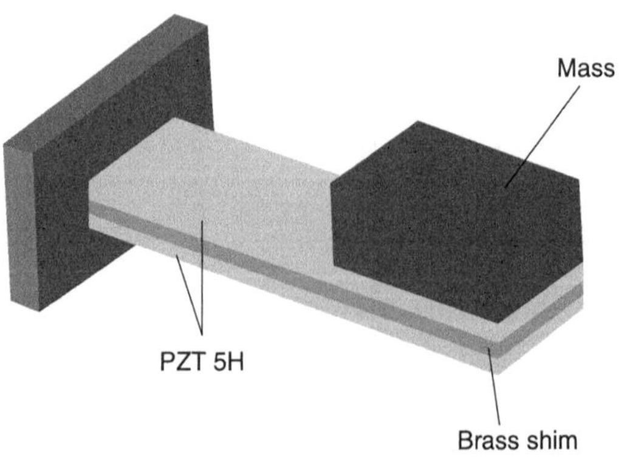

11.2 Cantilever piezoelectric vibration energy harvester.

Since energy harvesting fabrics may well be used to harvest energy from human motion, it is interesting to review piezoelectric harvesters designed for this application. An early example is a harvester based upon two piezoelectric shoe inserts, one located at the heel and one on the sole of the shoe (Schenk and Paradiso, 2001). The sole device consisted of eight layers of PVDF and associated electrodes fabricated on a plastic substrate, which operates in the d_{31} mode of operation. This device produced a power of 1.3 mW at a stepping frequency of 0.9 Hz. The heel generator was a PZT dimorph at the heel of the shoe, producing the power from the compression that occurs during walking. The compression-based device located at the heel of the shoe produces an average power of 8.4 mW at a step frequency of 0.9 Hz.

11.4.3 Piezoelectric energy harvesting on fabrics

The conversion of kinetic energy into electrical energy by incorporating piezoelectric materials into fabrics has begun to be explored but is at an early stage of development. Successful implementation will lead to a wide range of applications, such as energy harvesting from fabrics flapping in the wind, harvesting from floor coverings (carpets and rugs) as people walk across them, or simply harvesting the movement of a person via the deflection of their clothes. The following review summarises the research in this area, either directly on fibres or using processes that could be transferred to fabric substrates in the future.

One approach is to use piezoelectric fibres and weave them into the fabric, but this requires flexible piezoelectric fibres to be fabricated in the first place. One approach demonstrated was to coat Kevlar fibres with ZnO nanowires (Qin and Wang, 2008). ZnO nanowires have been grown radially on the surface of the fibre using a hydrothermal technique and have a range of diameter of 50 to 200 nm and a length of 3.5 μm (Fig. 11.3). To improve the binding of the nanowires and the mechanical robustness of the overall fibre, layers of tetraethoxysilane (TEOS) polymer were placed above and below the ZnO seed layer.

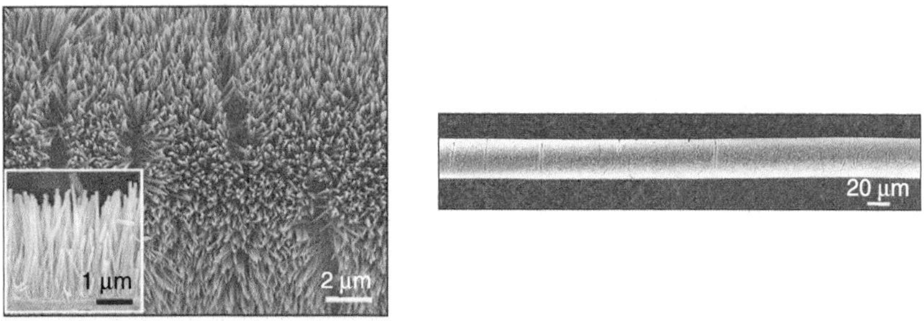

11.3 SEM images of a fibre coated with ZnO nanowires (left) and the distribution of the nanowires along the fibre (right) (Qin, 2008).

The fibres were initially tested by winding two fibres together and stretching them with one of the fibres having been coated with a 30 nm thick gold layer. The output current and voltage resulting from the testing process were 5 pA and 3 mV, respectively, the small output current being caused by the large internal resistance (250 kΩ) due to the cracks in the ZnO layer. Making a single yarn with 6 fibres with 3 of coated Au gave an improvement of between 30 and 50 times more than single fibre. This increase in power has been generated because of the increase of the surface contact area between the fibres. Also, depositing a conductive layer directly on the fibre before depositing the ZnO layer has reduced the internal resistance of the fibre to 1 kΩ and the output current has been increased from 4 pA to 4 nA. The scaled-up reported optimum power density of this yarn was of the range 20 to 80 mW/m^2, which indicates useful levels of power could be generated in this way given suitable input energy. The cost of the approach and the robustness of the fibres remain issues for further development.

High-quality piezoelectric materials have been realised on flexible substrates by transfer printing using a PDMS stamp (Qi *et al.*, 2010). Crystalline PZT ribbons are synthesised on an MgO host substrate and patterned to form individual PZT ribbons. The host substrate is then immersed in a liquid etchant, which loosens the ribbons without removing them entirely from the substrate. A flexible PDMS rubber piece is then bought into contact with the substrate and peeled back taking the PZT ribbons with it. This leaves a 'piezo-rubber' chip that is highly flexible and contains high-quality piezoelectric material. Piezo-Force Microscopy (PFM) was used to measure the d_{33} coefficient, which was found to be 101 pm/V. The performance and stetchability of the 'piezo-rubber' was improved by using buckled PZT ribbons (Qi *et al.*, 2011). Buckling the PZT ribbons is achieved by pre-stretching the PDMS before transferring the PZT ribbons. After transfer, the PDMS is released and the ribbons buckle. The measured d_{33} in the buckled region was 130 pm/V, an improvement of 30%. Whilst not applied directly to fabrics, this process could be used to transfer PZT material to fabrics with the appropriate coating pre-applied. Issues to be investigated would include the deposition of the required electrodes and the protection of the piezoelectric material after transfer.

Piezoelectric materials can be printed and high thick-film temperature inks have been screen printed onto stainless steel substrates for the purposes of energy harvesting (Zhu *et al.*, 2010). For smart fabrics, low temperature inks are required. Piezoelectric films have been demonstrated in the form of a piezoelectric glove designed to sense the movement of the hand (Edmison *et al.*, 2002). These films were not printed but were flexible laminated piezoelectric sheets from Measurement Specialties Inc., which were cut to size and attached to the textile glove. This work demonstrated the suitability of piezoelectric films for sensing motion, which is very similar to energy harvesting. If the piezoelectric film and electrodes could be printed directly onto the fabric then the textile can easily become functionalised and subsequent assembly into a finished garment would be much simpler. Flexible piezoelectric films have been printed onto fabrics as part of the Microflex EU FP7 project, which demonstrates the feasibility of this process, although the films have not been used as energy harvesters (Hansen and Zawada, 2011). At present, these printed films demonstrate a d_{33} of around 30 pC/N, which is sufficient to capture energy from the textile.

11.4.4 Discussion

Piezoelectric textiles will certainly be a feasible approach for energy harvesting in the future, but the optimum approach for achieving this has not been identified. The simplest approach would be to coat the fabric in a flexible low temperature material. Screen printable materials have been demonstrated, but these are very new and further development is required to improve flexibility and piezoelectric activity. In the case of woven fibres, the piezo nanowires are an exciting technology and the coating of fibres in this manner is very interesting. Whether this approach could be a cost effective, robust and practical solution when scaled up to make woven fabrics is doubtful. More conventional piezoelectric fibres can be fabricated using ceramic materials, but these are relatively inflexible and not compatible with requirements of smart textiles. One point is that fabrics are designed to be compliant and garments do not resist the motion of the wearer. In order to extract energy from fabrics, work has to be done to move the fabric and the more energy required the more work that must be done, that is the less compliant the fabric or garment will become. Care will have to be taken in the implementation of this technology, in order to avoid compromising the fabric properties and this is likely to require careful thought in the design of the application of the fabric (e.g. the garment).

11.5 Thermoelectric energy harvesting for smart textiles

Thermal energy can be converted into electrical energy using heat engines or thermoelectric generators (TEGs). Heat engines are bulky in nature and not compatible with textiles. TEGs are made up of a number of thermoelectric elements (thermocouples) connected electrically in series. When a temperature gradient exists across the generator, a voltage is generated. This voltage is explained by the thermoelectric theories that describe the energy transfer and conversion between thermal gradients and electrical energy in solid devices. Thermoelectric devices are planar in nature and, as such, may be included in smart textiles, although their full integration requires much further research.

11.5.1 Thermoelectric theory

There are three thermoelectric effects: Seebeck Effect, Peltier Effect and Thomson Effect, and these are linked by the Kelvin relationships. The Seebeck Effect describes the generation of an electrical voltage when a temperature difference exists between the ends of a thermoelectric device (Goldsmid, 1960). When a temperature difference exists between the junctions formed from two different conductor or semiconductor materials, for example materials *a* and *b* in Fig. 11.4, an open circuit voltage is produced named the Seebeck Voltage. The Seebeck Voltage occurs due to the flow of charge carriers in the different materials. In the absence of a temperature gradient, the charge carriers are uniformly distributed in the materials. However, when a temperature gradient exists, such as a high temperature (T_H) at one junction and a low temperature (T_C) at the other junction, the kinetic energy of charge carriers at the hot end will be greater than the energy at the cold end, resulting in a higher carrier concentration at the hot end. Hence, the charge carriers will diffuse from the high concentration end to the lower end. Also, the accumulation of carriers will build up an electrical field to further increase the rate of charge flow. The stable open circuit potential difference, Seebeck Voltage, is given by:

$$V = \alpha_{ab} \cdot \Delta T = \alpha_{ab} \cdot (T_H - T_C). \quad [11.3]$$

Here, α_{ab} is the Seebeck coefficient of the junction between the materials *a* and *b*. The value of α_{ab} is determined by:

$$\alpha_{ab} \cong \frac{V}{\Delta T} (\text{at } \Delta T \rightarrow 0). \quad [11.4]$$

For a small temperature variation, the relationship of thermoelectric voltage V and Seebeck coefficient α_{ab} is approximately linear. More correctly, the Seebeck coefficient is a physical property of the junction only between two different

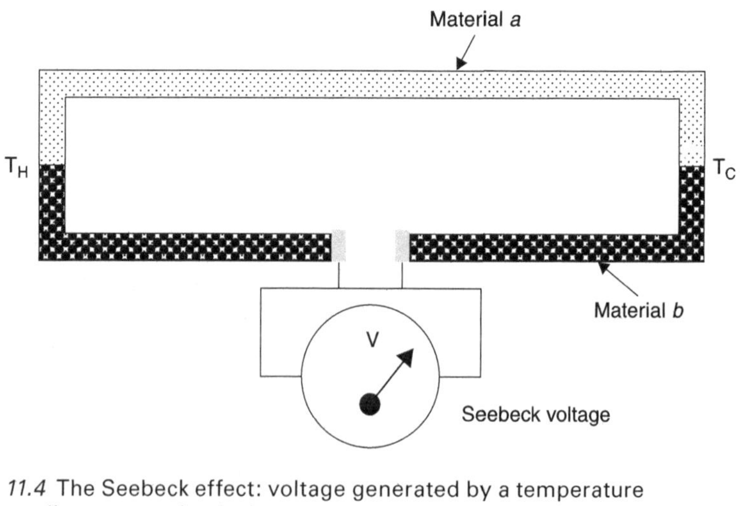

11.4 The Seebeck effect: voltage generated by a temperature gradient across the device.

materials and therefore it is also called the relative Seebeck coefficient. The unit of α_{ab} is V/K, or more often in μV/K, and the value can be either positive or negative, depending on the direction of the current between materials *a* and *b* at the hot junction. In typical applications, all TEGs work using the Seebeck mode.

Another main application of thermoelectric devices, refrigeration, is based on the Peltier Effect. Instead of applying a temperature difference to the thermoelectric junction, a voltage is applied across the junction. The current flow around the circuit will cause the junctions to absorb or dissipate heat, resulting in one junction heating up and the other cooling down. Reversing the direction of current flow will cause the heating and cooling effects to switch junctions. The transferred heat ($\mathbf{Q}_p$) is given by Eq. 11.5, where π_{ab} is the Peltier coefficient:

$$\dot{Q}_P = \pi_{ab} \cdot 1. \qquad [11.5]$$

The Thomson Effect relates to the heat dissipation and absorption in a single material, when a current flow and temperature gradient exist simultaneously. The unit heat absorbed or dissipated is calculated by:

$$\dot{Q}_T = \beta \cdot 1 \cdot \Delta T. \qquad [11.6]$$

Here, β is the Thomson coefficient. The unit is the same as the Seebeck coefficient, μV/K. In most cases, the Thomson Effect does not dominate in thermoelectric devices but cannot be neglected in a detailed calculation or when the temperature gradient is large.

The Seebeck, Peltier and Thompson coefficients are related to each other by the Kelvin relationships, as shown in the following equations:

$$\alpha_{ab} = \frac{\pi_{ab}}{T} \qquad [11.7]$$

$$\frac{d\alpha_{ab}}{dT} = \frac{\beta_a - \beta_b}{T}. \qquad [11.8]$$

11.5.2 Thermoelectric device structure

The most commonly used thermoelectric materials are semiconductors with several n- and p-type thermocouples connected electrically in series but thermally in parallel. Figure 11.5 shows the structure of a standard ceramic based thermoelectric module assembled from an array of thermocouples. The individual n- and p-type elements are connected by a conductive link and it is important that the resistance of these parts is as low as possible.

11.5.3 Evaluation of thermoelectric properties

The performance of the TEG is dependent on the combination of n- and p-type materials used in its construction. In order to evaluate thermoelectric materials, a

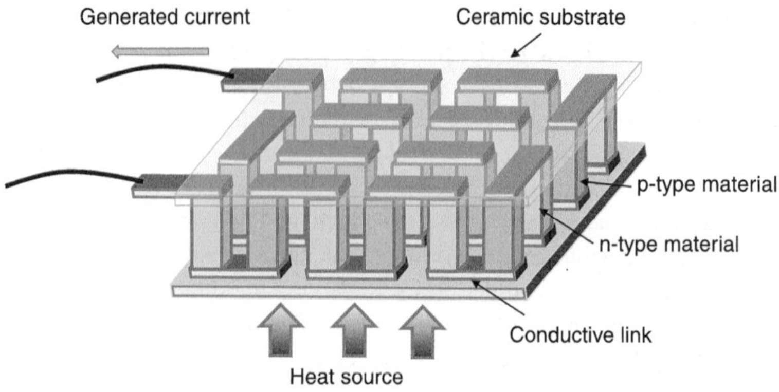

11.5 Ceramic thermoelectric module.

figure of merit in the specified operating temperature range is used. The figure of merit includes the Seebeck coefficient, thermal conductivity and electrical resistivity and this is given by:

$$z_c = \frac{\alpha_{np}^2}{R \cdot \lambda'}. \qquad [11.9]$$

Here, α_{np} is the Seebeck coefficient of the thermal junction between n- and p-type thermal materials, R is the equivalent electric resistance of the thermal junction, and λ' is the thermal conductivity of this junction. In order to gain a high value of figure of merit, the thermal junction should have a high Seebeck coefficient, low electric resistance and low thermal conductivity. In practice, however, there is a trade-off between these last two parameters.

11.5.4 Review of thermoelectric devices in smart fabrics applications

This review is focused on the choice of material, the structural design and the fabrication technologies used to, or potentially suitable for, the realisation of thermoelectric devices in smart fabrics applications. The review initially looks at existing ceramic thermoelectric devices (Fig. 11.5) in human applications, including a clothing based implementation. It then discusses new technologies, which could be used in smart fabric applications in the future.

Human applications of thermoelectric generators (TEGS)

The use of human body heat to power autonomous electronics systems has been demonstrated in the case of a wireless autonomous pulse oximeter (Torfs *et al.*, 2007). The autonomous sensor system is powered by a conventional ceramic thermoelectric module (Fig. 11.6). The 38 × 34 mm TEG can provide more than 100 μW with an ambient temperature of 22 °C, which is more than sufficient to power the oximeter, which requires 89 μW. In this system, a capacitor was used as a storage element for short-term energy buffering, to ensure the full energy autonomy.

In an example of a smart fabrics application, the same ceramic TEGs have also been incorporated into an energy harvesting shirt that uses both body heat and incident light to generate electrical power (Fig. 11.7).

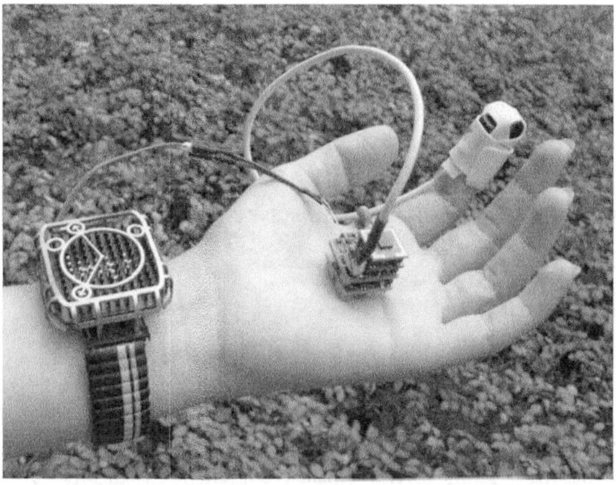

11.6 Body heat powered pulse oximeter. Left: 'watch'-style thermoelectric generator. Middle: electronics with wireless module. Right: commercial finger pulse oximeter.

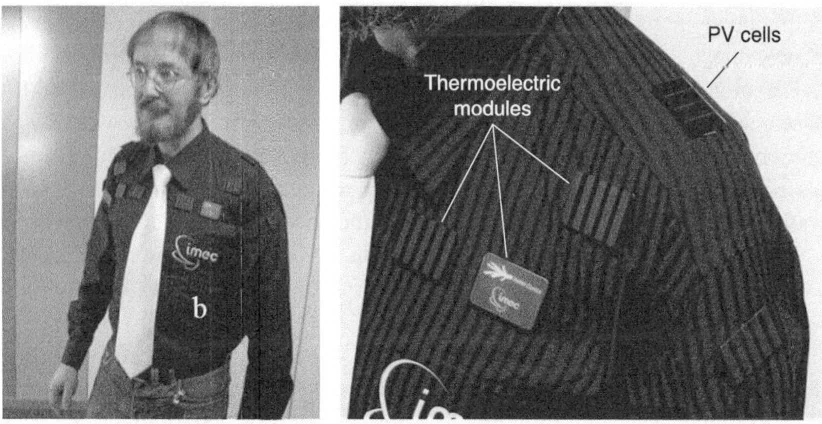

11.7 Left: electrocardiography system integrated in a shirt. Right: TE modules and PV cells.

The example system includes a wireless electrocardiography (ECG) incorporated into a standard long sleeved shirt (Leonov, 2009). In this system, 17 TEG modules, each 6.5 mm thick, with outer metal plates of 3 × 4 cm in size acting as radiators, have been integrated into the shirt by removing a rectangular section of fabric, thereby making room for each TEG. These modules occupy less than 1.5% of the total area of the shirt and a single TEG typically provides the 0.8 to 1 mW power at around 1 V into a matched load in a typical office environment. In this case, the TEGs are the main charging circuits, while the photovoltaic (PV) cells placed on the shoulders are the ancillary power charging circuits that provide the standby power for the system when the shirt is not worn for long periods of time.

Whilst this example demonstrates the feasibility of thermoelectric energy harvesting in garments, the implementation is not practical. The ceramic TEGs are rigid structures whose mass and lack of compliance will alter the feel of the shirt and will be unlikely to survive washing. In order to improve the practicality of thermoelectric energy harvesting in smart fabrics, the TEGs have to be either much smaller and/or flexible, and these are described in the following sections.

Micro thermoelectric generators

A considerable amount of research into MEMS micro-TEGs has been reported and a selection of significant devices is presented here. An early example was based on silicon thermocouples with n-type (doped with phosphorous) and p-type (doped with boron) thermoelements fabricated on a sapphire substrate (Rowe and Min, 1998). The undoped silicon between the active parts was removed by reactive ion etching (RIE) and electrical interconnections were realised by a thin aluminium layer. The implanted carrier concentration for both thermoelements was 5×10^{-25} m^{-3}, which gives a Seebeck coefficient of 530 μV/K. With a temperature difference of 10 K, the open circuit voltage per area was reported to be 0.55 V/cm^2 and the output power per area 0.11 μW/cm^2, which is less than for ceramic TEGs. This is largely due to the poor electrical conductivity of the silicon-based thermocouples.

Research in the SEIKO Corporation (Kishi *et al.*, 1999) has developed a Bi-Te-based micro-TEG device to act as the energy source of a wristwatch. The micromachining used a hot pressing process to form bismuth telluride compounds plates with nickel bumps on both sides. These plates were bonded to silicon wafers that contain the conductive links and the unwanted material was removed using a dicing saw. These thermoelectric modules were used to power the SEIKO Thermic watches, a limited edition range of watches powered by human body temperature.

Flexible micro-TEGs on plastic substrates have also been investigated and these are of interest because they could potentially be applied to smart fabrics. One example uses thin film Bismuth-Telluride p-type legs and n-type legs deposited onto flexible Kapton substrates (Stark and Stordeur, 1999). The different thin film thermoelements were fabricated by the Physical Vapour Deposition

(PVD) process combined with photolithography and wet etching. The Seebeck coefficients of the thin film p-type and n-type materials were around 200 μV/K and −70 μV/K, respectively, at room temperature 25°C, which gives a thermo power coefficient of 370 μV/K for a single thermocouple near room temperature. A 9 × 7.5 × 2.8 mm device produced 2.1 V and an output power of 5.8 μW (8.6 μW/cm^2) from a temperature difference of 10 K.

Polyimide substrates have also been used, with the thermoelectric substrate being corrugated and laminated between two flexible thermally conductive sheets (Itoigawa *et al.*, 2005). The thermoelectric sheet consists of copper and nickel acting as p-type and n-type thermoelectric material, respectively. The final open-circuit voltage for one thermocouple is 16 μV/K and the minimum bending radius of curvature is 9 mm. Structured SU-8 has been used to form flexible moulds for Bi-Te based flexible thermocouples deposited by electrochemical deposition (Glatz and Hierold, 2007). These can operate down to a radius of curvature of 7.5 mm and delivered 4.20 μW/cm^2 for a thermal gradient of 40 K.

Printed thermoelectric devices

During the last decade, printing technology has been explored for the manufacture of TEGs, because it is a low cost, simple and fast thick-film technology that is fully compatible with the textile industry. The following examples describe TEGs realised to date using printing techniques.

The earliest screen printed thick film thermocouples were printed on polyester (Duby and Ramsey, 2005). The printable paste formulation was based on Panipol M (a conductive poly-aniline), polyester and cellulose, binders in which the metal particulates were suspended. Various other additives, such as wetting agents and chemicals to penetrate and destabilise passive oxide layers, were also included in the formulation to improve the electrical properties of the thermocouples. NiCr and Ni/Panipol were chosen to be the n-leg and p-leg materials, respectively and the Seebeck coefficient of single legs were measured as 22 μV/°C. This performance was lower than expected, because forming the material into a paste will dilute its properties due to the addition of other ingredients. The binders used in the paste for screen printing are typically electric insulators, which decrease the electric conductivity of the printed film and, as a result, the figure of merit of the thermocouples will be decreased.

To mitigate against the dilution effect, active powders with a high Seebeck coefficient are required, such as BiSbTe materials. A coiled-up TEG with 900 thermocouples printed on top of polyimide (Kapton) substrate (Fig. 11.8), has been demonstrated (Weber *et al.*, 2006). For the ink preparation, Sb and $Bi_{0.85}Sb_{0.15}$ alloy materials were used for each leg and ethylene glycol was added as the binder, which is predicted to have a high Seebeck coefficient (Obata *et al.*, 1990). For a single screen printed thermocouple, the Seebeck coefficient was measured as around 97 μV/K. More recently, a similar planar geometry has been stencil

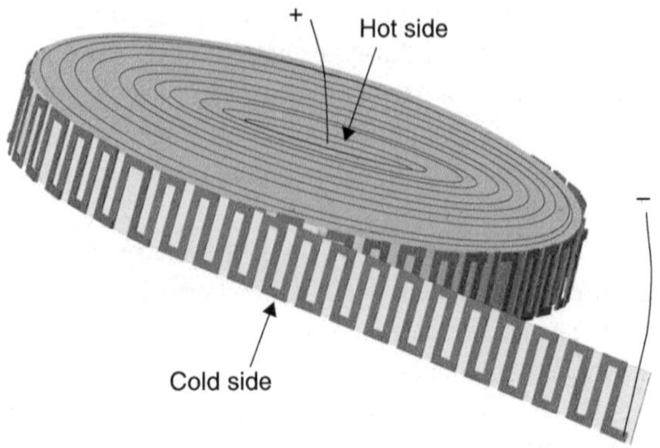

11.8 Schematic of the coiled-up TEG (Weber, 2006).

printed onto polyethylene naphthalate (PEN) flexible substrates (Navone *et al.*, 2010). The n-leg and p-leg were fabricated from $Bi_2Se_{0.3}Te_{2.7}$ and $Bi_{0.5}Se_{1.5}Te_3$, respectively, with the material processed in powder form and mixed with 2 wt.% polystyrene as the binder and 22 wt.% toluene to provide a suitable viscosity for printability and structural integrity. Thermolegs of dimensions 0.8 cm × 1.2 cm × 80 μm were printed onto the PEN substrate and laser-annealed using a pulsed 473 mJ/cm^3 Tamarack excimer laser. The Seebeck coefficient was found to decrease with increase in the number of pulses, whilst the electrical conductivity increases with the increase in pulses numbers. Hence, there is a trade-off between these factors, and the highest power factor reported in this article was 0.06 mWK^{-2}cm^{-1}.

The final TEG described here used dispenser printing to deposit n-type Bi_2Te_3 and p-type Sb_2Te_3 onto a flexible 63.5 μm thick polyimide substrate (Chen *et al.*, 2011). The thermoelectric materials come in a powder form and were mixed with an epoxy binder to produce the paste, which was printed onto the polyimide substrate, which included evaporated metal contacts. A 50-couple prototype with 5 mm × 640 μm × 90 μm printed element dimensions (Fig. 11.9) produced a power output of 10.5 μW (1.3 μA and 171.6 mV) for a 20 K temperature difference with a device area power density of 75 μW cm^{-2}.

11.5.5 Discussion

Although some wearable TEGs have been reported in the literature, none have been fabricated on a large area of clothes. The micro-TEGs presented here could be integrated into fabrics, but are still rigid devices that, despite their small size, may alter the feel of the fabric. Also, the fundamental amount of power delivered is a function of surface area, and therefore small device size, may have to be compensated for by a large number of devices. On the positive side, micro- and

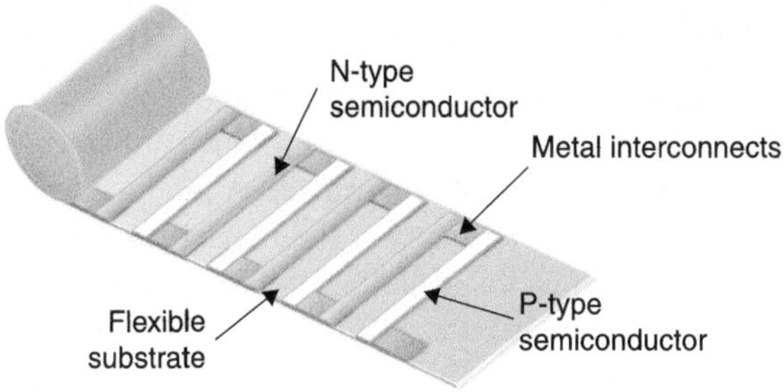

11.9 Planar dispenser printed TEG (Chen, 2011).

nanoscale fabrication techniques offer the ability to realise thermoelectric nanostructures, which reduce thermal conductivity without compromising other material properties and thereby improved device performance can be obtained.

The printing approach has obvious cost advantages with respect to large-area production and offers the possibility of realising flexible generators printed directly onto the fabric itself. However, the Seebeck coefficient of printable pastes is reduced compared to the bulk values due to the dilution of the material with the other ingredients. The choice of material has a significant effect on the device performance and the Bismuth-based semiconductor materials offer the best performance. Practical factors must also be considered in smart fabrics applications. In the case of printed devices and where the fabric is used as a substrate for the TEG, the thermal conductivity of the fabrics will affect the temperature gradient across the generator. Printed devices are also not as small as MEMS TEGs, since they are fabricated in a planar manner. Devices printed on flexible substrates are typically wound up in a coil form in order to maximise the density of thermocouples. Such an arrangement is not as small or efficient as the MEMS TEGs.

There is clearly an opportunity for thermoelectric energy harvesting to be applied in smart fabrics, but further development is required in both materials and fabrication processes in order to achieve a practical solution that does not overly compromise the feel and function of the fabric.

11.6 Solar energy harvesting for smart textiles

PV devices use the energy of photons (incident light) to generate electrical energy. Traditional PV cells are typically rigid and made from crystalline or polycrystalline materials using whole wafers 100 to 150 mm in diameter. Flexible PV cells have also been demonstrated using thin-film semiconductor processes or roll-to-roll printing techniques both on polymer substrates. These flexible devices can be thin film PV cells or dye-sensitised solar cells (DSSC).

11.6.1 Solar cell theory

Conventional semiconductor PV cells work when a photon with sufficient energy is absorbed by a semiconductor material (e.g. silicon), which promotes an electron into the conduction band resulting in a flow of charge. A p-n junction is required in the semiconductor material for this to happen. The effectiveness of the approach depends upon the wavelength of the incident light, which is linked to the energy of the photon, and the amount of light reflected off the surface of the material. The maximum efficiency that can be obtained is 33.7% from a single junction fabricated in silicon.

DSSC combine organic dyes and electrochemical cells comprising an anode, cathode and electrolyte (Hinsch *et al.*, 2001). The top electrode must be transparent to allow incident light to reach the photosensitive dye, which generates the electrons that then flow to the anode. The attraction of this approach is the relative simplicity of the fabrication process, which does not require expensive cleanroom equipment or high cost materials. Printing processes have been demonstrated for fabricating flexible DSSC, which are now commercially available (e.g. Solaronix and G24i).

The efficiencies of different solar cells can be compared using the fill factor shown in Eq. 11.10. The fill factor is determined by the ratio between the actually achievable maximum power (given by the current times the voltage at the maximum power point (mpp)) and the theoretical value (given by the short-circuit current (I_{sc}) times the open-circuit voltage (V_{oc})). Typical commercial solar cells have a fill factor of more than 0.7:

$$FF = \frac{I_{mpp} \times V_{mpp}}{I_{sc} \times V_{oc}} \qquad [11.10]$$

11.6.2 Example photovoltaic (PV) fabrics

The simplest form of PV fabric is achieved by simply sewing a flexible solar cell to the surface of the fabric. The fabric itself is not functionalised, the solar cell being fabricated on a plastic substrate, which is then attached to the textile. This basic approach has been used to realise energy harvesting jackets (e.g. Scott-E-Vest' solar jacket) and bags (e.g. Reware's Juice Bag) and is discussed in some detail by Shubert and Werner (2006).

In order to functionalise the fabric itself, PV fibres have been investigated that could then be woven together to form the fabric. The first example of this approach fabricates the PV layers around a flexible polypropylene (PP) core with a diameter of 0.59 mm (Fig. 11.10) (Bedeloglu *et al.*, 2010). The polymer mixture PEDOT:PSS is used as an anode. After chemical treatment to improve conductivity and adhesion to the PP-fibre surface, the initial fibre is dip-coated with the PEDOT:PSS mixture and dried for 3 hours at 50°C. Two different photoactive compounds, P3HT:PCBM or MDMO-PPV:PCBM, are then deposited onto the anode using

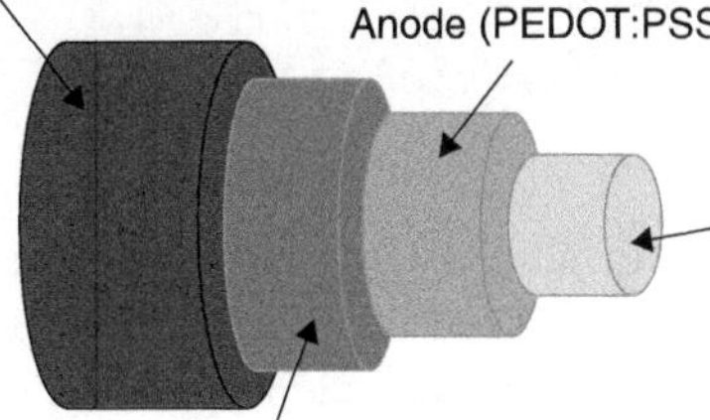

11.10 Schematic view of a coaxial photovoltaic fibre (Bedeloglu, 2010).

the same method. The cathode consists of an approximately 0.7 nm thin layer of lithium fluoride and 10 nm layer of aluminium of which semi-transparency is required for the light to reach the photoactive compound. These layers are deposited by thermal evaporation in near-vacuum at an evaporation rate between 0.01 and 0.2 nm per second. The MDMO-PPV:PCBM compound produced 300 mV and 0.27 mA/cm^2, giving a fill factor of 26% and an efficiency of 0.021%.

A second fibre-based development used multiple functional layers fabricated concentrically around a flexible polyimide-coated silica fibre (O'Connor *et al.*, 2008). The coatings (25 nm Mg, 20 nm Mg:Au (1:1), 65 nm Au, 25 nm copper phthalocyanine (CuPc), 40 nm C60, 7 nm tris(8-hydroxyquinoline) aluminium (Alq3), 6.5 nm Mg:Ag (7:1) and 5 nm Ag) were deposited using vacuum thermal evaporation. This PV fibre produced 390 mV and 2.85 mA/cm^2, giving a fill factor of 47% and an efficiency of 0.5%. Identical layers were also deposited on a planar glass substrate to allow comparison between fibre-based and planar geometry. The planar device performed significantly better demonstrating an efficiency of 0.82%.

11.6.3 Discussion

The efficiency of PV fibres lies well below the values that are achieved by conventional solar cells. The radial design does allow for the collection of photons from a larger range of angles than a planar device would allow. However, where light is coming from one direction, the radial design will mean the majority of the active surface is not facing the incident light and efficiency will be reduced. Absolute levels of energy harvested by a PV fabric can be increased by using more fibres, that is covering a larger area. Work is progressing on low temperature printed PV (Barr *et al.*, 2011), which may in the future enable these to be directly printed on fabrics. Functionalising the surface of the fabric in this manner should improve performance, since more of the PV structure will be facing the incident light and not wasted as is the case of the fibre.

11.7 Conclusion

Research into energy harvesting on fabrics is at an early stage and it is a challenging task to integrate this functionality without spoiling the feel of the fabric or the behaviour of the fabric when used in its application. There are three main sources of energy that may be exploited – kinetic, thermal and solar. In each case, energy harvesting can be incorporated into a fabric by integrating discrete devices in the fabric, such as MEMS kinetic energy harvesters, microscale TEGs and flexible plastic solar cells. However, it would be preferable if the fabric itself possessed the energy harvesting functionality. This can be achieved by functionalising either the fibres used to weave the fabric or the surface of the fabric itself using material deposition techniques. Standard processes commonly used in the textiles industry for the patterning of fabrics such as screen or inkjet printing can be used to deposit active materials onto textiles and by developing active materials that can be processed in a manner compatible with the limitations imposed by the fabric, for example maximum temperatures. Research into suitable materials is under way and, as these become increasingly available and optimised, achieving energy harvesting functionality in a fabric with minimum impact on the properties of the textile will become a practical proposition.

11.8 References

Barr M C, Rowehl J A, Lunt R R, Xu J, Wang A, *et al.* (2011), 'Direct monolithic integration of organic photovoltaic circuits on unmodified paper,' *Adv Mat*, 23(31), 3500–3505.

Bedeloglu A C, Demir A, Bozkurt Y and Sariciftci N S (2010), 'A photovoltaic fiber design for smart textiles,' *Text Res J*, 80, 1065–1074.

Beeby S and White N (Eds.) (2010), *Energy Harvesting for Autonomous Systems*. Norwood, MA, Artech House.

Beeby S P, Tudor M J and White N M (2006), 'Review paper: energy harvesting vibration sources for microsystems applications,' *Meas Sci Tech*, 17(12), R175–R195.

Chen A, McAdan D, Wright P K and Evans J W (2011), 'Dispenser-printed planar thick-film thermoelectric energy generators,' *J Micromech Microen*, 21, 104006.

Duby S and Ramsey B (2005), 'Printed thick-film thermocouple sensors,' *Electron Lett*, 41, 6–7.

Edmison J, Jones M, Nakad Z and Martin T (2002), 'Using piezoelectric materials for wearable electronic textiles,' *Proceedings of the 6th Symposium on Wearable Computers*, pp. 41–48.

Glatz W and Hierold C (2007), 'Flexible micro thermoelectric generator,' *20th IEEE International Conference on Micro Electro Mechanical Systems Technical Digest*, pp. 89–92.

Goldsmid H J (1960), *Applications of Thermoelectricity*. London, Wiley.

Hande A, Polk T, Walker W and Bhatia D (2007), 'Indoor solar energy harvesting for sensor network router nodes,' *Microproc Microsyst*, 31(6), 420–432.

Hansen K and Zawada T (2011), 'Microfabrication and printed electronics on flexible substrates,' *Piezo 2011*, Sestriere, Italy, available from: http://microflex.ecs.soton.ac.uk/externaldocs/Piezo2011_Microflex%20poster%20v3.pdf

Hinsch A, Kroon J M, Kern R, Uhlendorf I, Holzbock J, *et al.* (2001), 'Long-term stability of dye-sensitised solar cells,' *Prog Photovolt: Res Appl*, 9(6), pp. 425–438.

Itoigawa K, Ueno H, Shiozaki M, Toriyama T and Sugiyama S (2005), 'Fabrication of flexible thermopile generator,' *J Micromech Microeng*, 15, S233–S238.

Kishi M, Nemoto H, Hamao T, Yamamoto M, Sudou S, *et al.* (1999), 'Micro thermoelectric modules and their application to wristwatches as an energy source,' *Proceedings of the 18th IEEE International Conference on Thermoelectrics*, pp. 301–307.

Marzencki M, Ammar Y and Basrour S (2008), 'Integrated power harvesting system including a MEMS generator and a power management circuit,' *Sens Actuat A: Phys*, 145–146, 363–370.

Miles L W C (2010), '*Textile Printing*,' revised Second Edition, available from: http://www.digitallibrary.kcci.com.pk/handle/32417747/163

Navone C, Soulier M, Plissonnier M and Seiler A L (2010), 'Development of $(Bi,Sb)_2(Te,Se)_3$-based thermoelectric modules by a screen-printing process,' *J Electro Mat*, 39, 1755–1759.

Obata T, Kajikawa T and Kumashiro Y (1990), 'Characteristics of $(Bi,Sb)_2(Te,Se)_3$-based thick-film thermoelectric elements for power generation,' *Electr Eng Jpn*, 114(4), 213–219.

O'Connor B, Pipe K P and Shtein M (2008), 'Fibre based organic photovoltaic devices,' *Appl Phys Lett*, 92, 193306-1–193306-3.

Qi Y, Jafferis N T, Lyons K, Lee C M, Ahmad H and McAlpine M C (2010), 'Piezoelectric ribbons printed onto rubber for flexible energy conversion,' *Nano Lett*, 10, 524–528.

Qi Y, Kim J, Nguyen T D, Lisko B, Purohit P K and McAlpine M C (2011), 'Enhanced piezoelectricity and stretchability in energy harvesting devices fabricated from buckled PZT ribbons,' *Nano Lett*, 11, 1331–1336.

Qin X W T and Wang Z L (2008), 'Microfibre-nanowire hybrid structure for energy scavenging,' *Nature*, 451, 809–813.

Richards C D, Anderson M J, Bahr D F and Richards R F (2004), 'Efficiency of energy conversion for devices containing a piezoelectric component,' *J Micromech Microeng*, 14, 717–721.

Roundy S and Wright P K (2004), 'A piezoelectric vibration based generator for wireless electronics,' *Smart Mat Struct*, 13, 1131–1142.

Rowe D M and Min G (1998), 'Evaluation of thermoelectric modules for power generation,' *J Power Sources*, 73(2), 193–198.

Rowe D M, Morgan D V and Kiely J H (1989), 'Miniature low-power/high-voltage thermoelectric generator,' *Electron Lett*, 25, 166–168.

Schubert M B and Werner J H (2006), 'Flexible solar cells for clothing,' *Mat Today*, 9(6), 42–50.

Shenck N S and Paradiso J A (2001), 'Energy scavenging with shoe-mounted piezoelectrics,' *IEEE Micro*, 21, 30–42.

Stark I and Stordeur M (1999), 'New micro thermoelectric devices based on bismuth telluride-type thin solid films,' *Proceedings of the 18th IEEE International Conference on Thermoelectrics*, pp. 465–472.

Starner T (1996) 'Human powered wearable computing,' *J IBM Syst*, 35618–629.

Torfs T, Leonov V and Vullers R (2007), 'Pulse oximeter fully powered by human body heat,' *Sens Transd J*, 80(6), 1230–1238.

Weber J, Potjekamloth K, Haase F, Detemple P, Volklein F and Doll T (2006), 'Coin-size coiled-up polymer foil thermoelectric power generator for wearable electronics,' *Sens Actuat A: Phys*, 132, 325–330.

Zhu D, Beeby S P, Tudor J and Harris N R (2010), 'A Bimorph multi-layer piezoelectric vibration energy harvester,' *Proceedings of the PowerMEMS 2010*, Leuven, Belgium, pp. 335–338.

12

Signal processing technologies for activity-aware smart textiles

D. ROGGEN and G. TRÖSTER, ETH Zurich, Switzerland and A. BULLING, University of Cambridge, UK

DOI: 10.1533/9780857093530.2.329

Abstract: Garments made of smart textiles have an enormous potential for embedding sensors in close proximity to the body in an unobtrusive and comfortable manner. Combined with signal processing and pattern recognition technologies, complex high-level information about human behaviors or situations can be inferred from the sensor data. The goal of this chapter is to introduce the reader to the design of activity-aware systems that use body-worn sensors, such as those that can be made available through smart textiles. We start this chapter by emphasizing recent trends towards 'wearable' sensing and computing and we present several examples of activity-aware applications. Then we outline the role that smart textiles can play in activity-aware applications, but also the challenges that they pose. We conclude by discussing the design process followed to devise activity-aware systems: the choice of sensors, the available data processing methods, and the evaluation techniques. We discuss recent data processing methods that address the challenges resulting from the use of smart textiles.

Key words: wearable sensing, signal processing, activity recognition, context recognition.

12.1 Introduction: from on-body sensing to smart assistants

Sensors integrated into textiles open up the way to new kinds of functionalities, built right into our everyday clothing. As an example, a heart rate belt – a gadget commonly used by sports enthusiasts – may be realized in a shirt with conductive fibers serving as electrodes (Pola and Vanhala, 2007). Thus, the gadget's functionality becomes inherent to the clothing and available anytime and anywhere. This is the view promoted in wearable computing, as originally presented by Mann in 1996, which emphasized a shift in computing paradigm. Computing would no longer be separate from the user but would become an unobtrusive extension of our very bodies, providing us with additional ubiquitous sensing, feedback and computational capabilities (Fig. 12.1). In the figure, a number of sensors are integrated in the jacket, within the shoes, or inside other accessories. In a mature product, the integration is supposed to be unobtrusive. A see-through head display in the goggles provides the user with contextual information. With an internet connection, the on-body computer and sensors can infer the user's context, such as his activities or location, and his needs.

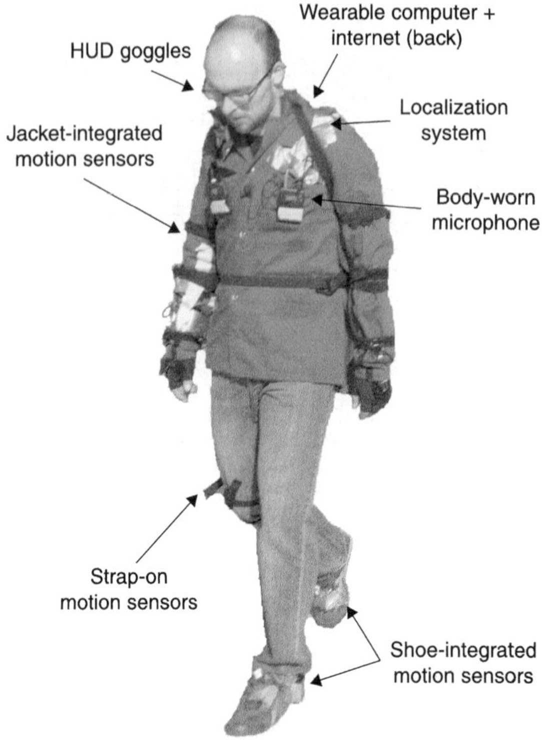

12.1 Representation of a prototype of a typical 'wearable computing' system.

In particular, Mann (1998) and Starner *et al.* 1998) were among the first to show that complex contextual information can be obtained by interpreting the data from sensors placed on the body. They underlined that this would lead to novel adaptive applications. These so-called 'context-aware' applications are capable of adapting their behavior, providing information right when needed, or delivering other forms of assistance when specific needs are detected. Context-awareness has been defined as 'any information that can be used to characterize the situation of a person, place, or object that is considered relevant to the interaction between a user and an application' (Dey, 2001). The 'context' of a user collectively refers to the user's activities (activity awareness) (Davies *et al.*, 2008; Lukowicz *et al.*, 2010), his emotional and cognitive states (cognitive-affective awareness) (Picard, 1997; Bulling *et al.*, 2010), his social surroundings and interpersonal interactions (social awareness) (Lazer *et al.*, 2009), or the physical surrounding (environmental awareness).

Pattern recognition techniques are used to infer these elements of context from sensor data. In many cases, a close proximity of the sensors to the body is required.

This is especially the case for activity awareness, where sensors must measure limb movement (Davies *et al.*, 2008; Lukowicz *et al.*, 2010), for physiological sensing, and for cognitive-affective awareness that also requires physiological measurements (Picard, 1997; Bulling *et al.*, 2010). Smart textiles offer such close proximity to the body and represent an ideal substrate for sensing technology to disappear into. Thus *smart-textile sensing* is a key enabling technology for a wide range of novel applications.

The goal of this chapter is to introduce the reader to the design of activity-aware systems, as they can benefit from smart-textiles sensing. We discuss the choice of sensors, the selection of data-processing techniques, and the evaluation methodologies. We also discuss data processing options to address the specific challenges posed by smart textiles.

In Section 12.2, we illustrate the breadth of activity-aware systems being investigated, by presenting a few examples from the literature. Many sensors have been considered in past work. Some of these sensors can readily be integrated into smart textiles, while others may be integrated in the future. Nevertheless, researchers are exploring the scope of possibilities enabled by a wide range of sensor kinds. In Section 12.3, we emphasize the benefits brought about by smart textiles for on-body sensing, and give an overview of commonly used sensors. We also mention the challenges posed to data-processing techniques resulting from smart-textile sensing. In Section 12.4, we introduce the basic principles of activity recognition and some common terminology. In Section 12.5, we explain how signal processing and machine learning techniques can be applied to sensor signals to infer activities or context. We discuss recent methods that address some of the challenges of smart-textile sensing and capitalize on their properties. The experimental aspects, such as the acquisition of datasets, their annotation, the evaluation of recognition methods, and the means available to prototype activity-recognition systems are covered in Section 12.6. In Section 12.7, we identify future trends, and provide sources of further information on the topic.

12.2 Activity-aware applications

Some of the most active application domains for activity-recognition technologies using body-worn sensing include healthcare and assisted-living, sports, industrial worker assistance, entertainment, or human-computer interfaces (HCI). We present a few recent works drawn from various fields, to illustrate the rich diversity of activity-aware systems.

Stiefmeier *et al.* (2008) present methods to recognize complex manipulative gestures performed by industrial workers in the car manufacturing industry. The authors show that it is possible for a jacket with built-in sensors to recognize activities such as checking the hood latch mechanism, checking the seat sliding mechanism, or checking the spacing between doors and car body. The authors present this system as an aid to workers who perform quality assurance tasks, as the system may automatically log all the checked parts without the need for an additional interface (e.g. mobile device operated with buttons, or paper checklists). Activity recognition in industry may also be used to improve the safety of workers in hazardous areas, such as detecting unsafe movements towards running machinery.

Dunne *et al.* (2008) present a garment that can measure seated posture. The authors motivate their work by the need to prevent musculoskeletal disorders, which may arise from an inadequate seated posture over extended periods of time. They argue that a garment-integrated system has the advantage to be always available to the user, and thus may offer reminders if the user is at risk. In Bächlin *et al.* (2010), the authors present a system capable of detecting the freezing of the gait of patients suffering from Parkinson's disease, as a way to provide bio-feedback to support these patients. The system is ideally integrated in trousers, belts or shoes. These are but a few examples of activity recognition in health-related domains (Park *et al.*, 2003; Pentland, 2004; Sixsmith *et al.*, 2007).

Gestural interfaces for HCI have been envisioned for many years (Myers *et al.*, 1996). Body-worn sensors provide gestural interfaces anywhere the user goes. For instance, Kallio *et al.* (2006) presented a way to interact with ambient devices by means of natural hand gestures with a sensor placed on the wrist. Nowadays, a wide range of motion-based interaction patterns exist for entertainment, for instance with the Nintendo Wii game console and Microsoft Kinect.

Other fields that benefit from activity recognition with body-worn sensing include, for example, sports (Bächlin *et al.*, 2009; Gravenhorst *et al.*, 2011; Strohrmann *et al.*, 2011), human-robot interaction (Martin *et al.*, 2010), or even crowd management (Roggen *et al.*, 2011). Mobile phones also offer a wide range of sensors suited for activity recognition and can complement smart textiles (Campbell *et al.*, 2010). In Section 12.7, we discuss future trends and point to conferences and journals where further application-oriented research outcomes can be found.

12.3 Sensing principles for activity recognition

A *sensor* is a device that converts one physical quantity (e.g. acceleration) into another physical quantity (e.g. voltage), which can be acquired by a computer. Typically, the sensor data will be acquired (or *sampled*) by the computer at a regular interval (the *sampling interval* and its inverse, the *sample rate*). A *sample* is the digital representation of the sensor signal. It refers to a value at a point in time. The *resolution* is the smallest incremental change in the physical quantity that leads to a change in sample value. It is limited by the acquisition electronics (usually an analog-to-digital converter) that uses a limited number of bits to represent a sample. Sensors are typically continuously sampled and provide a continuous stream of sensor data. The *data rate* indicates the number of bits of data acquired per unit time. It is the product of the sample rate by the number of bits used to represent samples. With a *calibration* formula, the value of the physical quantity can be computed from the sample. This formula may be a simple linear transformation, where the relation between the physical quantity and the sample is defined by a slope (the sensor *sensitivity*) and an *offset*. However, the relation may also be non-linear or time-dependent (e.g. *hysteresis*).

In Fig. 12.2, we illustrate the signal obtained from an acceleration sensor with a sample rate of 32 Hz and a resolution of 10 mG. The data is from a recording that is 170 s long at 32 Hz sample rate. The periodicity related to the user's gait is clearly visible. Finer distinctions can be made, such as between walking on a flat surface and walking up a staircase. This is particularly visible on the back sensor. Sitting, standing and lying also have distinctive signal patterns. The thigh sensor helps distinguish between sitting and standing. Lying is no different than sitting for the thigh sensor, but together with the back or ankle sensor, this posture can be distinguished. The signal values remain mostly constant, while the user retains the posture. Note that some data loss occurred around sample 5000 for the thigh and back sensors. The thigh sensor suffered from inadvertent rotation around sample

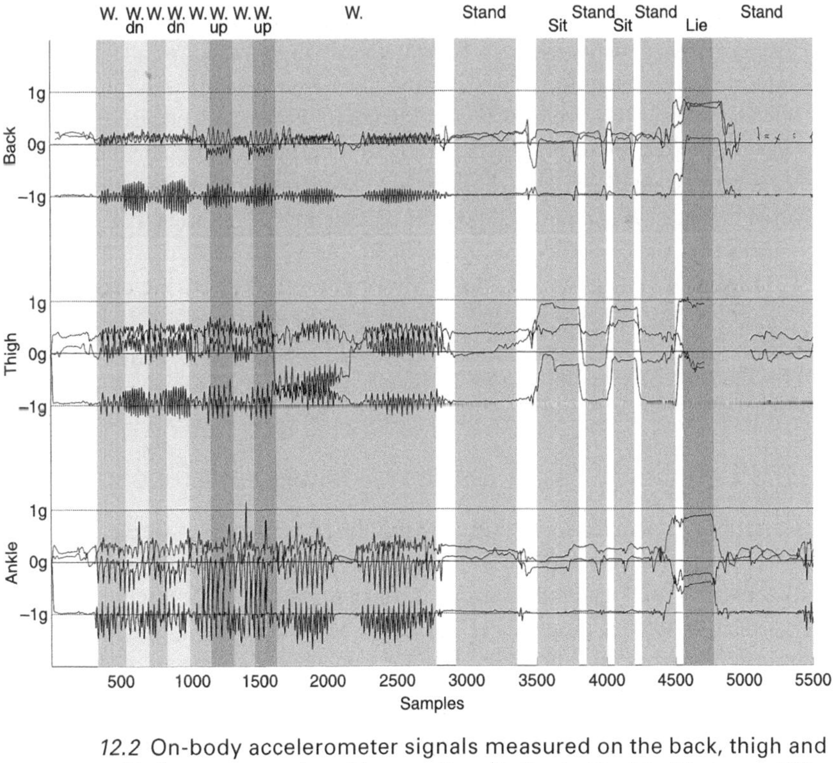

12.2 On-body accelerometer signals measured on the back, thigh and ankle for some modes of locomotion (indicated by W., W. dn, and W. up for walk, walk downstairs, and walk upstairs, respectively) and body postures.

2000. These issues are common with textile integrated sensors and wireless sensor nodes and the data-processing techniques must be robust to these situations.

12.3.1 Sensing requirements

The design of a recognition system starts by selection of a set of sensors based on the activities to be recognized. The sensors are selected according to a trade-off between information content, usability and power:

- *Characteristic signal patterns*: Activity recognition is based on the assumption that sensors placed on the body provide characteristic signal patterns when an activity is executed by the user. Thus, the selected sensors must measure signals that are characteristic of the activities that must be recognized. This influences the choice of the sensor modality and of its placement on the body.
- *Usability*: Since the sensors are on-body, they must meet usability requirements. A garment equipped with sensors should be comfortable, and in no way differ from a 'normal' garment. Consequently, the sensors must be small, unobtrusive and invisible. They must not hamper the user's freedom of motion, and must be wearable for extended periods of time without adverse effects.
- *Power*: Higher energy use calls for batteries of larger capacity and size, which adversely affect comfort. Thus, sensors must be chosen taking into account the energy use. This includes the energy use of the sensor, and also the energy required to communicate the data (e.g. to a wearable computer), and to run data processing algorithms. For instance, cameras are currently only rarely used in wearable computing, and even though they do not use much power by themselves, the computational requirements for video analysis are significant. Thus, sensor modalities that allow lightweight computations are preferred.

A major focus of the wearable computing community has been to look for sensor modalities that fit these requirements, and research in smart textiles has a key role to play.

12.3.2 Benefits of smart textiles for activity awareness

Hereafter, we use '*smart-textile sensing*' to refer to several approaches by which sensors are integrated within smart textiles. The ideal deployment vector for on-body sensors is to integrate the sensing element directly into the fiber with which the garment is woven (Tognetti *et al.*, 2006; Mattmann *et al.*, 2007). Another approach consists of using fiber-like substrates on which the sensors are deposited, and then to inter-weave this substrate between textile fibers (Kinkeldei *et al.*, 2011(a)). Highly miniaturized conventional sensing electronics can also be encapsulated into the garments (Harms *et al.*, 2009). Wireless communication or conductive fibers embedded within textiles can be used to interconnect the miniaturized electronics (Locher *et al.*, 2004).

Smart-textile sensing offers many benefits. A few of them include:

- *Close proximity to the skin*: This makes the smart-textile approach ideal to sense skin temperature, electric fields for electromyography (EMG), or electrocardiography (ECG). This is also a benefit to sense body movements, as the sensors can measure precisely the movements at any time, unlike camera systems that often suffer from image occlusion and limited field of view.
- *Integration of many multimodal sensors*: Smart textiles can integrate a large number of sensors of various modalities for a given context-aware application on a unique substrate (fiber, textile or garment). This simplifies deployment, as a single entity needs to be managed rather than a multitude of individual sensor nodes. Consequently, it also reduces the risk that a sensor is forgotten, swapped or placed at an inappropriate position during deployment.
- *Convenience and comfort*: Smart textiles allow for clothing that offers context-aware functionality and that is convenient and comfortable to wear.
- *Long-term use*: A consequence of convenience and comfort is to enable sensing applications on long timescales (weeks, months, or more). This is especially important in fields such as rehabilitation or assisted living, where trends must be assessed over long periods of time, or continuous assistance may be required day in and day out.
- *User acceptance and privacy*: Smart-textile sensing makes the sensors invisible to the outside. This enhances user acceptance: it preserves the user's privacy and does not draw the attention of others to the fact that one wears an assistive system.

However, capitalizing on these strengths is challenging, as Section 12.3.3 shows.

12.3.3 Challenges of textile sensing

There are several challenges to smart-textile sensing. Challenges cover technological aspects such as manufacturing, as smart textiles are still an active area of research. For activity awareness, we are especially concerned with data processing challenges. A few of those include:

- *Variability of sensor characteristics*: Variability in the production process can lead to different sensor sensitivity, offset or hysteresis. This can have adverse effects on subsequent data processing and must be compensated for, e.g. through a calibration process.
- *Deterioration of sensor characteristics*: Over long periods of time, sensor characteristics may be less stable with smart textiles than with conventional sensor nodes. This is due to the harsh conditions to which they are exposed, such as bending or washing (Schwarz *et al.*, 2011). This may lead to changing sensor characteristics after the initial calibration.

- *Sensor failure*: Sensors may fail altogether, e.g. due to cracks occurring in conductive materials (Kinkeldei *et al.*, 2011(b)). This requires the data processing methods to detect the faults, and ideally select replacement sensors at run-time.
- *Motion artifacts*: Motion artifacts occur when the user's own movements affect the placement of the sensor or its contact pressure to the skin. This can corrupt the sensor signal. Motion artifacts can affect, e.g. ECG measurements (changing skin-sensor impedance) or movement measurements (changing relative placement or orientation of a clothing-integrated sensor with respect to the body). Figure 12.3 shows an example of a motion artifact in the 'drink' gesture. As the user walks, we can see small oscillations in the 'drink' signal, which do not appear when the 'drink' gesture is executed when standing still. These oscillations are artifacts from the user's walking. The data processing methods must either be able to detect or compensate such motion artifacts. This is challenging, as the artifacts are usually difficult to model. The data was obtained from a recording 70 s long at 32 Hz sample rate. The drinking movement is a sporadic gesture, which must be identified in the continuous data stream. It is visible in the upper and lower arm sensors as a smooth

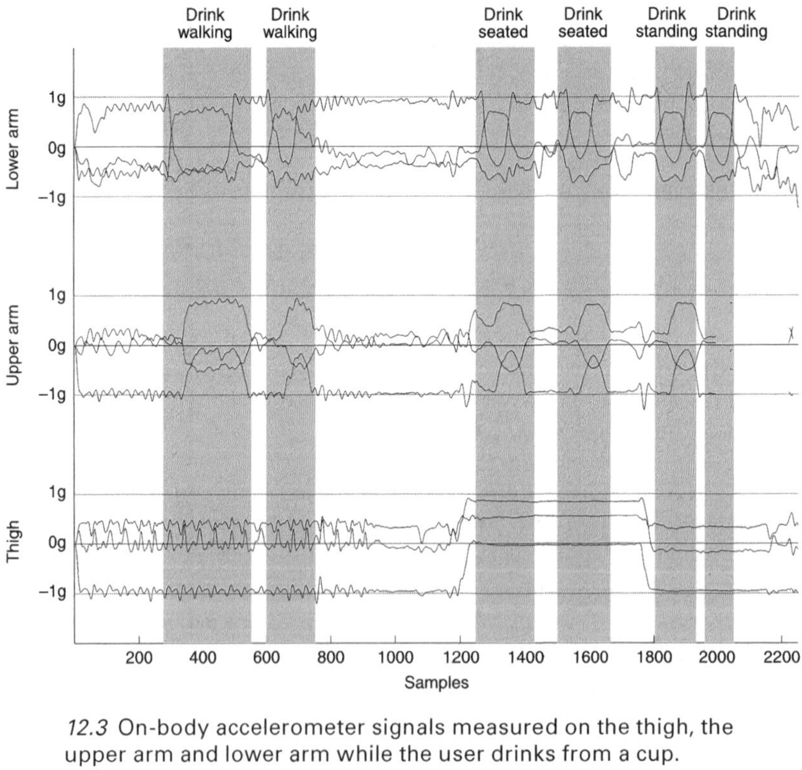

12.3 On-body accelerometer signals measured on the thigh, the upper arm and lower arm while the user drinks from a cup.

change in the acceleration readings. These reflect change in inclination of the arm as the user brings the cup to his mouth. The gesture is recorded while the user walks, is seated or stands. Periodic patterns due to the gait are superimposed on the readings of the arm sensors when the user walks. The readings of the sensors are smoother when the user is seated or standing still. The embedding of activities (drinking) within others (walking) must be taken into consideration when devising the recognition methods.

In Section 12.5, we present methods that can be used to address these challenges.

12.3.4 Typical sensor modalities

Developing sensors within fibers calls for significant investment in fundamental research, but has the potential to deliver ground-breaking functionalities. However, by using conventional electronics, researchers can tap into a large pool of well characterized and understood sensor modalities, some of which may not yet be available for integration into fibers. Researchers are investigating how to use existing as well as upcoming sensor modalities for activity-aware systems. A few typical sensor modalities used for activity recognition are indicated in Table 12.1, together with the possible means to deploy them on-body. Other sensor modalities are listed in Roggen *et al.* (2010(a)). We comment on a few common modalities below.

One of the most common modalities for activity recognition is the acceleration sensor. Accelerometers realized in MEMS technology are extremely small and low-power, and can measure the acceleration around three axes. They can easily be included into conventional but highly-miniaturized circuits and included in garments (Harms *et al.*, 2009). Many low-cost commercial solutions are available. When attached to a limb, accelerometers measure a static and a dynamic acceleration. The static acceleration stems from the Earth's gravity when the sensor is not moving. This can be used to determine the inclination of the sensor and thus the inclination of the limb with respect to the vertical plane (Ermes *et al.*, 2008; Randell, 2000; Harms *et al.*, 2009), to measure static body postures. The dynamic acceleration is due to the movement of the limb to which the sensor is attached. This can be used to recognize HCI gestures (Förster *et al.*, 2010) or modes of locomotion (Figo *et al.*, 2010; Randell, 2000). Inertial measurement units (IMUs) contain accelerometers, magnetometers and gyroscopes and sense the orientation of the device with respect to an Earth coordinated system. They are typically placed on each body segment and create a stick figure representing the user.

They can also be integrated into garments (Stiefmeier *et al.*, 2008) and are used when relative position and orientation of the limbs is relevant (e.g. the position of the hands with respect to the torso). Usually activity recognition performed on limb positions outperform the recognition performed on simpler sensors such as accelerometers (Zinnen *et al.*, 2009). Others have demonstrated that various

Table 12.1 Common sensor modalities for activity recognition from body-worn sensors, and the deployment possibilities for smart-textile sensing

Sensor	Observation	Smart-textile deployment
Microphone	Social interaction recognition, speaker identification, localization by ambient sounds, activity detection	No smart-textile equivalent, but commonly available on-body through mobile phones (Campbell, 2010)
Accelerometer, magnetometer, gyroscope, or inertial measurement unit	Movement patterns, orientation, or rotation of the body or limbs	No smart-textile equivalent, but flexible substrates, which can be woven into garments and on which bare-die sensors can be bound (Zysset, 2010), highly miniaturized conventional electronics encapsulated into garments (Harms, 2009)
Pressure	Pressure of clothing against the body, muscle activity, weight distribution, posture characterization	Textile pressure sensor used for muscle activity detection (Meyer, 2006) or posture recognition (Meyer, 2010)
Temperature	Health state (e.g. fever)	Textile integrated temperature sensor presented in Kinkeldei (2011)
Humidity	Sweat, physical activity	Textile integrated pressure sensor presented in Kinkeldei (2011)
Strain	Breathing (respiration belt), movement (elongation sensor in clothes)	Textile strain sensors for posture measurement in clothing (Mattmann, 2007; Giorgino, 2006; Tognetti, 2006), clothing-integrated strain measurement by optic fiber (Dunne, 2008)
ECG, EMG, EOG	Physical activity and health state (ECG), muscle (EMG) and eye (EOG) activation	Textile electrodes (Pola, 2007), dry electrodes (Lamparth, 2009; Bulling, 2009)

combinations of accelerometers, gyroscopes or magnetometers can in some cases be substituted for each other (Kunze *et al.*, 2010).

Sensorized textile fibers allow for truly unobtrusive garment-integrated on-body sensing. Mattmann *et al.* (2007) proposed the use of conductive elastic fibers as stretch sensors and showed that body postures can be detected based on the elongation of the fiber. Lukowicz *et al.* (2006) showed that pressure sensors can be used to measure muscle activity from which modes of locomotion can be inferred, or fatigue during sports exercise. They argue that the sensor is well suited for integration in garments, as force sensors can be realized as thin foils or even as textiles (Meyer *et al.*, 2006).

Smart textiles open the way to the development of new sensor modalities, which is pursued chiefly for two reasons: to further increase the usability of an existing system, or to enable the recognition of activities for which current modalities are not sufficient. Thus, new sensors are usually characterized in terms of the improvements they bring in usability and activity recognition performance. For instance, detecting swallowing without using implanted sensors is difficult. To address this, Cheng *et al.* (2010) developed a novel capacitive sensor, which may be integrated in close proximity to the neck in a shirt. This sensor can be used to detect drinking and eating from the movement of the throat during swallowing. As part of the evaluation procedure, the parameters of the sensor are characterized, its comfort evaluated, and the influence of body movements on the detection accuracy is quantified (Cheng *et al.*, 2010).

In order to enhance sensing capabilities, smart-textiles sensing can be combined with sensors that are readily available in commonly used gadgets. For instance, the user's location (e.g. using GPS or cellular network signals) is an important modality for context awareness. It does not lend itself well to be integrated into fibers but it is available in mobile phones. On-body microphones are also successfully used for activity recognition, as many human activities generate characteristic sounds (e.g. using a coffee machine, brushing teeth), and microphones are also available in mobile phones (Ward *et al.*, 2006). Mobile phones have a large number of other sensors to complement smart-textile sensing (Campbell *et al.*, 2010). Furthermore, they can be used as a computer to process the data from the smart-textile sensors and can realize the desired assistive function by providing user feedback (Want, 2009).

12.4 Principles of activity recognition

The purpose of activity recognition methods is to identify the patterns characteristic of an activity within the signals delivered by one or more sensors. The identification of a pattern indicates that the corresponding activity is likely to have been executed by the user. In this section we introduce the terminology commonly used in the field of activity recognition to distinguish different kinds of human activities and different kinds of recognition approaches.

12.4.1 Human activities

A wide range of different limb movements are commonly termed as 'activities', depending on their complexity or duration. Often, the terms 'action primitives' (or actions) and 'activities' are distinguished (Turaga *et al.*, 2008). Action primitives refer to simpler gestures (e.g. reaching for an object, raising a cup to the mouth to drink, toggling a switch), while activities are made up of a succession, interleaving or simultaneous execution of several action primitives spanning a longer duration. For instance, eating consists of using cutlery, bringing food to the mouth, chewing, swallowing, etc. To simplify, we will use 'activity' to refer to both action primitives and more complex sequences of primitives, unless the distinction is required.

Nevertheless, a precise definition of activity is elusive, as it involves perceptual aspects that are still being investigated (Blake and Shiffrar, 2007). These may lead observers of the same scene to describe it in different ways. Therefore activity-based computing usually defines activities from an application perspective: activities are sequences of body limb movements that – if they can be detected – can trigger useful behaviors in an application. Following this pragmatic approach, there is usually an agreement on the following broad categories of activities:

- *Body postures*: The subjects take a posture, which is static over a defined time period. Although static, the recognition of such postures is comprised in the broader umbrella of activity recognition. Examples of body postures are standing, lying or sitting (Randell, 2000; Ermes *et al.*, 2008) (Fig. 12.2). Also, finer distinctions in these postures can be made, such as seated and leaning forward or backward. Body postures may include particular limb configurations, such as when performing rehabilitation exercises (Mattmann *et al.*, 2007).
- *Modes of locomotion*: These activities relate to how the user moves around, e.g. by walking or running. Finer distinctions can also be made, such as walking upstairs or downstairs, or even jumping (Randell, 2000; Lester *et al.*, 2006; Figo *et al.*, 2010) (Fig. 12.2). Bicycling or driving may also be considered as modes of locomotion (Berchtold *et al.*, 2010).
- *Gestures*: The user performs a short gesture, such as toggling a light switch or bringing a cup to the mouth to drink. These gestures may be 'isolated' in the sense that the user returns to a rest position or marks a pause after the gesture is completed. This tends to be the case, e.g. in human-computer interaction scenarios with pointing gestures used to control a device (Martin *et al.*, 2010). Gestures may also be embedded among other user's activities. In this case, recognition tends to be harder, as the system has to disambiguate the gesture from other movements (Fig. 12.3).

Since activities are generally sensed with movement sensors (e.g. accelerometers), we can distinguish activities in terms of the nature of the resulting sensor signals:

- *Periodic nature*: This includes activities that exhibit periodicity (Fig. 12.2). Modes of locomotion usually exhibit such periodicity (e.g. walking, running, bicycling), but also some sports (e.g. step exercises, weight lifting). Such activities lend themselves to be detected in the frequency domain.
- *Static nature*: The activities are characterized by a signal that is relatively stable over time. This is typically the case with postures. These activities lend themselves to be detected by an analysis in the time domain (Fig. 12.2).

- *Sporadic nature*: The activities or gestures occur sporadically, interspersed with other activities or gestures. The detection of such activities usually requires a preliminary process of segmentation, which isolates the subset of data containing the activity, before doing a time or frequency analysis (Fig. 12.3).

In Fig. 12.2 and Fig. 12.3, we illustrate the sensor signals delivered by accelerometers placed on the body for the different activity categories introduced above. On longer timescales (e.g. days, weeks), human activities usually comprise a repetitive structure, where roughly similar activities are executed day after day. Such 'daily routines' are usually composed of a complex interleaving of lower-level activities (Blanke and Schiele, 2009). Thus, recognizing high-level activities builds upon the recognition of lower-level action primitives in a hierarchical manner (Hong *et al.*, 2009; Manzoor *et al.*, 2010).

12.4.2 Recognition system

Further distinctions can be made in the way that the machine recognition of activities operates. Activity recognition is done 'offline' when the system records the sensor data first, and the recognition is performed afterwards. Offline activity recognition is typically used for non-interactive or logging applications. For instance, the data can be collected over the day, and at night is uploaded to a server for analysis. Activity recognition is done 'online' when the system acquires sensor data and processes it on-the-fly. This is typically used for activity-based computing and interactive applications, where an immediate action must be carried out upon the detection of a specific activity.

Since sensors deliver continuous streams of data, the recognition system must identify at any time those parts of the data stream that are likely to correspond to the execution of an activity. This is referred to as 'continuous activity recognition'. When an activity is identified in the data sensor stream, it is said to be 'recognized' or 'spotted'.

Alternatively, in 'isolated activity recognition', the sensor data stream is 'segmented' at the start and end of each activity. In other words, the system gets to process only chunks of data that were acquired while the user performed exactly one activity, but the nature of the activity is unknown and must be determined. Thus, the system must 'classify' the sensor data into one of several predefined activity classes. The segmentation can be done by the experimenter when assessing classification performance during design phases. When the system is deployed, the segmentation can be done by an external system (e.g. cross-modality segmentation, where one sensor modality is used to segment the data stream of another sensor), or by a 'segmentation' algorithm. Continuous activity recognition combines the challenges of data segmentation with that of activity classification. Isolated activity recognition only deals with the problem of classification, and it is thus usually easier to address.

12.5 Signal processing and pattern analysis

The machine recognition of human activities consists in finding, in a stream of continuous sensor data, those points where the sensor signals are likely to be the result of the execution of an activity. Thus activity recognition is a problem of matching known patterns in large streams of data. To a large extent, the same principles also apply to the recognition of other elements of context, such as cognitive-affective states (e.g. detection of emotions (Picard, 1997) or of memory recall (Bulling *et al.*, 2010)), although the sensors used would be different. Despite the wide variety of sensors and activities, most of the methods used in the works mentioned previously can be cast into a common data processing architecture. We refer to it as the 'activity recognition chain' or ARC (Roggen *et al.*, 2011(b)) (Fig. 12.4). It is a roughly common processing structure that has emerged across most published work in activity recognition. It is based on streaming signal processing, machine learning and reasoning techniques (Bao and Intille, 2004; Ward *et al.*, 2006; Bettini *et al.*, 2010; Figo *et al.*, 2010). The raw sensor data is mapped to the occurrence of action primitive (events) with signal processing and machine-learning techniques. Here five sensors deliver data. Data fusion is illustrated at the feature, classifier and decision level. Optionally, symbolic processing can be used to infer higher-level activities from the occurrence of action primitives, usually with reasoning or statistical approaches.

The specific parameters and methods within the ARC are selected using a 'learning by demonstration' approach. In this approach, the user is instrumented with the selected sensors and put into a situation where he performs the activities of interest at design time. The sensor data is acquired with ground-truth annotations describing what the user performs or experiences. The resulting dataset is used to train the recognition system and test its performance. The training process consists of identifying the prototypical signal patterns corresponding to the user's activities, and then to select the methods and their parameters that best allows identification of these patterns in the sensor data stream. At run-time, the ARC essentially 'compares' the streaming sensor signals to prototypical activities. These prototypical activities are also called activity models (Section 12.6.1). The ARC identifies sensor patterns to match sufficiently the activity models to indicate that an activity has been 'spotted'.

The ARC illustrated in Fig. 12.4 shows the typical data processing steps from the input, which is the raw sensor signals, to the output, which is a series of events indicating the occurrence of an activity. First, sub-symbolic processing maps the raw sensor data (e.g. body-limb acceleration) to semantically meaningful action primitives (e.g. grasp). This is realized by streaming signal processing and machine-learning techniques. The outcomes of the sub-symbolic processing are events indicating the occurrence of action primitives.

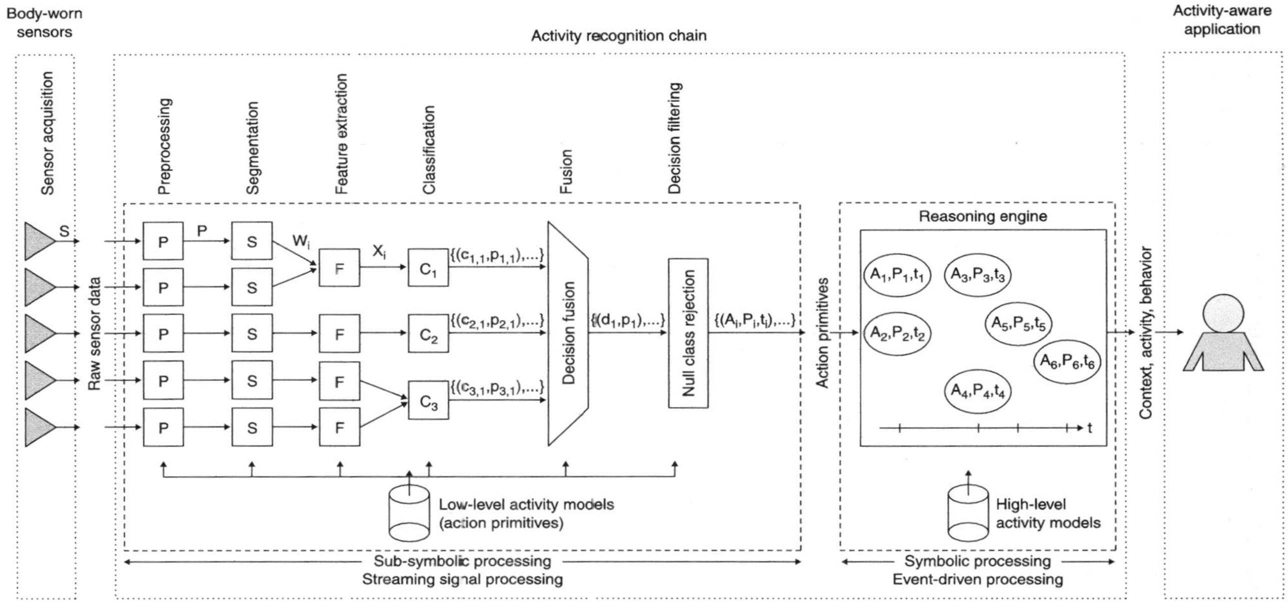

12.4 Processing steps used to infer activities from on-body sensors.

The ARC terminates at this stage when the activities of interest consist of simple gestures, for instance used for gestural interfaces (Kallio *et al.*, 2006). When more complex activities must be detected, symbolic reasoning can be used in a second stage to infer higher level activities. However, this is outside of the scope of this article.

Sub-symbolic processing ought to be robust to the large observed variability in sensor-signal to activity-class mapping resulting from variability in human behaviors. It must also be robust to variability in sensor characteristics and to deployment variability (e.g. displacement along limbs every time a garment is worn). This is especially important in a smart-textile approach to sensing, where the variability in sensor characteristics may be larger than for conventional sensor nodes, and may change over time due to the harsh conditions of use. In any case, the ARC (i.e. the specific methods that are applied and their parameters) is usually co-optimized with the sensors that are used, to maximize comfort, as well as to recognition performance.

The sub-symbolic processing stages are described below.

12.5.1 Sensor data acquisition

A time series corresponding to the sensor data is obtained. Since sensors can provide multiple values (e.g. an acceleration sensor provides a three-dimensional (3-D) vectorial acceleration), multiple sensors are jointly sampled and vector notation is used:

$$S = \{\vec{s}_0, \vec{s}_1, \vec{s}_2, \ldots\}. \qquad [12.1]$$

12.5.2 Signal pre-processing

Basic signal pre-processing is applied to the sensor data. The objective is to increase the signal-to-noise ratio. Typical pre-processing include calibration or normalization of sensor values, and various signal processing operations such as de-noising, bandpass filtering, discriminant analysis, etc. Sensor-specific processing can be performed. For instance, the l^2-norm of the values delivered by an accelerometer can be computed when the direction of the acceleration is not relevant. Other forms of pre-processing involve the conversion of the sensor values to other physical units. For instance, IMUs generally deliver quaternions indicating the sensor orientation. When several IMUs are placed on multiple limbs, it becomes possible to map the orientation of the IMUs to the spatial coordinate of the limb based on a body model (Zinnen *et al.*, 2009). This is a form of sensor fusion occurring at the signal level.

The outcome is pre-processed time series P:

$$P = \{\vec{p}_0, \vec{p}_1, \vec{p}_2, \ldots\}. \qquad [12.2]$$

12.5.3 Segmentation

The data stream is segmented into sections that are likely to contain an activity. Segment i is delimited by its start time t_i^s and end time t_i^e within the time series, yielding a segmented time series W_i:

$$W_i = \{\vec{p}_{t_i^s}, \ldots, \vec{p}_{t_i^e}\}. \qquad [12.3]$$

A common type of segmentation technique is the sliding window. In that case, the window has a fixed size w and we have $t_i^s = t_i^e - w$. The sliding window segmentation is typically used when the sensor signals are best analyzed in the frequency domain. This is the case for many periodic movements (e.g. walking, running, biking, rowing). A sliding window can also be used with static movements. In that case, it is used to increase the robustness of the subsequent feature extraction, in comparison to a sample by sample processing. In particular, it allows increased robustness to noise (e.g. by smoothing outliers resulting from communication errors or motion artifacts). Other segmentation methods include energy-based or rest-position based segmentation, when the user performs isolated gestures or returns to a rest position between gestures (Roggen *et al.*, 2010(b)). Other segmentation methods have been proposed in the context of time series analysis (Keogh *et al.*, 2001; Armstrong and Oates, 2007), online learning schemes (Kulic *et al.*, 2009) and hybrid segmentation-classification schemes (Deng and Tsui, 2000; Stiefmeier *et al.*, 2008).

12.5.4 Feature extraction

Features are computed on the identified segments. This is done to reduce data dimensionality, and to better discriminate activities of interest. The result is a feature vector $\vec{X}_i$, often represented as a point in a feature space:

$$\vec{X}_i = \Psi(W_i) \qquad [12.4]$$

A large set of features can be computed from the sensor signal. Knowing which must be computed is an explorative process. It benefits from past experience, 'educated guesses' supported by visual inspection of the data, or computational approaches (Guyon *et al.*, 2006). Evolutionary algorithms can also be used to explore a greater set of features than what is humanly possible to envision (Förster *et al.*, 2009(a)). A taxonomy of features commonly used with acceleration sensors has been proposed in Figo *et al.* (2010). It applies to a large extent to other sensor data (e.g. data from gyroscopes, magnetic field sensors, stretch sensors). Usually feature extraction attempts to generate an exhaustive set of candidate features. This set is then pruned to include only the most relevant features using feature selection heuristics (Guyon and Elisseeff, 2003).

12.5.5 Classification

The feature vector $\vec{X}_i$ is mapped into one output class (i.e. the user activity) c_i, using pattern classification or machine-learning methods (Duda *et al.*, 2000):

$$\vec{X}_i \to c_i, p_i \qquad [12.5]$$

Usually, classification also yields an indication as to the confidence in the classifier output. This is often a probability p_i with Bayesian approaches. Other classifiers may be calibrated to provide probabilistic outputs (Cohen and Goldszmidt, 2004).

12.5.6 Decision fusion

Decision fusion is one form of data fusion that combines the decisions of multiple classifiers into a common decision about the activity that occurred. Multiple classifiers are typically used with multi-modal sensors, when it is difficult to combine the sensor's data in the same feature vector (Sharma, 1998). In that case, the activities are first spotted in each sensor modality individually, then these individual decisions are combined. Generally multiple sensors are improving the recognition of complex real-world activities (Stiefmeier *et al.*, 2007). A detailed introduction to classifier fusion is found in Polikar (2006).

12.5.7 Null-class rejection

In principle, the segmentation stage should identify sections of the signal corresponding to an activity. In practice, segmentation may be inaccurate. For example, a data segment may cover a time span where 'no activity' took place. This is the case if a sliding window approach is used with sporadic activities such as short gestures. Null-class rejection attempts to discard these segments by assessing how likely it is that the data corresponds to an activity described by the classifier models. Thus, in cases where the confidence in the classification result is too low, the system may discard the classified activity i based on p_i, in the simplest case by comparison to a threshold (Lee and Kim, 1999; Sagha *et al.*, 2011). At this stage, the outcome is the detection of an activity A_i with likelihood p_i at time t_i.

12.5.8 Method and parameter selection

The specific methods for each of the elements outlined above and their parameters are defined at design time through an interactive or automated optimization process. The parameters subject to optimization include, for example, the window sizes, the thresholds to segment activities or reject the null class, the signal normalization parameters, the set of features to compute, and the classifier models. The objective of this optimization is usually to reach high activity recognition accuracy. This typically translates to increasing the separation and thus reducing the overlap between the distribution of the activity classes in the feature space. Power, computational requirements, and user comfort must also be taken into account when devising wearable activity recognition systems. Usually heuristics are available to find these parameters. Many methods exist for feature selection (Guyon and Elisseeff, 2003) and classifiers usually have a corresponding 'training' heuristic (Duda *et al.*, 2000).

For activities that are of periodic nature (e.g. walking, running, bicycling, rowing, hammering) a sliding window segmentation is generally used. The features that are used are then selected to capture the repetitive nature of the activity (e.g. zero or mean crossing rate, power spectrum, dominant frequency (Figo *et al.*, 2010)). With static postures, a sliding window approach is also commonly used with time-domain or statistical features (e.g. limb angle or angle between several limbs, means of acceleration). For periodic activities and static postures, the classifiers typically used include Support Vector Machines (Qian *et al.*, 2010; Bulling *et al.*, 2011(a,b)), decision trees, k-Nearest Neighbor or Naive Bayes classifiers (Randell, 2000). When the activities are sporadic (i.e. they are short and occur interleaved with other activities, which the system does not need to recognize) a sliding window of appropriate size (commensurate with the length of the activity), together with a null-class rejection, can be used (Sagha *et al.*, 2011). Alternatively, segmentation and classification techniques that take the temporal dynamics of the sensor signal into account can be used. These techniques include, for example, hidden Markov models (HMMs) (Starner *et al.* 1998; Deng and Tsui, 2000), dynamic time warping (Ko *et al.*, 2005; Stiefmeier *et al.*, 2008), methods based on feature similarities (Keogh *et al.*, 2001) or neural networks (Bailador *et al.*, 2007; Yang *et al.*, 2008). Further methods are mentioned in Figo *et al.* (2010).

In Fig. 12.5, we illustrate a set of activities carried out by an industrial worker during the quality assurance check of the car manufacturing process (Stiefmeier *et al.*, 2008). The data of an acceleration sensor integrated into the garment at the right wrist is depicted. We can note the variability in the gesture execution length and signal shape. Related activities, such as opening and closing the door (C5 and C6), show sensor signals with subtle differences. Unrelated activities, such as opening a door (C5) and checking the engine hood (C3), show markedly different signals. With simple statistical features, the sensor signals can be projected in a feature space where the activities form clusters suitable for classification. Here two possible feature spaces are represented. The features are the average value of the accelerometer reading on two or three axes of the sensor (this is an approximation of the average angle of the user's right lower arm). This yields a two-dimensional (2-D) and a 3-D feature space. Some activities are well separated, leading to accurate classification (e.g. C10 and C1 in the 3-D feature space), while others overlap as they are more similar (e.g. C5 and C6). In this example, the increase of the dimension of the feature space allows to better separate classes

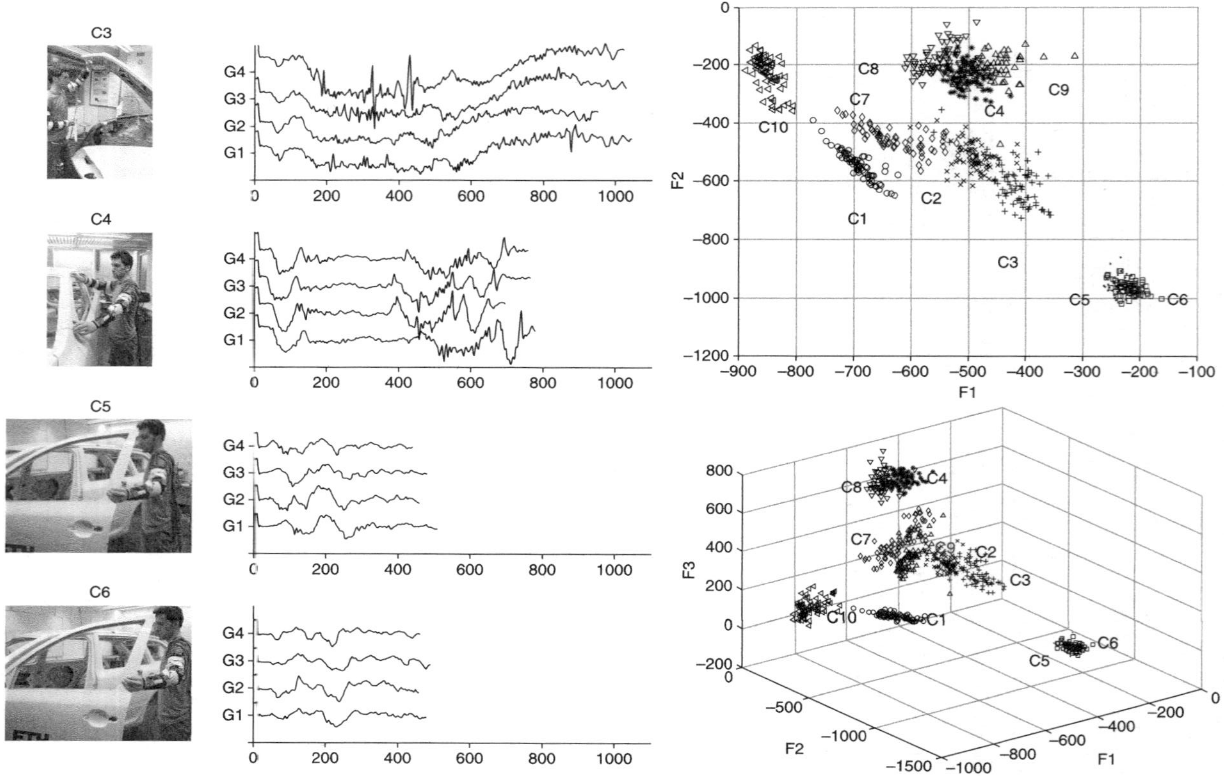

12.5 Four activities of a car assembly scenario are shown on the left: checking the engine hood (C3), the gap spacing between doors and car body (C4), the opening of the front door (C5), and the closing of the front door (C6). In the middle, the data of an acceleration sensor attached to the garment at the right wrist is shown for four repetitions (G1–G4) of the activity. After feature extraction, the sensor signals are projected into a feature space for pattern recognition. Here the projection in a two-dimensional (right up) and three-dimensional (right down) feature space is represented (sensor data from Zappi, 2008).

C1 and C7, which overlap in the 2-D feature space, but are well separated in the 3-D feature space. During the training of the recognition chain, the method and parameter selection aims at increasing the separation between the activity classes.

12.5.9 Addressing the challenges of smart textiles

In Section 12.3.3, we outlined a few of the challenges arising from the use of smart textiles for sensing, such as variability and deterioration of sensor characteristics, sensor failure and motion artifacts. Without precautions, these issues can adversely affect recognition performance once a system is deployed. We outline here a few approaches that can be employed to mitigate these effects. These approaches are also applicable when conventional sensor nodes are used.

Variability in sensor parameters is a result of the tolerances of the manufacturing process. One way to address this variability is to perform a *calibration* of the sensor (typically at production time) by characterizing and then compensating for differing sensitivity, offset and hysteresis prior to using it for activity recognition (Mattmann *et al.*, 2008; Meyer *et al.*, 2010). Sensor calibration may also be performed at the same time as user calibration (the process by which a generic recognition system is tuned for a specific user). Some recent user calibration methods were proposed in Ohmura *et al.* (2009) and Maekawall *et al.* (2011).

The variation of parameters over time may lead to a drift of activity classes in the feature space (concept drift). Addressing this needs a form of *online self-adaptation*. Self-adaptation methods were proposed for activity recognition in Förster *et al.* (2009(b)) and Bayati *et al.* (2011), to cope with sensor placement variability. These methods are related to expectation maximization and are also envisioned for adaptation to other forms of concept drift. Such methods must achieve an equilibrium between stability (the capacity to retain the current activity model) and plasticity (the capacity to adapt the model) by matching the speed of adaptation to the characteristics of the concept drift. These adaptation parameters can be found at design time from the characterization of the long-term properties of batches of sensors. Further details on an extension to the ARC to make it adaptive are found in Roggen *et al.* (2011(b)).

There is a thin line between a sensor whose parameters have significantly degraded, and a sensor which has failed. Since fiber-integrated sensors are often passive analog sensors (e.g. using resistive or capacitive sensing principles), they are not able to report whether they have failed, except obvious cases where sensing a high impedance indicates a loss of connectivity (e.g. when cracks form in a conductive material) (Kinkeldei *et al.*, 2011(b)). For activity recognition, it is important to estimate whether the data delivered by the sensor may come from a functional sensor, or whether the sensor is damaged and delivers data that should be discarded. The fields of *fault detection* and *anomaly detection* have investigated a wide range of methods, which can be used to assess whether the data delivered by a sensor are plausible for the given sensor. Many of these methods are reviewed in Betta and Pietrosanto (2000), Chandola *et al.* (2009) and Gage and Murphy (2010). Problematic sensors can be also identified when activities are recognized, by fusing the decisions taken by multiple ARCs operating on individual nodes (Sagha *et al.*, 2011(b,c)).

Motion artifacts are noise sources indirectly resulting from user movement. As the user moves, the pressure of the sensor against the skin may vary around a nominal value. Changing contact pressure against the skin is a common source of noise when measuring ECG or EMG from electrodes due to the variations in impedance induced by the movement. These motion artifacts can be identified using fault and anomaly detection methods, as mentioned above. As smart textiles can host a large variety of sensors unobtrusively, another approach is to use an additional motion sensor to perform *signal quality appraisal* (Schumm *et al.*, 2011).

Motion artifacts also affect movement sensors. A motion sensor may be displaced or become subject to rotations around its nominal position and orientation. This can occur when motion sensors are integrated into loose-fitting clothes. In that case, the orientation of a sensor may not reflect the orientation of the limb, but that of wrinkles of the clothing. This may be compensated by modeling techniques (Harms *et al.*, 2010(a)). Sensor displacement can also be compensated by a combination of multiple sensor modalities and a body model (Kunze and Lukowicz, 2008).

When the fibers themselves are functionalized, a large number of sensors may become available. The large amount of data may be challenging to process. *Feature and sensor selection techniques* can be used to reduce the amount of information that is subsequently processed. This is used, for instance, in Meyer *et al.* (2010) to select appropriate textile pressure sensing elements to monitor sitting behavior.

While having a large number of sensors can lead to increased computational requirements, it is also a unique characteristic of smart textiles that can be exploited. In order to do this, a typical practice followed with conventional sensor nodes must be overturned. With conventional nodes, the designer attempts to maximize user comfort by using only the minimal set of sensors sufficient to recognize the activities of interest. With smart textiles, the data processing paradigm rather shifts towards making the best use of a large amount of sensors readily available. This is achieved by relying on data fusion techniques. This in turn can address some of the challenges discussed above.

Setz *et al.* (2009) present a system to recognize emotions based on modalities, which could be integrated into textiles: ECG, EMG, EOG, galvanic skin response and respiration rate. Due to motion artifacts, these sensors deliver sporadically-corrupted data. By relying on the fusion of decisions taken by classifiers operating on each sensor modality individually, the authors could significantly improve the performance of the system, despite the motion artifacts. This approach can be chiefly generalized to other context recognition problems and other sensor modalities.

Zappi *et al.* (2007) showed that by fusing the decision of individual classifiers, the activity recognition performance can be significantly improved. They showed that the effect of unexpected sensor rotation, such as those that might occur if there are wrinkles in the clothing, can also be addressed. Finally, data fusion also improves the robustness to sensor failures. This work was extended to show that the sensor fusion approach can be dynamically adapted at run-time to achieve desired power-performance tradeoffs. This allows achieving a desired target performance while minimizing energy use (Zappi *et al.*, 2008). Furthermore, the system is able to replace at run-time faulty sensors by others to maintain the desired power-performance trade-off. Information theoretical principles can also be used to form optimal combinations of sensors for activity recognition (Chavarriaga *et al.*, 2011). Other examples of distributed fault-tolerant classification can be found in Wang *et al.* (2005).

In summary, one main drawback of smart textiles, compared to dedicated sensor nodes, is the higher variability in sensor characteristics due to manufacturing aspects, usage patterns and textile-embedding conditions. A promising way to handle these challenges is to capitalize on the characteristic of smart textiles: their capacity to embed a large number of sensors with high comfort. This requires shifting the activity and context recognition paradigm to a more prevalent use of data fusion techniques, rather than the traditional attempt to minimize the sets of sensors used. In this way, the variability issues can be compensated for by capitalizing on the large number of available sensors.

12.6 Experimental aspects

12.6.1 Data recording and ground truth annotation

A reference dataset comprising annotated activities (i.e. their start and end point is identified) is required to train and optimize the ARC, and to evaluate the recognition performance. Reference datasets can be recorded in a continuum from laboratory environments to daily life situations.

Recordings in the laboratory are typically done to ensure reproducible execution of activities without external influences. Laboratory set-ups are well suited to acquire annotations, as an experimenter can observe the scene and video footages can be made without privacy issues. Reference datasets can also be collected in daily life. This ensures a realistic execution of activities, which usually show a higher variability than in laboratory set-ups. Such recordings are more challenging to conduct. Experimenters must take care of technical issues (e.g. sensors that might be displaced while worn, or that run out of energy) and user aspects, in particular privacy issues. Annotation in daily life is also more challenging as video footage is often inappropriate (e.g. when recording data at the subject's home, or in the streets where other persons may appear in the footage). An intermediate approach is recordings in naturalistic environments. Such an environment is a laboratory environment designed to allow a set of highly realistic activities (Roggen *et al.*, 2010(a)).

Reference datasets need to capture the activity variability likely to be encountered when the system is deployed to ensure reliable training of activity models. The types of variability that must be captured include, among others, the inter-user and intra-user activity variability, and the sensor placement variability. Thus, ideally, datasets must include activities from multiple users, captured over multiple days or multiple deployments of the on-body sensors. The size of the reference dataset (number of repetitions of each activity) is a function of the expected variability, which can be assessed in a preliminary small-scale recording.

If multiple sensors are used but are not synchronized at the hardware level, offline synchronization is performed. Video footage assists greatly in this task. The subject is often asked to perform specific movement patterns at the start and end of the recording, such as clapping the hands (Bannach *et al.*, 2009; Roggen *et al.*, 2010(a)). This results in sound, a characteristic movement pattern visible in acceleration signals measured on body, and a clear visual identification of the moment of the clap (Roggen *et al.*, 2010(a)).

Annotation consists of identifying the starting and ending time of the activities of interest in the sensor data stream (see Fig. 12.2 and Fig. 12.3, for examples of annotated data streams). Data annotation can be done in real time by an expert observing the situation. However, this is difficult. Thus, annotation is preferably done offline using video footage documenting the data acquisition and synchronized with the sensor data recording. Short activities and transitions between activities are difficult to precisely identify, as human activities tend to be highly interleaved, and are also subject to different semantic interpretation. Labeling consistency is important. Thus, when possible, multiple experts should annotate the video footage. It allows quality control on the resulting annotations. When multiple persons share the work of labeling a dataset, it is recommended to arrange classroom sessions where it is explained to all persons how to perform the task.

We developed the concept of label tracks (Roggen *et al.*, 2010(a)) to represent a wide range of activities in a structured manner (Fig. 12.6). Hand activities are annotated by a pair of action tags and an object tag. An action tag indicates primitive activities, such as 'reach', 'open', 'move', 'grasp'. An object tag indicates the object that the action applies to. Another label track represents the modes of locomotion: 'walk', 'stand', 'sit', etc. Since the label tracks are independent of each other, new label tracks can be seamlessly added. In daily-life recordings, the annotations may be provided by the subject himself. This may occur in real time, *a posteriori* with a time diary (Van Laerhoven *et al.*, 2008), or by using 'experience sampling', which consists in asking the subject at regular intervals to annotate his activities (Froehlich *et al.*, 2007). Tools such as Anvil[1]

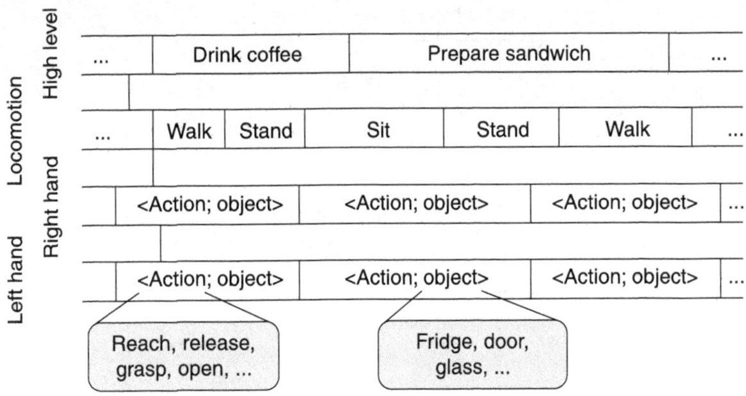

12.6 Activity labeling approach composed of multiple 'tracks', which indicate at each time step the ongoing activity of the user. This approach based on tracks allows independent labeling of different aspects of the user's activities in a scalable manner. Here we show a high-level activity track, modes of locomotion track, and two tracks for the action of the left or right hand.

can be used for annotation. Some groups are developing customized tools for data acquisition[2] and annotation (Bannach *et al.*, 2010).

In Roggen *et al.* (2010(a)) we describe a rich dataset of activities conducted in a kitchen environment. This dataset is annotated and available for use by the community. It has been used to organize an 'activity recognition challenge' within the System, Man, and Cybernetics conference in 2011, where several research groups took up the challenge of comparing their respective recognition methods (Sagha *et al.*, 2011(a)).

12.6.2 Performance evaluation

The evaluation of activity recognition systems is done along user aspects and objective performance and cost metrics, based on a reference dataset with ground truth annotations.

User aspects include user acceptance, comfort, etc. These subjective elements are assessed in user studies by means of questionnaires and expert interview (Gemperle *et al.*, 1998; Bächlin *et al.*, 2010; Franke *et al.*, 2011).

Depending on whether the activity recognition is continuous or isolated, different sets of performance metrics are used. For isolated activity recognition (i.e. when the sensor data streams are pre-segmented) the performance metrics are those used in machine learning (Duda *et al.*, 2000). The accuracy is commonly used to describe the performance. It indicates the number of correctly recognized activity instances divided by the total number of activity instances. The confusion matrix indicates for each activity class the distribution of the predictions made by the recognition system (Fig. 12.7(a)). The diagonal of the matrix contains the

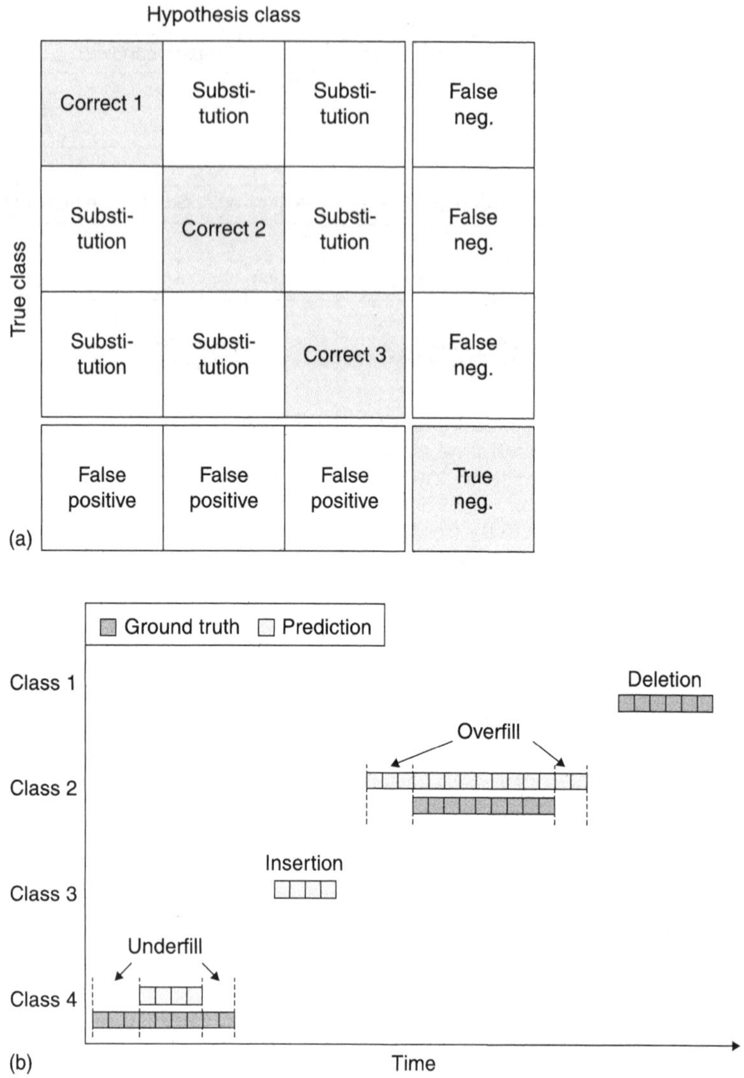

12.7 Performance metrics of activity recognition. (a) Confusion matrix. (b) Continuous recognition performance.

proportion of correctly recognized activity instances, while the off-diagonal elements denote the proportion of activity instances that have been confused with another activity. The off-diagonal elements can be summed up along the recognized activities column to obtain the false positives and along the true class row to obtain the false negatives.

In continuous activity recognition, the activities are interspersed by periods where no relevant activity occurs (null class). In that case, 'recognition accuracy', as defined above, is meaningless and metrics suited for continuous activity recognition must be used. From the confusion matrix, the following metrics can be derived: precision (true positives divided by positives), recall or sensitivity (true positives divided by true positives and false negatives) and specificity (true negatives divided by true negatives and false positives). Ward *et al.* (2011) introduced a *de facto* standard set of performance metrics specifically for continuous activity recognition (Fig. 12.7(b)). These metrics include 'insertions': activities that have been recognized but have not happened in reality; 'deletions': activities that happened in reality but have not been recognized; 'merges': when multiple activity instances are recognized as only one; 'fragmentations': when single activity instances is recognized as several instances. Finally, 'overfill' and 'underfill' describe how well the duration of an activity instance has been recognized.

The recognition latency is another important performance criterion, typically in gesture-based HFCs. Since power consumption affects battery size, the system evaluation should also include the energy required for activity recognition. Minimizing the number of recharges is especially important for continuous and long-term (week to month) activity monitoring (Van Laerhoven *et al.*, 2008). Additional ways in which recognition systems can be evaluated along 'cost and quality' metrics are presented in Villalonga *et al.* (2009).

Evaluation is typically conducted using cross-validation techniques to assess how the recognition system generalizes to a new situation. First, the reference dataset is partitioned into multiple folds. In a 'leave-one-X-out' cross-validation, all folds except one are used to train the recognition system. The left-out fold is used for testing. The process is repeated rotating the left-out fold until all folds have been used once for testing. Datasets may include recordings of multiple persons, on multiple days, and of multiple 'runs' containing repetitions of a set of activities. Folds are built differently to assess different aspects of generalization. Leave-one-person-out is used to assess generalization to an unseen user. This is done when designing a user-independent recognition system. Leave-one-run-out is used to assess a user-specific system. Since the way activities are executed may change over time, leave-one-day-out could be used to assess the robustness of the system over time.

12.6.3 Technical implementation, networking and prototyping

The technical implementation of activity recognition comprises elements such as databases for the storage of a dataset, the technical realization of networked sensing, and the implementation and optimization of the recognition algorithms.

Sharing datasets is identified as a key issue in the community to ensure reproducible machine-learning research (Sagha *et al.*, 2011(a)). This requires storing multimodal sensor data (including audio and video) in a standardized, synchronized and easy to retrieve manner, together with activity annotations. A set of open tools for data annotation, visual inspection and sharing of the dataset has been recently proposed to stimulate reproducible research (Bannach *et al.*, 2009).

Usually, there is no attempt to embed the context recognition processing in smart textiles. The dominant paradigm consists in using a mobile phone on the body as a 'wearable computer', which is running the recognition algorithms. Thus sensors are interfaced to the mobile device over a wireless gateway (Mattmann *et al.*, 2007). Prior to the gateway, multiple body worn sensors are interconnected by wires or wirelessly on the body. Wired connection may be employed to reduce the risks of data loss. This is possible in garments by hiding the wiring between textile layers (Harms *et al.*, 2009). Some textiles also offer built-in conductive wires (Locher *et al.*, 2004) and flexible substrate hosting sensors may also serve as interconnects (Zysset *et al.*, 2010). For greater flexibility, wireless interconnects can be used (Benini *et al.*, 2006). Usually the wireless sensors are directly connected to the wearable computer within one hop in a star network (Want, 2009). The typical wireless interconnections used include Bluetooth[3] (Roggen *et al.*, 2010(b)), Zigbee[4] (Calatroni *et al.*, 2009) or Ant[5] (Harms *et al.*, 2010(b)), all operating in the ISM band. Special care must be taken with wireless sensors to ensure the desired quality of data (Roggen *et al.*, 2010(a)). Several designs have been proposed for planar patch antennas based on textile materials (Locher *et al.*, 2006; Hertleer *et al.*, 2007). This allows the realization of sensing and communication with smart textiles.

Finally, software frameworks support the implementation of activity recognition algorithms and reduce the complexity of devising new algorithms. Frameworks focusing on the sub-symbolic network-oriented data processing include the CRN Toolbox[6] (Bannach *et al.*, 2008), TITAN[7] (Roggen *et al.*, 2011(c)) or SPINE[8] (Fortino *et al.*, 2009). Machine learning toolboxes, such as WEKA[9] or MOA[10], can be used to assist in the process of selecting features, training classifiers and assessing the system's classification performance. They also facilitate the comparative analysis of different methods.

12.7 Future trends

One major area of research is to devise systems that are portable across domains. For instance, rather than devising a recognition system for a specific garment, it should be able to generalize to any garment. The challenges result from the likely different sets of sensors and their placement offered in garments of different origins. Methods that are robust to this variability are investigated, among others, within the EU project OPPORTUNITY (Roggen *et al.*, 2009) and recent approaches are reviewed in Roggen *et al.* (2011(d)).

In past work, multimodal sensing was shown to increase activity recognition performance (Stiefmeier *et al.*, 2008). Smart garments have the capacity to hold a wide range of multimodal sensors in close contact to the skin. Thus, multi-modal data fusion for activity recognition is an important area of ongoing research.

The monitoring of human activities over a long period of time is extremely promising as a motivational tool to support persons that want to change their lifestyle. It is also relevant to quantify the progress of a rehabilitation procedure, or to assess health trends in the elderly. This requires sensors permanently on-body. Smart garments are ideally suited to fulfill this role. Translating laboratory prototypes of smart textiles to robust garments, which can be deployed to end-users, is key to pursuing this line of research.

Finally, smart textiles offer new possibilities for ambient intelligence, in particular for sensing on a large scale. Cheap industrial manufacturing means that many textiles found in our daily environments may be functionalized. A few examples illustrate the breadth of possible applications. Chair and seat coverings are made of textiles, which could sense human activities or physiological parameters at home, at work, in a car or in an airplane (Schumm *et al.*, 2011). Curtains are made of textile and could sense the amount of ambient and sun light, radiations, temperature and humidity for HVAC control. Carpets are found in many buildings and could offer user localization (Ueoka *et al.*, 2009) or gait pattern analysis to assess risk of falls in the elderly (Aud *et al.*, 2010). Power is less of an issue in ambient intelligence, as the sensing systems are deployed in a fixed environment. Thus, the aspects of miniaturization, computational limitations or comfort that we discussed previously, are less relevant or do not apply altogether. Besides this, the data processing methods that we introduced can be translated to context awareness in ambient intelligence environments with little differences, and the methodology to design and assess a context-aware system is essentially similar between ambient intelligence or on-body sensing.

12.8 Sources of further information and advice

We invite the interested readers to look for further information on activity-aware and context-aware systems in wearable computing and ambient intelligence in the following conference proceedings: *International Symposium on Wearable Computers; International Conference on Pervasive Computing*; and *International Conference on Ubiquitous Computing*. The following journals also cover the topic: IEEE *Pervasive Computing* Magazine; *Personal and Ubiquitous Computing* (Springer); and *Pervasive and Mobile Computing* (Elsevier).

There are a number of miniaturized sensors that can be used to prototype activity-aware smart garments. Sparkfun[11] and Phidgets[12] offer components for prototyping. The LilyPad[13] is a particularly interesting platform that offers components on a textile substrate, which is meant to be integrated into clothing. Commercial high-quality motion sensors include those from Xsens[14], Intersense,[15] MotionNode[16] or APDM.[17] Some hardware is also open source [18,19](Roggen *et al.*, 2010(b)).

A number of activity datasets have been made public and can be used to prototype recognition systems. A few well-known datasets include Intille *et al.* (2006), van Kasteren *et al.* (2008), Huynh *et al.* (2008), Tenorth *et al.* (2009) and Kawaguchi *et al.* (2011). Within the EU project, OPPORTUNITY, we collected a sensor-rich and activity-rich dataset comprising daily morning activities. This dataset has been used by multiple groups to benchmark activity recognition methods and can be obtained (Roggen *et al.*, 2010(a); Sagha *et al.*, 2011(b)).

12.9 Acknowledgments

We acknowledge the financial support of the Seventh Framework Programme for Research of the European Commission, under the project OPPORTUNITY with grant number 225938.

12.10 Notes

[1] http://www.anvil-software.de/
[2] http://wiki.esl.fim.uni-passau.de/index.php/Context_Logger_for_iPhone_OS
[3] http://www.bluetooth.com
[4] http://www.zigbee.org/
[5] http://www.thisisant.com/
[6] http://crnt.sourceforge.net
[7] http://code.google.com/p/titan/
[8] http://spine.tilab.com/
[9] http://www.cs.waikato.ac.nz/ml/weka
[10] http://moa.cs.waikato.ac.nz
[11] http://www.sparkfun.com
[12] http://www.phidgets.com
[13] http://web.media.mit.edu/~leah/LilyPad
[14] http://www.xsens.com
[15] http://www.intersense.com
[16] http://www.motionnode.com
[17] http://apdm.com
[18] https://sites.google.com/a/mis.tu-darmstadt.de/porcupine
[19] http://fiji.eecs.harvard.edu/CodeBlue

12.11 References

Armstrong T and Oates T (2007), 'UNDERTOW: multi-level segmentation of real-valued time series', *Proceedings of the AAAI*, pp. 1842–1843.

Aud M A, Carmen C, Abbott C, Tyrer H W, Neelgund R V, *et al.* (2010), 'Smart carpet: developing a sensor system to detect falls and summon assistance', *J Geron Nurs*, 36(7), 8–12.

Bächlin M, Förster K and Tröster G (2009), 'SwimMaster: a wearable assistant for swimmer', *Proceedings of the 11th International Conference on Ubiquitous Computing*, pp. 215–224.

Bächlin M, Plotnik M, Roggen D, Maidan I, Hausdorff J, *et al.* (2010), 'Wearable assistant for Parkinson's Disease patients with the freezing of gait symptom', *IEEE Trans Info Tech Biomed*, 14(2), 436–446.

Bailador G, Roggen D, Tröster G and Triviño G (2007), 'Real time gesture recognition using continuous time recurrent neural networks', *Proceedings of the 2nd International Conference on Body Area Networks*, available from: http://dl.acm.org/citation.cfm?id=1460247"article 15"

Bannach D, Amft O and Lukowicz P (2008), 'Rapid prototyping of activity recognition applications', *IEEE Perv Comp*, 7(2), 22–31.

Bannach D, Amft O and Lukowicz P (2009), 'Automatic event-based synchronization of multimodal data streams from wearable and ambient sensors', *Proceedings of the European Conference on Smart Sensing and Context*, pp. 135–148.

Bannach D, Kunze K, Weppner J and Lukowicz P (2010), 'Integrated tool chain for recording and handling large, multimodal context recognition datasets', *Proceedings of the 12th ACM International Conference Adjunct Papers on Ubiquitous Computing*, pp. 357–358.

Bao L and Intille S S (2004), 'Activity recognition from user-annotated acceleration data', *Pervasive Computing: Proceedings of the 2nd International Conference*, April, pp. 1–17.

Bayati H J, Millán J d R and Chavarriaga R (2011), 'Unsupervised adaptation to on-body sensor displacement in acceleration-based activity recognition', *Proceedings of the 15th International Symposium on Wearable Computers*, pp. 71–78.

Benini L, Farella E and Guiducci C (2006), 'Wireless sensor networks: enabling technology for ambient intelligence', *Microelectron J*, 37(12), 1639–1649.

Berchtold M, Budde M, Gordon D, Schmidtke H and Beigl M (2010), 'ActiServ: activity recognition service for mobile phones', *Proceedings of the 14th International Symposium on Wearable Computers,* pp. 1–8., available from: http://ieeexplore.ieee.org/xpls/abs_all.jsp?arnumber=5665868&tag=1

Betta G and Pietrosanto A (2000), 'Instrument fault detection and isolation: state of the art and new research trends', *IEEE Trans Instrum Meas*, 49(1), 100–107.

Bettini C, Brdiczka O, Henricksen K, Indulska J and Nicklas D, *et al.* (2010), 'A survey of context modelling and reasoning techniques', *Perv Mobile Comp*, 6(2), 161–180.

Blake R and Shiffrar M (2007), 'Perception of human motion', *Ann Rev Psychol*, 58, 47–73.

Blanke U and Schiele B (2009), 'Daily routine recognition through activity spotting', *Proceedings of Location and Context Awareness*, pp. 192–206.

Bulling A, Roggen D and Tröster G (2009), 'Wearable EOG goggles: seamless sensing and context-awareness in everyday environments', *J Amb Intell Smart Envir*, 1(2), 157–171.

Bulling A, Roggen D and Tröster G (2011a), 'What's in the eyes for context-awareness?', *IEEE Perv Comp*, 10(2), 48–57.

Bulling A, Ward J A, Gellersen H and Tröster G (2011b), 'Eye movement analysis for activity recognition using electro-oculography', *IEEE Trans Patt Anal Mach Intell*, 33(4), 741–753.

Calatroni A, Villalonga C, Roggen D and Tröster G (2009), 'Context cells: towards lifelong learning in activity recognition systems', *Proceedings of the 4th European Conference on Smart Sensing and Context*, pp. 121–134.

Chandola V, Banerjee A and Kumar V (2009), 'Anomaly detection: a survey', *ACM Comp Surv*, 41(3), 15:1–15:58.

Chavarriaga R, Sagha H and Millán J d R (2011), 'Ensemble creation and reconfiguration for activity recognition: an information theoretic approach', *Proceedings of Systems, Man, Cybernetics*. pp. 2761–2766.

Cheng J, Amft O, and Lukowicz P (2010), 'Active capacitive sensing: exploring a new wearable sensing modality for activity recognition', *Proceedings of the 8th International Conference on Pervasive Computing*, pp. 319–336.

Cohen I and Goldszmidt M (2004), 'Properties and benefits of calibrated classifiers', *Proceedings of Knowledge Discovery in Databases*, pp 125–136.

Davies N, Siewiorek D P and Sukthankar R (2008) 'Special issue: activity-based computing', *IEEE Perv Comp*, 7(2), 20–21.

Deng J and Tsui H (2000), 'An HMM-based approach for gesture segmentation and recognition', *Proceedings of the 15th International Conference on Pattern Recognition*, 3, 679–682.

Dey A K (2001), 'Understanding and using context', *Pers Ubiq Comp J*, 5(1), 4–7.

Duda R O, Hart P E and Stork D G (2000), *Pattern Classification*. New York, John Wiley & Sons.

Dunne L E, Walsh P, Smyth B, and Caulfield B (2008), 'Wearable monitoring of seated spinal posture', *IEEE Trans Biomed Circ Syst*, 2(2), 97–105.

Ermes M, Pärkkä J, Mäntyjärvi J and Korhonen I (2008), 'Detection of daily activities and sports with wearable sensors in controlled and uncontrolled conditions', *IEEE Trans Info Tech Biomed*, 12(1), 20–26.

Figo D, Diniz P C, Ferreira D R and Cardoso J M P (2010), 'Preprocessing techniques for context recognition from accelerometer data', *Perv Mobile Comp*, 14(7), 645–662.

Förster K, Brem P, Roggen D and Tröster G (2009a), 'Evolving discriminative features robust to sensor displacement for activity recognition in body area sensor networks', *Proceedings of the 5th International Conference on Intelligent Sensors, Sensor Networks and Information Processing*, pp. 43–48.

Förster K, Roggen D and Tröster G (2009b), 'Unsupervised classifier self-calibration through repeated context occurrences: is there robustness against sensor displacement to gain?', *Proceedings of the 13th IEEE International Symposium on Wearable Computers*, pp. 77–84.

Förster K, Biasiucci A, Chavarriaga R, Millán J d R, Roggen D and Tröster G (2010), 'On the use of brain decoded signals for online user adaptive gesture recognition systems', *Proceedings of the 8th International Conference on Pervasive Computing*, pp. 427–444.

Fortino G, Guerrieri A, Bellifemine F L and Giannantonio R (2009), 'SPINE2: developing BSN applications on heterogeneous sensor nodes', *Proceedings of the IEEE Symposium on Industrial Embedded Systems*. pp. 128–131.

Franke T, Pieringer C and Lukowicz P (2011), 'How should a wearable rowing trainer look like? A user study', *Proceedings of the 15th International Symposium on Wearable Computers*, pp. 15–18.

Froehlich J, Chen M, Consolvo S, Harrison B and Landay J (2007), 'My experience: a system for in situ tracing and capturing of user feedback on mobile phones', *Proceedings of the 5th International Conference on Mobile Systems, Applications and Services*.

Gage J and Murphy R R (2010), 'Sensing assessment in unknown environments: a survey', *IEEE Trans Syst Man Cyber, Pt A*, 40(1), 1–12.

Gemperle F, Kasabach C, Stivoric J, Bauer M and Martin R (1998), 'Design for wearability', *Proceedings of the 2nd International Symposium on Wearable Computers*, Los Alamitos, CA: IEEE Computer Society Press, pp. 116–123.

Giorgino T, Quaglini S, Lorassi F, Tognetti A and De Rossi D (2006), 'Experiments in the detection of upper limb posture through kinestetic strain sensors', in: *Proceedings of the XVI Congress of the International Society of Electrophysiology and Kinesiology, ISEK06*, pp. 4–12.

Gravenhorst F, Tessendorf B and Tröster G (2011), 'Towards a rowing technique evaluation based on oar orientation', *International Conference on Pervasive Computing*.

Guyon I and Elisseeff A (2003), 'An introduction to variable and feature selection', *J Mach Learn Res*, 3, 1157–1182.

Guyon I, Gunn S, Nikravesh M and Zadeh L (2006), 'Feature extraction, foundations and applications', *Series Studies in Fuzziness and Soft Computing*. Berlin, Physica-Verlag, Springer, available from: http://www.amazon.com/Feature-Extraction-Foundations-Applications-Fuzziness/dp/3540354875

Harms H, Amft O, Roggen D and Tröster G (2009), 'Rapid prototyping of smart garments for activity-aware applications', *J Amb Intell Smart Envir*, 1(2), 87–101.

Harms H, Amft O and G. Tröster G (2010a), 'Estimating posture recognition performance in sensing garments using geometric wrinkle modeling', *IEEE Trans Info Tech Biomed*, 14(6), 1436–1445.

Harms H, Amft O, Winkler R, Schumm J, Kusserow M and Tröster G (2010b), 'ETHOS: miniature orientation sensor for wearable human motion analysis', *Proceedings of the IEEE Sensors Conference*. pp. 1037–1042.

Hertleer C, Tronquo A, Rogier H, Vallozzi L and Van Langenhove L (2007), 'Aperture-coupled patch antenna for integration into wearable textile systems', *IEEE Anten Wireless Prop Lett*, 6, 392–395.

Hong X, Nugent C, Mulvenna M, McClean S, Scotney B and Devlin S (2009), 'Evidential fusion of sensor data for activity recognition in smart homes', *Perv and Mobile Comp*, 5, 236–252.

Huynh T, Fritz M and Schiele B (2008), 'Discovery of activity patterns using topic models', *Proceedings of the 10th International Conference on Ubiquitous Computing*. ACM New York, pp. 10–19.

Intille S, Larson K, Tapia E, Beaudin J and P. Kaushik P, *et al.* (2006), 'Using a live-in laboratory for ubiquitous computing research', *Proceedings of the International Conference on Pervasive Computing*, pp. 349–365.

Kallio S, Kela J, Korpipää P and Mäntyjärvi J, 'User independent gesture interaction for small handheld devices', *Int J Patt Recog Art Intell*, 20(4), 505–524.

Kawaguchi N, Ogawa N, Iwasaki Y, Kaji K and Terada T, *et al.* (2011), 'HASC challenge: gathering large-scale human activity corpus for the real-world activity understanding', *Proceedings of the 2nd Augmented Human International Conference*, available from: http://dl.acm.org/citation.cfm?id=1959853&dl=ACM&coll=DL

Keogh E, Chu S, Hart D and Pazzani M (2001), 'An online algorithm for segmenting time series', *Proceedings of the IEEE International Conference on Data Mining*, pp. 289–296.

Kinkeldei T, Zysset C, Cherenack K and Tröster G (2011a), 'A textile integrated sensor system for monitoring humidity and temperature', *Proceedings of the International Conference on Solid-State Sensors, Actuators and Microsystems*, pp. 1156–1159.

Kinkeldei T, Zysset C, Cherenack K, Noble W and Tröster G (2011b), 'Crack prevention of highly bent metal thin films in woven electronic textiles', *Eur Phys J Appl Phys*, 55(2), 1–5.

Ko M, West G, Venkatesh S and Kumar M (2005), 'Online context recognition in multisensor systems using dynamic time warping', *Proceedings of the Conference on Intelligent Sensors, Sensor Networks and Information Processing*, pp. 283–288.

Kulic, D, Takano W and Nakamura Y (2009), 'Online segmentation and clustering from continuous observation of whole body motions', *IEEE Trans Robot*, 25(5), 1158–1166.

Kunze K and Lukowicz P (2008), 'Dealing with sensor displacement in motion-based on-body activity recognition systems', *Proceedings of the 10th International Conference on Ubiquitous Computing*, pp. 20–29.

Kunze K, Bahle G, Lukowicz P and Partridge P (2010), 'Can magnetic field sensors replace gyroscopes in wearable sensing applications?' *Proceedings of the 14th International Symposium on Wearable Computers*. pp. 1–4.

Lamparth S, Fuhrhop S, Kirst M, Wagner G and Ottenbacher J (2009), 'A mobile device for textile-integrated long-term ECG monitoring', *World Cong Med Phys Biomed Eng*, 25(5), 278–281.

Lane N D, Miluzzo E, Lu H, Peebles D, Choudhury T and Campbell A T (2010), 'A survey of mobile phone sensing', *IEEE Comm Mag*, 48(9), 140–150.

Lazer D, Pentland A, Adamic L, Aral S and Barabási A. L, *et al.* (2009), 'Computational social science', *Science*, 323(5915), 721–723.

Lee H K and Kim J H (1999), 'An HMM-based threshold model approach for gesture recognition', *IEEE Trans Patt Anal Mach Intell*, 21(10), 961–973.

Lester J, Choudhury T and Borriello G (2006), 'A practical approach to recognizing physical activities', *Proceedings of Pervasive Computing*, pp. 1–16.

Locher I, Kirstein T and Tröster G (2004), 'Routing methods adapted to e-textiles', *Proceedings of the International Microelectronics and Packaging Society*.

Locher I, Klemm M, Kirstein T and Tröster G (2006), 'Design and characterization of purely textile patch antennas', *IEEE Trans Adv Pack*, 29(4), 777–788.

Lukowicz P, Hanser F, Szubski C and Schobersberger W (2006), 'Detecting and interpreting muscle activity with wearable force sensors', *Proceedings of the International Conference on Pervasive Computing*, pp. 101–116.

Lukowicz P, Amft O, Roggen D, and Cheng J (2010), 'On-body sensing: from gesture-based input to activity-driven interaction', *IEEE Comp*, 43(10), 92–96.

Maekawa T and Watanabe S (2011), 'Unsupervised activity recognition with user's physical characteristics data', *Proceedings of the International Symposium on Wearable Computers*, pp. 89–96.

Mann S (1996), 'Smart clothing: the shift to wearable computing', *Comm ACM*, 8, 23–34.

Mann S (1998), 'Humanistic computing: 'wearcom' as a new framework and application for intelligent signal processing', *Proc IEEE*, 86(11), 2123–2151.

Manzoor A, Villalonga C, Calatroni A, Truong H L, Roggen D, *et al.* (2010), 'Identifying important action primitives for high level activity recognition', *Proceedings of the European Conference on Smart Sensing and Context*, pp. 149–162.

Martin C, Steege E F and Gross H M (2010), 'Estimation of pointing poses for visually instructing mobile robots under real world conditions', *Rob Auto Syst*, 58(2), 174–185.

Mattmann C, Amft O, Harms H, Tröster G and Clemens F (2007), 'Recognizing upper body postures using textile strain sensors', *Proceedings of the 11th IEEE International Symposium on Wearable Computers*, pp. 29–36.

Mattmann C, Clemens F and Tröster G. (2008), 'Sensor for measuring strain in textile', *Sensors*, 8(6), 3719–3732.

Meyer J, Lukowicz P and Tröster G (2006), 'Textile pressure sensor for muscle activity and motion detection', *Proceedings of the 10th IEEE International Symposium on Wearable Computers*, pp. 69–72.

Meyer J, Arnrich B, Schumm J and Tröster G (2010), 'Design and modeling of a textile pressure sensor for sitting posture classification', *IEEE Sens J*, 10(8), 1391–1398.

Myers B, Hollan J, Cruz I, Bryson S, Bulterman D, *et al.* (1996), 'Strategic directions in human-computer interaction', *ACM Comp Surv*, 28(4), 794–809.

Ohmura R, Hashida N and Imai M (2009), 'Preliminary evaluation of personal adaptation techniques in accelerometer-based activity recognition', *Proceedings of the 13th IEEE International Symposium on Wearable Computers*, Late Breaking Results.

Park S and Jayaraman S (2003), 'Enhancing the quality of life through wearable technology', *IEEE Eng Med Biol*, 22(3), 41–48.

Pentland A (2004), 'Healthwear: medical technology becomes wearable', *IEEE Comp Mag*, 37(5), 42–49.

Picard R W (1997), *Affective Computing*. Cambridge, MA, MIT Press.

Pola T and Vanhala J (2007), 'Textile electrodes in ECG measurement', *Proceedings of Intelligent Sensors, Sensor Networks and Information Processing*, pp. 635–639.

Polikar R (2006), 'Ensemble based systems in decision making', *IEEE Circ Syst Mag*, 6(3), 21–45

Qian H, Mao Y, Xiang W and Wang Z (2010), 'Recognition of human activities using SVM multi-class classifier', *Patt Recog Lett*, 31(2), 100–111.

Randell C (2000), 'Context awareness by analyzing accelerometer data', *Proceedings of the 4th International Symposium on Wearable Computers*, pp. 175–176.

Ranganathan A, Al-Muhtadi J and Campbell R H (2004), 'Reasoning about uncertain contexts in pervasive computing environments', *IEEE Perv Comp Mag*, 3(2), 62–70.

Roggen D, Förster K, Calatroni A, Holleczek T and Fang Y, *et al.* (2009), 'OPPORTUNITY: towards opportunistic activity and context recognition systems', *Proceedings of the 3rd IEEE WoWMoM Workshop on Autononomic and Opportunistic Communications*. pp. 1–6, available from http://ieeexplore.ieee.org/xpls/abs_all.jsp?arnumber=5282442

Roggen D, Calatroni A, Rossi M, Holleczek T, Förster K, *et al.* (2010a), 'Collecting complex activity data sets in highly rich networked sensor environments', *Proceedings of the 7th International Conference on Networked Sensing Systems.* IEEE Press, pp. 233–240.

Roggen D, Bächlin M, Schumm J, Holleczek T, Lombriser C, *et al.* (2010b), 'An educational and research kit for activity and context recognition from on-body sensors', *Proceedings of the IEEE International Conference on Body Sensor Networks*, pp. 277–282.

Roggen D, Wirz M, Tröster G and Helbing D (2011a), 'Recognition of crowd behavior from mobile sensors with pattern analysis and graph clustering methods', *Networks Hetero Media*, 6(3), 521–544.

Roggen D, Förster K, Calatroni A and Tröster G (2011b), 'The adARC pattern analysis architecture for adaptive human activity recognition systems', *J Amb Intell Hum Comp*, pp. 1–18, available from: http://dx.doi.org/10.1007/s12652-011-0064-0

Roggen D, Lombriser C, Rossi M and Tröster G (2011c), 'Titan: an enabling framework for activity-aware "pervasive apps" in opportunistic personal area networks', *EURASIP Journal on Wireless Communications and Networking.*

Roggen D, Magnenat S, Waibel M and Tröster G (2011d), 'Wearable computing: designing and sharing activity-recognition systems across platforms', *IEEE Rob Auto Mag*, 18(2), 83–95.

Sagha H, Digumarti S T, Millán J d R, Chavarriaga R, Calatroni A, *et al.* (2011a), 'Benchmarking classification techniques using the Opportunity human activity dataset', Presented at *IEEE International Conference on Systems, Man, and Cybernetics*, Anchorage, USA, 9–12 October, available from: http://infoscience.epfl.ch/record/167935

Sagha H, Millán J d R and Chavarriaga R (2011b), 'Detecting anomalies to improve classification performance in an opportunistic sensor network', *Proceedings of the 7th IEEE International Workshop on Sensor Networks and Systems for Pervasive Computing, PerSens*. pp. 154–159.

Sagha H, Millán J d R and Chavarriaga R (2011c), 'Detecting and rectifying anomalies in Opportunistic sensor networks', *International Conference on Body Sensor Networks.*

Schumm J, Arnrich B and Tröster G (2012), 'Quality appraisal of an ECG signal measured in an airplane seat', *IEEE Perv Comp J*, available from: http://ieeexplore.ieee.org/xpl/tocresult.jsp?isnumber=5210084

Schwarz A, I. Kazani I, Cuny L, Ghekiere F, Hertleer C, *et al.* (2011), 'A study on the lifetime behaviour of elastic and electro-conductive hybrid yarns', *Proceedings of the AUTEX Conference*, pp. 843–845.

Setz C, Schumm J, Lorenz C, Arnrich B and Tröster G (2009), 'Using ensemble classifier systems for handling missing data in emotion recognition from physiology: one step towards a practical system', *Proceedings of the 3rd International Conference on Affective Computing and Intelligent Interaction and Workshops*, pp. 1–8.

Sharma R (1998), 'Toward multimodal human-computer interface', *Proc IEEE*, 86(5), 853–869.

Sixsmith A J, Gibson G, Orpwood R D and Torrington J M (2007), 'Developing a technology wish-list to enhance the quality of life of people with dementia', *Gerontech*, 6(1), 2–19.

Starner T, Weaver J and Pentland A (1998), 'Real-time American sign language recognition using desk and wearable computer based video', *IEEE Trans Patt Anal Mach Intell*, 20(12), 1371–1375.

Stiefmeier T, Roggen D and Tröster G (2007), 'Fusion of string-matched templates for continuous activity recognition', *Proceedings of the 11th IEEE International Symposium on Wearable Computers*, pp. 41–44.

Stiefmeier T, Ogris G, Roggen D and Tröster G (2008), 'Wearable activity tracking in car manufacturing', *IEEE Perv Comp*, 7(2), 42–50.

Strohrmann C, Harms H, Tröster G, Hensler S and Müller R (2011), 'Out of the lab and into the woods: kinematic analysis in running using wearable sensors', *Proceedings of the 13th ACM International Conference on Ubiquitous Computing*, pp. 119–112.

Tenorth M, Bandouch J and Beetz M (2009), 'The TUM Kitchen Data Set of everyday manipulation activities for motion tracking and action recognition', *Proceedings of the 12th IEEE International Conference on Computer Vision Workshops*, pp. 1089–1096.

Tognetti N, Carbonaro G, Zupone D and De Rossi D (2006), 'Characterization of a novel data glove based on textile integrated sensors', *Proceedings of the 28th Annual International Conference of the IEEE on Engineering in Medicine and Biology Society*, pp. 2510–2513.

Turaga P, Chellappa R, Subrahmanian V S and Udrea O (2008), 'Machine recognition of human activities: a survey', *IEEE Trans Circ Syst Video Tech*, 18(11), 1473–1488.

Ueoka R, Masuda A, Murakami T and Hirose M (2009), 'RFID textile and map making system for large area positioning', *Proceedings of the International Symposium on Wearable Computers*, pp. 41–44.

van Kasteren T, Noulas A, Englebienne G and Kröse B (2008), 'Accurate activity recognition in a home setting', *Proceedings of the 10th International Conference on Ubiquitous Computing*, pp. 1–9.

Van Laerhoven K, Kilian D and Schiele B (2008), 'Rhythm awareness in long-term activity recognition', *Proceedings of the 12th International Symposium on Wearable Computers*, pp. 63–68.

Villalonga C, Roggen D, Lombriser C, Zappi P and Tröster G (2009), 'Bringing quality of context into wearable human activity recognition systems', *Proceedings of the First International Workshop on Quality of Context*, pp. 164–173.

Wang T S, Han Y S, Varshney P K and Chen P N (2005), 'Distributed fault-tolerant classification in wireless sensor networks', 23(4), 724–734.

Want R (2009), 'When cell phones become computers', *IEEE Perv Comp*, 8(2), 2–5.

Ward J, Lukowicz P, Tröster G and Starner T (2006), 'Activity recognition of assembly tasks using body-worn microphones and accelerometers', *IEEE Trans Patt Anal Mach Intell*, 28(10), 1553–1567.

Ward J, Lukowicz P and Gellersen H (2011), 'Performance metrics for activity recognition', *ACM Trans Info Syst Tech*, 2(1), available from: http://dl.acm.org/citation.cfm?id=1889687

Yang J Y, Wang J S and Chen Y P (2008), 'Using acceleration measurements for activity recognition: an effective learning algorithm for constructing neural classifiers', *Patt Recog Lett*, 29(16), 2213–2220.

Zappi, P, Stiefmeier T, Farella E, Roggen D, Benini L and Tröster G, (2007), 'Activity recognition from on-body sensors by classifier fusion: sensor scalability and robustness', *Proceedings of the 3rd International Conference on Intelligent Sensors, Sensor Networks, and Information Processing*, pp. 281–286.

Zappi P, Lombriser C, Farella E, Roggen D, Benini L and Tröster G (2008), 'Activity recognition from on-body sensors: accuracy-power trade-off by dynamic sensor selection', *Proceedings of the 5th European Conference on Wireless Sensor Networks*, pp. 17–33.

Zinnen A, Blanke U and Schiele B (2009), 'An analysis of sensor-oriented vs. model-based activity recognition', *Proceedings of the International Symposium on Wearable Computers*, pp. 93–100.

Zysset C, Cherenack K, Kinkeldei T and Tröster G (2010), 'Weaving integrated circuits into textiles', *Proceedings of the 14th IEEE International Symposium on Wearable Computers*, IEEE Computer Society. pp. 1–8.

Part III

Product development and applications

13

Technology management and innovation strategies in the development of smart textiles

A. GARLINSKA and A. RÖPERT,
Interactive Wear AG, Germany

DOI: 10.1533/9780857093530.3.369

Abstract: This chapter presents an overview of smart-textile market economics. It discusses fundamentals of technology and innovation management in relation to the smart-textile community, especially in the environment of small and medium enterprises (SMEs). A few smart-textile market-related business models are discussed, showing the structure of the market (definitions, products, applications) and its dynamics (growth, trends, perspectives). Furthermore, a short overview of key challenges in the smart-textile business is presented.

Key words: smart-textile markets, technology and innovation management, smart-textile vertical and horizontal business models.

13.1 Introduction

This chapter presents an overview of the economics of the smart-textile market. It discusses fundamentals of technology and innovation management in relation to the smart-textile community, especially in the small and medium enterprise (SME) environment. A few smart-textile market-related business models are discussed, showing the structure of the market (definitions, products, applications) and its dynamics (growth, trends, perspectives). Furthermore, a short overview of key challenges in the smart-textile business is presented.

In Section 13.2, the fundamentals of technology management and marketing are reviewed, focusing on strategies and methods for new technologies in the SME environment of early markets. Furthermore, various innovation management approaches are discussed. Practical tips, based on the experience of innovation managers, as well as the authors themselves, are also presented. In Section 13.3, an overview of the market and its structure, as well as key trends and perspectives, are described. Then key challenges in smart-textile business are presented. In the last section, the authors outline the market structure and characterize the current smart-textile industry business models and provide closing conclusions.

The authors believe that the industrial structure combining electronics and textile worlds and the early market conditions create a very challenging non-mature competitive landscape, where companies are looking for strategic positioning in the technology and application segment. The understanding of the

Published by Woodhead Publishing Limited, 2013

drivers, challenges and market requirements for the specific segments should lead to clear value propositions in the delivery of both the competitive horizontal technologies and vertical applications. Different levels of maturity in both technology and solution segments should be taken into account while defining strategy and market approach. Strategic cooperations and alliances for solutions in specific segments will speed up the market innovation process and availability of market-ready products.

13.2 Fundamentals of innovation, technology and intellectual property management

We have found many different definitions and approaches to technology management, marketing and innovation management but no clear, unique and commonly-accepted views on the scope, boundaries and interfaces in the scientific literature (Fig. 13.1). As those of us working in SMEs have probably no time to read thousands of pages on different theories and views, we propose to concentrate on commonly understood pragmatic and clear definitions and strategies.

For us, these definitions are:

- Innovation management turns inventions into new products and commercial successes.
- Technology management builds the base for product development and maintenance and exploits the technological assets.
- Intellectual Property (IP) management protects the company's valuable assets.

In general, management means planning, organizing, implementing and controlling activities and in this manner can be applied to any discipline.

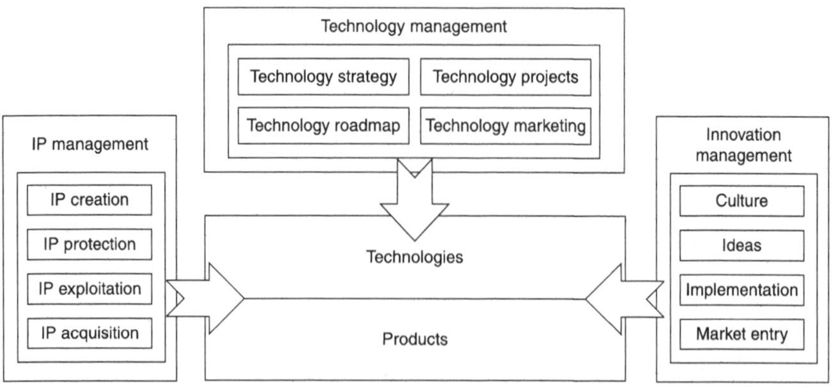

13.1 Major tasks of technology, innovation and IP management. (Source: Interactive Wear AG.)

13.2.1 Technology management

The major goal and task of technology management is to ensure that the company has the right technology base to provide competitive products. Technologies themselves can also represent a significant source of revenues. Having the right technology at the right time is the major competitive advantage for companies. However, the greater the progress of the new technology, the larger the gap for companies who have missed providing the appropriate technology base.

In many cases, a combination of technologies (e.g. smart phone = cell phone + PDA + wireless internet access + app market or YouTube = internet + cheap digital camera and camcorder) has made a real change in the business world and in our lives.

The smart-textile community is looking for answers on how the marriage between electronics and textile can deliver value to the products and how to design and market these products successfully. In addition to control and computation, electronics add actuating and sensing features to textile products; however, new organic materials are providing new possibilities for design and production of electronic components and products. There is a large potential on functions and possibilities on the one hand, but also some serious technological challenges on the other.

In this chapter, we will discuss the tasks of technology management and provide some ideas on strategies for SMEs.

Terms and definitions

In 1972, Rosenberg defined technology as tools, devices and knowledge that mediate between inputs and outputs and/or that create new products or services (Tushmann and Anderson, 1986). In 1987, the US National Research Council defined technology management as: 'Management of technology links engineering, science, and management disciplines to plan, develop, and implement *technological* capabilities to shape and accomplish the strategic and operational objectives of an organization.' Furthermore, the council listed the following key elements of industrial practice of technology management:

- the identification and evaluation of technological options;
- management of R&D itself, including determining project feasibility;
- integration of technology into the company's overall operations;
- implementation of new technologies in a product and/or process; and
- obsolescence and replacement.

In simple terms, technology management means to create and utilize the technological advantage for the company. The importance of appropriate technology management can especially be seen when, during technological changes, new companies grow very quickly and some already well-established companies get in trouble. 'In high wage countries, both the competivity of firms and more general welfare depend critically on the ability to keep up in innovative products and processes and in the underlying technologies' (Pavitt, 1990).

Technology management key tasks

Key deliveries of technology management to the organization are evaluation and protection of the value of the technology portfolio, definition of future needs (roadmap), roadmap implementation projects, and technology marketing programs. The technology management activities can be structured as follows:

Technology strategy

The goal of technology strategy is to define the role and value of technology in the company. How essential this understanding can be for the future of the company can be shown in many examples of strategic failures in judging new technologies. As a major producer of radios underestimated the advantages of the new transistor technology in comparison to vacuum tubes, Sony was able to gain world leadership in only eight years. At the beginning of the current century, a company called VIZIO, Inc. founded in 2002 by William Wang, with the idea that everyone deserves to own the latest technology, achieved first position in US sales of LCD-HDTVs in the first quarter of 2012, beating companies such as Samsung, LG Electronics and Sony.

The major goal is to understand how well your technology strategy supports the business strategy of your company. Technology strategy should provide a sound base for innovations. Major tasks are technology assessment, technology evaluation, IP management and the organizational set-up.

Helpful questions:

- *Strategic match*: How well does my current Technology Portfolio support the business strategy (are critical technological requirements clear)?
- *Competitive value*: Is the technology state-of-the-art or better? Is it unique and protected? How can we profit; protect, share or sell?
- *Organizational set-up and management quality*: Do we have a strategic technology plan? Who takes care of it?

Technology forecasting and road mapping

A technology roadmap is a forecasting and planning tool for the strategic development and application of new technologies. The major goal of the technology roadmap is to identify critical technologies and to provide a framework for coordination of future developments. In the smart-textile environment, there are some very helpful general roadmaps, such as 'A roadmap on smart textiles' (Schwarz *et al.*, 2010), a valuable general input for the smart-textiles community.

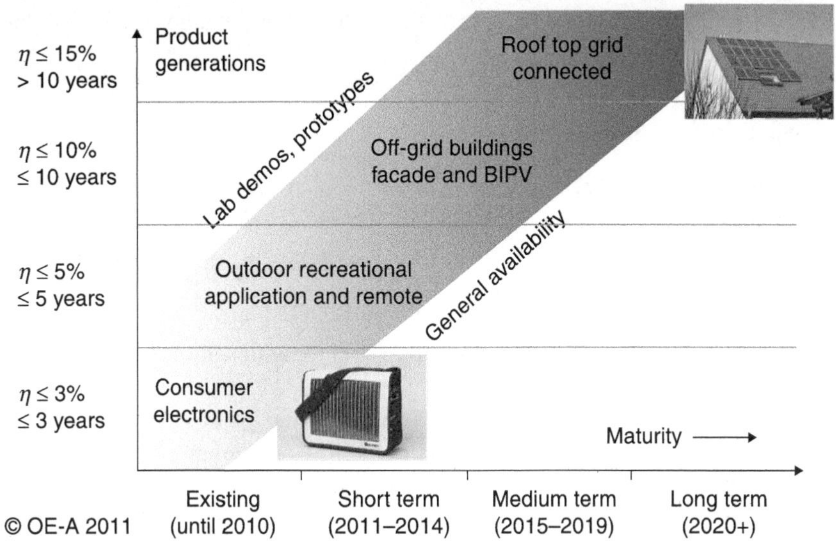

13.2 Organic photovoltaics roadmap.

Another good example of a smart-textile roadmap can be found in the overall roadmap for organic and printed electronics from the Organic Electronics Association OE-A (Hecker, 2011).

You will find many frameworks and examples available on the internet. The technology roadmap does not need to be complicated – it should visualize the plans. The easier it is to understand why and when we plan to implement our technology programs, the better is the basis for the next steps. The technology roadmap is usually more driven by need than by solution. For example, in our current roadmap, we address the major technology gaps in the process of development of smart-textile products. The major requirements are in the area of connectivity, housing and textile actuators technologies.

We find it very helpful to include the market and application focus in the technology roadmap. You can find an example above in the Roadmap for Organic Photovoltaics (Fig. 13.2). There are technology parameters on the left (efficiency and durability of photovoltaic modules), the time schedule on the right, and the applications in the middle.

Helpful questions:

- *Requirements match*: How well does the technology roadmap build the base to satisfy the market requirements? Does it address the needs of my organization to provide innovative products? How well does it support my product or solution roadmap?
- *Value check*: If we succeed with the implementation, how will it influence our market position?

- *Feasibility check*: Are resources available for implementation? What are the major risks and challenges?

Technology projects implementation

Implementing the defined roadmap means to plan and implement actions in technology development and/or acquisition. It has to be decided how the projects have to be implemented on our own, within a community, or by purchasing the development services or technology.

Technology projects are usually both exciting and risky. In comparison to projects for product development, they often include a higher level of uncertainty and new options. This can lead to more failures, especially if the risks and possible obstacles have been underestimated in the planning phase. We have gained valuable experience working with short feasibility studies before we started the implementation. We have focused on one side on major features to be delivered and, on the other, on key challenges to be addressed. We generally use feasibility studies to test possible options. As a result, within a short time we are able to exclude unattractive options (materials, methods, etc.) via testing prototypes or simulating the desired behavior. For example, domestic washing machines are sufficient tools as a first step to prove the feasibility of mechanical robustness of wearable electronics and conductive textile components.

Helpful questions:

- *Complexity level*: Are multiple organizations involved? Is my role and outcome clear?
- *Risk level*: What if the project task is not delivering the expected results? Is there a critical part of the development to be addressed separately?
- *Industrialization check*: Is the result stable, and ready to be implemented (produced) within the expected price range?
- *Project set-up*: Is the project properly prepared? Is a controlling schedule in place?

Technology marketing

There are some regional differences in defining and addressing technology marketing. The definition and understanding of the role of technology marketing has a major impact on the approach and the methods. Let us have a brief look at the differences.

Technology marketing in Europe is often related to marketing strategies of 'high-tech' products, while some American technology marketing centers focus on transferring the investment in research into commercial products and others emphasize the early-stage marketing strategies. According to the chapter on 'Technology Marketing' on the web pages of the Asian and Pacific Center for

Transfer of Technology (APCTT, 2012), technology marketing is 'a process involving the planning and executing of conceptualizing, pricing, distribution, and promotion of technology so that exchange activity can take place.' In technology marketing, process of the technology is seen as a product to be sold.

Behind all the disparities there are major common targets: how to efficiently transfer and transact technology, how to successfully transfer the technology into commercial products, how to market new high-tech products and how to address early markets.

For companies dealing with new technologies in early markets, it is very helpful to understand and implement marketing strategies in the difficult 'crossing the chasm' situation. Crossing the chasm means to overcome the gap between the early adopters of any new technology and the mainstream customers. In his famous book, Geoffrey A. Moore analyzes the specifics of marketing high tech products during the early phase of innovation (Moore, 1991). He has extended Everett Rogers' 'diffusion of innovation model' from 1962. His findings in this book and in his sequel, *Inside the Tornado* (Moore, 1995), has had a significant and lasting impact on high tech entrepreneurship.

13.2.2 Intellectual property management

Intellectual property (IP) is not easy to understand. Very helpful, especially to SMEs, is the World Intellectual Property Organization (WIPO, 2012), one of the 17 specialized agencies of the United Nations. To many, IP is a rather obscure legal concept that can only be properly understood and applied by those who are specially trained. You may be asking yourself why any SME and, more so, your SME, should pay attention to IP, or what benefits your SME could possibly draw from its use (WIPO, 2012).

According to WIPO, IP refers to creations of the mind: inventions, literary and artistic works, and symbols, names, images, and designs used in commerce. IP management comprises all methods and tools for protection and marketing of the company's proprietary know-how.

In comparison to large companies, many SMEs under-utilize the IP system. The reasons are mainly lack of awareness of the usefulness of the IP system combined with perceived high costs and complexity, as well as lack of easily accessible training and fast entry support. Sometimes IP management is limited in perception to patents, and the opportunities working with trade secrets are not evaluated. In some other cases, there is the question of how to start IP activities. To summarize, there is often uncertainty whether the investment in IP management will really pay-off. This question can be answered only case by case. Above all, however, this question must be asked and there must be enough knowledge about IP management to enable a qualified answer.

IP management can deliver a significant value to your SME. Properly done, IP management is the key to success of any innovation strategy. It can increase the value of your IP, support avoidance of unnecessary R&D spending, generate income through licensing, and reduce the risk associated with third-party IP. IP rights can enhance the value of your enterprise for financial reasons or in the event of a sale or a merger. In September 2008, adidas International, Inc. acquired Textronics, Inc. for a purchase price of US$ 35 million. Textronics, Inc. was a specialist in the development of wearable sensors for use in fitness and health monitoring. The net assets of Textronics, according to adidas publications (adidas Group Annual Report 2009), were solely trademarks and other intangible assets valued at US $9 million. In addition, adidas paid US$ 26 million goodwill arising on acquisition; a very high valuation of the trademarks, patents and know-how of the small smart-textile company.

The importance and choice of the IP system depends on the sector of the company's operations; for example, for science-based companies (e.g. electronics), a formal IP rights system is important. Many SMEs may choose informal IP protection methods, such as contracts, confidentiality agreements, circulation of tasks, etc. The formal and informal protection can be used in a complementary fashion during the lifecycle of a product or service.

How to start: first, after creating awareness for IP management within the company, it is essential to evaluate the strengths and weaknesses of the existing IP implementation. There are good introductions on IP management, both on WIPO and EU web pages. There are also helpful studies, tools and methods that every SME can use to understand how to get started and how to implement or improve the IP management within the company's strategy.

13.2.3 Innovation management

'Nothing is so embarrassing as watching someone do something that you said could not be done'. – Sam Ewing. Is there a magic formula to create and implement innovative ideas? In other words, is there an innovation management strategy that, regardless of the manufacturing organization size, market, and business cycle orientation, provides an innovative portfolio and process landscape? Although we have not found a silver bullet, at least our search for the answer has provided a number of useful strategies and techniques for managing innovation for SMEs.

SMEs rarely direct large research departments and central offices to execute significant innovation management tasks. Some are new businesses and just an innovative idea is the basis of their existence. The others have a more or less stable business and want to invest in new applications, products or processes. Examples are narrow-weaving companies, as well as embroidery companies who integrate metallic or carbon threads into their products. For example, Forster Rohner AG, a Swiss-based company with traditional embroidery products for more than 100 years, began in 2009 with activities in technical embroideries. In 2010, a first spinout, the Bionic Composite Technologies AG, was founded. While this company is ramping up first industrial productions with optimized processes for small- and medium-sized composite products, Forster Rohner, in 2012 created the business unit 'Textile Innovations'. The multi-disciplinary R&D innovation team identifies potential applications and together with partners in research and commercialization works on processes and solutions for volume production. Dr Jan Zimmermann, Project Manager of Technical Textiles, stated that 'Innovation should be part of the genetic code of an enterprise and has to be in line with the corporate strategy and management to be successful'. However, all SMEs want to take advantage of their best ideas and succeed in business with the changing market conditions and customer requirements.

In our search to find the best pragmatic and actionable strategies, we have studied the innovation management books and articles, analyzed innovation portals, listened to interesting lectures on innovation (Garry Kasparov, Matthias Horx) and have discussed the strategies with colleagues from research institutes and SMEs.

Innovation management terms and definitions

There are many definitions and views of innovation in science. Starting with French sociologist Gabriel Tarde in the 1890s, generations of scientists and managers have contributed to the picture of innovation that we have today. We have chosen the definition quoted by Paul Trott in his comprehensive work, *Innovation Management and New Product Development*: 'Invention, then, is the conception of the idea, whereas innovation is the subsequent translation of the invention into the economy' (US Dept of Commerce, 1967; Trott, 1998). The following simple equation helps to show the relationship between the two terms:

> Innovation = theoretical conception + technical invention + commercial exploitation (Trott, 1998).

We can simply say that invention is a new idea and innovation is the commercialization of the invention itself. Or, as Theodore Levitt said, 'Creativity is thinking up new things. Innovation is doing new things'. Innovation management is a set of tools and processes to ensure the ability of the organization to create and market innovative products and services. Furthermore, you have probably had some ideas about the differences between innovation management, new product development and technology management.

Innovation management supports both technology and product management in their abilities to develop and deliver to the market an innovative and accepted portfolio of products and technologies to build a sustainable value proposition in the marketplace. In this chapter, we describe how innovation management fulfills this task. We focus on pragmatic methods and models to help us set 'the conditions to be in place to ensure that the organization as a whole is given the opportunity to develop new products' (Trott, 1998).

Innovation management key tasks

There are thousands of pages written on innovation management theories, approaches, tools and methods. Despite the large amount of books, articles and portals, we have the impression that innovation management is still under development and there is no dominant method or approach. On the other hand, there are some common elements that we can rely on to support us in our innovation work.

In general, we propose to ask ourselves two essential questions about the innovation culture:

- Does our company's culture support new ideas and concepts?
- Do we have the right habits in place to make them successful?

If your company's culture is both open to new ideas and disciplined enough to implement them successfully, you are in the best starting position. If you are more creative than disciplined (and probably some projects were not successful), you will concentrate on finding a solution and work on realization methods. If you are very process-orientated and controlled, but may not be happy to find more innovative solutions in the market, you will probably be willing to think about open innovation or creativity techniques. As in sports, we should concentrate on our strengths and try to find a solution for our weaknesses.

Thomas Alva Edison (1876–1881) was one of the key personalities for an innovative culture. He managed to create a favorable space and climate for innovation. He was not only first in creating professional R&D labs in Menlo Park, he also managed to establish an informal environment with no rules for work but with easy communication between team members, great infrastructure and lots of fun. His team worked long hours (despite or because of the lack of a working hour clock) and produced substantial results. Thomas Edison said, 'I never did a day's work in my life, it was all fun'.

Creativity

Creativity is the source for future success. There are methods and tools supporting the organization in generating and collecting new ideas. Examples are very many creativity techniques such as brainstorming, associations, creative problem solving, improvisation, etc., which boost the generation of ideas. Very often companies concentrate on brainstorming and do not try to find new approaches. As the majority of techniques are not complicated to implement, we encourage trying some new ones, especially if brainstorming does not bring the expected results.

There is a huge potential in interaction with universities, communities, freelancers and others to access new sources of ideas. 'Not invented here' mentality is still very often an obstacle to creativity.

One of the most promising areas to generate innovations is the investment into a deep understanding of the requirements of the targeted customers:

> Innovation is fostered by information gathered from new connections; from insights gained by journeys into other disciplines or places; from active, collegial networks and fluid, open boundaries. Innovation arises from ongoing circles of exchange, where information is not just accumulated or stored, but created. Knowledge is generated anew from connections that weren't there before. – Margaret J Wheatley

How can smart-textile companies understand the requirements of their targeted customers? It is often not even clear who the potential customers are and what kind of applications can be targeted. We believe that these questions need to be discussed and to be tackled from the very beginning. We think it is worth looking for innovative customers early-adopters – who are willing to take risks in implementing new technologies. It is important to understand the advantages of smart-textile components in their solutions and applications. We work a lot with prototypes to visualize the application to the customer and create the basis for evaluation. We look at the smart-textile blogs and also follow discussions of user groups for end-users of our systems to anticipate feedback.

Sometimes ideas come and then go if we do not collect and keep them. We find it very important and helpful to have a 'treasure chest', where we can save our ideas and have them ready for later evaluation and decisions.

Feasibility

Right or wrong choices define the future value of our work. When we have an idea, it is time to challenge it and consider whether it is valuable enough and whether its implementation is practicable. This is depending on the level of innovation, sometimes a tricky task. You all know the different assumptions of the past that have lead companies not to follow the development:

- 'What use could the company make of an electric toy?' – Western Union, when it turned down rights to the telephone in 1878.
- 'Who the hell wants to hear actors talk?' – Harry M Warner, Warner Bros Pictures, 1927.
- 'Video won't be able to hold on to any market it captures after the first six months. People will soon get tired of staring at a plywood box every night.' – Daryl F Zanuck, 20th Century Fox, commenting on television in 1946.

As Einstein (*On Science*) said 'Imagination is more important than knowledge'. The majority of innovative ideas are not so difficult to imagine and evaluate, as the above-mentioned disruptive technologies. We can usually improve the quality of our choice by answering key questions on the innovation potential and practicability of a particular innovation:

- Is the idea desirable for the users? Why and what kind of product/technology will be replaced?
- Is it possible to implement? What are the major risks in the implementation?
- Is it viable in the marketplace? Can the product be delivered at a reasonable price? Are sales channels in place?

In this early phase, we used to working with feasibility frameworks, focusing on understanding the product advantages, potential and industrialization effort, as well as innovation level. We have gained valuable experiences working with demonstrators and looking at the product from the overall application perspective. We evaluate technological feasibility as well as economic aspects. For example, if you are working on an innovative heating material, but do not consider how to handle interconnects in a large-scale production, you will probably face delivery problems.

There are many tools – for example, Decision Memo, Pros and Cons or Nominal Group Technique – supporting the decision process; no one will guarantee success but almost all will give you valuable support for the evaluation and help you to limit the risks.

Implementation

The implementation phase is the basis for the future success of the innovation. Most of us tend to be very optimistic at the beginning of any project. However, as the project progresses, we realize that there are new issues to solve with no time left to do so.

How does the realization process of innovations differ from the standard development? What are the critical factors to be taken into account for the implementation? The focus of the development process depends on the innovation level:

- product development process for incremental and routine innovation;
- new product development + technology management + IP management for radical and improvement innovation.

Innovation level, incremental or routine innovation, does not usually generate big challenges and surprises. For example, we have continuously improved the method of integrating small electronics components such as LEDs or sensors, by using standard circuit boards for textile – cable interconnections combined with snap buttons and molds. This has optimized the robustness as well as the industrial look and feel of the components. The higher the innovation level the larger the uncertainty about the unknown challenges we may face during development. In some more advanced cases, we experienced that our supplier did not succeed in scaling up new production techniques for mass production, but even so, the prototypes and pre-series worked perfectly.

In the e-textiles world, you need to consider not only a (conductive) textile component but also the connectivity, integration, power supply and, last but not least, the complete system approach. Most of these issues are unknown to textile engineers, while traditional electronics engineers have no experience with the special requirements of the textile world:

- *Maturity of technologies*: which technology used in the realization process is new? Can I replace it if necessary?
- *Industrialization effort*: building a robust component (compound) ready for production;
- *Production aspects*: reproducibility, scalability;
- *Risk management*: 'what if' scenarios, if the realization process faces severe hurdles.

In the smart-textile environment, we face technical challenges in connectivity, isolation, mechanical and chemical stress. In addition, we are dealing with compound products containing multiple technologies. As SMEs usually have few resources for research and development, we may engage in cooperative projects, resulting in faster execution of the project and supporting the knowledge sharing.

Market introduction and dissemination

The success and payback will depend not only on the potential and quality of the innovation but also on market entry and implementation of market strategies. We will not describe the general theory in this chapter, but recommend reading Geoffrey Moore's exciting books about the challenges of marketing new technologies (Moore, 1991) and how evolving markets should be managed (Moore, 1995).

Strategies of innovation management

When we look at the history and classical strategies through the lessons that we have been taught from the famous innovation thinkers, we realize how valid their thoughts still are. In recent times, new approaches have become popular: Open Innovation and Distributed Innovation.

Open Innovation means that we open the development process of our products or technologies to external sources of ideas and solutions. Of course there is a broad spectrum of approaches to the project definition. We can be closer to the fully controlled innovation project by hiring an outside specialist or we can – be optimistic – putting the problem to be solved out in one of the portals and wait for suitable responses.

Openness starts in our minds. As Edison said, 'I readily absorb ideas from every source, frequently starting where the last person left off'. There is a big potential in building up on our common knowledge and not limiting ourselves to internal sources. Distributed Innovation is defined as a process of linking together numerous people with disparate expertise working in different institutions and countries, but united under a single project focused on rapid product development and deployment (Morey, 2009). Distributed Innovation is used by many cutting edge companies. European- or national-sponsored projects are often based on this principle.

For us, the most valuable aspect of Open and Distributed Innovation is to be part of a community and to have access to a broad knowledge base and to speed up the implementation process by task distribution. The communities and blogs are also an interesting source of feedback on our products and systems.

Strategies for small and medium enterprises (SMEs)

We believe that every company has to create its own way; it should match the culture, situation and way of operating. The former world chess champion Garry Kasparov said, 'You must also become intimately aware of the methods you use to reach your decisions' (Kasparov, 2007). We suggest the following guidelines:

- Be open for new ideas, encourage your organization to create and collect its own ideas and ideas from the 'outside'.
- Challenge your ideas – is it feasible, doable (possible to produce with reliability and market price)? Are there advantages that the customer is willing to pay for? Is the business model appropriate?
- Be clear about the strengths and weaknesses of your approach; focus on the strengths but be prepared to find a solution for the weaknesses.
- Create a link between the formal and informal, avoid bureaucracy, manage to incorporate informal results into your processes; e.g. all different forms of customer requirements (e.g. drawings are included in our certified process).
- Use networks – interact, compete and cooperate (e.g. in joint research projects), but care about your IP.
- Do not forget the industrialization and commercialization efforts – a prototype is not a product! You will find some of our showcase prototypes not on the market but in a museum. It is great if this happens as an exception, but it should not become a rule.

13.3 Business models for smart textiles

13.3.1 Business models

In the previous chapter, we looked at technology management and marketing, innovation and IP management. While technology in general is a common source for innovations, we need a market, for innovations should become real products. A market is defined as a network between sellers and buyers to exchange products and services for money. Sellers are normally established as companies or businesses. A business model is a description of the operation of a business including all components of the business, the functions of the business, the participants and the revenues and expenses that the business generates. The phrase 'business model' appeared for the first time in the late 1950s.

With the beginning and after the dotcom bubble, the interests in, and discussions on, business models were very strong. In our opinion, this was related to the new options of de-construction and re-construction of value chains through the new information and communication techniques (web 1.0). The stunning growth and availability of cheap information and communication technology enabled companies to work in networks, because coordination and transaction costs decreased substantially. This led to industrial boundaries becoming more and more open. A good overview can be found in the thesis of Alexander Osterwalder (2004). Henry Chesbrough mentioned that usually companies struggle hard to manage innovations that do not match their experience (Chesbrough and Rosenbloom, 2002). This is especially true for the textile companies dealing with electronics components.

A good model to work with is described by Andrews (1971):

- value proposition, i.e. the value created for customers by the product;
- value chain, i.e. the structure of creation and distribution of the product;
- value network, i.e. suppliers and complementors, customers and competitors;
- market segment, i.e. customers and applications;
- competitive strategy;
- cost structure and profit potential.

The value chain describes the value added from raw materials until end consumer products. If a company controls all the steps – or at least most of the steps – it is called 'vertically integrated'. A company that sells one product to different market segments is utilizing a 'horizontal' business model.

Horizontal market players focus on high-volume products or (patented) technologies and/or processes. For example, fabric manufacturers try to leverage cost advantages in mass production due to roll-to-roll processes. The drawback for the horizontal market players: there is no end product. And if they do not find a vertical end product where the new technology will fit, there will be no pull for the technology.

13.3.2 E-textile business models

Of course, we cannot define a general business model for smart-textile vendors. Looking at the value proposition, e-textiles could enhance or improve basic functions of textiles as there is climate regulation (i.e. heating, cooling, and moisture management) or create new functionalities (i.e. sensing and energy harvesting).

From our experience, smart-textile companies should carefully review their business model. Helpful questions are:

- What kind of USP do you offer?
- Is it know-how, a component, a system, a product?
- To which segments do your customers belong?
- Which complementary products and or partners do you need, to offer the right level of product or service to your customer?

The material provider offers common or specific material such as printed circuits, roll-to-roll, highly flexible wires for weaving and knitting, flexible printed circuit boards, etc. The material providers need a pull from the next level of the value chain, the component manufacturer. When we analyze the value chain of a system with incorporated conductive textile, we will find:

- *Base material*: e.g. conductive yarns, polymers or inks.
- *Textile material*: e.g. the woven smart fabric heating fabric on roll.
- *System components*: e.g. a heat-pad, cut and provided with interconnects.
- *Integrated sub systems*: e.g. a heating system including a heat pad, controller, interconnection and batteries.
- *Products*: e.g. heated gloves or LED jackets.

Of course, it is relatively easy to design a value proposition for a niche market or even for a single customer. However, most of the textile companies looking into the smart-textile markets are targeting mass markets.

Forming alliances is an interesting and appropriate way to work on innovations such as smart textiles. Alliances are usually installed to market systems with the potential to address several market segments. Ideally, the partners of an alliance contribute not only components or technologies that build a complete product but also provide complementary roots to markets.

Vertical model supplier

There are some companies that offer a vertical product based on a core of smart-textile components. In most cases, a lack of customers for the base technology forces the companies to develop the full vertical product and sell it. The possibilities of offering goods via the internet help these companies to establish a sales channel without large investments.

WarmX® is a subsidiary of a German knitwear company that offers heated underwear with elastic silver-coated yarns (Fig. 13.3). They developed their own control electronics and offer the products to distributors and end users. After two years of development and one year of field tests, in 2006 WarmX® introduced its first products. Since then, it has had more than five years of slight but constantly growing business, using the internet as the main sales channel. WarmX® has established itself as a commercial vendor of elastic heated underwear products

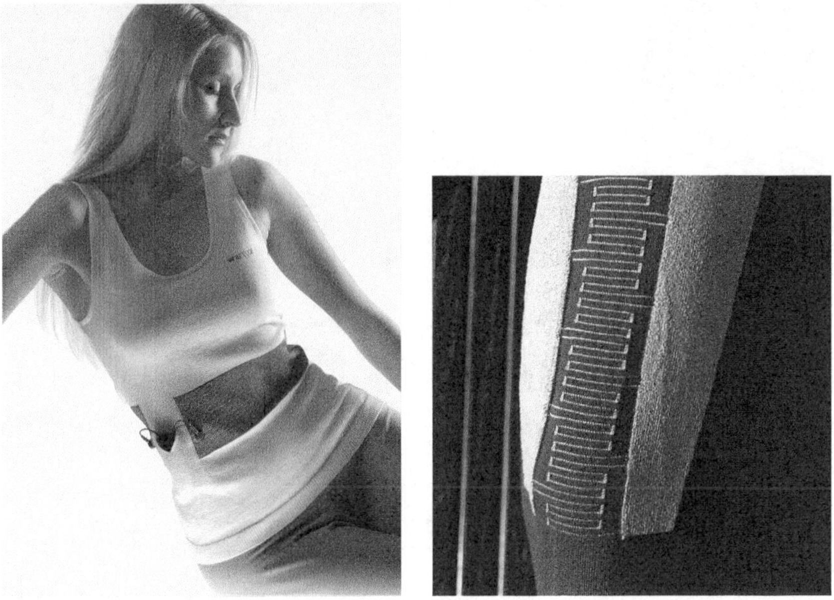

(a) (b)

13.3 WarmX®: Heated underwear shirt/detail knitted heating wires: (a) front; (b) back. (Reproduced with permission from WarmX®.)

(Fig. 13.4). Christoph Müller, CEO of WarmX®, recognizes that the main barrier for the sale of his products is the need to explain the benefits and properties of the products, due to the fact that this kind of product is not widely known to potential consumers.

Future-Shape, with its roots in the Infineon Research Labs, has developed special bonding techniques to build large-area capacitive sensor systems, which could be integrated into any floor covering. The SensFloor® is based on a smart-textile underlay with built-in microelectronic components and capacitive proximity sensors (Fig. 13.5). Future-Shape has developed a radio module on a flexible substrate, which is specially designed for integration into textile surfaces and can be connected to as many as eight sensor areas. Products like the SensFloor® mats are offered to senior residences and hospitals, shops and architects, as well as to the end user.

AiQ Smart Clothing Inc. is an example of building a vertical company within a textile group, with subsidiaries in different steps of the value chain (Fig. 13.6). AiQ is a subsidiary of TexRay Industrial Co. Ltd., a publicly listed global fabric and apparel company based in Taiwan, with global operations across North America, Asia and Africa. AiQ offers a complete vertical integration of wearable technologies to its customers, by combining and installing multiple key components from its upstream affiliate, King's Metal Fiber Technologies. AiQ

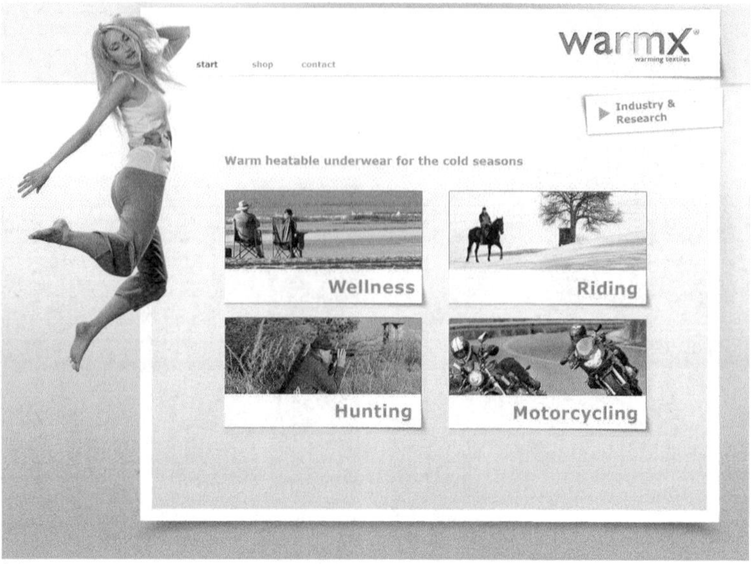

(a)

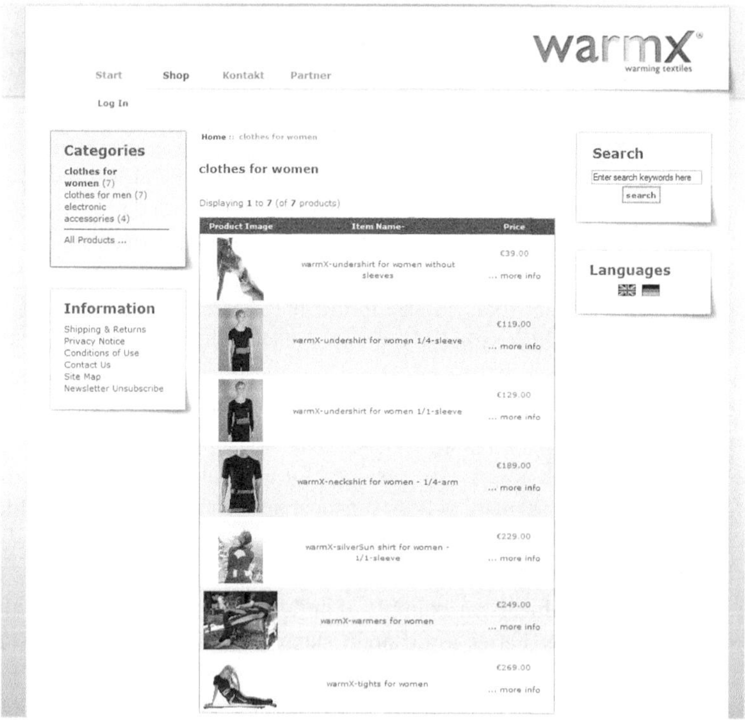

(b)

13.4 Typical for a vertical product supplier: (a) WarmX® is proposing a value added product for special target segments: (b) the niche customers can order the products in the online shop.

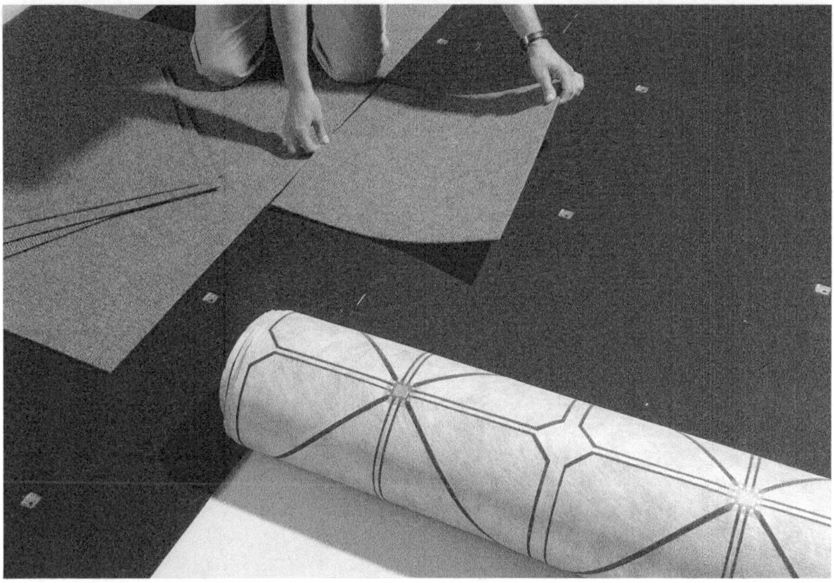

(a)

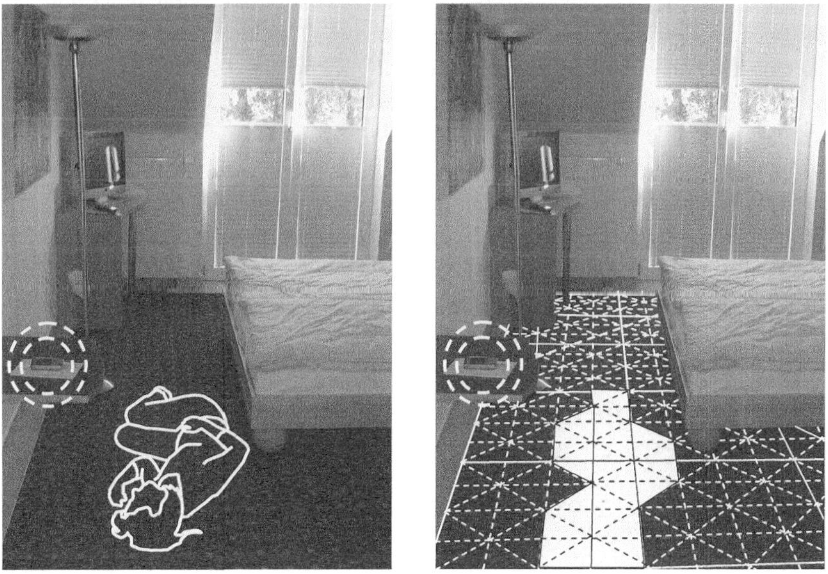

(b)

13.5 Future-Shape's Sensfloor®: underlay with: (a) textile sensors; and (b) bonded electronics, downfall pattern.

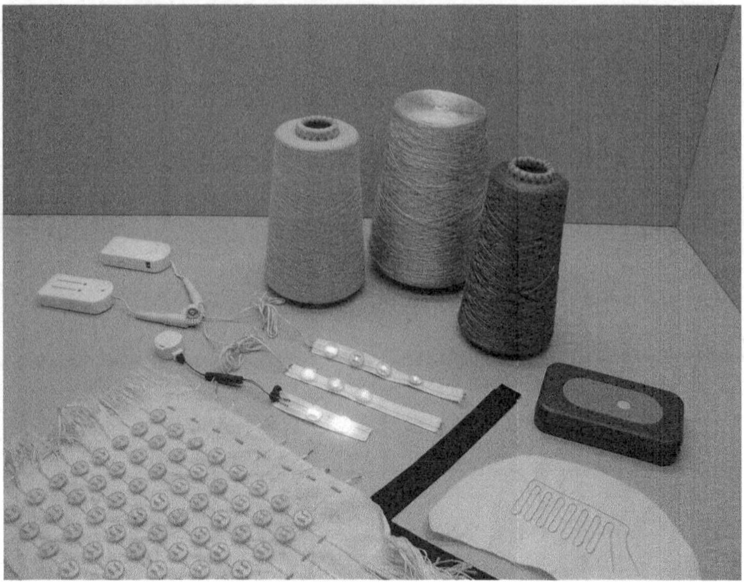

13.6 AIQ LED buttons and other components from the group.

specializes in developing wearable systems that are fully compatible with its downstream manufacturer and parent company, Tex-Ray. The company has developed a range of e-textile systems: the AIQ Biker Jacket with LEDs to indicate turns, and LED safety garments, heated garments, bio-monitoring shirts and touch screen compatible gloves.

Other examples of vertical approaches have been seen from Cute Circuit, famous for their 'hug-shirt', designing and marketing LED-garments, and also Textronics Inc., which launched the Numetrex sports bra when sports clothing companies were reluctant to build products with their sensors. After being acquired by adidas in 2008, products of Textronics Inc. submerged in the adidas universe.

13.4 Opportunities and challenges in the e-textiles business

13.4.1 Smart textiles terms and definitions

A definition of the smart-textile market is very challenging. An exact and clear definition does not exist. An often quoted definition is 'smart textiles are materials and structures that sense and react to environmental conditions or stimuli, such as those from mechanical, thermal, chemical, electrical, magnetic or other sources'. In the e-book, *Smart Fabric Technologies and Applications* (Intertech Pira, 2011(a)), smart textiles are defined to:

> . . . provide feedback to the wearer, improve safety and supply power. It covers a wide range of fibres and raw materials, as well as the connection methods necessary to integrate transistors, circuits and, ultimately, components such as sensors, LEDs and power supplies.

'Smart' is a buzzword that describes somehow unexpected and useful properties of well-known materials or products. In some research studies, the phrase 'smart fabrics and interactive textiles' (SFIT) is used. However, the expression of 'wearable electronics systems', 'wearable electronics' or even 'wearable' is used for electronic components or systems integrated into textiles. But from the exact meaning, also a mobile phone or a torch could be categorized as 'wearable'.

In this chapter, we would like to define the scope of smart textile as 'conductive textiles' or 'e-textiles', which has the capability to transport an electrical current. In our opinion, this term is even more general and descriptive than the phrase of 'electro-active fabrics'. There is also the huge number of innovative fabrics that, for example, improve the fabrics' moisture or temperature control in a very smart way but without electrical control or stimulus. E-textile systems are built with conductive textile components such as electrodes for monitoring, resistive or capacitive sensors, or resistors for heating elements.

Instead of using the broad term 'wearable electronics', we will focus on the system aspect of an electronic solution integrated into textile. Within the scope of smart-textile solutions, we would also like to point out that one of the main differences to simple mobile – hence wearable – devices like mp3-players, is the usage of a distributed architecture, flexible material and special textile compliant integration methods and techniques.

The e-textile market is strongly fragmented and unclear. While a variety of ongoing research results foster high expectations in the mid- and long-term future, the technology maturity level of most of the advanced applications is still very low.

13.4.2 E-textiles market

The textile industry itself, generally used to producing high quantities, expects to produce conductive material in a roll-to-roll process. Their customers, like the garment brands or automotive industry, aim to integrate sub-systems and treat them as trims or dashboards. In between, we find the assembly lines and integrators who need knowledge of all related technologies.

We will not conduct a complete market analysis, instead referring the reader to Section 13.6. Most of the studies expect strong growth for the smart-textile market. They are looking at the textile part of the value chain or at vertical products such as automotive seat heating systems or sensing and monitoring products. In the smart-textile business, we know of only one segment that shows sustainable growth and a noticeable market size, the segment of automotive seat heaters that use heat pads built of stainless steel or copper wires, or carbon fibers. This segment is constantly growing, as seat heating as a comfort function is in demand for all car classes, from luxury to compact.

13.4.3 E-textiles business drivers

What are the business drivers to invest in conductive textiles? The textile and clothing industry in Europe and North America has been an important part of the overall economy in the last century. Like in other industry segments, value added products and services are still growing, especially in the area of technical textile production. Mass production is moving from north to south, and from west to east. This is one of the major reasons why the European Community (EC) is allocating significant amounts of money for research projects in the field of smart textiles and especially e-textiles.

In 2006 there were 220 000 companies in the textile industry in the EC. These companies employed 2.5 million people and generated a turnover of €190 billion, which accounted for 3% of the total manufacturing value added in Europe (EC's Directorate General for Enterprise and Industry website).

Technical textiles find their way into more and more application fields, from innovative materials for traditional markets of garments and home textiles into automotive, architecture and other industrial applications. The European electronics industry exported €124 billion while importing €224 billion. A strong segment is measuring devices, which had a positive trade balance (Sectore Fiche: EU Trade in Electronics, 2011).

In the electronics industry, the market of embedded systems is increasing steadily. Driven by automotive and consumer electronics, components such as microprocessors and micro-electromechanical systems (MEMS) are becoming increasingly powerful and are available in mass production. Sensing and monitoring networks are becoming a state-of-the-art application in several industries. IDC analysts expect market growth of intelligent connected embedded systems from US$800 million in 2011 to US$2.3 billion in 2015 (IDC, 2011). Reynette Au, executive partner for GTIA L.L.C, said:

> Today's smart embedded technologies do more than just control and contribute features to a system. These technologies enable systems to connect, offer security and manageability, and open the door to new applications and innovation that are fundamentally changing the way we work, live, create and consume. How products in the Intelligent Systems category evolve and get to market will be an extremely dynamic process. It will require new levels of cross-industry collaboration with new business models emerging. This will make the ecosystem critical and it will be an exciting and value rich environment.' (IDC, 2011)

Furthermore, we see two major trends that will foster special application segments of smart textiles:

1. Increasing interest in electric vehicles lead to a growing demand for pure electrical heating systems beyond seat heaters in the automotive industry.
2. The replacement of traditional light bulbs with LEDs will allow the design and construction of completely new lighting systems, especially in conjunction with textile constructions (interior and exterior architecture).

13.4.4 Technology push versus market pull

During the hype from the years 2005 to 2008, smart textiles were used as a marketing tool to gain brand attention. Successful examples are Infineon, O'Neill and Zegna. After dipping into the trough of disillusion in 2008 to 2010, we see a constant growth of interest in this technology reaching a plateau of productivity in the current decade. Although the wearable electronics products became very popular in the hype period, the real business was in the automotive industry: with a 25% growth from 61.7 million cars in 2009 to 77.6 million cars in 2010, this market is still the most interesting for the smart-textile industry. Furthermore, due to growing demand, the number of heated seats per car is also increasing. As the e-mobility gets more and more focus, the automotive industry is looking for purely electrically driven alternative heating concepts as the engine heat waste used to warm up the car will not be available any more. Therefore, we expect that in the short term the automotive heating applications will be the largest business segment in the e-textile market.

13.4.5 Two worlds apart – textile and electronics businesses

Developing, manufacturing and marketing of products in the textile and the electronics industries are completely different. We will compare the garment industry and the consumer electronics industry at a high level. The main difference in the product characteristics is the analog, continuous and flexible world of textiles versus the digital, discrete and precise world of electronics. Garment makers address their customers each season with new styles, whilst the product itself stays more or less the same – a jacket with pockets, sleeves and maybe a hood. But it comes to the market in a variety of styles, fabrics and sizes. In the electronics industry, product innovation is driven by integrating new technologies, adding functionalities and increasing performance. A smart phone of today is a mobile phone, a pocket computer, a camera and an internet network device, more powerful than a mainframe computer of 25 years ago.

The product properties influence the development processes. The garment development process is clearly defined by schedule. A product manager of a garment brand is responsible for dozens to even hundreds of products, to be launched each season (twice a year, sometimes even every quarter). Product changes seem to be simple; no matter how the button is designed, from which company, material or size, a garment designer does not need to care about the integration. However, in the electronics industry, the integration of new components is always complicated; building a prototype, testing and measuring the electrical values, power consumption, etc. In the next step, the developer will program the microcontroller, integrate the components and peripherals and perform the integration testing. After successful implementation and integration, certification tests are necessary before the product can be produced.

Looking at the production processes, tailoring is still a labor-intensive job. Textile productions, fabrics and trims are automated bulk production comparable with mass production of semiconductors. But putting the electronic parts together is an automated and very precise process. The lead time to do all this correctly is much longer than in the garment industry.

Last but not least, sales people at the point of sales are used to handling textile products, but not electronic devices. In addition, the service processes are not designed and organized to handle electronic component returns. The bottom line for these differences is that today we do not have appropriate sales channels for e-textile products.

13.4.6 Economics from production to retail

We also find major differences between the garment and consumer electronics industries in the area of sales. A typical margin structure in the garment industry shows a ratio of four to seven from the cost of manufacturing to the retail prices. A brand company sells the garment to a distributor and takes a margin of 25 to 50%. A distributor and a dealer each take a margin of around 50 to 100%. Compare this with the electronics industry, where the ratio from production to retail in consumer electronics varies typically between 50 and 150%.

Now imagine that an electronic product manufactured for €20 is available in a store for €49. A garment brand produces a jacket for €20 and sells it for €99 in the store. If this brand integrates that electronic system into the jacket, it would be offered in the store for €199 and the customer would pay twice as much for the integrated electronic than just for the device in the electronics store.

13.4.7 Regulations, certifications and markings

Beside the differences in the commercial price calculations in the garment and electronics industries, there are also significant differences in the effort required to market e-textile products.

Legal issues also have to be taken into account. Textile and electronic products need to comply with different norms. Garment manufacturers know about the region- or country-specific norms of material properties required for products with skin contact. There are especially restrictions for use of chemicals, like the European Union (EU) regulation REACH (Registration, Evaluation, Authorization and Restriction of Chemicals). Some garment brands define their own criteria for suppliers of fabrics and trims. Usage of voluntary marks, such as the Oekotex 100 or the EUR Eco-label, are well-known. Markings are commonly in use for material compounds and care procedures.

Electronics products have to be compliant with safety norms and regulations for electro-magnetic compatibility (EMC) and several standards, depending on their usage. Although standardization of these norms is ongoing, most of them have specific national or regional versions. Look at all the marks of a power supply (e.g. from a laptop) to gain an impression. Fortunately, most of the wearable devices do not need to comply with these safety norms, as voltages below 25 or 60V DC are rated as safe, even to children.

On the material level, compliance with the EU Directive on Restriction of Hazardous Substances (RoHS) is mandatory within the EU and part of the CE declaration. REACH is of course, also applied as to textiles, but the importance is not so high.

Standardization on the interfaces, communication, as well as power supplies, is widespread. Some of the standard interfaces that are used within wearable systems are Bluetooth and USB. Some vendors offer compliance marks with their products (e.g. made for iPod™).

13.5 Conclusion

We see ongoing development in the technological areas, as well in materials and (textile) production, electronics hardware and software, which will foster the development of new smart-textile applications. However, there will be ongoing increasing demand to develop portable, connected devices, which will educate users to handle electronics devices as self-evident as clothes. A breakthrough of wearable electronics or smart textiles, as predicted by numerous market studies in the past, will depend on overcoming the following multiple obstacles in the mid-term: Technological challenges are:

- textile look and feel for electronics (robust, lightweight, flexible, elastic);
- connectivity and textile integration;
- lack of standards and inter-operability.

Business challenges are:

- complexity of products;
- channels (know-how in point-of-sales, service processes, margin systems);
- compliance and certifications.

Acceptance challenges are:

- use of batteries (bulky, limited performance and lifetime);
- fear of health impacts (burning batteries, radiation);
- care (lifecycle compliance, demand of green products).

Product managers, designers and developers of smart-textile solution should keep in mind that they are dealing with an electronic system and partner up with companies that complete their own experiences. The conditions to do so are very good, as the smart-textile community is dominated by SMEs, working in flexible networks and alliances.

Short term, we are expecting a combination of market driven projects and offerings from vertical organized alliances and segment specialists. Long term, there is a need to develop horizontal product vendors, in order to offer components and products at lower prices, supplying stable mass production.

13.6 Sources of further information and advice

Fundamentals of innovation, technology and IP management. Enterprise Europe Network: develop your business in new markets, source or license new technologies, access EU finance and EU funding: http://portal.enterprise-europe-network.ec.europa.eu/

European Commission on Innovation and SMEs: http://ec.europa.eu/research/sme/leaflets/en/

Transnational funding opportunities for European SMEs: http://www.innovation-for-sme.eu/

Online Knowledge Center for creating value with innovation management: http://www.innovationmanagement.se/

Paul Trott (1998) was building links between innovation management, new product development, technology marketing and IP management. His work has created a consistent base for the understanding of how to execute innovation as a management process.

Robert Burgelman, Clayton Christensen, Steven Wheelwright: *Strategic Management of Technology and Innovation*, a book with concepts and case studies.

We have found a helpful framework focusing on SMEs in the technology management process (SME Innovation, 2012).

Guidelines on Technology Management for SMEs can be found on the website of Ten^3 (1000ventures.com), with focus on technology audit, business and technology strategy definition and a formulation of a technology strategy.

Centre for Technology Management of University of Cambridge: http://www.ifm.eng.cam.ac.uk/ctm/

A comprehensive information framework for SMEs is provided by the World Intellectual Property Organization (WIPO, 2012).

The EC Centre for Strategy and Evaluation Services has issued a study on working IPR for SMEs (EC IPR Enforcement Group).

There are many great thinkers, writers, inventors and entrepreneurs who created the base in terminology, models and methods. We can choose only a few of them.

We found interesting notes in a blog about the inventor and entrepreneur Thomas Alva Edison (Don Mangum, 2011).

From Tarde (1843–1904), we have learned the meaning and importance of imitation and the role of social networks in the innovation diffusion (Tarde, 1890).

Joseph Alois Schumpeter (1883–1950) was an Austrian-Hungarian-American economist and political scientist. He identified innovation as the critical dimension of economic change. He was one of the business cycle fathers and pioneer of innovation management. He analyzed the impact of organization size on innovation potential, evaluating the dilemma of large organizations (powerful but often bureaucratic) versus SMEs (flexible and fast but limited in resources) (Pol, 2006).

Peter Ferdinand Drucker (1909–2005) was an influential writer and management consultant. He created a base and many systematics in our management reality. His thoughts on innovation, simple, market-driven and built on our strengths, are still relevant today. We have learned from him that 'innovation is work' (Luebke, 2010).

Geoffrey Moore (Moore, *Crossing the Chasm*, 1991) has explained how and why technology companies succeed. He has enhanced the Rogers' diffusion of innovation (Moore, 1995).

Market studies on smart textiles

Several analysts and research organizations, as well as institutes, have worked on smart-textile studies. Their studies predict tremendous market growth and a bright future for the smart-textile market.

VDC stated in the first market study on SFIT (VDC, 2003) that 'the technologies and solutions reported here will be as ubiquitous as are cotton, polyester, computers and cell phones today'. At that time, VDC expected a common annual growth rate (CAGR) between 11% and nearly 30%, with a potential billion dollar business in 2008. Four years later, in the third update of the study in 2007, VDC's estimates showed a growth rate of less than 7% from 2003 to 2006 with even higher CAGRs of up to 50% for the segment of conductive textiles (VDC, 2007). In parallel to the SFIT study, VDC launched a wearable electronics study with emphasis on the electronic part, especially communication and computing, sensing and monitoring, and infotainment solutions (VDC, 2007).

It was reported in the BCC research study of 2007, that the US market for smart textiles was worth an estimated \$70.9 million in 2006. BCC expected this market to reach \$78.6 million in 2007 and \$391.7 million in 2012, at a CAGR of 37.9% between 2007 and 2012 (BCC Research, 2007). Furthermore, the projected rapid growth of military, biomedical, and vehicle safety and comfort applications for smart textiles was expected to have a major impact on both the size and structure of the market. Sales of conductive fabric products were expected to more than double each year between 2007 and 2012.

Frost and Sullivan (2010) stated that still:

> a wide gap exists between the end consumer and the product manufacturer's awareness. As a result, we do not see many big players into smart textiles. However, in the near future, the textile industry is expected to realize its true potential due to a recent surge in the demand for luxury and aesthetic products. This would open new avenues for smart textiles and a gradual growth in this sector will be seen.

In 2011, Global Industry Analysts forecasted the worldwide market of SFIT to reach US$1.8 billion by the year 2015 (Global Industry Analysts, Inc., 2011) (Fig. 13.7). They stated that:

> The market for smart fabric and interactive textile technologies, although in embryonic stages, is poised for tremendous growth. Despite slow adoption rates in several markets, demand for SFIT technology is expected to escalate, driven by emergence of advanced products and new application areas with significant potential. Years of refining technology innovation have helped smart fabrics progress steadily towards achieving successful commercialization in key end-use industries, such as biomedical and consumer electronics.

In the study *The Future of Smart Fabrics to 2021* (Intertech Pira, 2011b), Adrian Wilson evaluates smart fabrics as 'currently extremely fractured in their actual and potential end-use market applications, as well as the products that are being provided. There are few transparent supply chains and widely varying prices'.

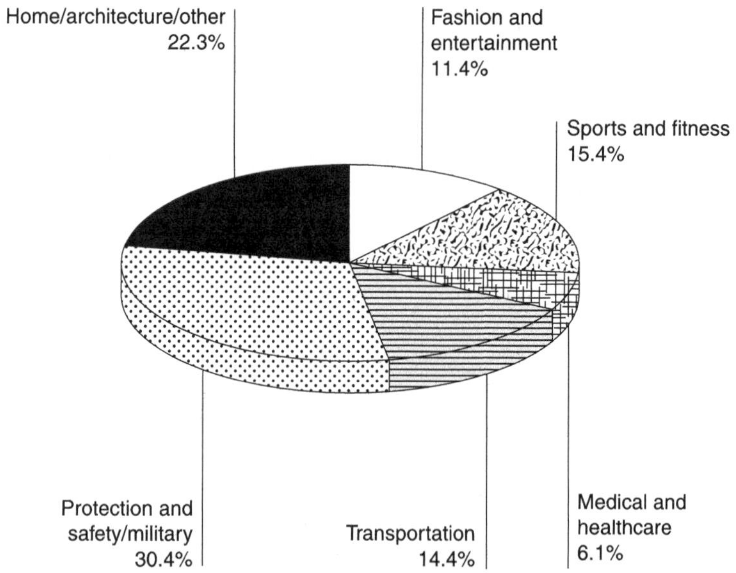

13.7 Smart fabric end use markets 2011. (Reproduced with permission from Smithers Apex, formerly Intertech Pira.)

Intertech Pira, now established as Smithers Apex, estimated the market for smart fabrics at €188.15 million in 2011 and forecast a CAGR of almost 23% to 2016, and nearly 30% for 2016 to 2021.

According to the study, external drivers contributing to this growth include:

- Sensors and other electronics continue to get smaller in size and weight, as well as cheaper, even as their capabilities increase.
- The dramatic increase of the popularity and capabilities of smart phones will result in increased use of smart fabrics, for sensing and monitoring applications, but also in fashion and entertainment.
- By as early as 2016, there are expected to be around 300 million body-worn wireless sensor-based gadgets on the market, with Bluetooth Low Energy (BLE) technology having a major impact.
- Many emerging product packages, including smart fabrics and related products, are rapidly proving to have more value than the purpose for which they were originally intended, largely as a result of advances in software, apps and the internet.
- One key issue in the development of truly smart fabrics has been the lack of microconductors that exhibit the intrinsic qualities of textiles. This is now being successfully addressed.
- Smart fabrics continue to secure funding for new research and development projects.

13.7 References

Andrews K R (1971), *The Concept of Corporate Strategy*. Dow Jones- Irwin.

APCTT. Available from http://www.apctt.org [accessed 19 February 2012].

BCC Research (2007), *Smart and Interactive Textiles*. Norwalk, CT, BCC Research.

Chesbrough H and Rosenbloom R S (2002), 'The role of the business model in capturing value from innovation', *Ind Corp Change*, 11(3), 531.

Don Mangum J (2011), 'Edison and Innovation Blog', in *Innovation in Space*, available from http://norwellconsulting.com/innovationblog/2011/06/28/innovation-in-space/ [accessed 19 February 2012].

EC IPR Enforcement Group, *EC IPR Enforcement Group Report*, available from http://ec.europa.eu/enterprise/newsroom/cf/_getdocument.cfm?doc_id=4053 [accessed 5 March 2012].

European Commission's Directorate General for Enterprise and Industry website, available from http://ec.europa.eu/enterprise/sectors/textiles/index_en.htm [accessed 19 February 2012].

Frost and Sullivan (2010), *Smart Textiles – Assessment of Technology and Market Potential*. San Antonio, TX.

Global Industry Analysts, Inc (2011), *Smart Fabrics and Interactive Textiles: A Global Strategic Business Report*. San José, CA, Global Industry Analysts, Inc.

Hecker K (2011), *Organic and Printed Electronics*, Fourth Edition. Frankfurt, VDMA Verlag.

IDC (2011), *Intelligent Systems Transforming the Embedded Industry*, available from http://www.idc.com/getdoc.jsp?containerId=prUS23026311 [accessed 24 February 2012].

Intertech Pira (2011a), *Smart Fabric Technologies and Applications*. Leatherhead, UK, Intertech Pira.

Intertech Pira (2011b), *The Future of Smart Fabrics to 2021*. Leatherhead, UK, Intertech Pira.

Kasparov G (2007), *How Life Imitates Chess*. USA, Bloomsbury.

Luebke R (2010), *Innovation Excellence*, available from http://www.innovationexcellence.com/blog/2010/07/30/peter-drucker-on-innovation/ [accessed 19 February 2012].

Moore G (1991), *Crossing the Chasm*. New York, Harper Collins.

Moore G (1995), *Inside the Tornado*. New York, Harper Collins.

Morey J (2009), *Distributed Innovation*, available from http://www.adb.org/documents/events/2009/CCEWeek/Presentation-Jessica-Morey-Innovation.pdf [accessed 19 February 2012].

National Research Council (1987), *Management of Technology: The Hidden Competitive Advantage*. Washington, DC, National Academic Press.

Organic Electronics Association (2011), *OE-A Roadmap for Organic and Printed Electronics*, Fourth Edition. Frankfunt a.M., OE-A, VDMA e.V.

Osterwalder A (2004), *The Business Model Ontology: A Proposition In A Design Science Approach*. Lausanne: Universite De Lausanne – Ecole Des Hautes Etudes Commerciales.

Pavitt K (1990), 'What we know about the strategic management of technology', *Calif Mgmt Rev*, 32, 3, 61 pp.

Pol E (2006), *An Introduction to Economics with an Emphasis on Innovation*. Melbourne, Australia, Thomson.

Schwarz A, Van Langenhove L, Guermonprez P and Deguillemont D (2010), 'A roadmap on smart textiles', *Text Prog*, 42(2), S 88.

Sectore Fiche: EU Trade in Electronics (2011), available from http://trade.ec.europa.eu/doclib/docs/2011/july/tradoc_148053.pdf [accessed 19 February 2012].

SME Innovation. Available from http://www.smeinnovation.com/tm_concept.htm [accessed on 19 February 2012].

Tarde G (1890), *Les lois de l'imitation*. Paris, Alcan.

Trott P (1998), *Innovation Management and New Product Development*. Financial Times Management.

Tushmann M L and Anderson P (1986), 'Technological discontinuities and organizational environments', *Admin Sci Quart*, 31(3), 440.

VDC (2003), *Smart Fabrics and Interactive Textiles: Global Market Oppotunity Assessment*. Natick, MA, Venture Development Corporation.

VDC (2007a), *Smart Fabrics, Interactive Textiles, and Related Enabling Technologies*, Third Edition. Natick, MA, Venture Development Corporation.

VDC (2007b), *Wearable Electronics Systems*, Third Edition. Natick, MA, Venture Development Corporation.

WIPO *Program Activities/Small and Medium-Sized Enterprises*, available from http://www.wipo.int/sme/en/ [accessed 5 March 2012].

1000ventures.com. Available from http://it4b.icsti.su/1000ventures_e/business_guide/sme_tech_mgmt_ics.html [accessed 19 February 2012].

14
Improving the sustainability of smart textiles

S. H. W. OSSEVOORT, Lucerne University of Applied Sciences and Arts, Switzerland

DOI: 10.1533/9780857093530.3.399

Abstract: Smart textiles do not necessarily imply a less sustainable option to ordinary textiles, as long as the product offers better user value, user attachment and longevity. This chapter discusses the difference between ordinary sustainable methods based on saving energy and resources and methods that tackle excessive consumption, such as user involved design to enhance product durability. It discusses the theoretical model of user involved design through a practical example of the development of a smart lightweight tracking tent and concludes with a set of general guidelines for developing sustainable smart-textile products.

Key words: user involved design; smart-textile durability; smart-textile recycling; sustainable design guidelines.

14.1 Introduction

Smart textiles are the result of joint research activities in the field of nanotechnology, microelectronics, information technology and textile technology. It is the interdisciplinary nature that makes smart textiles one of the most promising areas of innovation.

The development of smart textiles is of vital importance for the textile industry. Due to globalisation, most traditional textile companies have lost their core activities to low wage countries. Smaller companies have merged into massive industries, which try to survive in a highly-competitive environment. Despite these mergers, Chinese companies have taken the lead, since they are able to offer an attractive low priced 'one-stop textile marketplace.' At the moment, more than a quarter of the production of textiles and clothing is in China, a market share that will increase along with China's growing domestic market (Allwood *et al.*, 2006). In order to remain competitive, smaller traditional textile companies have had to specialise, a process in which smart textiles offer a promising way in which to enter a high tech niche (*Economist*, 2005). The development of smart textiles involves necessary improvements in order to remain competitive (Keenan *et al.*, 2004):

- new production technologies (process technologies, automation);
- new information and communication technologies (ICT);
- new materials and products (multi-functional textiles and garments);
- innovation and research and development (R&D).

In order to realise the necessary technological developments, many companies have increased investment, which has put a corresponding increase on the industry's financial burden. We are reaching a point in which smart textiles should deliver their promise. Innovation for the sake of change is not a sustainable practice.

14.1.1 The sustainability paradox

There are serious doubts about the sustainability of smart textiles (Allwood *et al.*, 2006; Köhler *et al.*, 2011). It is not strange to see where this distrust originates; both the textile and the microelectronic industry do not have the best records on sustainability.

The textile industry, as one of the oldest industries, has introduced many environmental and social problems. The most important raw materials, cotton and polyester, are responsible for severe environmental problems due to the excessive use of water, energy, insecticides, and chemical treatments and improper waste disposal after use. As part of the clothing industry, the textile industry has inspired unbridled consumerism, which leaves no space for durable products.

The microelectronic industry has a similar reputation. It is responsible for polluting our environment with heavy metals and toxic chemical compounds. For the production of components, there is a great need for rare earth metals and minerals (i.e. coltan, gallium or indium), which has lead to many international and national conflicts (SVTC, 2004). The industry has created a highly digital and automated world, bringing forth social changes, which have not always been received positively.

14.1.2 Creating a sustainable future

Although the word 'sustainable', derived from Latin *sustinere*, *tenere*, to hold (*Collins English Dictionary*, 2009) may imply a rigid endurable state, it actually stands for a repetitive cycle, which can only be achieved by looking forward. According to the Brundtland Commission (1987), sustainable developments meet the needs of the present, without compromising the ability of future generations to meet their own needs. In order to be sustainable, the industry should never lose sight of its future aim. There are plenty of examples from the past in which a mindless search for innovation has led to environmental catastrophes.

One example from the textile industry is the development of aniline dyes. Up until the nineteenth century, textile dyes were a rare costly material, mostly extracted from minerals and plants. Their natural origin implied an uncertainty, since quality and supply depended on natural parameters. With the mechanisation of the textile industry, natural dyes became an unreliable source, which called for an alternative. The answer came in 1856, when an 18-year-old chemistry student named William Henry Perkin (1838–1907) accidentally created the first aniline dye distilled from coal, named *Paris* or *Emerald Green*. The common availability of the raw material and the brightness of its colour were the base for its success and soon many aniline dyes where developed. However, nobody wanted to admit, despite common suspicion, to the poisonous nature of the green dye, due to its arsenic compound; the wheel of progress had run too fast to be slowed down. Arsenic dyes remained in use for over 60 years, causing many casualties in our society (Chase and Scott, 2010).

Also, today, we encounter examples in which the innovation potential blinds us from future pitfalls. The development of smart phones and tablet PCs has been highly competitive. In a bid to be the first on the market, producers have adopted indium tin oxide (ITO), a sintered mixture of indium (In_2O_3) and tin oxide (SnO_2), as a transparent conductive layer for their capacitive touch-screen panels (Grossman, 2007). There are not many alternatives to ITO, particularly given its ability to function in humid conditions. As a consequence, the popularity of smart phones and tablet PCs has created a worldwide shortage of indium deposits. Currently, research is being carried out to find prospective replacements such as carbon nanotube conductive coatings (Kaempgen *et al.*, 2005). If no suitable alternative is found, recycling will become a necessity in order to avoid a severe material shortage by the year 2040, but that might prove to be too difficult to be a viable option (Kapilevich *et al.*, 2009).

To make matters worse, ITO has caused interstitial pulmonary diseases amongst those working with the material (Chonan *et al.*, 2007), whilst further research has proven ITO to be a new toxicological entity (Lison *et al.*, 2009). There are still many uncertainties regarding the health and environmental risk of any nanomaterial. Caution is in place, certainly for textile applications at large scale and volume. Let us learn from these mistakes and aim for a smart textile future, where technological developments improve our health, comfort and well-being, without interfering with our natural environment.

14.2 Sustainable production of smart textiles

It is in the interest of any company to minimise the environmental impact of production by reducing waste, energy consumption and material usage. These measurements of efficiency do not only support the environment but also have a positive impact on business revenue.

Through a lifecycle analysis (LCA), an intensive analysis of the consumption of energy and materials before, during and after production, companies can strive for increased efficiency. A practical LCA method to measure the eco-efficiency is the MIPS method (Material Input per Service unit), which takes into account the materials required to produce a product or service (Wuppertal Institute, 2012).

Although efficiency is a way forward, a company can only reach truly sustainable production if it assumes responsibility for the resources to the end user, from the collection of raw materials to the end life of a product. This may be problematic for smart-textile producers, since their products are intermediate goods, disconnected from the end user. Like other textile producers, they import yarn and outsource textile treatments; their end product is one of the many supply chains available for smart product concepts, amongst others such as textile fasteners, electronic components, sensors, batteries or displays. Therefore, the sustainability of smart-textile products depends on the sustainability of its various components.

14.2.1 Transparency

Where do the raw materials come from? What happens during the production process and who takes responsibility at the end of a product's lifecycle? Each supplier is liable for its contribution and has to be open about the environmental impact, early examples of which can be seen in the car industry. Car manufacturers demand certain liabilities from their suppliers in terms of packaging materials; otherwise they will be sent back. As a result, the car industry has increased the need for reusable packaging material.

Despite these liability schemes, complex products lack the transparency needed to offer truly sustainable production (Shedroff, 2009). In our globalised, price-driven and competitive economy, many material resources remain deliberately unidentified. This lack of transparency can only change through the enactment of new laws. However, real changes can only come from consumers' demands and the sustainability policies of the companies.

14.2.2 Closing the loop

Most products are destined to become waste at the end of their lives, a process in which valuable resources are lost forever. It is of benefit to the environment to reduce the amount of waste; however, true sustainable production means closing the loop of a product's lifecycle and eliminating the concept of waste.

Minimising waste

Recollection schemes are a good way to minimise waste. They will become common practice since the European directive for Waste Electrical and Electronic Equipment (WEEE Directive, 2003) has enforced manufacturers to re-collect electrical and electronic equipment for disassembly. However, this directive does not explicitly address smart textiles (Köhler *et al.*, 2011). As long as smart textiies are a niche market for professional and medical clothing applications, they may find their way back to the original manufacturer, but this will change when smart textiles become part of the mass market of today's clothing industry. Ruled by the concept of fashion, smart-textile products will become subject to the notion of change, which generally produces waste (Hethorn and Ulasewicz, 2008).

Eliminating waste

Most products endure a lifetime from 'cradle to grave,' in which their original material is lost indefinitely. McDonnell and Braungart (2002) initiated an approach to carry products from cradle-to-cradle, thereby eliminating waste. They observed that nature lacks the concept of waste; any remainder from a natural process serves as a nutrient. In other words, waste equals food. The cradle-to-cradle approach makes a clear separation between technical and biological nutrients. Biological nutrients are of natural origin and can be composted after use. Technical nutrients may be inorganic or synthetic materials, as long as they are non-toxic, have no negative effects on the natural environment and can be recycled indefinitely, thereby staying out of the natural environment. Since smart-textile materials are mostly of synthetic origin, the challenge is to exclude any toxic materials and create smart materials, which can be recycled indefinitely.

14.3 Recycling, a necessity

14.3.1 Textile recycling

In general, textiles are well suited to recycling, although the quality of the fibres limits the field of application for recycled materials to insulation, protection, cleaning or filling materials. Even though this process involves degrading (down-cycling), it gives 95% of all textiles and clothes (which are handed in for recycling) in the UK a second life.

Despite the possibility to recycle, the majority (85% in the UK) of textiles and clothes end up in landfill sites, partly due to the fact that it is all too easy for consumers to throw unwanted products in a rubbish bin (Textile Recycling Association, 1999; Allwood *et al.*, 2006).

This was very different about 200 years ago, when the demand for raw materials for the textile industry grew beyond its supply. The shortage in wool initiated the development of the pulling process in 1815, which involves breaking down woollen textiles into their constituent fibres (shoddy), so that they can be re-spun into fresh thread (Cupit, 1996). Nowadays, the recycling industry for wool has almost disappeared. The open market economy has pushed the availability of cheap raw materials to unsustainable levels, which makes recycling a less attractive option. Besides, the diversity of materials and their blending compositions is so large that only experienced personnel are able to separate textiles by their material composition. These difficulties make textile recycling both difficult and costly.

Another problem is that currently about 54% of textiles are of synthetic origin (Gherzi Consultants, 2011), which is hard to break down for re-spinning. The separation process of synthetics could potentially be improved through electrostatic and chemical separation, but the two processes are expensive and might not be feasible on a large scale (Allwood *et al.*, 2006). Smart textiles, with their multiple material compositions, will certainly make the process of recycling more complicated.

14.3.2 Smart textile recycling

The term 'smart' has been used to describe functional textiles with engineered properties. This includes anything from curtains that light up in the dark to odour-free socks or concrete reinforcements. The ability to recycle smart-textile products depends on the materials used and the level of integration between the textile and technological components (Mecheels *et al.*, 2004; Tang and Stylios, 2006; Cho *et al.*, 2010).

The lowest level of integration between textile and technology is a pocket, designed to contain a piece of technology. A well-known example is the watch-pocket on a pair of Levi's jeans introduced in 1903 (Fig. 14.1).

14.1 Watch-pocket on a pair of Levi's jeans.

14.2 Smart object built inside the button of a shirt.

The logical next step is using the textile surface to mount electronic hardware in an ordinary way: stitched, attached with Velcro™, buttoned, zipped, clipped or pinned. The autonomous sensor button, an independent senor node (Bharatula *et al.*, 2004) is a good example of a smart object built inside the button of a shirt (Fig. 14.2). It can be stitched onto a textile surface, but should be removed when the product is disposed of. Not removing the electronic hardware will almost certainly disturb the recycling processes of the textile materials (Wäger *et al.*, 2010; Köhler *et al.*, 2011). Small RFID tags have already caused problems in established recycling processes of non-electronic goods.

Even when removed, electronic hardware is very difficult to recycle. An average electronic device may contain as many as 40 different materials in very small quantities. Although most harmful materials have been restricted in a European directive for the Restriction of Hazardous Substances (RoHS), the most common way of recycling, shredding products into smaller entities, does not separate the basic materials. Experiences with the disposal of contemporary electronic waste (e-waste) give reason to expect severe environmental and social impacts worldwide (Köhler, 2008).

Further integration of electronic hardware into textiles has lead to the development of conductive textile materials (Table 14.1). Whilst, in theory, the electronic hardware can be removed and reused in future applications, it does not count for the conductive textile materials, which are infinitesimal metal depositions, such as silver, gold, copper and nickel on a polyamide or nylon yarn. The layer of these metal depositions is too thin to be recycled in a mechanical

Table 14.1 Overview of conductive (electric, magnetic and optical) textile materials

Material	Application	Producer
Metals:		
Conductive black polyester (Ni), Cu, Ni – Sn	Self cleaning, antibacterial and UV protection	http://www.shguilian.com
Stainless steel threads	Magnetic wave shielding	http://www.bekaert.com
Silver plated fabric: (Ni) Cu, (Ni), Ag	Conductive fabric, EMR shielding, wound healing, odour control or anti-microbial	http://www.x-staticfiber.com
Gold Plated fabric: (Ni), Cu, (Ni), Au	Conductive fabric, EMR shielding	http://ajinelectron.co.kr
Electro Magnetic Fabrics: (Ni), Cu, Ni – Co or (Ni), Cu, Ni – Fe		http://www.chinabaicheng.com
Shape Memory Alloys Cu – Zn – Al – Ni and Ni –Ti	Deformable fabrics, ergonomic support, thermo regulation suits	http://www.dynalloy.com
Polymers:		
Polyaniline (PANI)	Microwave and Radar Absorption, Electrostatic Discharge, Resistive or Microwave Heating and Pressure Sensors	http://www.eeonyx.com http://www.fibrontech.com
Polypyrrole (PPy)		http://www.eeonyx.com http://www.fibrontech.com
Poly(3,4-ethylenedioxythiophene (PEDOT)	Conductive coating for Electro Luminescent Panels and Antistatic agent	http://www.heraeus.com http://www.agfa.com
Shape memory Polymers (PU, PET and PEO)	Deformable fabrics, ergonomic support, thermo regulation suits	Laschuk and Souto (2008)
Carbon nanotubes	Conductive coating	http://nanoshel.com
Carbon fibres	Durable clothing, improved heat absorbency	http://nanoshel.com
Optical fibres (PMMA and PS) or silica		http://www.lumigram.com http://www.starscape.co.uk

way. According to McDonough and Braungart (2002), they should be classified as monstrous hybrids, material amalgamations that make recycling and cradle-to-cradle impossible.

The notion of product-smartness has recently shifted toward more active and 'intelligent' functions (Laperre, 2010). New, sophisticated smart textiles, will be able to sense, transmit signals, process data or control the behaviour of the textile. Various new materials will be introduced such as shape memory materials, to adjust the texture of fabrics depending on temperature change or phase change materials, to create a cooling effect or store excess heat. Although the improved efficiency and functionality of novel materials can bring environmental benefits, the risk of introducing these new materials into our waste stream, and ultimately our environments, is not well enough understood.

Recent research uncovered that textiles can lose more than 1900 fibres per washed item (Browne *et al.*, 2012). As the human population grows and people use more synthetic textiles, the environmental problems become more eminent.

This is certainly the case with the introduction of nanoparticles in textile materials. The behaviour of the nanoparticles of any given material may be very different from the actual material properties. For instance, in its natural form, gold is famously inert. However, at a particle size of 2 to 5 nm, gold becomes highly reactive. The chemical composition of these two materials is identical: it is the different physical size of bulk materials and nanoparticles that accounts for their very different chemical properties (RCEP, 2008). The introduction of engineered nanomaterials into textile products will ultimately be followed up by a large-scale production. With this future vision in mind, risk assessment is extremely important before any unintended release of engineered nanomaterials might harm human and environmental health (European Commission, 2011; Som *et al.*, 2011). It emphasises that smart-textile developers have to make design decisions under conditions of uncertainty for which it is important to understand where we have choices and where we do not (Allenby, 2009).

14.4 Product durability

Smart textiles are often introduced as a potential development to make the textile world more sustainable. The driving forces behind these claims are based on textile care, textile durability and feel good fabrics.

- Textile care

Within the lifespan of a piece of clothing, washing, drying and ironing make up the largest consumption of energy and water (Fletcher and Goggin, 2001).

Cleaning: Ideally textile materials should be self-cleaning or stain repellent to reduce the necessary washing cycles. New nanomaterials, such as Nano-Sphere (Schoeller, 2011), have been developed with a special finish that repels water and dirt (Lai, 2003). Unfortunately, users wash their clothes not only because they are dirty but just as much to get the fresh feeling of a newly laundered item (Black, 2008). Therefore, this new material would only be effective if consumers change their habits and refrain from washing to refresh their clothing.

Ironing: Another approach is a new textile finish, which eliminates wrinkles, such as Nano-Tex (Brutten, 2011). Here we have to take into account that ironing is a formal habit. We might question the need for ironing as Dame Vivienne Westwood advocates.

- Textile durability

The lifecycle of clothing is relatively short. One of the causes is the poor quality of textile materials, which is an effect of mass production in combination with price competition. The development of self-repairing textiles is able to counteract this movement and produce textiles that last longer. However, such developments would only be possible if people were willing to pay more for their clothing and change their collection less often.

- Feel good fabrics

Our clothing has a unique function, which is to make us feel good about ourselves. Comfort and well-being are important drivers for happiness. In general, smart clothing can create a positive feeling to which the design has a central role (Black, 2008).

The claims above are focused on the user side of the smart-textile products. These are just as important as the sustainable aim to create smart textiles, which are efficient and recyclable. There is nothing more wasteful than a product that does not work or does not meet the expectations of the user. So the first question we need to ask is what is the added benefit of smart textiles and do these benefits weigh against the inevitable negative consequences for our environment?

In reality, manufacturing products that last longer does not necessarily pose a threat to a healthy economy. Reducing the need for rapid product replacement leads to new services, possibilities to upgrade and above all, it allows products to become prone to aesthetic ageing which will ultimately attract loyal customers (Hinte, 1997). Whilst improving the physical quality is a precondition, creating durable products requires more. According to Jonathan Chapman (2009), if we limit ourselves to the physical durability, we will simply end up with durable waste.

To improve the product's lifetime, it is important that the product represents more than its functionality, which can be accomplished through its design. Design provides meaning; it offers the possibility to attach oneself to a product.

Increased product durability implies a prolonged lifetime that is out of tune with universal economic principles of growth. Products that live longer cause a reduction in sales volume and minimise the option for product innovation. Yet in the area of sustainability, consumer ability to gather and apply the knowledge necessary to make the right choice is diminished by the established system of production. This system, coupled with the desire of consumers to pay the least possible price for the product, has led to over-consumption and waste (Hethorn and Ulasewicz, 2008).

14.4.1 Improving durability

It is very difficult to compare products based on sustainable criteria such as energy consumption, material usage, reusability or recyclability. Even for products of the same category, it is difficult to judge the environmental impact, because most sustainable criteria are complex and may not produce comparable results (Shedroff, 2009).

Comparisons are much easier when the product lifetime is considered. Products that last a long time are more likely to be the best sustainable choice, with the exception of products that consume a substantial amount of energy during usage, such as cars, fridges or light bulbs (Hinte, 1997). The lifetime of these energy-hungry products may be challenged when they become less efficient in comparison with new technologies.

In general, extending the product lifetime is an important approach toward reducing the environmental burden. However, if product life is such an important issue, why are products replaced at ever-increasing rates?

There are several factors that influence the end of a product's life (Walker, 2006; Shedroff, 2009). These factors vary from disposability, wear, non-reparability, functional obsolescence, technological obsolescence and aesthetic (psychological) obsolescence (Ossevoort, 2010). Most of these are directly linked to the product's physicality with the exception of aesthetic obsolescence. The latter is a decisive factor for textile materials.

14.4.2 User involved design

A way to challenge aesthetic obsolescence is by actively involving the user in the design process. Traditionally, designers work from their own experience, observe problems, and generate an accepted solution. The challenge lies within accepting the different points of view of the user. After all, a good designer interprets user needs into a desirable solution.

User involved design is an important topic in the design community these days. It is more diverse than a user test at the end of the product development stage. The latest addition in user involved design is 'crowd funding,' in which design initiatives are presented on the internet to create a group of followers, who make a pledge to purchase the product. Only if a design initiative gets enough followers, will it be produced and distributed.

Ideally users are involved in every stage of the design process, from the first observations to the final use inclusively. For each stage, there are several methods available to support meaning directed to the user involvement (Fig. 14.3).

For each of the key stages, I have identified a general approach:

- *Understand*: In this approach, the designer tries to understand the mechanism of creating value and to apply his/her knowledge (indirect), or a designer engages in observing the user and applies the outcome (direct).
- *Inspire*: Instead of generating a project from technological, market or business opportunities, the user will inspire a new idea.
- *Involve*: Rather than trying to understand, the designer invites the user to play an active role in the design process.
- *Motivate*: An object or service is not brought into completion until it is actually used. Motivate is an approach in which the designer deliberately creates space for the user to complete the design within its use.
- *Engage*: In this approach, the designer tries to establish a link between the user and object/service, which is durable over time.

For each approach, there are several design research methods available, which will be further explained through the example below.

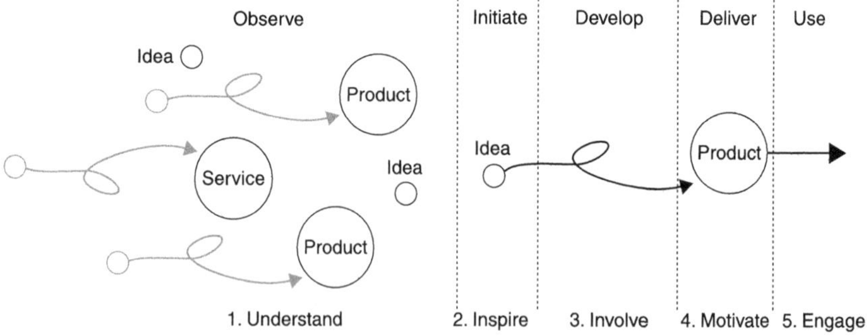

14.3 User involvement in each stage of the design process.

14.5 Sustainable design approach for a smart-textile product, an example

Products that satisfy the end user will have more chance to become a sustainable solution than products that support no particular value system. Unfortunately, users can only articulate product value in terms of existing products, which makes it difficult to predict the value of new products such as smart-textiles products.

In 2004, I was invited to the Hewlett Packard Labs in Bristol, to attend a workshop on smart-textile applications for teenagers. Amongst a group of researchers, we were convinced about the technological challenges that smart textiles offer. We presented some state-of-the-art working examples of high tech fashion to a group of teenagers. Afterwards, we ran a session with these teenagers from whom we expected to receive new ideas and insights. Instead, the outcome of the creative session was merely a direct translation of the technology we had presented onto their most desired pieces of clothing. Was there any reason we could have expected otherwise?

Of course, the user should be central within the design process; however, it is a classical mistake to expect the end user to be able to visualise their needs and wishes. The process of visualising needs and wishes into an inspiring new product concept remains the task of the designer. I will explain how the user can be involved in every stage of the design process with a short design example, a smart tent.

14.5.1 Initial design

Imagine the design of a tent, in which the upcoming smart-textile technologies offer potential for innovation. It is tempting to focus on new materials to increase the performance, such as weight reduction or the addition of new features, such as solar panels or thermo-electric materials (Hewitt *et al.*, 2012) to charge electric equipment. However, rather than asking the question, 'How could we apply smart-textile technology to improve a tent?' designers should ask: 'How could we improve the pleasure of spending the night outside?', which automatically calls for a deeper understanding of the end user.

Understanding

Of course, there are many types of tent, which provide different solutions for different users. There are tents for outside parties, families or single persons, tents to exclude mosquitoes, to shelter from the sun, or even tents within a hospital environment to protect patients from airborne infections. Let us focus in this example on a lightweight outdoor tracking tent for two people.

There are different reasons why one would buy a lightweight outdoor tracking tent, which also implies that there are various types of users or user profiles. Designers normally improvise a range of user profiles to challenge their design ideas.

Unfortunately, most improvised user profiles are too shallow to spur new ideas; they remain a reflection of the designer's own cultural background (Shedroff, 2004).

Instead, try to map the whole range of possible user groups, which includes unintended or extreme users. In our tent example, one should not only look at people hiking with a rucksack but also observe people who would like to take a tent to shelter whilst fishing or people who visit an outdoor music festival. Unlike the average user, unintended and extreme users give the best ideas for user driven innovation.

Once it is clear who might be the end user(s) of the product, it is time to make things real. Understanding users means talking with them, giving them certain tasks and getting involved. Methods like individual interviews and rapid ethnography (Table 14.2) prove to be useful at this stage.

Inspire

Now we understand the variety of users, we could ask what they expect from their tent. However, asking such a direct question like: 'What do you expect from your tent?' would not produce a useful answer. People judge objects through their value(s). If the most important value for a tent is security, we should ask what users expect from a secure outdoor place.

Multiple values can be retrieved from the stories users tell about their product. Words such as saves time, clear, flexible, convenient or easy will indicate a certain value (Kelly and Littman, 2001). Next to rapid ethnography, methods such as cultural probes (Gaver *et al.*, 1999) and personifications (Table 14.2) are helpful to unleash user stories.

In our tent example, the product value could be described as: 'A portable personal space to enjoy the outside with a certain amount of comfort.' This description gives designers a good base to discover new ways to increase or add product value.

Involve

We know that users expect a tent to provide privacy and protection against cold, windy and wet circumstances. However, we may have picked up on user stories, which challenge these values. Some users may have experienced a lack of privacy due to light projections when using flashlight or due to the poor sound insulation of the tent fabric. Other users may have experienced a lack of comfort through the dark, hot and humid atmosphere inside the tent.

Methods like role-plays or designer games (Table 14.2) are useful at this stage to analyse the problems, synthesise solutions or discuss concept directions. For example, a particular story about a user who could not find his or her tent during an outdoor music festival may be the inspiration to create a tent with a tracking device or an audio/visual feedback like a car in a parking lot.

Table 14.2 Methods for user involvement

Approach	Method	Further reading
1. Understanding	Individual interviews	Burns (1989)
	Direct observation	Coleman (2007)
	Rapid ethnography	Norman (2005) (http://jnd.org)
2. Inspire	Cultural probes	Gaver *et al.* (1999) (http://www.gold.ac.uk)
	Personifications	Ossevoort (2002)
	User scenarios	http://www.designcouncil.org.uk under: design methods Moggridge (2007)
3. Involve	Workshops	Laurel (2003)
	Role-play	http://designgames.com.au/role-plays
	Designer games	Flex and the Innovationlab (http://www.designgame.nl)
	Focus groups	http://www.designcouncil.org.uk under: design methods
	Think aloud protocols (TAP)	Dumas and Loring (2008)
	Rapid prototyping	Brown (2009) http://designthinking.ideo.com
4. Motivate	User intervention	Verbeek and Slob (2006)
	Personalisation	van Hinte (1997)
	Co-creation	Winsor (2006)
5. Engage	Sustaining enjoyment	Pieter Desmet (http://studiolab.ide.tudelft.nl)
	Renewing experiences	Brunner *et al.* (2009)
	Multisensory product design	Schifferstein and Hekkert (2008)
	Emotional design	Norman (2005)

A versatile outcome of stories, ideas and concepts can be narrowed down by the likelihood of finding a smart-textile solution. In our example, a tent surface is ideal for generating electricity, creating a self-regulating climate textile or a one-way see-through layer. A further selection of concepts could take place in close discussion with the actual users.

Motivate

Imagine the best-valued concept is a tent with a solar fabric roof, which provides an indirect low-level task light inside the tent (Fig. 14.4). The main reasons are that flashlights are much too focused, may blind the other person in the tent, challenge the privacy by casting shadows on the tent and are reliant on batteries that may run out. The light could be a simple panel of electroluminescent material or LEDs, which shine through the inside tent layer.

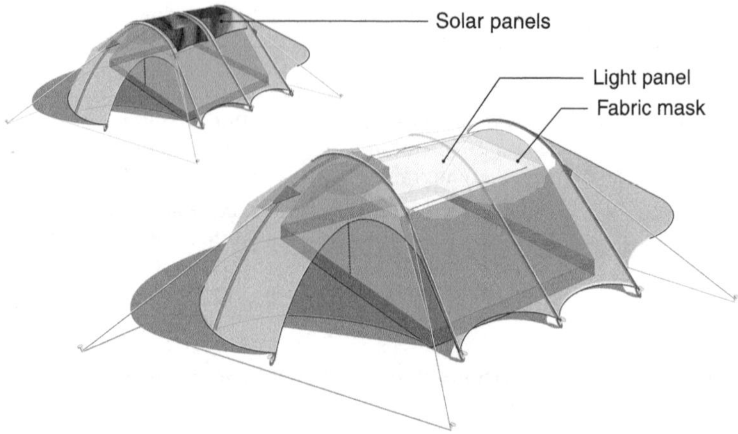

14.4 Simple panel of electroluminescent material or LEDs, which shines through the inside tent layer.

14.5 A fabric mask to resemble a starry night, plant-like motifs, and an indoor ceiling.

The designer could further improve the user-product relationship by motivating the user to intervene in the design. The light panel could shine through a fabric mask, which could be selected by the user to resemble a starry night, plant-like motifs, or an indoor ceiling (Fig. 14.5).

Engage

Engaging the user is especially important when the product is interactive. Interactive products behave in a particular way, which may become casual and uninspiring. Enjoyment and pleasure are important factors, which help to sustain the user-product relationship in order to increase the product durability.

14.5.2 Product development

Material

If weight were not an issue, the best sustainable tent material would be canvas (cotton), a renewable material, which lasts up to three times longer than a nylon.

This poses the first challenge to a smart lightweight tent, the life expectancy of the material. Due to the UV exposure, a nylon tent only lasts about 12 to 30 weeks outdoors. Any smart materials utilised, such as solar panels and conductive thread, would ideally last no longer unless they could be separated and reused, which has some serious implications for the design.

A tent with the solar fabric roof would ideally have a detachable solar panel made from a more UV-resistant material such as polyester. Other important sustainable goals are using less material, mono-materials and renewable materials.

Production

The use of renewable resources or reusable parts and reduction of waste (or zero waste) are the best sustainable achievements in production. For lightweight products, such as tents, local manufacturing is not an issue.

Any electronic component, which exists as multiple material combinations, is ideally reused and should be kept separate from the normal waste disposal. The tent would include rechargeable batteries, which should be placed in a pocket or (well marked) inside a tent pole, to allow the user to separate them for recycling (Fig. 14.6).

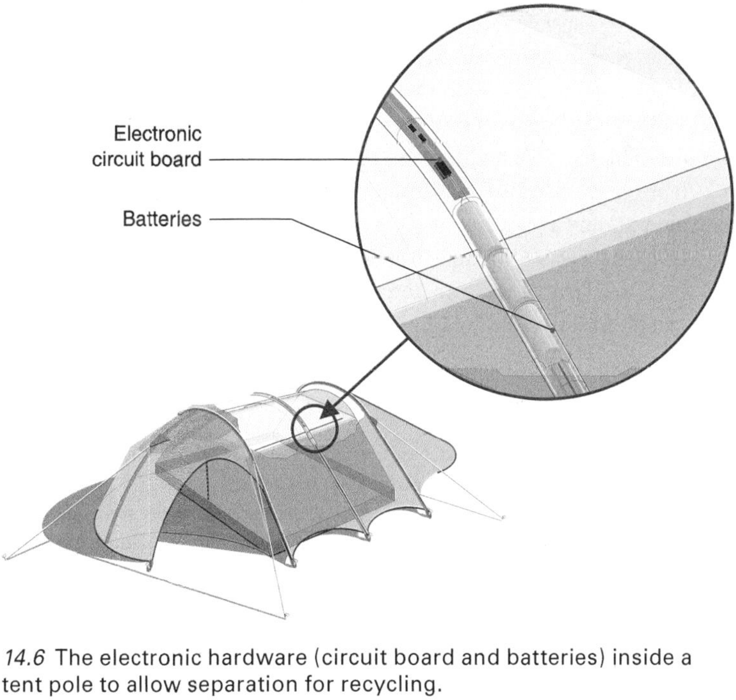

14.6 The electronic hardware (circuit board and batteries) inside a tent pole to allow separation for recycling.

14.6 General guidelines for the design of sustainable smart-textile products

Making smart-textile products truly sustainable is not a viable aim, nevertheless it is important to do whatever is necessary to reduce the environmental burden. For smart textiles, like most electronic products, the best results can be obtained by creating user commitment and a closed loop recycling system. I have created a general list of guidelines, which are important to consider for smart-textile products:

- *Simplicity*: Use smart textiles as much as possible as a carrier for simple and coarse sensors, such as resistive or capacitive sensors. Obtain precision through multiple sensors and extrapolation of data.
- *Surface*: Make use of the vast amount of surface textiles on offer. There is no need to create infinitesimal sensors and electronic components for textile surfaces.
- *Reliability*: Ensure reliability through multiple data paths, parallel lines and fail safe circuitry. Nothing can be more frustrating than a smart-textile product that fails to work properly.
- *Material*: Focus on a minimal use of materials, or even better mono-materials. There are some good examples of clothing, which consist of 100% polyester including Velcro™ connections and zippers, as well clothing made of 100% organic materials such as woollen jumpers.
- *Energy*: A sustainable future can only exist when it is independent of fossil fuels. Smart textiles in building façades, agricultural tarpaulins or outdoor clothing offer the possibility to harvest energy. Other applications, in which smart textiles may offer new possibilities, are membranes for reverse osmotic installations (Kim and Logan, 2011) or materials for energy storage.
- *Recycling*: Because electronic components contain (rare earth) metals, minerals and a plurality of other synthetic materials, they are best kept outside the normal waste stream. If possible, separate any (micro) controlling circuit from the smart-textile material or provide an easy way to separate them (design for disassembly).

14.7 References

Allenby B (2009), 'The industrial ecology of emerging technologies; complexity and the reconstruction of the world,' *J Ind Ecol*, 13, 168–183.

Allwood J M, *et al.* (2006), 'Well dressed? The present and future sustainability of clothing and textiles in the United Kingdom,' Report from the Institute for Manufacturing, University of Cambridge, UK, available from: www.ifm.eng.cam.ac.uk/sustainability/projects/mass/uk_textiles.pdf [accessed 12 February 2012].

Bharatula N B, Ossevoort S, Stäger M and Tröster G (2004), 'Towards wearable autonomous microsystems,' *Perv Comp, Lecture Notes in Comp Sci*, 3001, 225–237.

Black S (2008), *Eco Chic, the Fashion Paradox*. London, Black Dog Publishing.

Brown T (2009), *Change by Design: How Design Thinking Transforms Organizations and Inspires Innovation*. New York, Harper Business.

Browne M A, Crump P, Niven S J, *et al.* (2012), 'Accumulation of microplastic on shorelines worldwide: sources and sinks,' *Envir Sci Technol*, 45, 9175–9179.

Brundtland Commission (1987), *Our Common Future*. Oxford, Oxford University Press.

Brunner R, Emery S and Hall R (2009), *Do You Matter? How Great Design Will Make People Love Your Company*. New Jersey: Pearson Education Inc.

Brutten M (2011), 'Nano-Tex introduces Fortify DP: an innovation in wrinkle-free technology,' *Nano-Tex company report*, 29 September.

Burns C (1989), 'Individual interviews,' in Robson S and Foster A, *Qualitative Research in Action*. London, Edward Arnold.

Chapman J (2009), 'Design for (emotional) durability,' *Design Issues*, 25, 29.

Chase L and Scott S H (2010), 'Two nerdy history girls' blog: "A Deadly Shade of Green," ' blog post Friday, 4 June, available from: http://twonerdyhistorygirls.blogspot.com/ [accessed 20 November 2011].

Cho G, Lee S and Cho J (2010), 'Review and reappraisal of smart clothing,' in: *Smart Clothing Technology and Applications*. Raton, FL, CRC Press.

Chonan T, Taguchi O and Omae K (2007), 'Interstitial pulmonary diseases in indium processing workers,' *Eur Resp J*, 29, 317–324.

Coleman R (2007), *Design for Inclusitivity: A Practical Guide to Accessible, Innovative and User-centred Design*. Hampshire, UK, Gower Publishing Limited.

Collins English Dictionary (2009), *Complete and Unabridged*, 10th Edition.

Cupit M J (1996), 'Opportunities and barriers to textile recycling,' *AEA Technology: Recycling Advisory Unit*, Report 0113, Oxford.

Dumas D and Loring B (2008), 'Moderating usability test, principles and practices for interacting,' Chapter 6: *Interacting during the Session*. Burlington, MA, Morgan Kaufmann.

Economist (2005), 'The great stitch-up restricting Chinese textile exports will only rebound on America and Europe,' *The Economist*, 26 May.

European Commission (2011), 'Towards responsible nanotextiles and coatings: a new risk approach,' *Science for Environment Policy document*, EU DG ENV.

Fletcher K T and Goggin P A (2001), 'The dominant stances on ecodesign: a critique,' *Des Issue*, 17, 5–25.

Gaver B, Dunne T and Pacenti E (1999), 'Design: cultural probes,' *Interactions*, 6, 21–29.

Gherzi Consultants (2011), 'The Cellulose Gap, Zürich,' available from: www.lenzing.com [accessed 12 February 2012].

Grossman L (2007), 'The Apple of Your Ear,' *TIME Magazine*, 12 January.

Hethorn J and Ulasewicz C (2008), *Sustainable Fashion: Why now? A Conversation about Issues, Practices and Possibilities*. New York, Fairchild Books.

Hewitt C A, Kaiser A B, Roth S, Craps M, Czerw R and Carroll D L (2012), 'Multilayered carbon nanotube/polymer composite based thermoelectric fabrics,' *American Chemical Society publication*, Nanoletters, E-pub 8 February.

Hinte van E (1997), *Eternally Yours: Visions on Product Endurance*. Rotterdam, 010 Publishers.

Kaempgen M, Duesberg G S and Roth S (2005), 'Transparent carbon nanotube coatings,' *Appl Surf Sci*, 252, 425–429.

Kapilevich I and Skumanich A (2009), 'Indium shortage implications for the PV and LCD market: technology and market considerations for maintaining growth,' *Photovoltaic Specialists Conference (PVSC)*, 34th IEEE.

Keenan M, Saritas O and Kroener I (2004), 'A dying industry – or not? The future of the European textiles and clothing industry,' *Foresight*, 6, 313–322.

Kelly T and Littman J (2001), *The Art of Innovation: Lessons in Creativity from IDEO, America's Leading Design Firm*. New York, Doubleday.

Kim Y and Logan B E (2011), 'Miocrobial reverse electrodialysis cells for synergistically enhanced power production,' *Envir Sci Tech*, 45, 5834–5839.

Köhler A R (2008), 'End-of-life implications of electronic textiles: assessment of a converging technology,' *University essay from Lunds universitet/Internationella miljöinstitutet*, available from: http://www.lunduniversity.lu.se [accessed 18 February 2012].

Köhler A R, Hilty L M and Bakker C (2011), 'Prospective impacts of electronic textiles on recyling and disposal,' *J Ind Ecol*, 15, 496–511.

Lai S C S (2003), 'Mimicking nature: physical basis and artificial synthesis of the Lotus-effect,' *Report (0020370)*, Universiteit Leiden, August.

Laperre J (2010), 'Jaarverslag 2010: Textiel Overschrijdt Grenzen, Centexbel, Gent, available from *http://www.centexbel.be* [accessed 20 February 2012].

Laschuk T and Souto A (2008), 'Incorporation of SMA technologies in fashion underwear apparel,' poster *Ambience 2008*, Boras, Sweden.

Laurel B (2003), *Design Research: Methods and Perspectives*. Cambridge, MA, MIT Press.

Lison D, *et al.* (2009), 'Sintered indium-tin-oxide (ITO) particles: a new pneumotoxic entity,' *Oxford J Life Sci Med Tox Sci*, 108, 472–481.

McDonough W and Braungart M (2002), *Cradle-to-Cradle: Remaking the Way We Make Things*. New York, North Point Press.

Mecheels S, Schroth B and Breckenfelder C (2004), *Smart Clothes Brochure*. Bönnigheim, Germany, Hohensteiner Institute.

Moggridge B (2007), *Designing Interactions*. Cambridge, MA, MIT Press.

Norman D A (2005), *Emotional Design*: W*hy We Love (or Hate) Everyday Things*. New York, Basic Books.

Ossevoort S H W (2002), *Wearable Dreams*. Interaction Design Institute, Ivrea (internal publication).

Ossevoort S H W (2010), 'Product durability for the experience society,' *DeSForM* (Design and semantics of form and movement), pp. 129–134.

RCEP (2008), *Royal Commission on Environmental Pollution, Novel Materials in the Environment: The Case of Nanotechnology*. London, RCEP.

Schifferstein H N J and Hekkert P (2008), *Product Experience*. New York, Elsevier.

Schoeller (2011), '*Nano-Sphere*, dirt repellent nanocoating,' available from: http://www.schoeller-tech.com/ [accessed 20 February 2012].

Shedroff N (2004), 'The InfoDesign interview by Dirk Knemeyer,' available from: http://www.informationdesign.org/special/shedroff_interview.php [accessed 12 February 2012].

Shedroff N (2009), *Design is the Problem, the Future of Design must be Sustainable*. New York, Rosenfeld Media.

Som C, Wick P, Krug H and Nowack B (2011), 'Environmental and health effects of nanotextiles and façade coatings,' *Envir Intern*, 37, 1131–1142.

SVTC (2004), 'Poison PCs and toxic TVs, California's biggest environmental crisis that you never heard of,' *Silicon Valley Toxics Coalition report*, available from: http://svtc.org [accessed 20 February 2012].

Tang L P S and Stylios G K (2006), 'An overview of smart technologies for clothing design and engineering,' *Int J Clot Sci Tech*, 18(2), 108–128.

Textile Recycling Association (1999), 'Destination of post-consumer textiles,' available from: http://www.wasteonline.org.uk/ [accessed 8 February 2012].

Verbeek P and Slob A (2006), *User Behavior and Technology Development: Shaping Sustainable Relations between Consumers and Technologies*. Dordrecht, Springer.

Wäger P, Schluep M and Müller E (2010), 'RoHS Substances in mixed plastics from waste electrical and electronic equipment,' *Empa WEEE Forum Final Report*, 17 September, St Gallen.

Walker S (2006) *Sustainable by design: Explorations in Theory and Praxis*. London, Earthscan.

WEEE Directive 2003/108/EC of the European Parliament and of the Council amending Directive 2002/96/EC on waste electrical and electronic equipment (WEEE), 8 December, available from: http://ec.europa.eu/environment/waste/weee [accessed 20 February 2012].

Winsor J (2006), *Spark: Be More Innovative Through Co-Creation*. Evanston, IL, Agate Publishing.

Wuppertal Institute, MIPS Material Input Per Service unit. Documents on the method can be downloaded from: http://www.wupperinst.org/en/publications [accessed 10 March 2012].

15
Medical applications of smart textiles

S. COYLE and D. DIAMOND, Dublin City University, Ireland

DOI: 10.1533/9780857093530.3.420

Abstract: Smart fabrics and interactive textiles are a relatively new area of research, with many potential applications in the field of biomedical engineering. Sensors integrated into textiles in contact with the body have the ability to capture many useful physiological signals. This chapter discusses some of the signals that can be measured using smart textiles and examples of the application of this technology. Future trends in this area of research are presented and issues regarding technology development and its uptake are highlighted.

Key words: smart textiles, wearable sensors, biomedical engineering, physiological measurement, wearable chemosensors, textile sensors, functional fabrics, connected health.

15.1 Introduction

Medical engineering aspires to improve human well-being and promote health through the creation of devices for diagnosis, treatment and cure of medical conditions. Some devices are designed to analyse the physiology of the human body, such as ultrasound imaging or the familiar stethoscope. Other devices function as assistive tools, where the body's system has weakened, such as hip replacements or orthotic limbs. The development of new materials and the miniaturisation of electronic devices have resulted in great advances in this largely multidisciplinary field. Medical engineering combines the skills from engineering, scientific and medical communities. A surge in multidisciplinary research in the 1960s and 1970s led to many advances in modern healthcare, based on breakthrough discoveries in biomedical engineering (Fagette, 1997; Bronzino, 2000). Developments in microelectronics and material science have helped to miniaturise many of these technologies. The first electrocardiograph, to monitoring heart function, developed by Einthoven at the beginning of this century, weighed about 600 pounds and required 5 people to operate it (Street, 2008). In the 1960s, portable Holter monitors were being used in clinical settings to allow 24 h monitoring of patients. Today, with the advances in smart textiles, it is possible to measure heart rate through the clothes we wear and link up this information to wristwatches or mobile phones.

Smart fabrics and interactive textiles are a relatively new area of research, with many potential applications in the field of biomedical engineering. The ability of smart textiles to interact with the body provides a novel means to sense the wearer's physiology and respond to the needs of the wearer. Wearable sensors and smart fabrics in principle have the ability to sustain the health and wellness of the wearer. Clothing that we don for hours each day has the capacity to continuously monitor the physiology of the wearer. Physiological signals, such as heart rate, breathing rates and activity levels, are useful indictors of health status. These signals can be measured by means of textile-based sensors integrated into smart clothing. Smart garments have the potential to offer a personalised healthcare solution, by making the wearer more aware of their health status. This would thereby encourage individuals to take a more active role in their own healthcare.

The advantage to medical practitioners is a more thorough insight into their patients' well-being in the home setting. This provides a full picture rather than snapshot clinical visits, which may miss important events. Wearable sensing devices can give doctors quantitative measures of how their patients are progressing at home. Doctors rely on patients' accounts of how they feel. Patients often tend to misreport the severity of their condition. Smart clothing has the ability to keep a digital record of the patient's physiological response since their last clinical visit, allowing doctors to make a more accurate diagnosis. Similarly, in rehabilitation, it is difficult for therapists to ensure that patients are complying with prescribed exercises. Smart garments sensing body movements have the potential to guide wearers through their exercises, while also recording their individual movements and adherence to their prescribed programme. This detailed information gives a quantitative measure of improvement since the last clinical visit. Through long-term monitoring, medical staff have a broader window into the person's condition and can provide more personalised care. This extends from lifestyle and health status management to individualised medicines and treatment.

15.2 Monitoring of body parameters

Most of the existing wearable technologies are based on physical sensors, such as electrodes, thermistors and accelerometers. These sensors respond to physical changes in their environment, for example electric fields, heat and movement. Textile electrodes can be used to detect electric signals from the body, such as electrocardiography (ECG) from the heart and electromyography (EMG) from skeletal muscles. Textile strain gauges and pressure sensors can detect body movements, such as breathing movements and foot pressure. Accelerometers pick up speed of movement and depending on placement, can focus on a specific limb or determine the body's general activity level. An emerging field in the area of wearable sensors is in wearable chemo-sensors. These devices have the potential to measure many more variables relating to the person's well-being and safety. Chemical sensors have an active surface, which reacts or interacts with a sample at the molecular level, for example through the use of immobilised receptors that selectively bind a particular target species, and in so doing generate an observable signal (Janata, 2009). Wearable chemo-sensors can monitor the composition of

Table 15.1 Physiological signals that may be measured using textile based sensors

Physiological measurement	Textile-integrated sensors	Signal source	Typical sensor placement
Breathing patterns	Piezo-resistive stretch sensors, inductive plethysmography, impedance plethysmography, optical fibres	Expansion and contraction of ribcage during breathing	Thoracic-abdominal region
Heart activity	Woven/knitted electrodes	Electrical activity of heart	Thoracic region
Muscle activity	Woven/knitted electrodes	Electrical activity of muscles	Skin surface overlying relevant muscles
Blood oxygen saturation	Optical sensing components, plastic optical fibres	Light absorption of haemoglobin in blood	Regions with good blood perfusion, e.g. finger tip, earlobe
Blood pressure	Features of the photoplethysmography (PPG) signal	Arterial pressure pulsations	Finger, wrist and earlobe
Body movement, posture	Piezo-resistive strain/ pressure sensors, accelerometers, gyroscopes, optical fibre sensors	Body kinematics	Dependent on motion to be analysed
Electrodermal activity	Woven electrodes	Skin electrical conductivity	Fingertips
Composition of body fluids	Electrochemical sensors, colorimetric pH fabric	Composition of sweat, saliva, urine	Fluidic sampling system necessary

body fluids, such as sweat, saliva, tears and urine. There are also potential biomarkers from gaseous samples in breath and perspiration. Table 15.1 lists some physiological parameters that have been widely studied using textile-based sensors.

15.2.1 Breathing

Breathing is closely related to our physiological and psychological state. Breathing patterns can be monitored by measuring changes in thoracic volume as the ribcage expands and contracts with each breath. This movement can be detected using wearable strain gauge sensors or using textile electrodes for electrical impedance or inductance plethysmography.

Textile-based strain sensors have been demonstrated using stretch fabrics modified with inherently conductive polymers (Rovira Carlos, 2011) or carbon loaded rubbers (Tognetti *et al.*, 2005). Knitting with conductive yarns is another approach to creating textile piezo-resistive sensors (Loriga *et al.*, 2007). Stretch of the textile sensors leads to a change in conductivity of the material. The use of multiple sensors allows not only the breathing rate to be measured, but also gives an indication of the amplitude of breathing, for example deep breathing manifests as a large change in signal amplitude. A respiratory monitoring vest, developed by the Adaptive Sensors Group in Dublin City University in collaboration with Shimmer Research (www.shimmer-research.com), is shown in Fig. 15.1.

Electrical inductance plethysmography integrates two conductive wires into a garment, one around the ribcage and the other around the abdomen. Motions of the chest wall cause changes of the self-inductance of the two loops. Impedance pneumography uses two or four textile electrodes placed at the thorax. It involves injecting a high-frequency and low-amplitude current through the electrodes and measuring the trans-thoracic electrical impedance changes. This technique has been used in projects such as myHeart and WEALTHY (Paradiso *et al.*, 2005(b); Paradiso and DeRossi, 2006).

An alternative approach to monitoring breathing is using textile-integrated optical fibres. The OFSETH (http://www.ofseth.org/) project aims to develop new

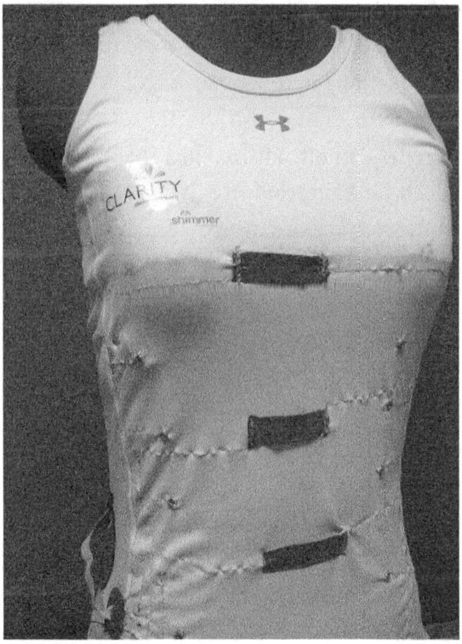

15.1 Respiratory monitoring shirt with integrated piezo-resistive sensors. Data is acquired and wirelessly transmitted using a Shimmer™ device.

monitoring devices for various vital parameters by embedding optical sensors into textiles using textile processes optical fibres. To monitor breathing, fibres were stitched onto textile in a sinusoidal shape. The fibres were illuminated with a laser and light detected with photodiodes. The curvature of the bends affected the light attenuation through the fibre. Analysis of the bending of these fibres was used to recognise the breathing movements of the wearer's upper body. This technology has the advantage of being compatible with MRI scanning (D'Angelo *et al.*, 2008).

15.2.2 Heart activity

ECG records the electrical activity of the heart from the skin surface. With each heartbeat, an electrical signal spreads from the top to the bottom of the heart. As it travels, the signal causes the heart to contract and pump blood. Analysis of the timing and morphology of the electrical signal can help to diagnose heart problems. ECG is typically measured using silver chloride electrodes coupled to the skin with gel. In clinical settings, 12 electrodes are used, which allows the heart's activity to be viewed from 12 different angles; 3 and 5 lead portable systems, such as a Holter monitor, are used when continuous monitoring is needed, but these require wires and adhesive electrodes to be placed on the body.

Flexible conductive yarns, fully metal yarns, or natural/synthetics blended with conductive fibres have been knitted into garments to develop textile electrodes (Catrysse *et al.*, 2003). A study by Paradiso *et al.* (2005(a)) shows good correlation between fabric electrodes and traditional silver/silver chloride electrodes. A major issue with bio-potential electrodes is the need to use a gel to provide a good contact between the sensor and the skin. This restricts the ease of use and can cause skin irritation. In recent years, much effort has been focused on the development of 'dry' electrodes. Dry electrodes are well suited for long-term monitoring, although they are subject to more noise interference, particularly from motion artefacts and power line interference (Xu *et al.*, 2008; Merritt *et al.*, 2009).

15.2.3 Muscle activity

EMG is a method of measuring electrical activity from muscle. Surface EMG (sEMG) technology allows information regarding the overall muscle function and condition to be collected from the surface of the skin in a non-invasive manner. Intramuscular EMG involves the use of a needle electrode to penetrate directly into the muscle, giving a more localised measurement about muscle fibre activity. This method is primarily used as a diagnostic tool to detect very subtle changes of electrical muscle activities to determine underlying disease processes. For wearable applications, surface EMG is more useful and feasible and has applications in biofeedback relaxation and in rehabilitation. Traditionally, disposable metallic (Ag/AgCl) EMG electrodes with conductive paste are glued to the skin upon the muscles of the person under the test. An EMG device can give a patient feedback on how hard he or she is exercising an injured muscle and provide an evaluation of muscle function to monitor improvement with therapy. A study by Finni *et al.* (2007) showed that the signals from textile electrodes are in good agreement with traditionally measured surface EMG signals. In the textile electrodes, the shape and size of the conductive area and the inter-electrode distance are different and much larger than in the traditional Ag/AgCl electrodes. The larger conductive area is not so sensitive to slight differences in electrode positioning in longitudinal studies as compared to the traditional electrodes. However, the larger area is not muscle-specific but collects data from entire muscle groups, which can be an advantage in conditions where information on precise motor control is not required.

Another application of EMG is in the control of prosthetic devices. Textile electrodes embedded into a garment are a straightforward means of electrode placement. A recent study demonstrated that e-textiles can be used as recording systems for highly accurate EMG control of hand and wrist functions for active prostheses. A sleeve with high-density electrode grids covering the upper and lower arm was used to record EMG signals (Farina *et al.*, 2010).

The ConText project, funded under EU FP6, involved the development of a vest that measured a sEMG signal with textile integrated sensors, without direct electrical skin contact. Textile electrodes were embroidered into a garment and functioned as capacitive transducers measuring through a textile layer (Linz *et al.*, 2007). The possibility of avoiding direct skin contact reduces skin irritation problems. When using traditional electrodes, the skin is usually prepared by shaving, abrading and cleansing, which can cause irritation.

15.2.4 Blood pressure

Blood pressure is an indication of the amount of work that the heart has to do to pump blood around the body. A traditional blood pressure meter, known as a sphygmomanometer, comprises an inflatable cuff to restrict blood flow and a mercury or mechanical manometer to measure the pressure. A stethoscope is used to listen to the pulse to take a pressure reading when the heart muscles contract (systolic pressure) and when the heart muscles relax (diastolic pressure). Automated electronic devices are beginning to replace the traditional manual sphygmomanometers and this technology has led to portable ambulatory monitoring devices (Pickering *et al.*, 2005). A wrist-worn device available for home use is available from Omron Healthcare (http://www.omronhealthcare.com). The Omron Wrist Blood Pressure Monitor takes periodic blood pressure readings and stores the readings to memory with a time stamp. Blood pressure varies at different times during the day, and night-time levels are important for diagnosis.

Blood pressure can also become elevated due to anxiety. This can lead to 'white coat hypertension,' where anxiety in the doctor's surgery causes a reading that is higher than normal. Therefore, to get a true picture of the patient's blood pressure, ambulatory monitoring is necessary. Ideally, this should be carried out in a discrete manner. Earlier systems made a loud sound when the cuff was being inflated or deflated. Harry Asada's group, at the Massachusetts Institute of Technology, have been developing wearable blood pressure monitors based on features of the pulse signal (Asada *et al.*, 2010). Photoplethysmography (PPG) is an optic signal related to the volumetric pulsations of blood in tissue, which in turn is related to arterial pressure pulsations (Reisner *et al.*, 2008). Rather than using the traditional occlusion cuff, Asada *et al.* (2010) created a device that uses two in-line PPG sensors. One sensor was placed against the ulnar artery at the wrist and one was placed against the digital artery of the little finger. This measured the peripheral pulse transit, which can be correlated with important arterial blood pressure. The device is being commercialised by CardioSign, a company launched by a former group member, Philip Shaltis (Trafton, 2009). Another device has been developed by Tatara *et al.* (2007), to monitor blood pressure. The device is small, lightweight and designed to be worn on the ear. It has the advantage of being unaffected by arm movement. Although these devices are wearable, they are not textile-based. However, the ability of optical fabrics to detect PPG signals (Selm *et al.*, 2010) imply the potential for smart-textile blood pressure monitors.

15.2.5 Blood oxygen saturation

Blood oxygen saturation is a measurement of the amount of oxygen attached to the haemoglobin (Hb) cells in blood. Pulse oximetry is a simple non-invasive method of monitoring the percentage of Hb in blood, which is saturated with oxygen. In a fit, healthy individual, this is usually above 95%, although it varies with degree of fitness and current altitude. A pulse oximeter works by measuring the absorption of light through body tissue with a high perfusion rate of blood, usually at the finger or earlobe. Hb has a different absorption spectrum, depending on whether it is oxygenated (oxy-haemoglobin) or deoxygenated (deoxy-haemoglobin). Oxygen saturation (SpO_2) is estimated by measuring the absorption of two different wavelengths of light through the tissue. LEDs are typically used as light source and photodiodes as light detectors. These optical components may be placed in a transmission mode configuration, on either side of the tissue, or else in reflectance mode on the same side of the tissue. Photonic textiles using OLEDs or woven polymer optical fibres (POFs) offer an alternative to conventional LEDs, to create a textile-based pulse oximetry system. Rothmaier *et al.* (2008) have demonstrated such as system, using a cotton glove with woven POFs positioned at a fingertip of the glove.

The disadvantage of using fingertip or earlobe probes is that they may interfere with the user's daily activities. A pulse oximetry sensor using optical fibres integrated into fabric has been demonstrated to measure blood oxygen saturation at the sternum (Solà *et al.*, 2007).

15.2.6 Body movement

There are many variations of textile-based and wearable sensors to monitor movement of the body. The configuration depends on the motion under analysis and the application. Monitoring body movement can give an indication of general activity, which may be useful in monitoring the well-being of isolated individuals or motivating patients to maintain a daily activity level in the case of obesity or cardiovascular disease. Measuring specific motions has applications in rehabilitation therapy. Monitoring characteristics of the body's movement is also useful in assessing conditions affecting motor skills, such as Parkinson's disease.

There are two main approaches in smart garments for monitoring biomechanics. One area is in the use of inertial sensors together with advanced algorithms, the other area is the development of textile-based sensors (De Rossi and Veltink, 2010).

Fabric-based sensors, which respond to stretch or pressure, can be used to capture kinematic movements of the body. Fabric stretch sensors can detect movement of the limbs by placing sensors across joints, such as the knee or elbow. The sensors may also be placed on the back, to examine posture of the spine and upper body position and gestures. Danilo De Rossi's group, at the Interdepartmental Research Centre 'E. Piaggio,' University of Pisa, in collaboration with Smartex Srl (www.smartex.it), have developed various garments to measure biomechanics. Conductive loaded rubbers have been integrated with textiles as sensing elements and as conductive tracks to data acquisition units (Carpi and De Rossi, 2005). An upper limb kinesthetic garment, which detects the posture of wrist, elbow and shoulder, has been developed for post-stroke patient rehabilitation. A photo of this garment and a conceptual diagram of such kinematic garments are shown in Fig. 15.2. This garment can be integrated into a healthcare service, which allows patients to continue the rehabilitation training at home without the help of physicians, after the intensive rehabilitation period (Tognetti *et al.*, 2005).

Gerhard Tröster's group, in ETH, Zurich, have also developed a prototype garment that can recognise upper body postures using textile strain sensors. A strain sensitive fibre was made using a mixture of a thermoplastic elastomer and carbon black particles. The fibre was extruded using a capillary rheometer. It was integrated into the back region of a tight fitting garment, using conductive thread and conductive epoxy (Mattmann *et al.*, 2008). A further study with this smart garment involved the identification of exercise execution on a gym machine to measure the intensity of training and dynamics of movement (Mattmann *et al.*, 2007).

These sensor garments offer potential aids to physiotherapists and their patients. At present, therapists monitor and regulate exercises performed by patients during therapy sessions and encourage patients to continue self-exercise programmes.

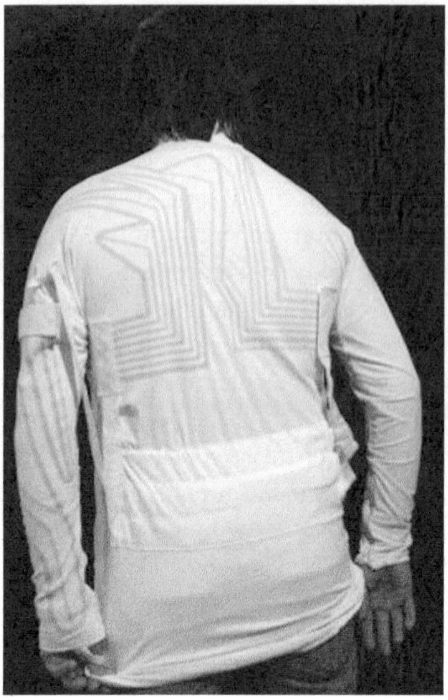

(a)

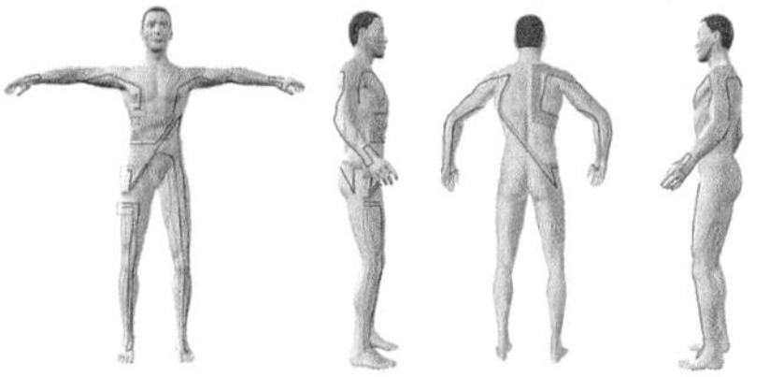

(b)

15.2 (a) Post-stroke rehab garment. (b) Conceptual diagram for the design of kinematic sensing garments developed by Interdepartmental Research Centre 'E. Piaggio', University of Pisa and SMARTEX Srl. (Photo courtesy of Professor Danilo De Rossi.)

Patients need to be kept motivated between therapy sessions, to carry out their prescribed exercises. Lack of adherence to prescribed exercise is a common problem and delays recovery. It is very easy for subjects to become distracted and unmotivated in performing the exercises at home. In some cases, debilitating conditions may lead to depression and lack of self-motivation, leading into a cycle that stalls progress. Smart garments have the potential to inform therapists about what exercises have been carried out and whether they have been performed correctly. At the same time, they can be connected to a user interface providing guidance and feedback to patients.

Another important clinical application of biomechanical sensors is in monitoring and aiding management of conditions, such as Parkinson's disease. Currently, clinical visits are inadequate to sample the severity of Parkinsonian symptoms, because symptoms vary in response to medication dosage and can affect a patient's perception of their own motor status. Therefore, there is a need to gather objective measures of the severity of symptoms over time, to reliably assess the effectiveness of medication adjustments (Bonato, 2010).

Fall detection is another key application of wearable kinematic sensors. Approximately 28 to 35% of people aged 65 and over fall each year, increasing to 32 to 42% for those over 70 years of age (World Health Organisation, 2007). The frequency of falls increases with age and frailty level. Approximately one-quarter of falls result in physical injury and incur high costs in terms of quality of life and to health and social services. Falls may also result in a post-fall syndrome, which includes dependence, loss of autonomy, confusion, immobilisation and depression, which will lead to a further restriction in daily activities. Typically accelerometer-based devices are used to detect falls, the principle being that the fall has a different pattern of motion compared to other activities. While accelerometers are low-cost and relatively easy to position on the body or in a walking stick, the classification of falls is a complex task, which can often lead to many false alarms. Yu (2008) gives a survey of fall detection for elderly patients, focusing on identifying approaches and principles of the existing fall detection methods. Perry *et al.* (2009) have also published a survey of fall detection methods, with evaluations based on their own results. Their study found acceleration to be a critical role in elderly mobility monitoring, and should be used in fall detection. Compared to visual and acoustic sensors, accelerometers consume less energy and are much easier to integrate into wearable mobility monitoring devices. The Technology Research for Independent Living (TRIL) Centre has a central research focus on fall detection and prevention. The centre aims to elucidate the factors contributing to falls and to use this information to develop assessment tools to identify those at risk of falls (http://www.trilcentre.org/falls-prevention research).

15.2.7 Electrodermal activity (EDA)

Emotional reactions often cause increased sweat gland activity on the palms of the hands and the soles of the feet. The baseline level of skin conductance varies from person to person and depends on both patient physiological state and autonomic regulation. This baseline varies slowly over time, indicating the general activity of the perspiratory glands that may be affected by factors such as temperature. A faster phasic change or 'peak' in skin conductance can occur as a result of a specific stimulus or event. This response appears between 1.5 and 6.5 seconds after the stimulus. Electrodermal activity (EDA) is generally measured by placing silver-silver chloride electrodes on the first and second fingers of either one or both hands. One electrode emits an electrical signal, whilst the other primarily acts as a receiver, thus a circuit is established that passes across the skin. Prototype gloves and socks have been developed to measure the EDA response in a wearable manner. The MARSIAN smart glove (Axisa *et al.*, 2004) was developed to measure the activity of the autonomic nervous system by measuring physiological parameters of the skin. Two prototypes were developed, the first based on Kapton/copper foil and the second was a hairnet structure with standard silver/silver chloride electrodes embroidered into it. The European Union PSYCHE and MONARCA projects have also developed garments measuring EDA, as part of a system to support the diagnosis and treatment of bipolar disorder patients (Gaetano Valenza, 2010; Kappeler-Setz *et al.*, 2010). The MONARCA project has shown the integration of the sensor system in the shoe or sock as a promising approach. EDA peaks at the foot occurred around 0.5 seconds later than the EDA peaks at the hand. The context of the user, such as the outside temperature and clothing, needs to be taken into account when measuring and analysing EDA traces, as this affects the evaporation of sweat from the skin. In the PSYCHE project, the hands and feet have been used for measuring EDA. Another study by Poh *et al.* (2010) has demonstrated a wrist-worn sensor that measures EDA from the forearm.

15.2.8 Composition of body fluids, sweat analysis

Wearable chemo-sensors can be integrated within a garment to access body fluids (Coyle *et al.*, 2010a). Blood is the most reliable diagnostic medium; however, as it requires invasive techniques for sampling, it is typically sampled at specified time intervals. Therefore other body fluids, which can be accessed more easily through non- or minimally-invasive means, must be considered for continuous analysis. Possible samples include urine, saliva, sweat, interstitial fluid and wound exudate. Of all these body fluids, sweat is the most accessible within a garment structure. Sweat is a filtrate of blood plasma and contains many substances of interest, such as sodium, chloride, potassium, bicarbonate and calcium (Shirreffs and Maughan, 1997) (Fig. 15.3). Sweat analysis offers valuable physiological information. The sweat test, measuring sodium and chloride concentration levels, is the gold standard technique for the diagnosis of cystic fibrosis (CF). Sweat composition is highly variable among individuals and is also affected by environmental conditions and activity levels.

Analysis of sweat loss and sweat composition can offer valuable information regarding hydration status and electrolyte balance, which is vital for our well-being. Analysis of sweating patterns during the day and night-time may be useful for monitoring patients with conditions such as diabetes and hyperhydrosis (Haider and Solish, 2005; Asahina *et al.*, 2008).

The most widely-used methods of sweat analysis to date involve the collection of sweat by using patches or pouches and then analysis afterwards using a standard measurement technique (Brisson *et al.*, 1991). The BIOTEX project developed a textile-based system to collect and analyse sweat, by using a textile-based sensor

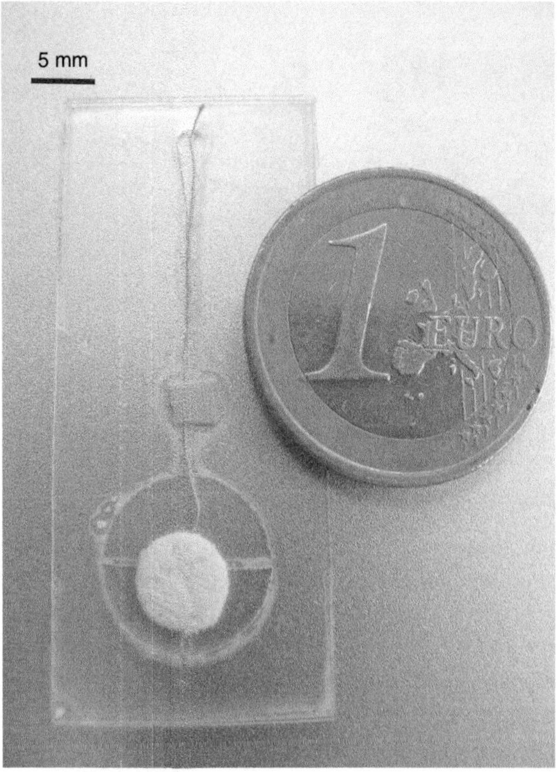

15.3 Microfluidic chip for analysis of sweat pH. (Photo courtesy of CLARITY, Dublin City University.)

capable of performing chemical measurements. This work demonstrated the use of a fabric-based patch to measure sweat pH, sodium and temperature and a textile-based sweat rate sensor utilising humidity sensors (Coyle *et al.*, 2010a, Salvo *et al.*, 2010). The sweat rate sensor (Fig. 15.4), integrated two humidity sensors in a textile structure at two distances (0.5 and 1.5 cm) from the skin. Sweat rates were estimated from the gradient of humidity measured by these devices. Recent work by Benito-Lopez *et al.* (2010) has involved a miniaturised microfluidic sensing platform for real-time sweat analysis. Figure 15.3 shows a microfluidic chip used to collect sweat samples for real-time analysis of sweat pH. The fluidic channel contains a pH sensitive material, which changes colour, depending on sweat pH.

Yang *et al.* (2010), at the University of California, San Diego, have demonstrated printed biosensors in clothing. Electrochemical sensors were printed directly onto the elastic waist of underwear that offers tight direct contact with the skin. The group plans to incorporate chemically-selective layers, which can detect substances, such as ethanol or lactate.

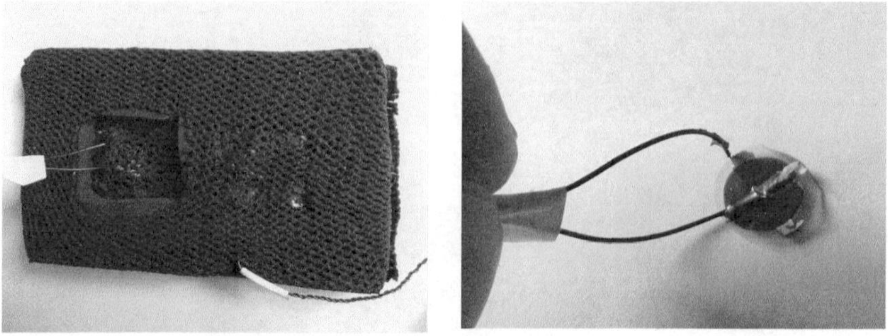

(a) (b)

15.4 (a) Sweat rate sensor developed at University of Pisa during the BIOTEX project. (b) Humidity sensor integrated into textile pockets of the sensing device at different distances from the skin. (Photo courtesy of Dr Fabio Di Francesco.)

Another approach to sweat detection has been demonstrated by Vaughan and Scully (2008). Polymer optical coatings to detect moisture were developed to clad POFs. A POF sensor has been integrated into a garment that can be worn during daily activity and is unobtrusive. The plastic optical fibre sensors were designed to continuously monitor human perspiration in response to changes in environmental conditions, physical and mental activity, medical condition or psychological stress.

15.3 Challenges in medical smart textiles

There is no doubt that wearable technologies have much to offer to the medical sector, such as continuity of care from early detection through to home therapeutic and rehabilitation aids. They have the potential to introduce more personalised care and to encourage people to take a more active role in their health and well-being. This will hopefully help to reduce the demand on future healthcare systems and result in more efficient care. In order for this technology to become widely accepted, there are a number of challenges that must be overcome.

15.3.1 Wearability

Wearable technologies should be soft, flexible and washable, to meet the expectations of normal clothing. Washing is an important factor in terms of product lifecycle. The first wearable computers were bulky and rigid and were really more 'portable' rather than 'wearable' systems. With increased miniaturisation of electronic devices and the advances in functional materials, wearable technologies are becoming more integrated into the textile structure. By adding function to the fabric and creating a fabric sensor, the smart garment can retain its normal tactile properties. The design of smart garments and sensor integration depends on the application, the users' needs and what measurements are most useful (McCann *et al.*, 2005). A smart garment must be suited to the user, for example a smart jacket for a neonate must be designed for minimum stress when dressing (Chen *et al.*, 2010).

Ideally, all the components that constitute a smart garment, including power, sensors and electronics, would be textile-based and washable. In reality, there are limitations with these components. Advances in flexible batteries (Liu *et al.*, 2012) and textile transistors (Barbaro *et al.*, 2010) show promise for the future; however, until such a time, electronic modules need to be sealed or removable. They should also be packaged without sharp edges, which could be hazardous, and should be positioned in an ergonomic manner. Sensors have the issue of calibration. In particular, chemical on-body sensors require an active surface to react. One solution to avoiding calibration is to use low-cost sensing elements, which can be replaced. An example of this is a sweat sensing device, where the microfluidic chip can be easily replaced, while the optical and electronic components are re-useable (Curto *et al.*, 2012).

Fabric sensors can be easily integrated into garments and connected by using embroidery or weaving techniques using conductive threads. While there have been developments in flexible antennae, the wireless electronic components are generally conventional electronic devices. These must be packaged carefully to ensure user safety and comfort. Methods of interconnection between flexible textile conductive traces and conventional electronics include crimping, soldering and embroidery. There is a pressing need for a common set of standards defining methodologies, specifications and best practices for the various smart-textile applications and products. Due to the young nature of smart textiles, standards to evaluate the performance and compatibility in healthcare are missing and it remains to be seen if available medical or electronics standards can be applied (Schwarz *et al.*, 2010). A European task group (CEN/TC 248), led by Belgian CentexBel, was established in 2006 to develop guidelines for the development of standards for smart textiles. There is a need for international standards and regulations, a need for common certification of smart textiles and garments, and a common certified validation protocol and quality control during production (Lymberis and Paradiso, 2008).

15.3.2 Ease of use

Portable monitoring devices often involve wires and may require electrodes to be positioned and adhered. What could be simpler than putting on a 'smart' shirt, which has inbuilt sensors in the right places to monitor vital signs? The goal of smart garments is to enhance everyday clothing and to increase its functionality for the benefit of the wearer. The garment needs to be straightforward to use and a suitable user interface is needed. Data must be presented to the user in a useful manner, without overloading the user with too much data. Much of the research in wearable technologies and personal health systems has focused on monitoring patients; however, one of the largest challenges still remains and that is in the design of methods of user feedback. This needs to be designed to suit its purpose, for example, for a respiratory feedback system for children with respiratory illness, an avatar game has been developed to provide assistance and motivation (Mitchell *et al.*, 2010). The system was designed to be accessible, requiring a web-browser to run a Flash application. Another project carried out by CLARITY, DCU and NUIM, was a stroke rehabilitation glove, which was designed to provide physiotherapists and patients with animations of hand movements. An oedema glove was used to integrate the sensors; such a glove would be familiar, as it is typically used to reduce swelling of the hand in stroke patients (Coyle *et al.*, 2010b).

There are many signals that can be measured from the body; however, the big question is how to use this information in a way to benefit the user? In order to do this, multiple data sources need to be carefully processed into a useful index to give an overall picture or to highlight changes and unusual physiological events. New algorithms are needed to do this, taking into account the user context. Data should be transmitted and stored automatically with minimum effort from the wearer. In two separate blood pressure studies, it was found patients fabricated readings when asked to record their blood pressure measurements. The patients were not aware that the readings were stored to memory cards during the studies (Pickering *et al.*, 2005).

15.3.3 Acceptance by medical profession

Sensor systems become redundant if patients or clinicians do not want to work with them. A review of patients' and clinicians' preferences for non-invasive body-worn sensor systems was carried out by Bergmann and McGregor (2011). The key user preferences were that a body-worn sensor system should be compact, embedded and simple to operate and maintain. It also should not affect daily behaviour nor seek to directly replace a healthcare professional.

The integration of personal health monitoring systems in healthcare implies their interconnection with existing health information systems and electronic health records. Therefore it is essential to ensure interoperability with current- and future-generation communication infrastructure – broadband, fixed and wireless (Gatzoulis and Iakovidis, 2007). Studies need to be carried out to help convince the medical community of the benefits of this technology. These studies need to investigate the impact on quality of life and it must be demonstrated that this technology can help to reduce hospitalization rates and quality of care.

15.3.4 Ethics

Physiological data of users are a highly personalised and private source of information. Important medical and lifestyle trends can be gleaned by analysing the physiological data from large populations. Concerns about patient privacy, trust, product liability and negligence in handling medical data are affecting the adoption rate and usage of current electronic health systems (Sarabdeen, 2012). There is currently an inadequate level of legal protection or unawareness of availability of laws and regulation that addresses the electronic healthcare system privacy. With the increased use of information systems in society, user privacy is becoming an issue of major social concern. Data containing personal information requires security in storage, transmission and third-party use. It is vital to ensure that encryption and security methods are in place. Computer misuse, common to the internet, is a potential risk.

The sharing of patient data between different healthcare professionals, departments and other information systems could affect the confidentiality of a patient's data or it could compromise the integrity and timeliness of the treatment of the data (Whitehouse and Duquenoy, 2011). These threats are extremely difficult to address technologically and enforce internationally (Duquenoy and Whitehouse, 2010). Patient confidentiality must be preserved and the users' informed consent must be given to use this data. Given that a user may have little knowledge and lack of control of the information system handling the data, we need to ask to what extent and in what way, the particular user is 'informed' of the system (Duquenoy and Whitehouse, 2010). Another potential ethical impact on users is compatibility – if the user is tied in to one particular system platform and its applications, this raises issues of affordability and opportunity for equal access. Reliability is another concern, particularly where systems are life-critical. Users need to be made aware of how reliable the systems are.

15.4 Trends and applications of medical smart textiles

Given the range of physiological signals that smart textiles can sense, there are many scenarios where they can be applied. Cardiovascular diseases represent the leading cause of deaths worldwide and much of the research in wearable and portable eHealth systems has focused on these diseases. The non-invasive monitoring capability of these systems concerns not only the prevention of cardiovascular diseases (e.g. myocardial infarction and stroke) but also their management, as in the case of chronically-ill patients (Gatzoulis and Iakovidis, 2007). Since the mid-1990s, the European Commission (EC) has developed innovative systems and services for personal health monitoring. A number of collaborative research projects (AMON, MOBIHEALTH, WEALTHY and MYHEART, to name a few) have been funded by the Fifth and Sixth Research Framework Programme of the EC (FP5, FP6). Recent projects, part of the Seventh Research Framework Programme (FP7), have brought a focus onto mental health, including projects such as PSYCHE.

One of the first commercial smart garments that became available was the Lifeshirt® developed by Vivometrics (Heilman and Porges, 2007). The shirt contained respiratory function sensors woven into the shirt around the patient's chest and abdomen, a three-lead, single channel ECG and a three-axis accelerometer to record subject posture and activity level. The Smart Shirt by Sensatex was developed around the same time, based on technology from Georgia Tech (http://www.gtwm.gatech.edu/), funded by the US navy. The shirt contained optical and conductive fibres integrated into the garment to monitor vital signs. The optical fibres allowed the detection of a bullet penetration and its location for law enforcement and military applications. The shirt was designed to be used for physiological monitoring in various scenarios, including chronically-ill patients, elderly persons living alone, athletes and infants.

Exmovere Holdings have now developed a garment based on patents licensed from Sensatex and Georgia Tech. The Exmobaby garment is a baby sleep garment to measure vital signs. Thanks to embedded electrocardiogram, skin temperature, moisture and movement sensors, the Exmobaby system wirelessly transmits a baby's vital sign data to a PC located within 100 feet. Smartlife Technologies (http://www.smartlifetech.com), based in Manchester, also offer a monitoring shirt called the SmartLife HealthVest®. The garment is created with integrally knitted ECG electrodes, respiratory sensors and conductive pathways. Doubled covered elastomeric yarn is used for the base structure and silver coated yarn for the sensors and conductive pathways. A three-dimensional (3-D) image of the knitted garment is shown in Fig 15.5. Zephyr Technologies (http://www.zephyr-technology.com/) have developed a BioHarness™ to measure critical vital signs (ECG, heart rate, breathing rate, skin temperature) and contextualise the information with the individual's physical activity using an accelerometer (activity and posture). The device is available in strap or clothing form and can link to a smart phone, PC or web portal. The device is designed for use by first responders, military, personal trainers and professional sports teams. The device was used to monitor the 33 Chilean miners that were trapped in the San Jose mine in 2010.

Underarmour have teamed up with Zephyr to develop the E39 compression shirt, which measures heart rate, breathing rate and G Force information. It combines with the Omnisense software to develop automatic tests for fitness and vertical and broad jumps, and is being used by the National Football League (NFL) in the US.

Smart textiles are also available as consumer sportswear, which can monitor parameters such as heart rate and running statistics linking wirelessly to electronic devices such as iPhones, as with the Nike™ + products or smart phone android apps, as with the adidas Micoach system.

While much focus has been placed on monitoring applications and using textiles to gather information from the body, it is also possible to use smart textiles to provide therapeutic benefits to the wearer. Smart textiles are capable of providing heat to the body to treat conditions such as arthritis, muscle pain and spasm, joint stiffness or Raynauds Syndrome. Fabric heating systems are

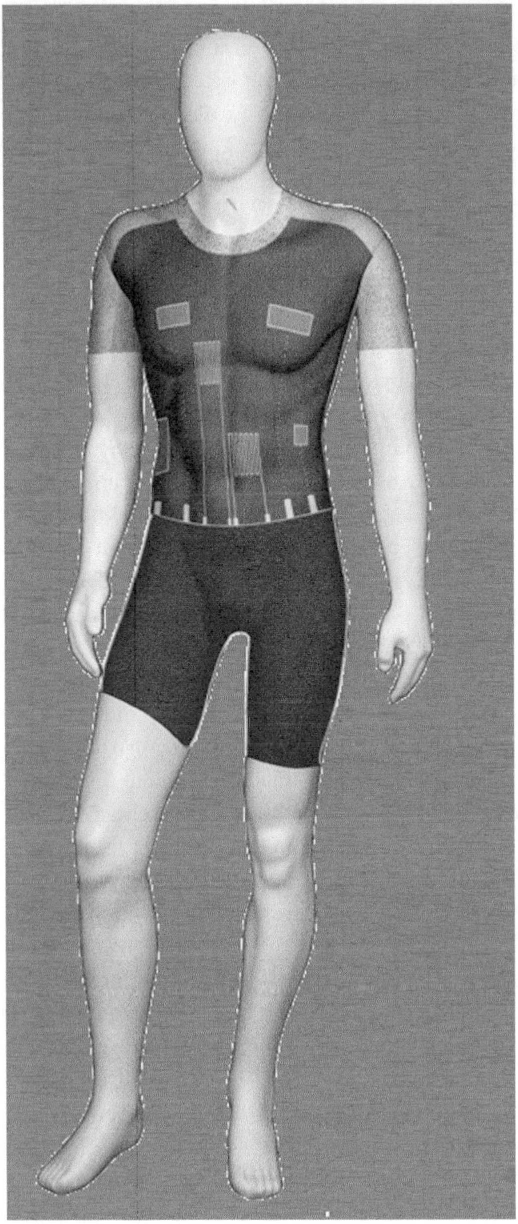

15.5 SmartLife™ has developed softsensor™ systems aimed for the improvement of lifestyle. The HealthVest® is created with integrally knitted ECG electrodes, respiratory sensors and conductive pathways. (Image courtesy of Smartlife Technologies.)

developed by EXO^2 (http://www.exo2.co.uk/). The EXO^2/FabRoc® heating system delivers infrared heating to the affected areas, which are typically the knees, hips, lower back, hands and neck. The company also develops heated clothing for outdoor, equestrian and motorcycle use. Another company developing heated garments is Gerbing's Heated Clothing. This technology was originally designed for use on motorbikes with a 12 V power supply and they have extended their product range to be used with portable 7 V battery packs. They moved from integrating copper wires into garments to Microwire™ technology, using bundles of hundreds of microscopic stainless steel strands.

Wound care is one of the most lucrative medical device markets and there has been much effort in developing textiles to promote rapid healing of wounds in order to obtain functional and cosmetic results (Gupta *et al.*, 2010). Smart textiles can help to provide the optimum environment, such as optimum moisture levels and gas diffusion. A system for monitoring changes in pH and inflammatory proteins has been developed by Pasche *et al.* (2008), as part of the European BIOTEX project, with potential use in the supervision of skin grafts and ulcer treatments. Another EU funded project, Lidwine, focused on the development of multifunctional medical textiles to prevent and treat decubitus wounds. Target applications include an antibacterial textile for wound care, integrated with medication depots, including an active circulation support bandage with a contractile cuff. Textiles, given their flexible nature, have the advantage of being easy to apply to any location of the body. A thin and flexible luminous textile embroidery was developed by Selm *et al.* (2007), using plastic optical fibres. The textile diffuser has novel biomedical application in the field of cancer treatment, where the luminous embroideries are used to apply light energy to cancerous tissue during photodynamic therapy (Selm *et al.*, 2007). Philips are developing a phototherapy blanket, emitting blue light, to treat neonatal jaundice. At present, babies are placed beneath overhead lamps, whereas a soft blanket would be in close range with the baby's skin, which should improve the efficiency of the treatment. This approach also allows interaction with parents and caregivers, without interrupting the baby's treatment. Figure 15.6 shows the woven electronic platform developed by Philips, which integrates light emitters into a textile.

While electrodes may be integrated as detectors, another possibility is to use the electrodes to stimulate muscles. Electrodes for functional electrical stimulation (FES) have been integrated into fabrics, to provide actuation stimuli to muscles of the spinal cord of injured and stroke subjects, in order to generate or improve lost motor function, for example, for walking or hand gripping movements (Kirstein *et al.*, 2003). Electro-active polymers represent suitable candidate flexible components for the embedding of 'artificial muscle' functions into garments (Carpi and De Rossi, 2005). In the future, robotic and wearable technologies or smart fabrics could be combined to deliver therapeutic interventions (Bonato, 2010).

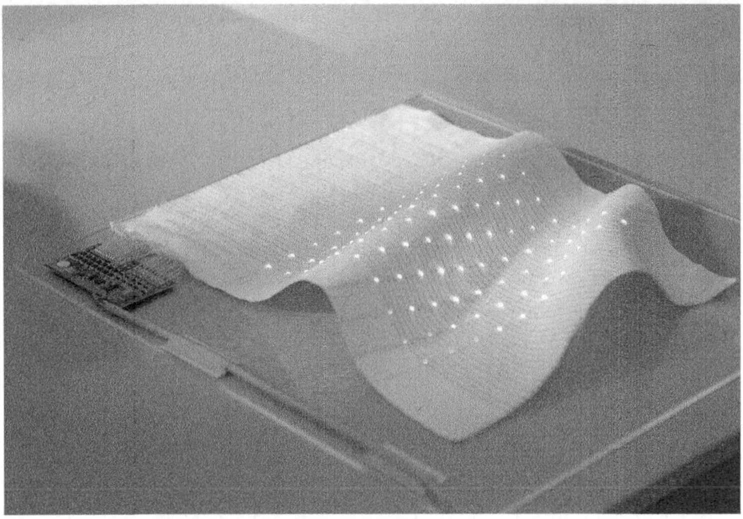

15.6 Philips Lumalive 'Woven electronics' fabric platform. (Image courtesy of Philips.)

15.5 Conclusions

The stages of treatment for most medical conditions include prevention, immediate care, rehabilitation and long-term support. Smart textiles have a role to play in each of these stages in treatment and prevention of illness. Integrating smart garments into our lives is a natural step, linking into our smart phones and interconnecting electronic devices. Through effective user feedback applications, there is potential to motivate individuals, promoting wellness and a healthy way of living. In the case of illness, smart garments can offer support to the medical community by providing a fuller picture of their patients' health and enable remote monitoring to reduce the frequency of clinical visits. In rehabilitation, a smart garment may help the patient take an active role in their recovery and prevent future relapse. In the future, smart textiles may have therapeutic functionalities, providing flexible and adaptable means of care.

15.6 References

Asada H H, Shaltis P, McCombie D and Reisner A (2010), 'Wearable blood pressure sensor and method of calibration', *US Patent 7674231*.

Asahina M, Yamanaka Y, Akaogi Y, Kuwabara S, Koyama Y and Hattori T (2008), 'Measurements of sweat response and skin vasomotor reflex for assessment of autonomic dysfunction in patients with diabetes', *J Diab Complic*, 22(4), 278–283.

Axisa F, Gehin C, Delhomme G, Collet C, Robin O and Dittmar A (2004), 'Wrist ambulatory monitoring system and smart glove for real time emotional, sensorial and physiological analysis', *Proceedings of the 26th Annual International Conference of the IEEE EMBS*, San Francisco, CA.

Barbaro M, Caboni A, Cosseddu P, Mattana G and A B (2010), 'Active devices based on organic semiconductors for wearable applications', *IEEE Trans Inf Technol Biomed*, 14(3), 758–766.

Benito-Lopez F, Coyle S, Byrne R and Diamond D (2010), '"Sweat-on-a-Chip": analysing sweat in real time with disposable micro-devices', *Proceedings of the 9th Annual IEEE Conference on Sensors*, 1–4 November.

Bergmann J H M and McGregor A H (2011), 'Body-worn sensor design: what do patients and clinician's want?' *Ann Biomed Eng*, 39(9), 2299–2312.

Bonato P (2010), 'Wearable sensors and systems: from enabling technology to clinical applications', *IEEE Eng Med Biol Mag*, 29(3), 25–36.

Brisson G R, Boisvert P, Péronnet F, Parrault H, Boisvert D and Lafond J S (1991), 'A simple and disposable sweat collector', *Eur J Appl Physiol*, 63, 269–272.

Bronzino J D (2000), *The Biomedical Engineering Handbook*. Boca Raton, FL, CRC Press.

Carpi F and De Rossi D (2005), 'Electroactive polymer-based devices for e-textiles in biomedicine', *IEEE Trans Info Tech Biomed*, 9(3), 295–318.

Catrysse M, Puers R, Hertleer C, Van Langenhove L, Van Egmond, *et al.* (2003), 'Fabric sensors for the measurement of physiological parameters', *Transducers '03, 12th International Conference on Solid-State Sensors, Actuators and Microsystems*, Boston, MA.

Chen W, Bouwstra S, Bambang Oetomo S and Feijs L M G (2010), 'Intelligent design for neonatal monitoring with wearable sensors', in Somerset, V. S. (Ed.), *Intelligent and Biosensors*. InTech. pp. 1–24

Coyle S, Lau K, Moyna N, Diamond D, Di Francesco F, *et al.* (2010a), 'BIOTEX: Biosensing textiles for personalised healthcare management', *IEEE Trans Info Tech BioMed*, 14(2), 364–370.

Coyle S, Mitchell E, May G, O'Connor N E and Diamond D (2010b), 'Textile sensors for personalized feedback', *Proceedings of the ECIR 2010 Workshop on Information Access for Personal Media Archives*, Milton Keynes, UK.

Curto V F, Coyle S, Byrne R, Angelov N, Diamond D and Benito Lopez F (2012), 'Concept and development of an autonomous wearable micro-fluidic platform for real time pH sweat analysis', *Sens Actuat B*, DOI:10.1016/j.snb.2012.02.010 (in press).

D'Angelo L T, Weber S, Honda Y, Thiel T, Narbonneau F and Luth T C (2008), 'A system for respiratory motion detection using optical fibers embedded into textiles', *Proceedings of the Engineering in Medicine and Biology Society. EMBS 2008 30th Annual International Conference of the IEEE.*

De Rossi D and Veltink H (2010), 'Wearable technology for biomechanics: e-textile or micromechanical sensors?' *IEEE Eng Med Biol Mag*, May/June, 37–43.

Duquenoy P and Whitehouse D (2010), 'A 21st century ethical debate: pursuing perspectives on ambient intelligence', in: Duquenoy P, Zielinski C and Kimppa K (Eds.), *The Information Society: Emerging Landscapes*, New York, Springer. pp. 293–314.

Fagette P (1997), 'Tracking the historical development of biomedical engineering: the 1960s and 1970s', *IEEE Eng Med Biol Mag*, 16, 164–175.

Farina D, Lorrain T and Negro F (2010), 'High-density EMG E-textile systems for the control of active prostheses', *Proceedings of the 32nd Annual International Conference of the IEEE EMBS*, Buenos Aires, Argentina.

Finni T, Hu M, Kettunen P, Vilavuo T and Cheng S (2007), 'Measurement of EMG activity with textile electrodes embedded into clothing', *Physiol Meas*, 28, 1405–1419.

Gaetano Valenza A L, Scilingo E P and De Rossi D (2010), 'Towards a smart glove: arousal recognition based on textile Electrodermal Response', *Proceedings of the 32nd Annual International Conference of the IEEE EMBS*, Buenos Aires, Argentina, 31 August–3 September.

Gatzoulis L and Iakovidis I (2007), 'Wearable and portable e-Health systems: technological issues and opportunities for personalized care', *IEEE Eng Med Biol Mag*, 26(5), 51–56.

Gupta B, Agarwal R and Alam M S (2010), 'Textile-based smart wound dressings', *Ind J Fibre Text Res*, 35(2), 174–187.

Haider A and Solish N (2005), 'Focal hyperhidrosis: diagnosis and management', *Can Med Assoc J*, 172(1), 69–75.

Heilman K J and Porges S W (2007), 'Accuracy of the LifeShirt® (Vivometrics) in the detection of cardiac rhythms', *Biol Psych*, 75(3), 300–305.

Janata J (2009), *Principles of Chemical Sensors*, Second Edition, London, Springer.

Kappeler-Setz C, Schumm J, Kusserow M, Arnrich B and Tröster G (2010), 'Towards long-term monitoring of electrodermal activity in daily life', *Ubicomp 2010*.

Kirstein T, Lawrence M and Tröster G (2003), 'Functional electrical stimulation (FES) with smart textile electrodes', *International Workshop on a New Generation of Wearable Systems for E-Health*, Pisa, Italy.

Linz T, Gourmelon L and Langereis G (2007), 'Contactless EMG sensors embroidered onto textile', *Proceedings of the 4th International Workshop on Wearable and Implantable Body Sensor Networks*, Aachen, Germany.

Liu Y, Gorgutsa S, Santato C and Skorobogatiy M (2012), 'Flexible, solid electrolyte-based lithium battery composed of $LiFePO_4$ cathode and Li_4Ti5O_{12} anode for applications in smart textiles', *J Electrochem Soc*, 159(4), A349–A356.

Loriga G, Taccini N, Pacelli M and Paradiso R (2007), 'Flat knitted sensors for respiration monitoring', *IEEE Elect Mag*, Fall 2007, 5–8.

Lymberis A and Paradiso R (2008), 'Smart fabrics and interactive textile enabling wearable personal applications: RandD state of the art and future challenges', *Proceedings of the 30th International Conference of IEEE Engineering in Medicine and Biology Society*, Vancouver, BC.

Mattmann C, Amft O, Harms H, Clemens F and Tröster G (2007), 'Recognizing upper body postures using textile strain sensors', *Proceedings of the 11th International Symposium on Wearable Computers (ISWC07)*, Boston, MA.

Mattmann C, Clemens F and Tröster G (2008), 'Sensor for measuring strain in textile', *Sensors*, 8(6), 3719–3732.

McCann J, Hurford R and Martin A (2005), 'A design process for the development of innovative smart clothing that addresses end-user needs from technical, functional, aesthetic and cultural view points', *Wearable Computers, 2005. Proceedings of the 9th IEEE International Symposium*, 18–21 October, pp. 70–77.

Merritt C R, Nagle H T and Grant E (2009), 'Fabric-based active electrode design and fabrication for health monitoring clothing', *IEEE Trans Info Tech BioMed*, 13(2), 274–280.

Mitchell E, Coyle S, Ward T, O'Connor N E and Diamond D (2010), 'Breathing feedback system with wearable textile sensors', *Body Sensor Networks (BSN 2010)*, 7–9 June, Biopolis, Singapore.

Paradiso R, Belloc C, Loriga G and Taccini N (2005a), 'Wearable healthcare systems, new frontiers of e-textile', in: Nugent C D, Mccullagh P J, Mcadams E T and Lymberis A

(Eds.), *Studies in Health Technology and Informatics*, vol. 117, *Personalised Health Management Systems: The Integration of Innovative Sensing, Textile, Information and Communication Technologies*.

Paradiso R, Loriga G and Taccini N (2005b), 'A wearable healthcare system based on knitted integrated sensors', *IEEE Trans Inf Technol Biomed*, 9(3), 337–344.

Paradiso R and DeRossi D (2006), 'Advances in textile technologies for unobtrusive monitoring of vital parameters and movements', *Proceedings of the 28th IEEE EMBS Annual International Conference*, New York, 30 August–3 September, pp. 392–395.

Pasche S, Angeloni S, Ischer R, Liley M, Luprano J and Voirin G (2008), 'Wearable biosensors for monitoring wound healing', *Adv Sci Tech*, 57, 80–87.

Perry J T, Kellog S, Vaidya S M, Jong-Hoon Y, Hesham A and Sharif H (2009), 'Survey and evaluation of real-time fall detection approaches', *Proceedings of the 6th International Symposium on High-Capacity Optical Networks and Enabling Technologies (HONET) 2009*.

Pickering T G, Hall J E, Appel L J, Falkner, B E, Graves J, *et al.* (2005), 'Recommendations for blood pressure measurement in humans and experimental animals', *Hypertension*, 45, 142–161.

Poh M Z, Swenson N C and Picard R W (2010), 'A wearable sensor for unobtrusive, long-term assessment of electrodermal activity', *IEEE Trans Biomed Eng*, 57(5), 1243–1252.

Reisner A, Shaltis P A, McCombie D and Asada H H (2008), 'Utility of the photoplethysmogram in circulatory monitoring', *Anesthesiology*, 108(5), 950–958.

Rothmaier M, Selm B, Spichtig S, Haensse D and Wolf M (2008), 'Photonic textiles for pulse oximetry', *Opt Exp*, 16 12 973–12 986.

Rovira Carlos W T, McCoy A, Coyle S, Corcoran B, Stroiescu F, *et al.* (2011), 'Web-based wearable sensor streaming for respiratory monitoring applications', *Proceedings of the IEEE Sensors*, Limerick, Ireland.

Salvo P, Di Francesco F, Costanzo D, Ferrari C, Trivella M G and De Rossi D (2010), 'A wearable sensor for measuring sweat rate', *IEEE Sens J*, 10(10), 1557–1558.

Sarabdeen J (2012), 'Legal issues in e-healthcare systems', in: Wafta M K (Ed.), *E-Healthcare Systems and Wireless Communications: Current and Future Challenges*. IGI Global.

Schwarz A, Van Langenhove L, Guermonprez P and Deguillemont D (2010) 'A roadmap on smart textiles', *Text Prog*, 42(2), 99–180.

Selm B, Rothmaier M, Camenzind M, Khan T and Walt H (2007), 'Novel flexible light diffuser and irradiation properties for photodynamic therapy', *J Biomed Opt*, 12, 034024.

Selm B, Gürel E A, Rothmaier M, Rossi R M and Scherer L J (2010), 'Polymeric optical fiber fabrics for illumination and sensorial applications in textiles', *J Intell Mat Syst Struct*, 21, 1061–1071.

Shirreffs S M and Maughan R J (1997), 'Whole body sweat collection in humans: an improved method with preliminary data on electrolyte content', *J Appl Physiol*, 82, 336–341.

Solà J O C and Krauss J (2007), 'On the reliability of pulse oximetry at the sternum', *Proceedings of the 29th Annual International Conference of the IEEE Engineering in Medicine and Biology Society*, Lyon, France.

Street L (2008), *Introduction to Biomedical Engineering Technology*. Boca Raton, FL, CRC Press.

Tatara N, Koizumi H, Mino S, Hayashida S, Aihara K, *et al.* (2007), 'A novel blood pressure monitoring device for ubiquitous healthcare services', *Proceedings of the 29th Annual International Conference of the IEEE EMBS*, Lyon, France.

Tognetti A, Lorussi F, Tesconi M, Bartalesi R, Zupone G and De Rossi D (2005), 'Wearable kinesthetic systems for capturing and classifying body posture and gesture', *Proceedings of the Engineering in Medicine and Biology 27th Annual Conference*, Shanghai, China.

Trafton A (2009), 'Wearable blood pressure sensor offers 24/7 continuous monitoring. Device could help diagnose hypertension, heart disease', *MITnews*, available from http://web.mit.edu/newsoffice/2009/blood-pressure-tt0408.html.

Vaughan J and Scully P J (2008), 'Optical sensing for perspiration', *Photon 08: Optics and Photonics Conference*, Heriot Watt University, Edinburgh. Bristol, Institute of Physics Publishing.

Whitehouse D and Duquenoy P (2011), 'e-Health and ethics: theory, teaching, and practice', (Festschrift in honour of Bradley), in: Haftor D and Mirijamdotter A (Eds.), *Information and Communication Technologies, Society and Human Beings: Theory and Framework*. Hershey, PA, IGI Global.

World Health Organisation (2007) *WHO Global Report on Falls Prevention in Older Age*.

Xu P J, Zhanga H and Tao X M (2008), 'Textile-structured electrodes for electrocardiogram', *Text Prog*, 40(4), 183–213.

Yang Y L, Chuang M C, Lou S L and Wang, J (2010), 'Thick-film textile-based amperometric sensors and biosensors', *Analyst*, 135(6), 1230–1234.

Yu X (2008), 'Approaches and principles of fall detection for elderly and patient', *HEALTHCOM 2008*.

16
Automotive applications of smart textiles

M. WAGNER, Daimler AG, Germany

DOI: 10.1533/9780857093530.3.444

Abstract: Smart textiles have become a new topic in vehicle engineering. New applications in the automotive industry seem to have potential and may give access to entirely new system approaches. This chapter presents some of them and focuses especially on the potential of measuring physiological parameters such as heart rate, electrodermal activity and others. Investigations on prototype systems of car seats and steering wheels will be discussed.

Key words: smart textiles, car seats, capacitive coupled electrocardiogram, textile multilayer electrode, vital parameters, sensor steering wheel, INSITEX.

16.1 Introduction

Smart textiles have become a new topic in vehicle engineering, although they are not yet being used in mass-produced vehicles. Why is it important for car firms to get involved in this new technology? Cars are products with a long lifecycle. It can take between 25 and 30 years to develop a new car design from the start of research up to its production and the organization of spare parts supply. New technologies are generally not sufficiently reliable to be used immediately in automobiles and so the consumer goods industry is way ahead because of its far shorter production cycle. However, it is precisely this relatively long production cycle that makes it indispensable for the automobile industry to keep in touch with and make use of new technologies. The potential change in consumer attitudes also has to be taken into account.

Futurology analyzes trends in consumer attitudes and markets. One of those researching future trends, M. Horx from the German 'Zukunftsinstitut' [1] has, for example, published a list of megatrends [2] comprising individualization, urbanization and behaviour, which he has called 'the new aging' (Fig. 16.1).

Some social trends important for the automobile market are individualization and the increase in average age of populations (and therefore customers) in Europe and the US. It is also important to note a change in behaviour amongst younger consumers; they like to have access to a car when they need to be mobile, but they do not need to own the car. It is also clear that the modern consumer is far more conscious of his/her health in all areas of daily life. As the job market requires high mobility, the customer spends more and more time in his/her car. 'Health on wheels' and 'wellness on wheels' are new areas of research where new applications are needed. Prototypes of smart-textile systems show that there are new

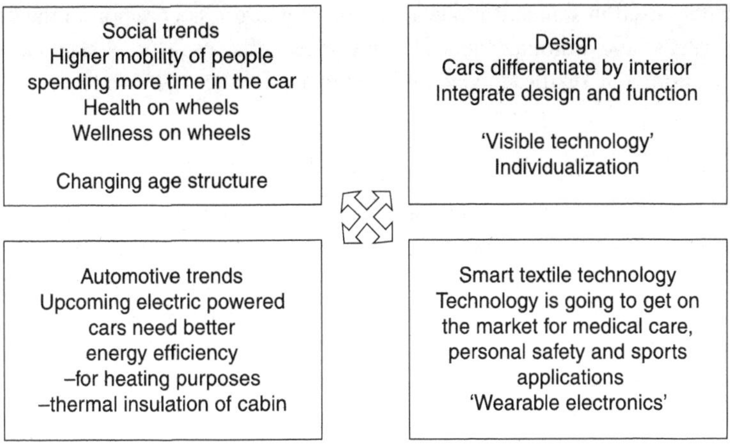

16.1 Social and technological trends (source: Daimler AG/Manfred Wagner).

opportunities for applications in cars to meet such needs. Sections 16.3 and 16.4 will deal with these applications more closely.

Even the design of car components proves this trend to shape the interiors of the car to our own taste [3]. This is especially apparent for the seats of a car, where mainstream vehicles have followed the example of luxury cars. Smart textiles may enable 'visible design', the combination of functionality and design. Currently in demand are health and protective clothing applications based on smart-textile technology. Some of these products and prototypes will be considered in Section 16.3. Section 16.5 points out that there is at present a mismatch between the requirements of the automotive industry and the state-of-the-art of smart textiles. Qualification standardization is necessary and has already started. The irreversible trend in purely electrically operated cars opens up the appliance of smart textiles in lightweight construction, heat insulation and power efficiency. Some of these appliances with regard to automobiles will be considered in Section 16.6.

16.2 The use of textiles in vehicles

Textile elements can be used in both the interior and the exterior (e.g. canvas tops) of cars; however, most of the applications are located in the interior. Typical textile components are seats, seatbelts, fabric covers for roofs, floors, side walls and sometimes parts of the dashboard. The textile surface is responsible for the sensory and aesthetic appearance of the interior and therefore the feeling of quality experienced by the customer. This is especially true where passengers have direct contact with the textile surface, for example with the seat and armrest. The seat is the component with the most contact with the passenger. Fabric-covered seats are widely used in standard models, though they are less common in the luxury car market where customers choose leather-covered seats. More detailed information on automotive textiles in general is available in other research [4].

16.2.1 Classification of textile intelligence

The integration of microelectronic devices into textile products has taken place during the last decade. The terms used for these new textile systems are 'intelligent textiles' or 'smart textiles'. Unfortunately, the terms have different meanings for different people. The draft of the new standard CEN/TR 16298 (DIN SPEC 60298):2012-02 *Textiles and textile products – Smart textiles – Definitions, categorisation, applications and standardization needs* includes a classification of textile systems (Table 16.1) [5].

16.2.2 Potential applications of smart-textile technology in vehicles

Car interiors are well-suited to the first applications of smart textiles in vehicles. On the one hand, textiles have already been in use for some time in car interiors and research into their handling and testing is available. However, requirements concerning environmental conditions are much lower in this field than in exterior applications. It is common knowledge that applications relating to heating, lighting and sensor functions will have the highest initial potential for car applications. Interior components that can be combined with textile sensors based on smart-textile technology are discussed below. More detail will be presented in Sections 16.3 and 16.4.

Table 16.1 Classification of textile intelligence

System	Characteristics	Example
Textile material	Material made of textile fibres, for use together with another textile or non-textile products or stand-alone	
Functional textile material	Textile material modified by structure, treatment or characteristic of the textile fibre	Spacer fabrics, Ag coating
Smart-textile material *or* intelligent textile material	Functional textile material which can interact actively with the environment (sensor/ actor)	ECG T-shirt

Seat

Seats and steering wheels are the components that are in contact with the driver throughout the driving experience and therefore the seat is one of the most popular components in which to integrate sensors for measuring the driver's condition.

Seatbelt

Sensors might be integrated in the seatbelt too; for example, breathing activity can be measured by the deflection of the seatbelt. This is not a major task for textile sensors, because this signal can be taken by the belt retractor. The integration of textile electrocardiography (ECG) sensors into seatbelts is becoming of interest, because the position of the seatbelt is always near the heart. Seatbelts as safety-related components are submitted to a strict test and qualification process, and therefore sensor integration needs intensive coordination in development and production. New extended test methods with safety requirements have to be defined.

Roof and floor

Integration of capacitive matrix structures for recognizing the position of the passengers can be envisaged, as can lighting as a design element and for ambient illumination.

Steering wheel

The steering wheel is, along with the seat and the seatbelt, a component that is in contact with the driver all the time. Therefore the steering wheel is also a candidate for integration of sensor functions. Steering wheels usually have a surface of wood, leather or plastic and are therefore not a typically textile component. Conductive yarns in the seams of a leather cover can provide a contact electrode to measure vital parameters [6].

The process flow of making a smart-textile system is shown in Fig. 16.2. Some special components, technologies and processes are necessary. Some of them have to be adapted to vehicle requirements or new developments have to be undertaken.

First, a conductive element is needed. This may be a yarn, as the one shown in Fig. 16.2, which is a multi-filament yarn [7] with a combination of conductive and insulating threads. Making a yarn conductive is usually done by plating it with a conductive layer such as Ag or Cu. A metal wire, a carbon fibre or a special polymer fibre could also be used. The integration of this conductive element in a fabric by a textile process such as, for example, weaving, delivers the base material for the application. Depending on the application, this basic fabric contains an arrangement of parallel conductive elements, a matrix arrangement, or even

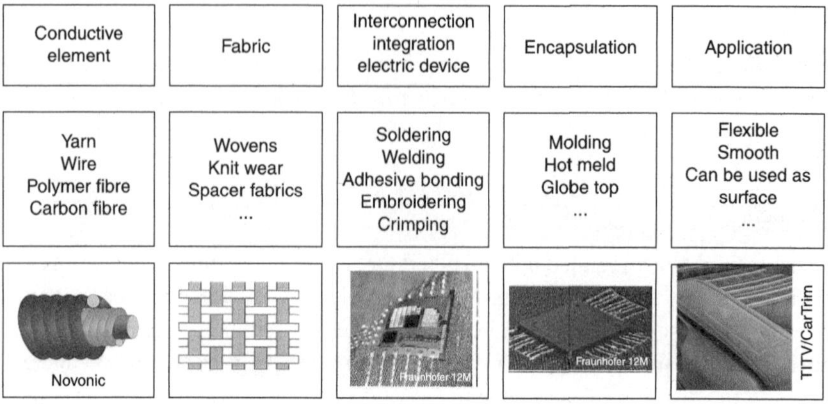

16.2 Smart-textile process (source: Daimler AG, Manfred Wagner with use of pictures of [45], [8], [34]).

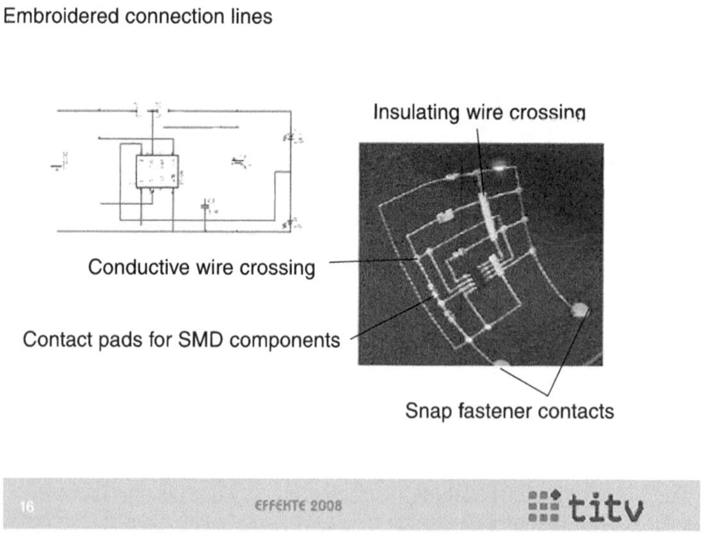

16.3 Textile circuit board [45] (source: TITV).

conductive planes. For producing electrode structures, spacer fabric technology is one of the most popular processes. The next step is connecting electronic devices, for example microsystems, by appropriate interconnection technologies. This may be done by soldering, welding, adhesive bonding or by special textile techniques such as embroidering with conductive yarns or fibres. Through embroidering, a fully textile circuit board can be produced (Fig. 16.3).

Crimping is a new interconnection method in textile technology. A suitable crimping method was developed by Fraunhofer IZM [8] in the INSITEX project [9]. For automotive environments in particular, interconnection technology is one of the most critical topics. As discussed in more detail in Section 16.5, the standard automotive test schedule applies cycled mechanical loads to the textile, and the interconnection must withstand these tests without failing. Reliability and durability over the component's lifetime is an automotive requirement. The typical lifetime requirement is defined as up to 15 years.

Environmental conditions also produce the need to shield the electronic components by encapsulation. Different processes are known from the electronic industry as moulding, hot meld and globe top encapsulation. In addition, the aimed localization of the system in the car may demand some more requirements to the design of the smart-textile component caused by, for example, temperature, vibration or humidity. The requirements and boundary conditions of a defined application are typically detailed in a specification book.

16.3 Smart-textile applications and their potential for use in cars

In this section, some developments of smart-textile products, which have the potential to be used in vehicle applications, will be discussed. The chosen selection is the view of the author and makes no claim to be complete.

16.3.1 ECG T-shirt and textile electrode

Long-term monitoring of heart disease patients was one motivation to begin the development of textile capacitive electrodes and, in the European project ConText [10], contactless sensors have been developed for the purpose of measuring electromyography EMG and ECG signals. EMG measurement aims to monitor muscle activity for sport and medical applications. The capacitive electrode is connected by a textile conductor path to a miniaturized pre-processing unit, which transmits the data wirelessly to a computer for further processing. The knowledge gained here was a pre-condition of the applications discussed in Section 16.4.

16.3.2 Switches and touch pads

Textile switches were first incorporated into jackets and gloves for on/off and volume control, and even keypads have been demonstrated (Fig. 16.4). TITV [45] with partner CarTrim [11] has demonstrated a car seat with textile switches incorporated directly into the seat for motion control, and newer developments use spacer fabrics as the distance layer. Pushing the button connects two electro-conductive ribbon structures.

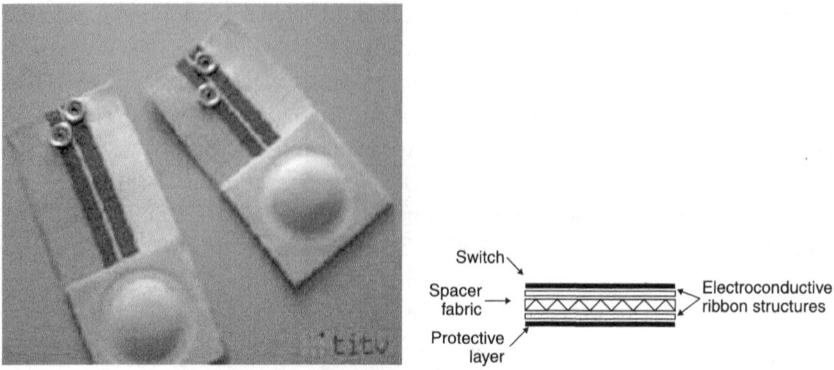

(a) (b)

16.4 (a, b) Textile switches (source: TITV [45]).

Another recent development is a woven two-dimensional (2-D) touchpad sensor and a one-dimensional (1-D) slide sensor using soft capacitor fibres. [12] A special conductive polymer-based capacitor fibre was used to build a fully woven 2-D touchpad sensor and a 1-D slide sensor. Smart phones and notepads have been the motivation to develop the touchpad and build a demonstrator pad of 10 cm by 15 cm. Finger touches or swipes modify the capacitance of the fabric when an alternating current is passed through the fibres. An automotive use of this device should be of high interest and press releases indicate that OEM is interested in the technology [13].

16.3.3 Lighting

There are different ways to light textile material, one of which is to combine textile material with incorporated conductive fibres with LEDs. Several R&D institutes have demonstrated the feasibility of this technology [14] and Philips/NL has started a commercial business with lighting textiles for home applications.

The fibre itself can be the light emitting device, as TITV demonstrated with an electroluminescent thread constructed of high conducting ELITEX®-yarns coated with electroluminescent paste. This technology can be used for creating lighting areas or lines. If the lifetime is extended to ten years or more, the use of such fibres for ambient illumination will be of high interest.

In the BMBF project, Texoled [15] researchers tried to boost integration on new levels, whereby the optical fibre itself is the light-emitting device. A stack of layers is deposited on the fibre, which builds an LED structure. The goal is to use this fibre in textile processes such as weaving to form a lighting fabric.

At the Geneva Motor Show in 2012, concept cars with light applications were exhibited [16].

16.4 Prototypes of smart-textiles applications in vehicles

At present there is no smart-textile system in use in car production. The first prototypes of potential vehicle applications are shown in funded research projects [9, 17, 18] or by cooperation of research institutes and OEMs/suppliers [16, 20]. The BMBF [19] funded project INSITEX, 'Active Passenger Safety by Intelligent Textile Applications' [9], as the first project with an OEM participation, developed the vision of an intelligent cabin (Fig. 16.5). INSITEX focused on applications that may enhance passenger safety and can be realized by smart-textile systems.

Its goal was to apply knowledge from other smart-textile businesses, such as sport and medicine, to estimate the driver's condition. By measuring physiological parameters, such as heart rate (HR), electrodermal activity (EDA) [20], and skin temperature and oxygen saturation of the blood (SpO_2) [21], it should be possible to estimate attention, drowsiness and/or stress of the driver. Furthermore, the goal was to measure seating position and passenger weight classification according to FMVSS 208 [22] by a textile sensor system.

In Fig. 16.5, possible sensor locations are shown. In principle, all textile elements that are in contact or near the driver/passenger can be equipped with textile sensors. The location of the sensor also depends on the intention of the measurement:

- monitoring of the driver's condition without conscious interaction;
- conscious action of the driver or passenger to use vital signal measurement. This may be for testing health and fitness conditions or controlling another comfort system therewith.

Numerous medical investigations show the relationship between the condition of a subject and physiological parameters, such as electrical brain activity (α-signals),

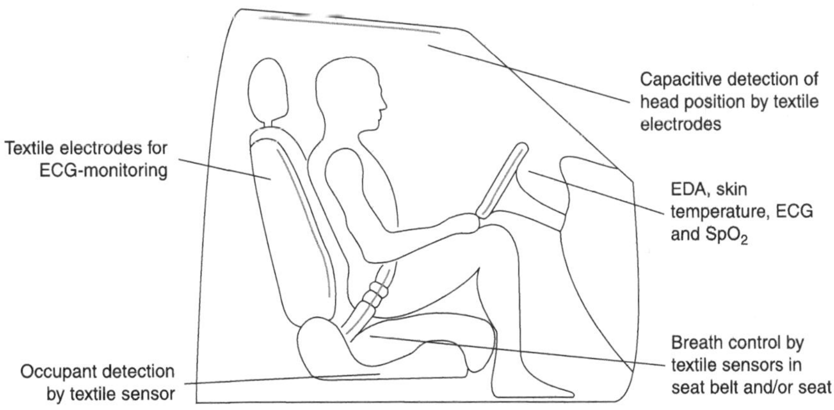

16.5 Intelligent cabin (source: INSITEX).

HR, HR variability (HRV), EDA [20], skin temperature (T_{skin}), oxygen saturation of the blood SpO_2 [21] and some other parameters not considered here. Parameters, such as eyelid movement [23], are indicative of drowsiness of the subject, but are much better measured by optical rather than by textile sensors and will therefore not be discussed here.

EEG (electroencephalogram) is the most reliable method to measure drowsiness and α-signals, as a measure of brain activity, are directly correlated with drowsiness [24, 25]. The measurement of the electrical activity of the brain by EEG requires electrodes directly attached to the head of the subject to be able to detect these weak signals. Often these sensor arrangements are incorporated into a sensor cap worn by the driver, which is not an object for everyday use when driving a car. However, as test equipment in a driving simulator, it is helpful. It is possible to correlate the test driver's reactions on given traffic situations to the α-signals measured in parallel.

ECG [26] is also a well-established measurement method used in clinical as well as in mobile applications. We have to distinguish here between only measuring the HR or the HRV and real clinical applications. Measuring the HR needs only a reliable detection of the R-peak of the QRS signal. This is possible by a 2-electrode arrangement and HRV is the variability of the HR. Time periods between R-peaks can vary, which results in a heart frequency distribution. The width of this distribution is a measure of HRV, which is known as a good measure for stress [27]. Clinical application needs more than two electrodes to characterize the whole QRS complex and can diagnose possible heart diseases from this information.

EDA uses conductive electrodes to measure the EDA of the skin. In practice, this means measuring the conductivity of the skin with contact electrodes, which is mainly influenced by the humidity of the skin. Sweating, for example, changes the skin humidity, and is a physiological process correlated with stress situations [29]. Most lie detectors work on this principle.

T_{skin} can be estimated by all known temperature measuring methods. Temperature is also related to the condition of the person and therefore a patent exists that correlates T_{skin} to the sleepiness of a subject [30]. Nevertheless, the temperature of surfaces in a car interior is heavily influenced by environmental impacts, such as sunshine and other heat sources from car components. Skin temperature measurement in an actual car environment is not as easy as in a laboratory.

SpO_2 is a standard measurement in clinical applications. The signal is typically taken by a finger clamp applied to the person being diagnosed. The saturation of blood with oxygen varies with heart activity. Each time the heart ventricle is pumping blood enriched with oxygen in the lung, the saturation of oxygen is at its maximum. Between the contractions of the heart muscle, the saturation with oxygen decreases. The grade of saturation can be detected by an optical differential measurement. Therefore it is possible to calculate from the SpO_2 measurement the pulse, which means the HR of the subject. A runtime measurement, which uses the

R-peak as the start trigger and the arrival of the maximum oxygen saturation at the point of SpO_2 measurement as the stop trigger, can additionally provide a value of relative blood pressure.

Beyond the measurement of these vital parameters, the measurement of seat occupancy detection (OCS) by textile sensor arrangement was one of the objectives of the INSITEX project. OCS is well-defined by the FMVSS 208 [23] US standard. The weight of a subject must be recognized and different cases such as no load, child, and f05 percentile have to be distinguished. In addition, the detection of the position of the seat occupant is required for different airbag scenarios. Hence the activation of the airbag should be inhibited when the passenger's position is near the dashboard.

Three prototypes have been built in the INSITEX project to demonstrate the potential of textile sensors:

- contactless ECG measurement by capacitive coupled sensors in a car seat (Fig. 16.6);
- steering wheel with vital sensors;
- a textile seat occupancy detection sensor for weight and position recognition.

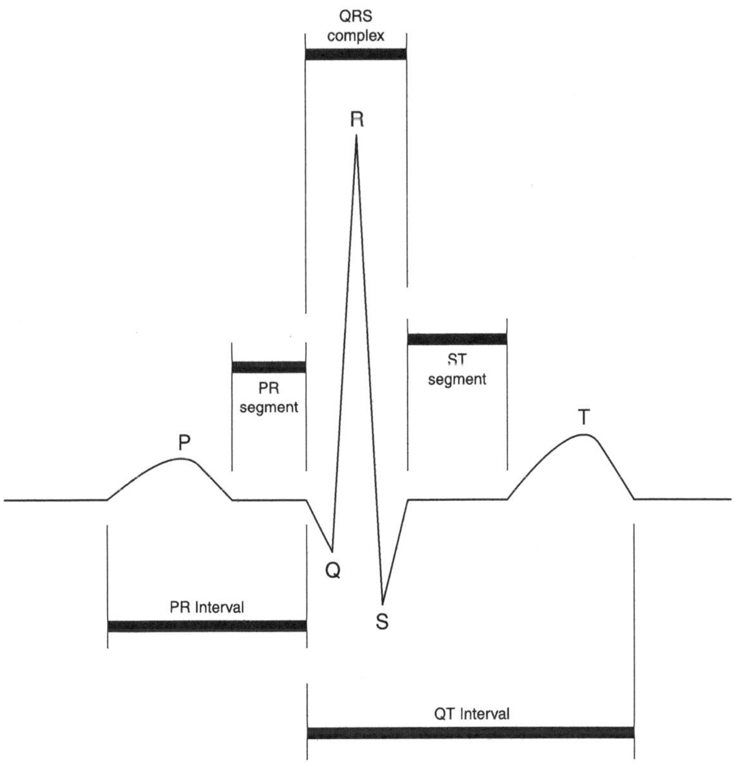

16.6 ECG signal with QRS complex [28].

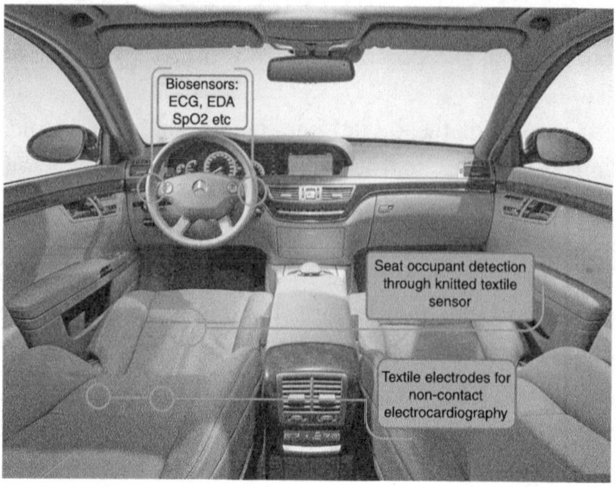

16.7 Realization of the 'intelligent cabin' vision (Mercedes S-Class) (source: Daimler AG).

Figure 16.7 shows the localization of the prototype systems in a real car environment.

16.4.1 ECG-seat with textile electrodes

Measuring the ECG of the subject is an approach to monitor the driver's condition and to detect awareness, drowsiness or stress. HR and HRV can be calculated from ECG measurement data. Medical research has found that there is a relationship between HRV, stress and drowsiness [31]. ECG is usually measured by a contact electrode method, coupling the electrodes by gel to the skin of the patient. Contact electrodes are obviously not suitable in a car for daily use and so a contactless ECG measurement method has been established through research projects for long-term monitoring of heart disease patients and persons working with special safety clothes (e.g. firefighters) [32].

INSITEX made the approach to apply a capacitive coupled ECG measurement method to car interiors and implement it in a car seat.

16.4.2 Measurement principle

Muscle activity generates an electric potential on the skin. The potential distribution of the electric field generated by the heart activity is shown in Fig. 16.8. The maximum amplitude is more than 2 mV.

The signal is measured by a differential potential measurement with a two-electrode arrangement. Both capacitive coupled electrodes are integrated into the backrest of the car seat (Fig. 16.9) and a driven circuit method is used to reduce noise.

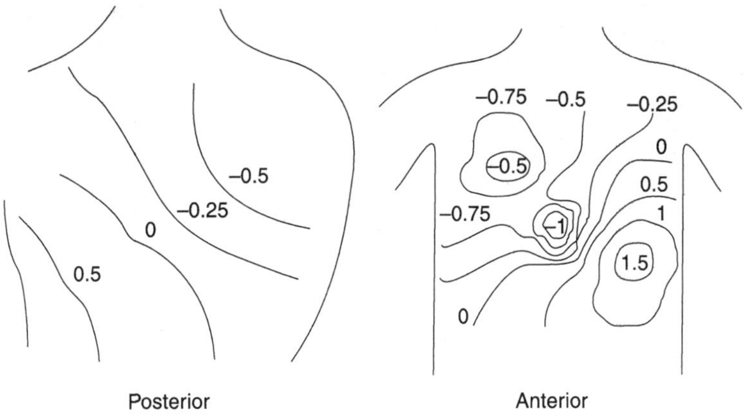

16.8 Potential distribution caused by heart activity [46].

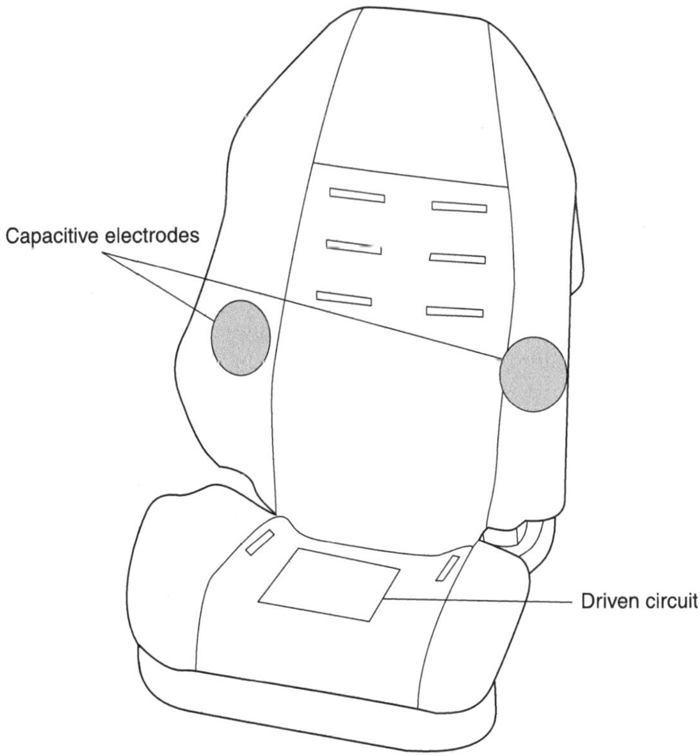

16.9 Positions of electrodes in car seat [33].

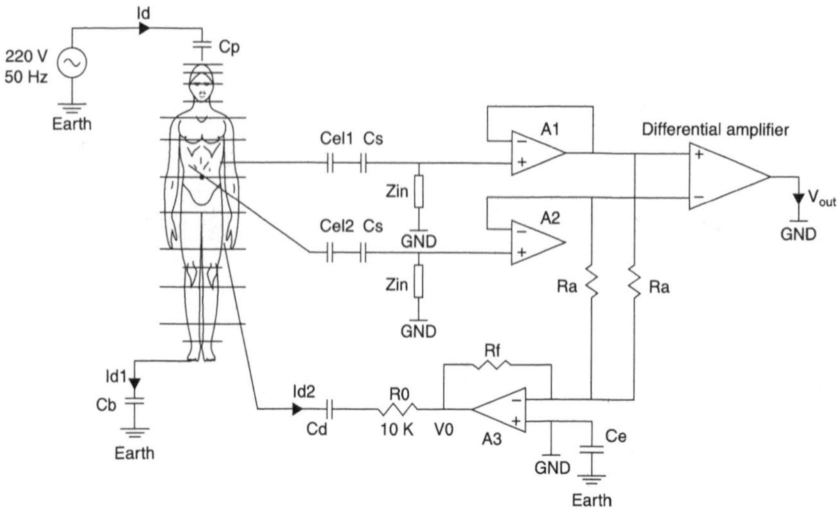

16.10 ECG measurement set-up [47] (source: INSITEX Bhavin).

Figure 16.10 shows the electrical circuit of the measurement set-up. The electrodes are formed by Cel1 and Cel2, which are similar to parallel plate capacitors. One plate is built by a shielded textile electrode and the other plate is the body surface. The dielectric layer between the plates is given by a textile shielding layer of the textile electrode and the clothes of the subject. The input signal is amplified by a high impedance differential amplifier with |ZQ| of more than 100 GΩ. For noise reduction, a driven seat concept with a third electrode in the seat cushion is used.

Quality and stability of the ECG signal depends on size, shape and position of the electrode. Despite Fig. 16.8 demonstrating that positioning the electrode near the heart is favourable, Chamadiya [33] found that taking into account robustness against movement artefacts and the differing height of people, a position in the lumbar region is the best compromise. Movement artefacts are caused by the relative movement of the person to the seat.

A key feature in capacitive coupled ECG measuring is the electrode. In INSITEX, the project partners with textile production and development competences [34] developed a multilayer electrode based on spacer fabric technology (Fig. 16.11). The conductive yarn used was e-blocker© [35]. Other textile electrodes with stacked laminated layers are also suitable. The conductive element here is a conductive paste printed to the corresponding layer [36].

The measurement results with textile electrodes are as good as those with metallic electrodes, which is an excellent result, because metallic electrodes are not usable by reason of haptics. Textile electrodes have the advantage of being permeable to air. Sensitivity against humidity is still a problem and is presently

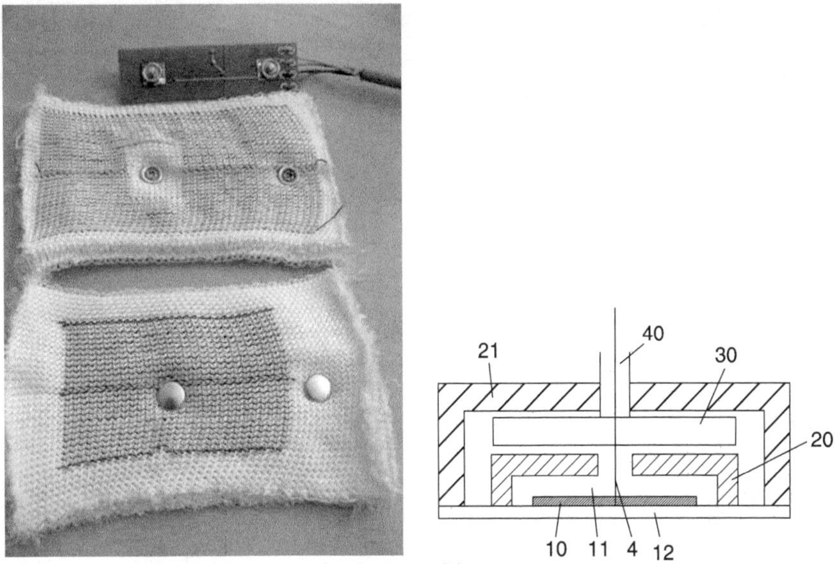

(a) (b)

16.11 (a) Prototype of textile multilayer electrode [35] and (b) cross section of multilayer electrode [36] (source: INSITEX (hto), Daimler AG (schematic).

being worked on. A seat cover with textile electrodes has been built and tested on real roads, as well as in the laboratory. Figure 16.12 shows the prototype in a Mercedes C-class.

Tests on roads show a good detectability and stability of the signal. Robustness of the set-up against distortions coming from the car's electronics, such as navigation, mobile phones and servomotors, were astonishingly good for a prototype. Some signals, such as the pwm signal of seat heating with a frequency of 24 Hz, were within the signal window. This distortion can be removed from the measured signal and is therefore no problem.

Driving on highways as well as on city roads has been performed. The most severe problem to be solved regarding signal quality is that of moving artefacts. Movements of the car body, for example on poor roads, or movements of the driver, can be the cause. Electrically this is a displacement of the plates of the capacitor and a resulting peak in the measured signal.

Correction of movement artefacts should be possible by measuring movements of the car body as well as the driver. These signals can then be used to distract them from the distorted ECG signal. Figure 16.13 shows ECG signals measured with the set-up shown in Fig. 16.12 on real roads for different driving situations.

Further investigations with diverse persons and different clothes have been performed. Although the probability of being able to measure ECG is high, the quality varies as for all measurements of physiological parameters. Clothes with

(a) ECG seat cover

(b) Driver seat electrode (on seat cushion)

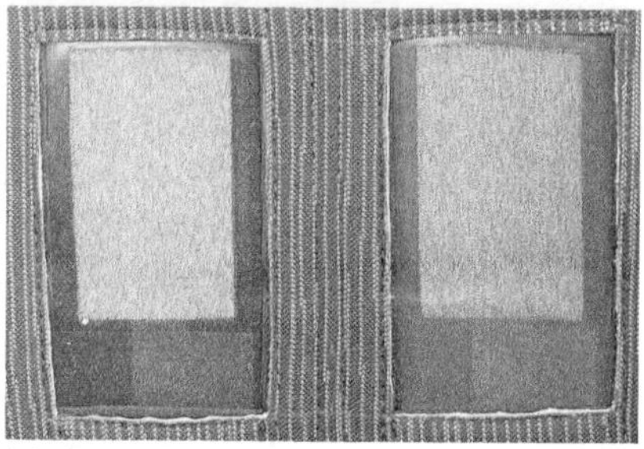

(c) Textile electrodes

16.12 (a–c) ECG seat cover. Installed in a Mercedes C-class (source: Daimler AG).

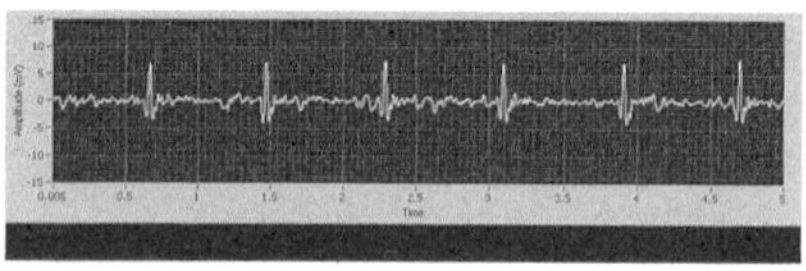

(a) Highway

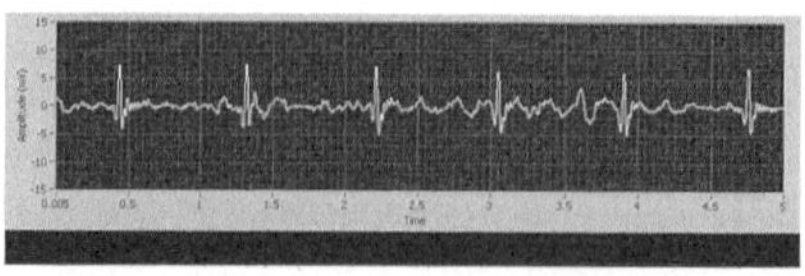

(b) City road

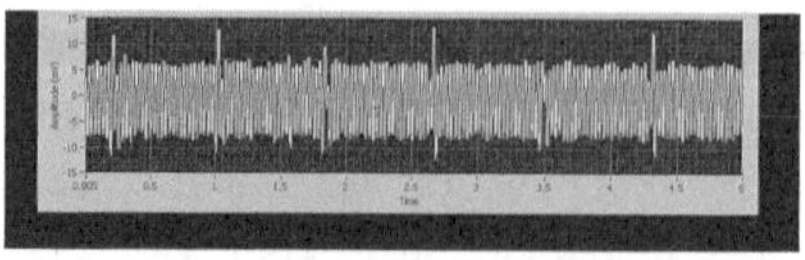

(c) Distortion by seat heating PWM signal

16.13 (a–c) ECG measurements on real road [48] (source: Daimler AG).

metallic coatings or surfaces that can be electrostatically charged are problematic for physical reasons.

16.4.3 Steering wheel with vital sensors

Another prototype of INSITEX is a steering wheel equipped with different vital signal sensors. The available integrated sensors and the block diagram are shown in Fig. 16.14.

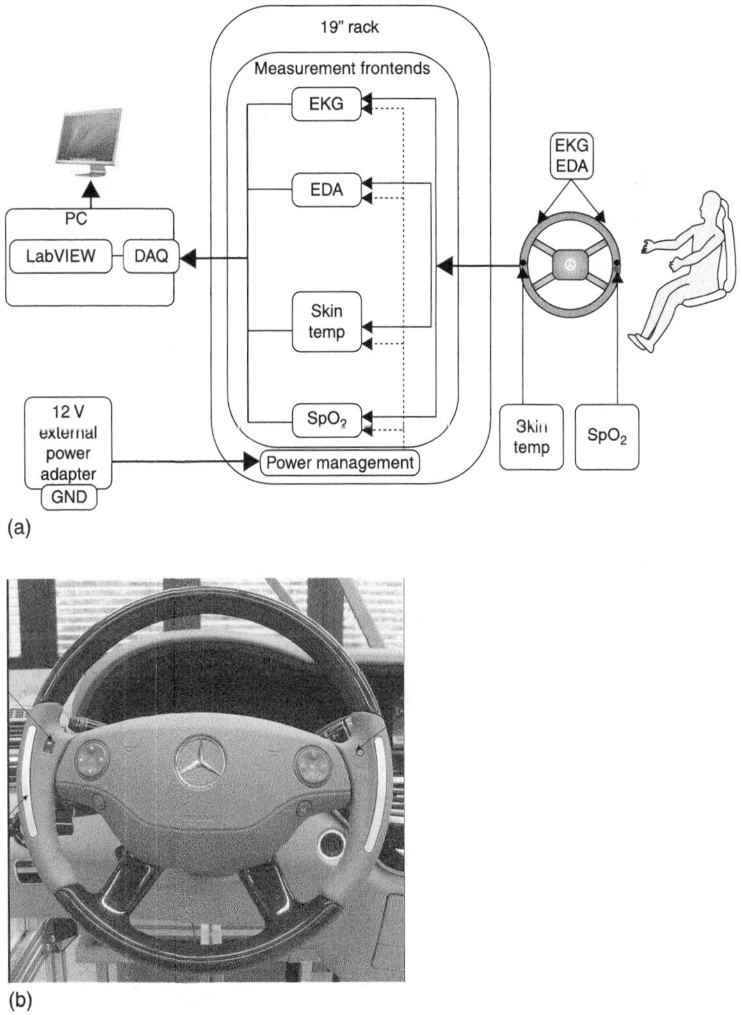

16.14 (a) Block diagram (source: FZF Karlsruhe Germany) and (b) prototype of steering wheel (Mercedes S-class) with vital signal sensors [49] (source: Daimler AG).

EDA is measured by two electrodes on the front and back side of the steering wheel. A weak constant current is applied and the resulting voltage is measured. The measurement can be taken on both sides of the wheel. Skin temperature is measured by a commercial thermopile. Oxygen saturation uses a commercial sensor with a two-wavelength differential principle.

A contact ECG is measured by two electrodes on the left and right side of the steering wheel, when they are contacted by the hands of the driver. By a runtime measurement, as described above, a relative blood pressure value can also be measured. This value can be calibrated for the individual person. This prototype has no textile component but, as mentioned above, it will be possible to build a 'textile' steering wheel [6] (Fig. 16.15). The prototype steering wheel can be used for test drives on test roads.

ECG measurement integrated in a car seat is also shown by other groups. The focus of the Smart Senior project was on disease detection and autonomous failsafe operation of the car [17]. Ford/RWTH Aachen have demonstrated a sensor array in the backrest of the car seat, which selects the best electrode pair for the evaluation of HR [37]. In the published results, they do not use textile sensors. For their experimental set-up, they also found a good detection probability of ECG signal.

16.4.4 Occupant detection and classification system with textile sensor

The third prototype built in INSITEX is a textile sensor able to measure weight and position of the passengers on the seat according to FMVSS208. The sensor is

16.15 Steering wheel mounted in the car and with displayed measurement values (source: Daimler AG).

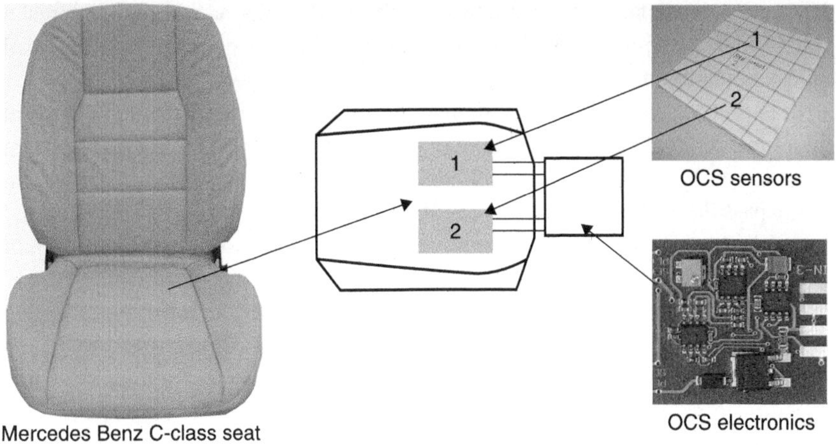

16.16 OCS with textile sensor [50] (source: IG bauerhin AG).

working on an inductive principle. By applying a pressure to the sensor, the inductive coupling is changed and a signal is generated. The leading partner for the textile system was IG bauerhin AG. The sensor [38] is based on spacer fabric technology. The special assembly is a development of IG bauerhin AG and patent registered. The use of two sensors in the prototype (Fig. 16.16) allows a rough estimation of the passenger's seat position.

The seat was tested according to automobile test standards and shows very good performance in the environmental and durability tests. Textile solutions for occupant detection and classification have also been the goal of the SeatSen project [18]. The SeatSen approach is based on a textile circuit board assembled with microelectronic devices in a matrix array. A special interconnection technology was developed in the project.

Beyond the above discussed applications of smart textiles for monitoring driver/passenger conditions, a set of other applications are of high interest and many really promising heating and lighting applications have been demonstrated [39], [16]. Textile switches and touchpads were mentioned in Section 16.3.

16.5 Key safety and quality requirements

All components used in the automobile industry are subject to a strong qualification process. This is necessary because of the long lifespan of a car and the complexity of its automotive systems. Qualification ensures a high level of quality and safety. Procedure and environmental tests are part of the component specification book. Each OEM has its own procedure, but all are developed from national or international standards and institutions such as IEC, DIN or SAE to name just a few, which are responsible for the standards or the national derivation of international norms.

In particular, environmental tests may be a hurdle for smart-textile systems at the moment. Upper and lower limits of the parameters of temperature, humidity, vibration and mechanical shock are defined in the specification book. The system has to pass these tests over the whole parameter range, otherwise it is not accepted for serial use. EMC qualifications and resistivity against liquids is also part of the test procedure. Seat systems, for example, have to perform the 'Coke' test, which means functions should not be affected when a soft drink is spilled on the seat.

Table 16.2 shows some typical values of the environmental parameters, which depend strongly on the location of the component in the car. Hence, the requirements for cabin electronics are typically lower than for systems located near the combustion engine.

The introduction of new technologies, like smart-textile technology, faces the problem that no appropriate test specifications are available. Sometimes existing norms that cover only a part of the system can be extended to cover the new item. Otherwise a new norm has to be defined. It may also be that the merging of two technologies generates new features that are, for example, safety relevant.

Testing of electronic components is widely covered by the IEC 60068 [40], but there is no textile system considered. Fortunately there is already an initiative for the smart-textile systems of the Standardization Institutes. The draft of the CEN/TR 16298:2011 [5] contains definitions for 'intelligent' and 'smart' systems. These definitions were used in Section 16.2.1. The draft defines furthermore what should be covered by the test norms to be developed. Adequate test norms for specific products, such as textile systems for vital signal measurement or intelligent seat heating, have to be created. This is an important and indispensable requirement, in order to use these systems in serial cars.

Table 16.2 Typical environmental conditions for car components (depends on system and OEM)

Parameter	Value
Temperature	−40 to 105°C
Humidity	Up to 85%
Vibration	Depends on location, typically 20 g (RMS)
Mechanical shock	Typically 50 g
Lifetime	Up to 15 years or 8000 h hours of operation

16.6 The impact of electric vehicles on smart-textiles applications

The fact that resources of fossil fuels are limited has boosted the development of alternative technologies in the last few years. The concept of electrically powered vehicles covers battery electrical cars as well as vehicles powered by a combination of fuel cells and Li-ion accumulators. Hybrid vehicles that use combustion engines with a supporting e-motor and a high voltage battery are considered as the link from the past to the fully alternative powertrain future. Battery capacity nowadays is sufficient to drive 150 to 200 km before recharging is necessary. A typical value for the capacity is approximately 18 kWh [41]. Battery power is not fully available for the powertrain, because all the comfort systems are drawing energy from the battery, so especially when systems for heating in winter or cooling in summer are activated, the range of the car is much shorter than the specified value. The need for energy optimized systems is obvious.

Table 16.3 gives some values of energy consumption of selected systems.

Air conditioning is the comfort system with the highest energy consumption. Heating, headlights and other systems are in the range of 100 to 200 W. The automotive industry discusses three main approaches to increase driving range and save waste energy:

- lightweight construction;
- thermal insulation of the cabin;
- energy efficient design of all components.

Smart textiles can contribute to each of these points and compete with new energy efficient system approaches. Standard solutions are often cost-optimized but not energy-efficient.

Table 16.3 Energy consumption of car components [43, 44]

Component	Power [W]	Fuel consumption [l/100 km]
Headlight, low mode	110	0.15
Electric air conditioning		up to 2.1 *
Seat heating	100 **	
Steering wheel heating	50	
Rear-window heater	150	0.12–0.17
100 kg additional weight		0.4

*depends strongly on temperature gradient cabin <> outside and car

**depends on car and number of heated seats

16.6.1 Lightweight construction

Each kg, which has to be moved, additionally needs additional energy (Table 16.3). Obviously it makes sense to enforce the trend to lightweight construction. Fibre compound materials are the materials of choice here. Textile materials as fabrics are usually not used in vehicle body construction, but new compound materials also use textile fibres. Textile fibres in construction are already in use in the building industry.

The potential of smart-textile systems is the possibility of integrating sensors and actors, for example, in insulating layers. The system weight of heating, cooling and ambient lighting may decrease this way. Spacer fabrics in particular, with their multifunctional properties, and the opportunity to produce different functions in one production flow can succeed here. Another feature of textiles, which should not be underestimated, is the potential to be a design as well as a functional element. Here again is the opportunity to reduce weight if, for example, a surface layer covering the functional layer can be omitted.

16.6.2 Thermal insulation of the cabin

Mass-produced cars on the current market are rarely thermally insulated. Thermal insulation of houses is, meanwhile, a standard, and even a legal restraint. Therefore in seasons when the temperature gradient from the cabin to outside or vice versa is high, a lot of energy is wasted and therefore thermal insulation and heat reflecting outer surfaces will be of great importance in future. The smart for vision concept car (Fig. 16.17), presented at the IAA 2011 in Frankfurt/Main, Germany, demonstrates the potential for using innovative materials in the automotive sector [42] – E-textiles for efficient temperature management; advanced composite materials for weight reduction; super absorbent non-wovens and flexible moulded foam systems for more comfort – the possibilities are endless. In conflict to lightweight construction, thermal insulation increases the weight on first glance but in summary there is a win in efficiency.

Here again combining design and function is an advantage.

16.6.3 Energy efficient design of all components

The comfort component with the highest energy consumption is electrical air conditioning, followed by heating (Table 16.3). Established heating applications are seat heating, heating of the steering wheel, panel heating and heating the rear and front windows.

So how could smart-textile heating be used to advantage? The heating of a body from temperature T1 to temperature T2 follows the physical laws of thermodynamics and needs the same energy regardless of the heating method. Obviously the advantage of smart textiles is the special feature to be functional as well as a surface element. In addition, as mentioned above, it can cover design

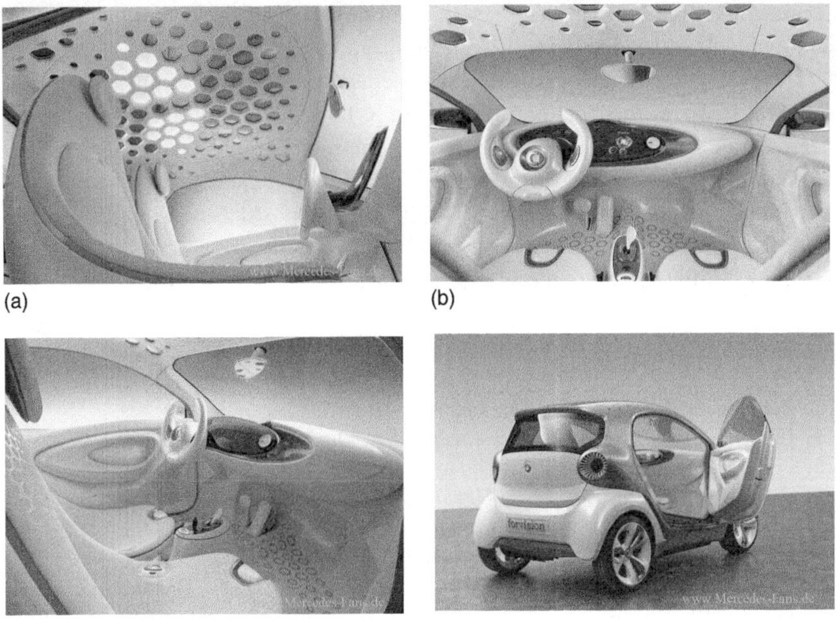

(a) (b) (c) (d)

16.17 Smart concept car 'for vision' with (a) solar roof, (b) and (c) thermal insulation and (d) heat reflecting paint and window coating

requirements. This means textile heating can be the surface itself or very close to the surface. Therefore no bulk material in the seat has to be heated, which would otherwise cause energy losses. Also the shape of the heating element can easily be adjusted to the body shape of the average passenger, which means no waste heating of non-contacted surface areas. A second way is to make heating 'intelligent' by dividing the heating area into segments and switching off segments where no body contact is detected.

The integration of contact sensors (e.g. capacitive devices) and temperature sensors is possible and needs no additional external wiring. These considerations are also valid for panel heating. If in future light-emitting textile fibres or devices are available, ambient light applications may also have energy saving potential. The growing electro-mobility market offers opportunities for new technologies. Smart textiles stand a good chance by using their very special features to compete successfully with other technologies.

16.7 Future trends

Smart-textile technology has left the research lab and found its way into medical-, sportswear- and safety-clothes products. New applications in the automotive industry seem to have potential and may give access to entirely new system approaches. Textile sensors, heating fabrics, lighting devices and other employments of conductive textile materials are under investigation.

The evaluation of new textile sensor systems for driver monitoring and passive safety systems in funded R&D projects (e.g. INSITEX) and other publications demonstrate the feasibility in the actual car environment.

Future car concepts, such as e-drive, will strongly require energy saving and intelligent components, for example for heating purposes. Smart-textile systems have the potential to succeed here if problems related to long-term stability are solved. Fulfilling the automotive requirements is just the challenge.

16.8 References

1. Zukunftsinstitut, Kelkheim, Germany and Wien, Austria.
2. Horx M (2011), *Das Megatrend Prinzip*, DVA, Munich, Germany.
3. Friedrichs J (2012), *Innovative Seating*, Darmstadt.
4. Fung W and Hardcastle M (2000), *Textiles in Automotive Engineering*, Series in Textile No. 13. Cambridge, UK, Woodhead Publishing, Ltd.
5. CEN/TR 16298 (2011), German version: DIN CEN/TR 16298.
6. Chamdiya B and Wagner M (2010), 'Lenkrad für ein Fahrzeug', Daimler AG, *Patent DE 10 2010 053 354.*
7. Patented yarn of Zimmermann GmbH and Co. KG/Novonic, Germany, Weiler-Simmerberg.
8. Fraunhofer IZM, Berlin/Germany, available from www.izm.fraunhofer.de/
9. BMBF funded project: INSITEX: Aktive Insassensicherheit durch intelligente technische Textilien, Förderkennzeichen: 16SV3463, 2007–2010.
10. European funded project, ConText: Contactless sensors for body monitoring incorporated in textiles (FP6-2004-IST-027291).
11. TITVs Car Trim AG, Plauen, Germany
12. Gorgutsa S *et al.* (2012), *Smart Mater Struct*, 21, 015010.
13. *Stroke your car seat to pump up the volume* – tech – 9 January 2012 – available from *New Scientist.mht*
14. ETH Zürich, TITV (to name some of them).
15. BMBF funded project Texoled, Förderkennzeichen: 16SV3450, 2007–2010.
16. http://www.talk2myshirt.com/blog/archives/5445
17. BMBF funded project 'Smart Senior' http://www1.smart-senior.de/
18. BMBF funded project SeatSen, Förderkennzeichen:16SV3457, 2007–2010.
19. BMBF: German Ministry of Education and Research.
20. http://en.wikipedia.org/wiki/Galvanic_skin_response
21. http://en.wikipedia.org/wiki/oxygenation_(medicine)
22. FMVSS 208, available from http://www.nhtsa.gov/Driving+Safety/Occupant+Protection
23 Hargutt, V (2003), 'Das Lidschlussverhalten als Indikator für Aufmerksamkeits- und Müdigkeitsprozesse bei Arbeitshandlungen', in: VDI-Gesellschaft Fahrzeug- und Verkehrstechnik (Hrsg.), *Reihe 17: Biotechnik/Medizintechnik* (VDI-Fortschritt-Bericht, Nr. 233). Düsseldorf, VDI-Verlag.
24. http://en.wikipedia.org/wiki/Electroencephalography
25. Schmidt E A. 'The objective assessment of fatigue during a monotonous daytime drive and the driver's verbal judgement', Dissertation, University Düsseldorf, Germany.

26. http://en.wikipedia.org/wiki/Electrocardiography
27. http://de.wikipedia.org/wiki/Herzfrequenz
28. http://en.wikipedia.org/wiki/File:SinusRhythmLabels.png
29. http://en.wikipedia.org/wiki/Galvanic_skin_response
30. Patent of Bosch GmbH, Germany.
31. www.fh-schmalkalden.de
32. Ottenbacher J and Heuer S (2009), 'Motion artefacts in capacitively coupled ECG electrodes', *World Congress on Medical Physics and Biomedical Engineering*, Munich, September.
33. Chamadiya B (2008), Master Thesis, 'Investigation of capacitive coupled electrocardiography for car seat integration', Böblingen, Universität Lübeck.
34. Moll C, 'Insitex partners with textile R&D and production competences', Zimmermann/Novonic and Fraunhofer IZM.
35. E-blocker is a product of Zimmermann GmbH and Co. KG/Novonic, Weiler-Simmerberg, Germany.
36. Chamadiya B and Wagner M. Patent multilayer electrode DE102010023369A1, Daimler AG.
37. Eilebrecht B, Artzek T W, Lem J, Vogt R and Leonhardt S (2011), 'Capacitive electrocardiogram measurement system in the driver seat', *ATZ 03*, 113, 50–55.
38. The sensor construction is a patent of IG bauerhin AG, Gründau, Germany.
39. Article 'Ein Faden heizt ein', Automobil Industrie, 5.5.2008
40. DIN IEC 60068.
41. Daimler AG, Smart eV, battery capacity, data sheet.
42. Smart for vision, Conceptcar, presented on IAA (2011), available from http://www.mercedes-fans.de/galerie/id=1325
43. ADAC, German Automobile Club.
44. University of Duisburg-Essen, Öko-Globe Institut, Germany.
45. TITV: Textilforschungsinstitut Thüringen-Vogtland e.V., Greiz/Germany, available from www.titv-greiz.de
46. Anderson J and Allen J. Apparatus for body surface mapping. *US patent US2003/0040677 A1*.
47. Kim K, Lim Y and Park K (2005), 'An application of electrically non-contact ECG measurement on a chair: heart rate variability during computer game involving mental activity', *Proceedings of the 2005 IEEE Engineering in Medicine and Biology 27th Annual Conference*, Shanghai, China, 1–4 September.
48. INSITEX (2011), Final REPORT, TB-Hannover, Germany.
49. Daimler AG and INSITEX partner, FZI, Karlsruhe, Germany.
50. IG bauerhin AG, Gründau/Germany, INSITEX partner.
51. FZI (Forschungszentrom informatik), Karlsruhe, Germany. INSITEX partner.

17
Architectural applications of smart textiles

A. RITTER, ritter architekten, Germany

DOI: 10.1533/9780857093530.3.468

Abstract: This chapter starts with an introduction about key themes and important factors in modern architecture. It then gives some general informations about smart materials, which can be divided into the categories of property-changing smart materials, energy-generating and exchanging smart materials, and matter-exchanging smart materials. Then some examples are given of existing smart textiles for architectural applications, which include light-emitting metal mesh, self-cleaning membranes, self-healing textiles, shape memory interior textiles and innovative actuators, made through the use of smart textiles. The chapter ends with future trends for exterior textiles, actuators and interior textiles.

Key words: property-changing smart materials, energy-generating smart materials, energy-exchanging smart materials, matter-exchanging smart materials, adaptive architecture, kinetic architecture, smart building textile, smart polymer actuator.

17.1 Introduction: key themes in modern architecture

There are a number of factors that are important when considering materials for modern architecture, including:

- sustainability
- adaptability
- intelligence.

Sustainability in the architectural context refers to building materials and products that are preferably natural, fully recyclable, or both. Furthermore, it requires that buildings are able to absorb energy and matter (e.g. heat and water) and to generate energy and use storage facilities, so that they no longer require extraction of gas and electricity from local distribution grids. These requirements generally lead to buildings consisting of materials and products made from natural materials such as wood, clay and straw. To be energy and matter self-sufficient, they may need to move (e.g. rotate to track the sun), as well as to include components that can concentrate and/or collect energy and/or collect matter like some smart materials do.

Adaptive or smart materials and structures have components that are able to respond to external and/or internal stimuli by, for example, changing colour and/or creating new geometries or shapes (Fig. 17.1(a,b), Fig. 17.14(a–c), Fig. 17.16(c)) [1–19]. Smart textiles, which are adaptive, could include shape memory alloy wires (Fig. 17.8(a,b)).

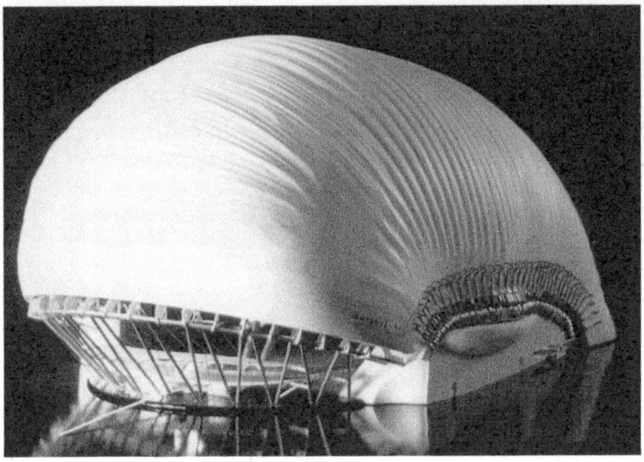

(a)

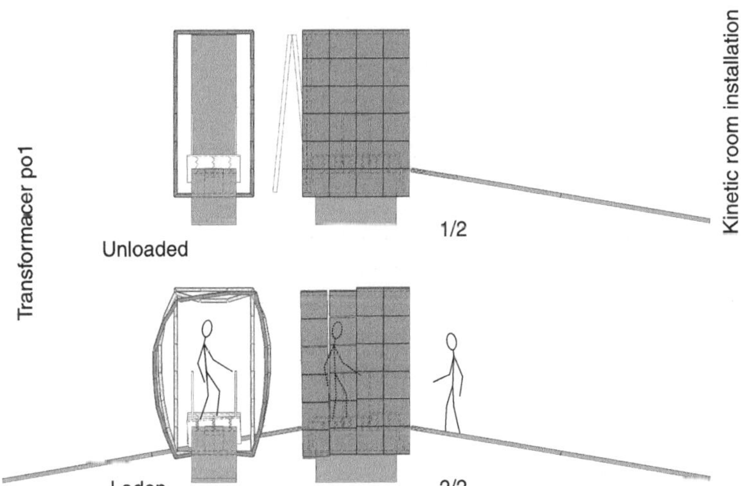

(b)

17.1 (a) Airship hangar *morphyt-nt* with an innovative inflatable, contractible/expandable three-layer membrane. (b) Human weight-controlled/operating building installation called *Transformacer p01*. (Source: Axel Ritter, 1995, 2002.) (Drawings by Axel Ritter, 2002/modified 2011/2012.)

Intelligent architecture is becoming increasingly important, moving away from devices that work using the electricity grid and towards structures able to convert mechanical energy into electrical energy. One such approach consists in weight controlled mechanical structures able to generate and regulate, for example, the light and electrical power in a home, based on the number of persons in it. Another such application links mobile phones with home automation.

It is now possible to create smart buildings that can react directly or indirectly to their environments. Leading on from this, it is now possible to consider buildings that are able, for example, to heal themselves (e.g. the bentonite-based textiles shown in Fig. 17.7(a,b)), able to clean the air in rooms, or convert harmful substances to harmless substances.

17.2 Smart materials

Depending on physical and/or chemical stimuli, such as light, temperature, pressure, electric or magnetic fields, such as a chemical environment, smart materials are able, for example, to change shape or colour, generate an electrical current, or absorb and store water. Some are able to react with their environment (e.g. convert harmful into harmless substances [3,4,13,14]). Smart materials fall into several categories:

- property-changing smart materials;
- energy-generating and exchanging smart materials;
- matter-exchanging smart materials.

17.2.1 Property-changing smart materials

Shape-changing smart materials for example shape memory alloys, can and are able to change their shape in response to temperature (Fig. 17.2(a)). Such materials might be used in switching operations, such as a sun shade opening or closing in response to the amount of sunlight (Fig. 17.8(a,b)). Other materials can change their optical properties, such as thermochromic paints and electro-optic functional layers. These may be used in switchable glass systems to control the light entering a room and/or on textiles. With adhesion-changing smart materials, adhesion forces at their surfaces depend on parameters such as light, temperature, and electrical or magnetic fields. Examples include photo-adhesive smart materials operating with photo-sensitive titanium dioxide (TiO_2). An example of their use includes membrane-coated buildings, which are self-cleaning and air-improving in response to light conditions (Fig. 17.6(a,b)) [3,4,25].

17.2.2 Energy-generating and exchanging smart materials

Electricity generating smart materials include both well-established silicon-based solar cells, and the lesser-known dye solar cells that mimic the natural process of photosynthesis (Fig. 17.2(b)). The use of these, together with thermo-piezo and mechano-electrically operating systems, make possible the generation of CO_2-neutral electricity independent of the grid system. Applications include textile membranes embedded with piezoelectric elements that generate electrical

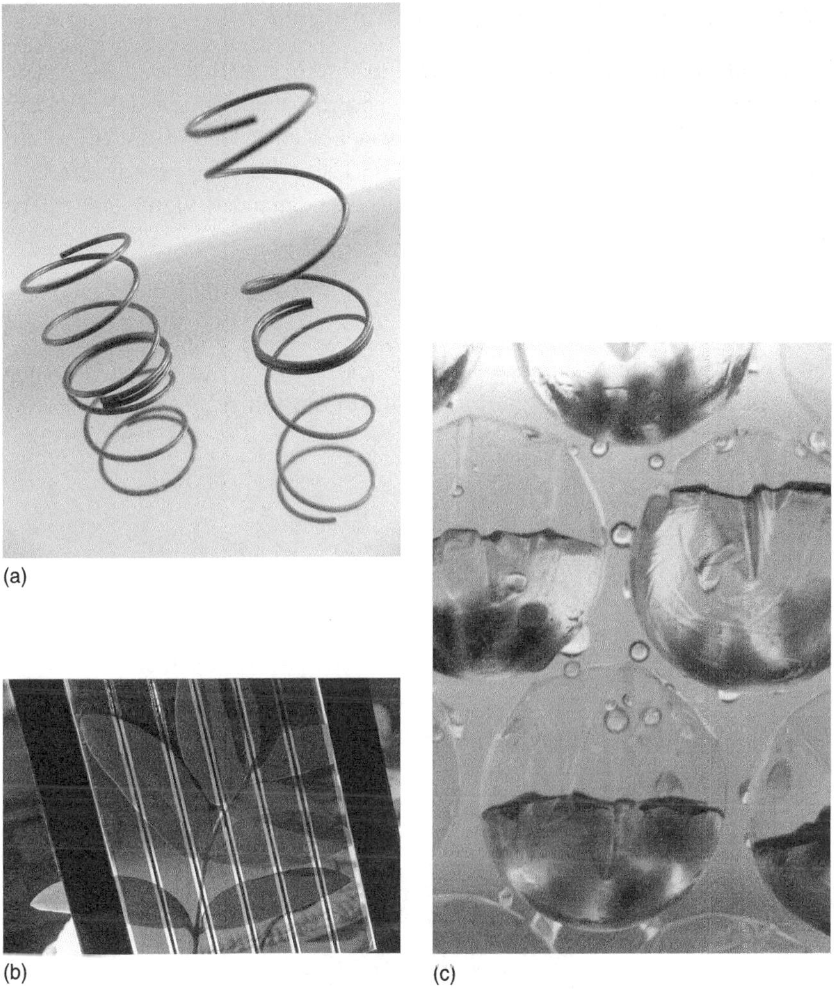

17.2 The three different types of smart materials: (a) example of property-changing smart materials, in this case shape memory alloy (SMA) springs showing a two-way effect (photo by Axel Ritter 2011); (b) example of energy-generating and -exchanging smart materials, in this case a dye solar cell (DSC) (photo courtesy of [20]); (c) example of matter-exchanging smart materials, in this case a functioning demonstrator of a *hydroabsorber foil* to control light transmission and improve the climate of rooms/buildings. (Source: Axel Ritter, 1995.) (Photo by Axel Ritter, 1995.)

voltage in response to wind-influenced mechanical pressure, which could be sufficient to switch a display operating with e-ink.

Other smart materials of this type are so-called phase change materials (PCMs), such as salt hydrate (Fig. 17.4b), which can cool rooms passively.

17.2.3 Matter-exchanging smart materials

These include smart materials that are able to absorb, store and/or release matter. Through ad- and absorbents it is possible to absorb water, store it and then release it through different textiles in which the water is embedded. Examples of use include passive cooling effects in façades. These materials are available in different forms, such as powders and granulates. In granulated form, they can be placed into the cavities of membranes (Fig. 17.2(c)).

17.3 Applications

Following are some examples of passive and active in smart-textiles usable components (Fig. 17.3(a–c) and Fig. 17.4(a–c), respectively) and examples of architectural applications.

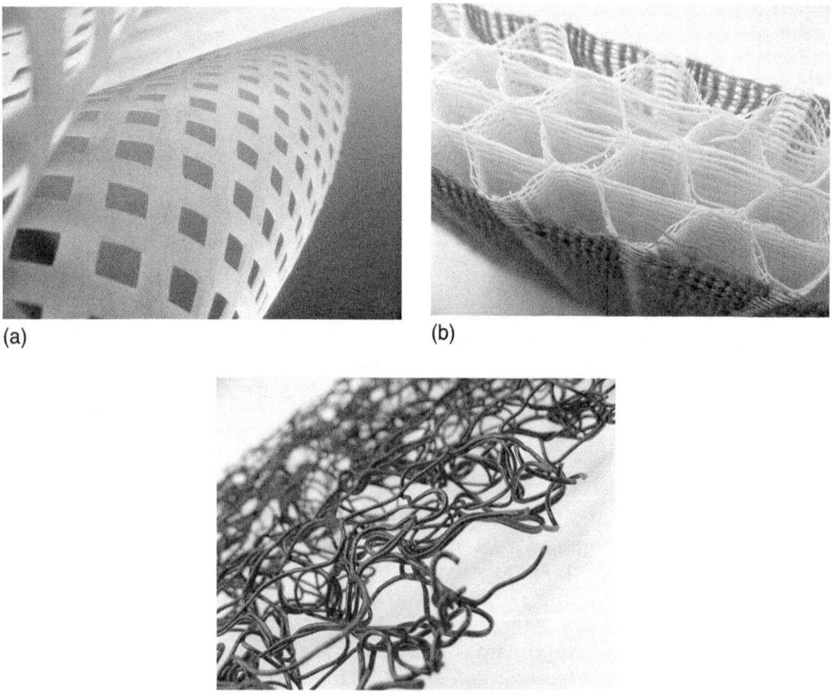

(a) (b) (c)

17.3 Examples of passive components, which could be used in smart textiles: (a) a square hole raster equipped membrane, made from coated polyether sulphone high tenacity (PES HT) yarn normally used as an architectural device for sun protection [21]; (b) a three-dimensional multilayer structural fabric cushion (unknown manufacturer); (c) a three-dimensional plastic mat normally used to protect against surface erosion [22, 23]. (Photos by Axel Ritter, 2011.)

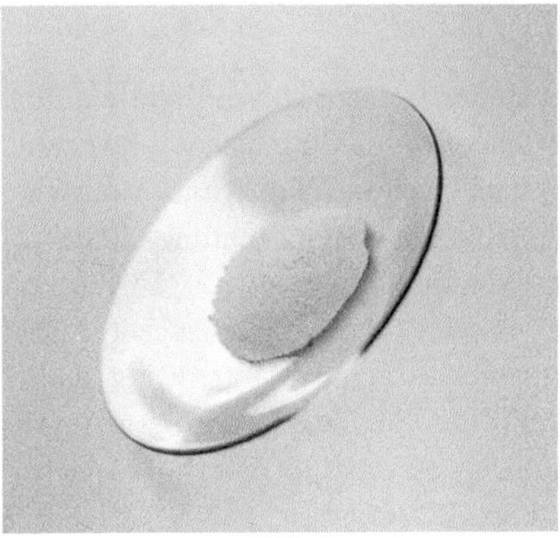

(a) ((TiO_2))

(b)

(c)

17.4 Examples of active components, which could be used in smart textiles (a) titanium dioxide (b) special phase change material (PCM); (c) water-swellable polyacrylate. These and other active components could be combined, e.g. with the three-dimension multilayer structural fabric cushion to create air-purifying and latent heat-storing textiles. (Photos by Axel Ritter, 2006, 2012, 2010.)

17.3.1 Light-emitting metal mesh

A relatively well-known smart textile for exterior use is a special light-emitting metal mesh (Fig. 17.5(a–c)). Originally designed for building façades, this could also be used for complete colour changes of both façades and interiors. Modifying this type of material might involve application of organic light-emitting diodes, as opposed to the synthetic light-emitting diodes commonly used (Fig. 17.12(a,b)), and the replacement of ordinary metal components with fluorescent and/or phosphorescent metals in various components to achieve interesting and/or energy-saving multi-colour and lighting effects. Another development is a filigran metal mesh with interwoven phosphorescent fibres. Able to deform repeatedly, such a material is able to glow in the dark (Fig. 17.11(a,b)).

17.3.2 Self-cleaning membranes

Self-cleaning membranes refer to textile products able to clean their own surfaces by using the photocatalytic effect. One such membrane consists of a two-dimensionally (2-D) woven fabric made of polyester or glass fibre covered on both sides with polyvinylchloride, with an additional protective adhesive layer and an outer photocatalytic layer embedded with photo-sensitive TiO_2 (Fig. 17.6(a,b)). Another such product is made of glass fibre covered on both sides with polytetrafluoroethylene. A third operates with TiO_2 and the insulation material aerogel.

17.3.3 Self-healing textiles

One example of self-healing textiles are geotextiles, including the natural product bentonite on or between two textile layers (Fig. 17.7(a,b)). This material is water swellable and is able to expand and close cracks, etc. Products using bentonite include 'water stop' strips which, for example, isolate concrete parts of a basement against incoming water.

17.3.4 Shape memory interior textiles and other interior textiles

Shape memory interior textiles are able to regulate light and heat transmission, depending on the level of sunlight. They use shape memory alloy wires such as Nitinol. In this case, the material is responsible for the reduction or enlargement of the translucent areas of the textiles. These products are still in the prototype phase (Fig. 17.8(a,b)). Another smart textile using shape memory alloy wires is the temperature controlled hook and loop fastener (Fig. 17.9(a,b) [3,4]).

Other examples of products for the interior use are clay boards with embedded microcapsulated PCM and glass textiles (Fig. 17.10) (PCM, Section 17.2.2).

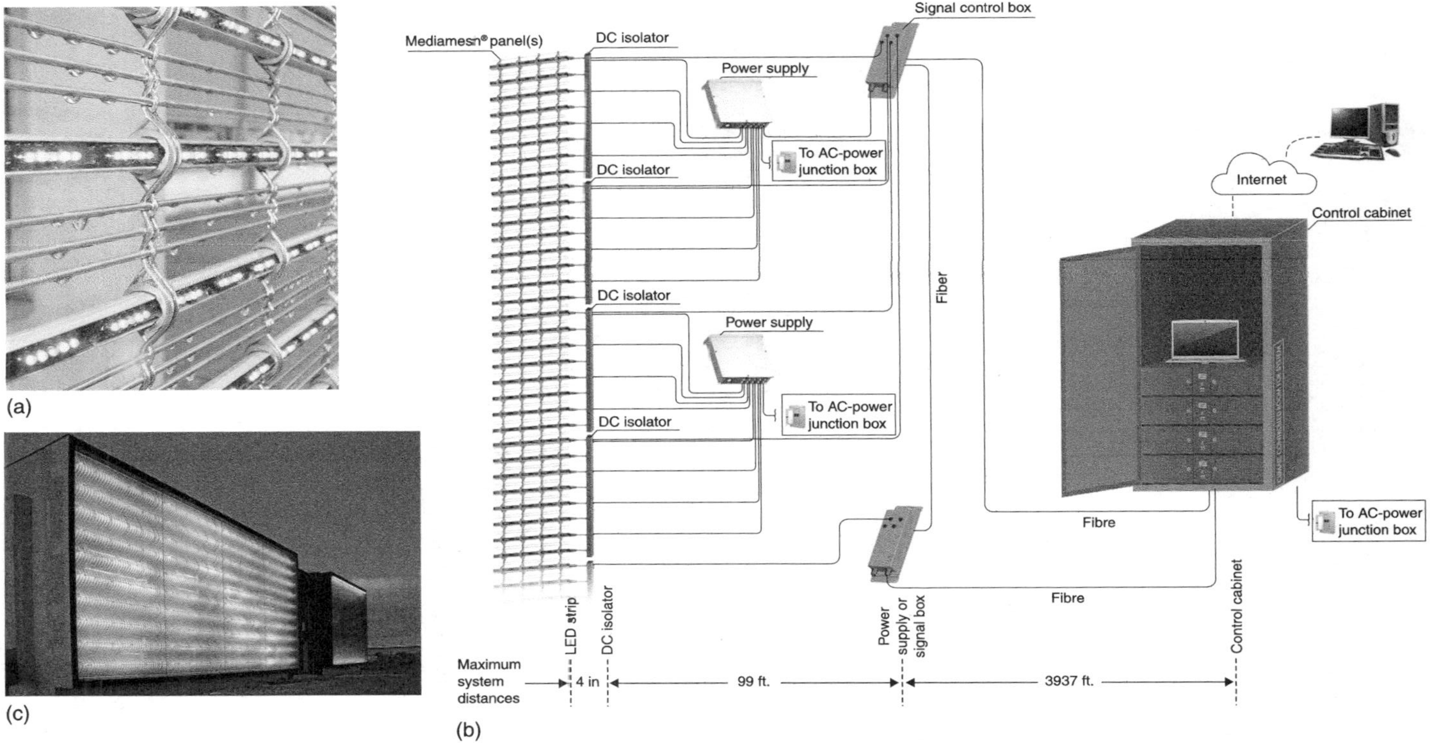

17.5 (a) A smart textile consisting of a metal mesh and LEDs (*Mediamesh*). (b) Wiring diagram of a new version of this textile, with six instead of five LEDs per pixel. (c) Application of similar textiles on the *Atelier a Torcé* (architect: Batir, France). (Photos courtesy of [24].)

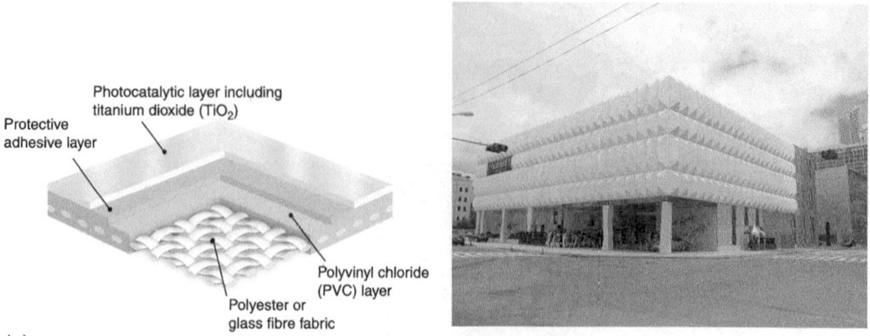

(a) (b)

17.6 (a) PVC-based membrane covered with a photocatalytic layer that has embedded TiO_2. (b) *Busan Bando* Model House (architect: Design Tomorrow Inc.), using a similar membrane. (Graphic and photo courtesy of [26].)

(a) (b)

17.7 Geotextile with water-swellable bentonite: (a) textile partly activated; (b) Shanghai Airport under construction using this textile (architect: Murphy/Jahn). (Sources: [22, 23].) (Left photo by Axel Ritter, 2012; right photo, courtesy of [22].)

17.3.5 Actuators

Figure 17.13(a,b) show an actuator consisting of eight hydroactive strips that are able to change their length, depending on the humidity of the surrounding air. One end is fixed on the body of the drive, whilst the other end is mounted on a mechanical device that transmits movement to a valve regulating air transmission between indoors and outdoors. Normally, the actuators are found on the top of window frames as part of a humidity controlled, energy self-sufficient, ventilation system. Using such strips in series (as opposed to in parallel) might enable new kinds of fibres for large-scale actuators, for example to be used in the positioning

(a) (b)

17.8 Shape memory interior textiles: (a) detail of a shape memory interior textile with embedded SMA wires; (b) proposal for a room divider using such a smart memory interior textile. (Textiles and photos by Yvonne Chan Vili, 2003.)

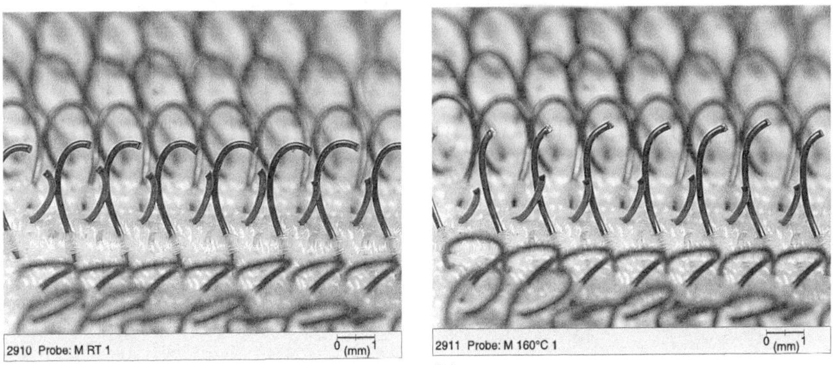

(a) (b)

17.9 Hook and loop fastener with nickel-titanium alloy hooks: (a) closed normal state; (b) open in response to a certain temperature. Textile developed and prototype built by EADS (A. Schuster, 2001). (Photos courtesy of [27].)

of building elements such as sunshades. This could also lead to large-scale fabrics operating like pinecones, opening and closing in response to humidity. Besides humidity transmission, this could also potentially be used to regulate light and/or noise transmission, filtration of polluted air, etc.

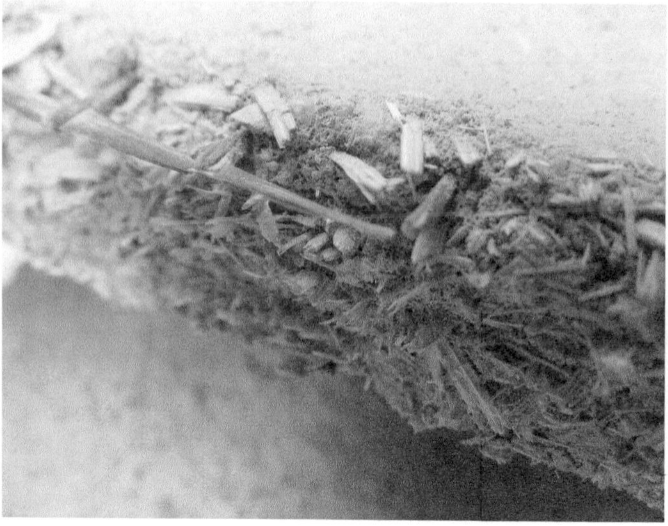

17.10 Detail of clay boards with embedded microcapsuled PCM and glass textiles. (Source: [28].) (Photo by Axel Ritter, 2011.)

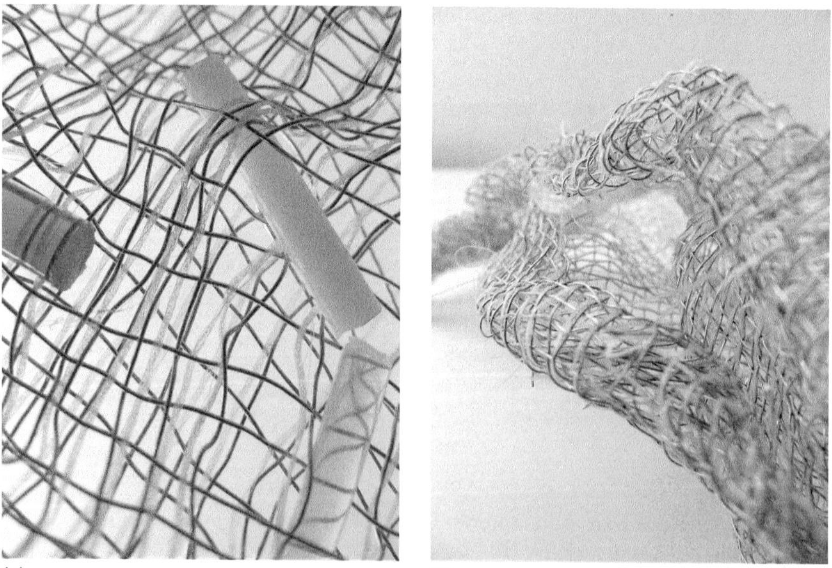

(a) (b)

17.11 Modified metal mesh with embedded phosphorescent fibres and additionally fixed fluorescent and phosphorescent sticks made from plastic polymethylmethacrylate (PMMA) and polyamide (PA): (a) close-up detail; (b) proposal for a building envelope. (Textile developed by Hannaliisa Johnson, 2005. The modification and deformation arrangement to simulate a building envelope was developed by Axel Ritter, 2011. (Photos by Axel Ritter, 2011.)

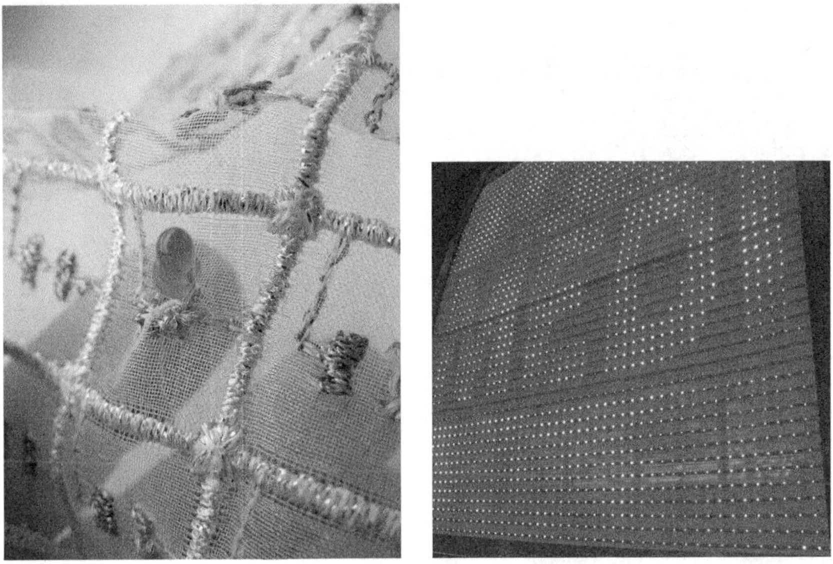

(a) (b)

17.12 Two different types of smart textiles with rows of linear orientated LEDs: (a) detail of a textile with a conductive yarn and a connected LED (modification and photo by Axel Ritter, 2011); (b) media screen consisting of a smart textile with rows of linear orientated LEDs as a room divider. (Photo courtesy of [29].)

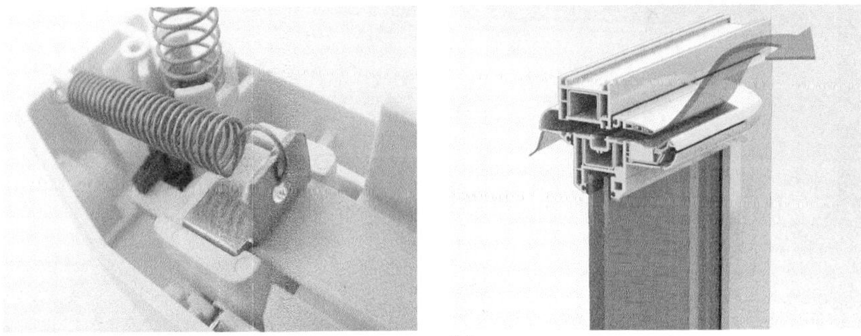

(a) (b)

17.13 Hygro-active strips and actuators: (a) details of an actuator (drive) consisting of eight strips; (b) rendering of a sectioned system mounted on a window frame. (Source: [30].) (Photo (a) by Axel Ritter, 2011.)

Figure 17.14(a,b), a new development by the author of this chapter, shows an expandable braided sleeve with an inner polymer-based strip able to swell with increase in water content. Depending on the absorption or desorption of water, the strips expand or contract and simultaneously, and the sleeve expands and contracts

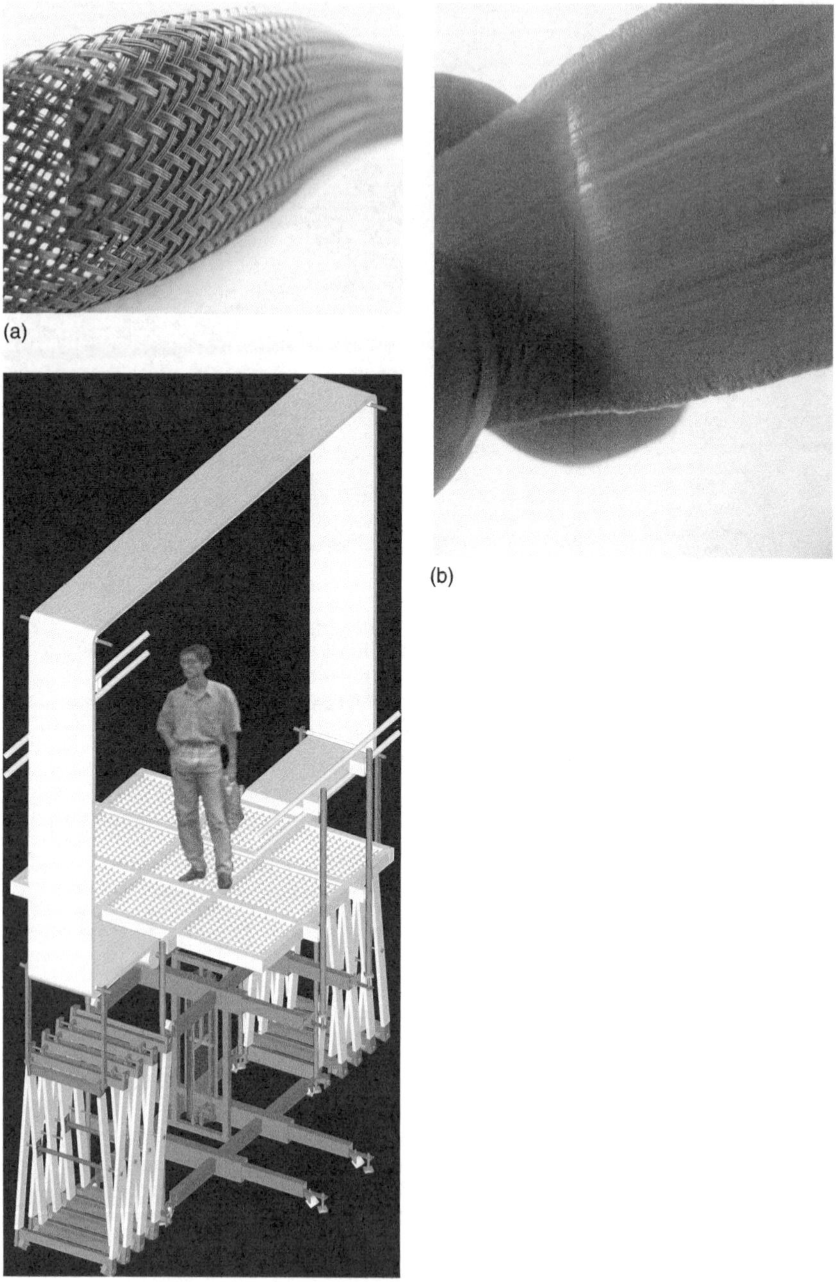

(a) (b) (c)

17.14 New actuators: (a) detail of *Antriebsstrang II*, a smart-textile prototype developed and built by Axel Ritter 2004; (b) detail of a water-swellable polymer (polyacrylate-based) strip, partly activated; (c) kinetic building structure using such an actuator developed by Axel Ritter 2007. (Photos and graphics by Axel Ritter, 2011, 2009, 2007.)

accordingly. Figure 17.14(c) illustrates a suggested energy and matter self-sufficient kinetic building structure using smart textiles in a large-scale actuator. The actuator is located under a weight-sensitive platform. Platform and actuator movements influence the space-forming textile structure, which changes its geometry accordingly.

In 2009, a group of scientists at the Technical University of Darmstadt developed new actuators using the technique of 'thermobiometals'. These consisted of several different thermo-sensitive composite strips made, for example, from PVC as the active components, and differently-orientated carbon fibres as the passive components, the two combined with conventional two-component adhesives. A proposed architectural application using these materials is shown in Fig. 17.15(a,b). Similar considerations were already made around 1995 by the author and another student at the University of Stuttgart.

17.4 Future trends

17.4.1 Exterior textiles and actuators

One concept is to create a building skin that behaves like human skin. A demonstrator developed by the author consists of a membrane that can change colour and structure, depending on temperature and humidity, and is able to cool down the temperature of the façade due to changes in temperature and absorbed rainwater (Fig. 17.16(a,b)). Figure 17.16(c) shows a possible architectural application. Another concept is the *_glowingSkin II* project, which is a multifunctional luminescent building and vehicle skin (Fig. 17.17(a,b)). This complex system is able, through the use of light tubes embedded into a flexible heat isolation material and an inner carrying layer made, for example, of strong polyester fabric, to harvest light energy and so save the requirement to use electricity from a grid.

Other proposals include conventional materials covered in gelatine or agar, serving as a food carrier for bacteria, algae, etc., which might be used for environmental monitoring, colour changes, air purification, etc. (Fig. 17.18). Also proposed is the use of nanometer-sized robots able to create different types of textiles with enhanced properties (Fig. 17.20).

Not far from practical application are new smart-textile actuators developed by the author (Fig. 17.21(a–c)). These operate with super-absorbent polymers (Fig. 17.4(c)) that are able to swell quickly. Instead of using one single surrounding expandable textile, this actuator also has an inner elastic fabric to ensure the water-swelling components can not leave the inside. Applications include sun blinds, etc. and it should also be possible to realize large-area applications such as building skins that are able to cool and change their structure depending on the amount of water absorbed.

17.15 Thermoactive strips and actuators consisting of PVC with carbon fibres. (a) detail of a test carrier, called *Blume* (photo by Axel Ritter, 2011), (b) proposal for an autonomously opening and closing device of a smart building skin [31]. (Other photos courtesy of [31].)

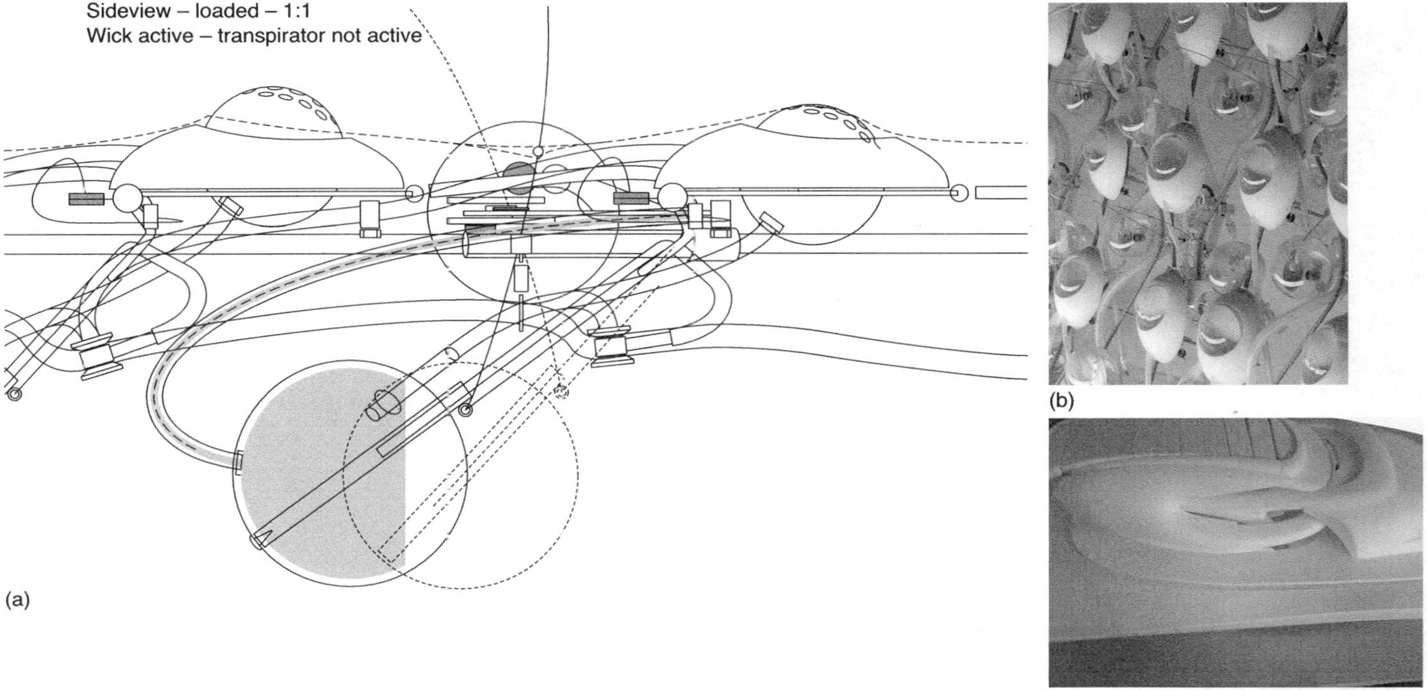

(a)

(b)

(c)

17.16 (a) Drawing of a section of *Polyreagible Mechanomembrane* showing different stages of construction. (b) Detail of the whole water-absorbing, -storing and -releasing system with thermobimetal and hygrobicomposite spiral springs as part of the smart component equipment. (c) Architectural application of the system at the *Futurologisches Zentrum*. (All content by Axel Ritter, 1995, 2007, 2012.)

(a) (b)

17.17 Multilayered, multifunctional luminescent skin called *_glowingSkin II*, consisting of an inner phosphorescent textile layer, with high power LEDs and a flexible outer layer made of silicone with embedded tiny flexible light tubes: (a) rendering; (b) a functioning demonstrator. (System developed/demonstrator built by Axel Ritter, 2007.) (Photo by Axel Ritter, 2007, 2008.)

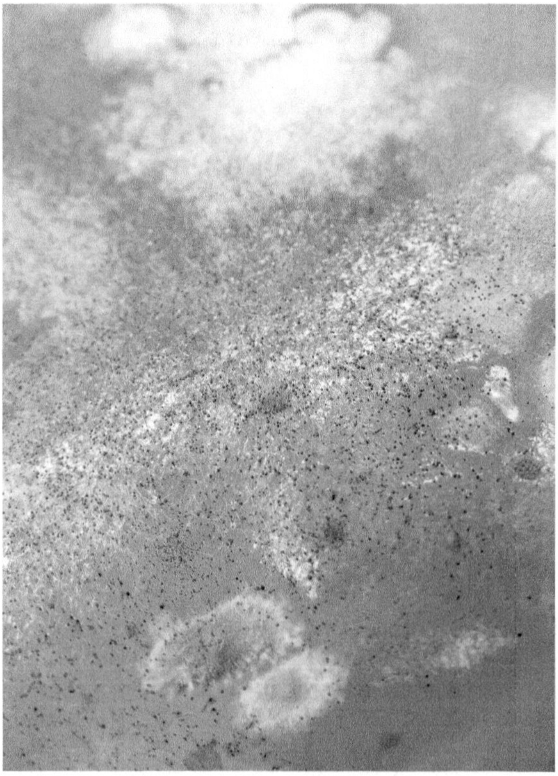

17.18 Detail of a green gelatine surface with mold fungus an actual population of mold fungus in a room. (Concept, realization and photo by Axel Ritter, 2011.)

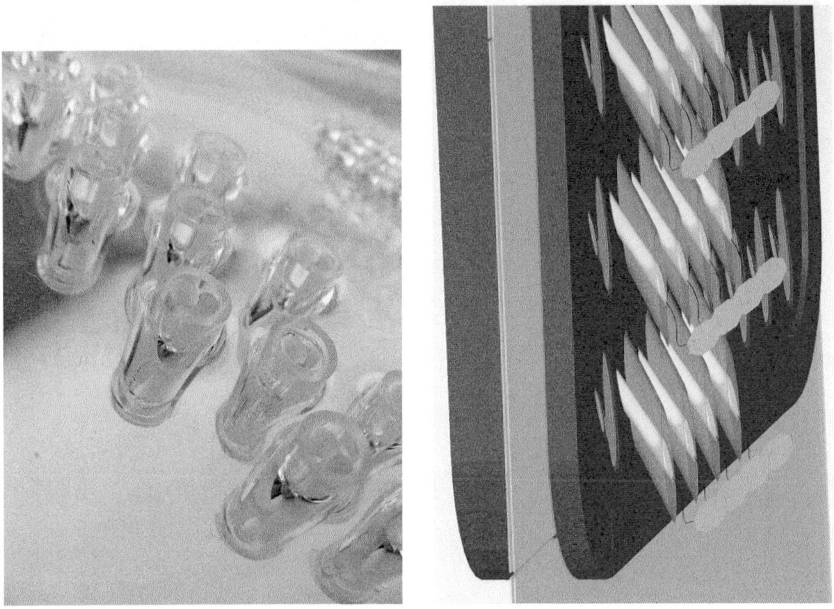

(a) (b)

17.19 Kinetic luminescent skin called *_glowingSkin I*: (a) detail of a functioning demonstrator; (b) proposal for an application as a room divider. (System developed and built by Axel Ritter, 2007). (Concept, realization and photo by Axel Ritter, 2011, 2007.)

17.20 Proposal for a flat structure constructed with *nanobots* [3,4]. (Courtesy of [32].)

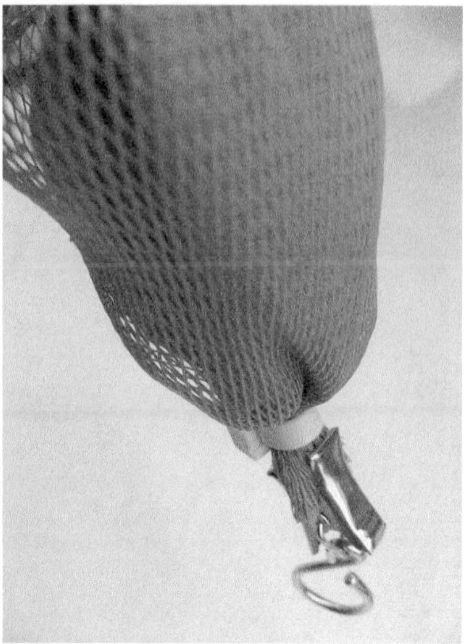

(a)

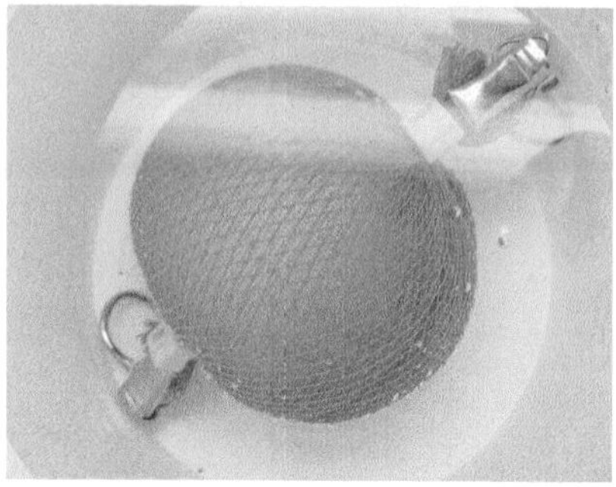

(b)

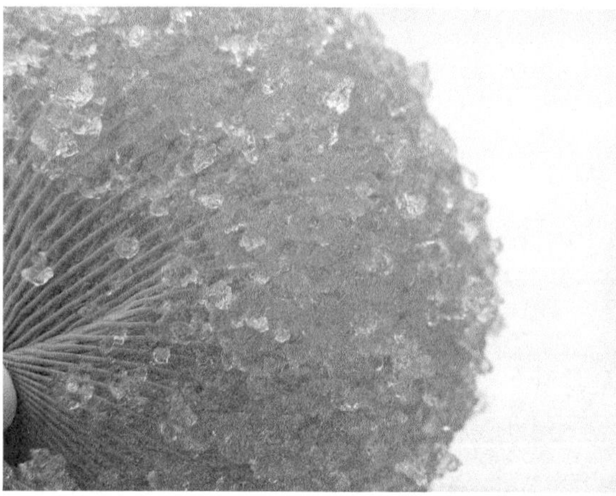

(c)

17.21 New actuator II: (a) detail of *Antriebsstrang II* equipped with a superabsorber 2012, when not activated and at normal length; (b) the entire device in contact with water, which has activated it and the device is nearly fully contracted; (c) detail of the activated surface. (Smart textile/demonstrator developed and built by Axel Ritter, 2012.) (All photos by Axel Ritter, 2012.)

17.4.2 Interior textiles

Figure 17.19(a,b) shows the *_glowingSkin I* project of the author, where a stable frame is embedded into a textile fixed, for example, to the floor and to the ceiling of a house. In the frame there is a translucent insulation element with fluorescent and phosphorescent sticks embedded in a matrix of silicone. The use of these sticks, surrounded by transparent tubes, and with the use of heat-sensitive Nitinol wires (fixed on small mirrors), forms a kinetic structure that is able to create different light effects energy self-sufficiently. Such an arrangement can also be used for room dividers (Fig. 17.19(b)), as ceiling elements, and also for façade elements. Nanotechnology may also be used to transform room interiors according to the wishes of the occupants. For example, a textile wall might convert to a more rigid wood-like wall, with an accompanying change of perfume from flowers to forest. Walls, and if wanted/needed, other building elements, might also change their geometry with this technology (Fig. 17.20).

17.5 References and further reading

1. Ritter A (2007), 'Material world: advanced materials have many potential applications in design and architecture, both practical and futuristic', *Engineering*, 248(1), 56–57.
2. Ritter A (2008), 'Advanced Materials: Neue Materialien für das Interior-Design', *medAmbiente*, 2, 30–31.
3. Ritter A (2006), *Smart materials in architektur, innenarchitektur und design*. Basel, Birkhäuser.
4. Ritter A (2006), *Smart Materials in Architecture, Interior Architecture and Design*. Basel, Birkhäuser.
5. Ritter A (2008), 'Innovative Building Materials: green smart materials – an international perspective', *Green Building Fest*, Toronto, Canada, 9 October. Sustainable Buildings, Canada.
6. Ritter A (2009), 'Green smart materials in architecture and interior architecture', *Green Building and Design Conference 2009, Green Materials*, Melbourne Convention Centre, Australia, 9 September. Centre for Design, RMIT University, Melbourne.
7. Ritter A (2009), 'Designing Zero and Plus Energy Buildings through the use of Smart Materials', *Solar Architecture and Urban Planning, Conference Proceedings of the 4th Energy Forum*, 2–4 December, Bressanone, Italy. Munich, Bozen, EF Economic Forum Ltd.
8. Ritter A (2009), 'Progettare Edifici Zero Energy e Energy Plus utilizzando Materiali Intelligenti', *Architettura and Urbanistica Solare – Documentazione, 4° Energy Forum*, 2–4 December, Bressanone, Alto Adige, Munich/Bozen: EF Economic Forum, Ltd.
9. Ritter A (2009), 'Smart Materials: Der Einsatz von Materialien mit Wechseleigenschaften in der Architektur als ein Ansatz zur Lösung aktueller Klima- und Energieproblematiken', *Montagsgespräche – Positionen: Junge Architekten gesucht*, Domforum, Köln, 14 December. Bund Deutscher Architekten (BDA), Cologne.
10. Ritter A (2010), 'Powered by Nature: Energie- und materieautonome Räume', *medAmbiente*, 2, 22–23 April.
11. Ritter A (2010), 'Desarrollado por la naturaleza: mediante el uso de materiales inteligentes/powered by nature: zero and plus energy buildings through the use of smart materials', *Proceedings of the European Conference on Energy Efficiency and Sustainability in Architecture and Planning/Jornadas Europeas Sobre Eficiencia Energetica y Sostenibilidad en la Arquitectura y el Urbanismo*. Universidad del Pais Vasco, UPV/EHU/España, June.

12. Ritter A (2010), 'Powered by nature: Energie- und materieautonome Gebäude durch den Einsatz von Smart Materials und adaptiven Konstruktionen', *Proceedings of the Bauhaus.SOLAR* (incl. CD-ROM) *3. Internationaler Kongress Technologie – Design – Umwelt*, 10–11 November, Messe Erfurt.
13. Ritter A (2011), 'Materialet inteligjente ne arkitekture, interier dhe projektim', (original title, text in Macedonian), Ars Lamina, Skopje, Macedonia.
14. Ritter A (2011), 'Smart materials in architektur, innenarchitektur und design', (title and text in Serbian), Ars Lamina, Skopje, Macedonia.
15. Ritter A (1995), 'Flexible Tragstruktur aus Glasfaserkunststoff', *Arch +*, 129, 136.
16. Ritter A (1996), 'Es' bewegt sich. – Reagible Tragstruktur aus Glasfaserkunststoff', *AIT Spezial – Intelligente Architektur*, 4, 70–71.
17. Ritter A (2002), 'Lektionen aus der Natur: Polyreagible Mechanomembran', *Intelligente Architektur*, 36 16–17.
18. N. N (2002), 'Polyreagible Mechanomembran', *Wallpaper*, 54, 67.
19. Ritter A (2012), 'Smart and kinetic! Entwerfen und Konstruieren von energie- und materieautarken kinetischen Räumen und hydroaktiven Polymer-Aktoren unter Einsatz von Smart Materials, adaptiven und gewichtsreagiblen Konstruktionen/Design and construction of energy and matter self-sufficient kinetic spaces and hydro-active polymer actuators under the use of smart materials, adaptive and weight-sensitive constructions', PhD thesis, Universität Duisburg-Essen/Campus Essen, Germany.
20. Dyesol, Australia.
21. Serge Ferrari SAS, France.
22. NAUE GmbH & Co. KG, Germany.
23. *Geotechnik mit Geokunststoffen*, product information of NAUE GmbH & Co. KG, Germany.
24. GKD – Gebr. Kufferath AG, Germany.
25. *TiO_2 Photocatalytic Membrane* product information of Taiyo Kogyo Group, Japan.
26. Taiyo Kogyo Group, Japan.
27. Schuster A EADS Deutschland GmbH, Germany, EADS Innovation Works.
28. Lebast, Germany.
29. LEUROCOM electronic displays, GmbH, Germany.
30. Aereco GmbH, Germany/France.
31. Hauschild M (TU Darmstadt, Germany), Karzel R (FH Köln, Germany), cand Arch Ljubas A cand Waidelich Tim – 1st phase: cand Arch Saskia Mayer cand Arch Frauke Hausi (TU Darmstadt, Germany).
32. J. Storrs Hall, USA.

Index